BAYESIAN INFERENCE AND DECISION TECHNIQUES

Essays in Honor of Bruno de Finetti

STUDIES IN BAYESIAN ECONOMETRICS AND STATISTICS

Editors

ARNOLD ZELLNER

JOSEPH B. KADANE

Volume 6

NORTH-HOLLAND–AMSTERDAM · NEW YORK · OXFORD

BAYESIAN INFERENCE AND DECISION TECHNIQUES

Essays in Honor of Bruno de Finetti

Edited by

PREM K. GOEL
Ohio State University

ARNOLD ZELLNER
University of Chicago

1986

NORTH-HOLLAND–AMSTERDAM · NEW YORK · OXFORD

ISBN: 0 444 87712 6

Published by:

ELSEVIER SCIENCE PUBLISHERS B.V.
P.O. Box 1991
1000 BZ Amsterdam
The Netherlands

Sole distributors for the U.S.A. and Canada:

ELSEVIER SCIENCE PUBLISHING COMPANY, INC.
52 Vanderbilt Avenue
New York, N.Y. 10017
U.S.A.

PRINTED IN THE NETHERLANDS

INTRODUCTION TO THE SERIES

This Series on Bayesian Econometrics will include monographs reporting significant research on the theory and applications of Bayesian analysis in Econometrics. Research on Bayesian methodology and applications of it has progressed remarkably in the past few decades thus making it appropriate to institute this Series at this time. It is anticipated that the Series will serve as a focal point for bringing significant Bayesian research to the attention of researchers, decision-makers, students and others.

The Editors

Bruno de Finetti 1906–1985

(Photo: courtesy of Massimo De Felice)

CONTENTS

Introduction to the series v

Acknowledgements xi

1. *Prem K. Goel*, Introduction 1

I. ON DE FINETTI'S CONTRIBUTION

2. *Ludovico Piccinato*, de Finetti's logic of uncertainty and its impact on statistical thinking and practice 13

3. *Edwin T. Jaynes*, Some applications and extensions of the de Finetti representation theorem 31

II. FINITE ADDITIVITY

4. *Bruce M. Hill and David Lane*, Conglomerability and countable additivity 45

5. *Joseph B. Kadane, Mark J. Schervish and Teddy Seidenfeld*, Statistical implications of finitely additive probability 59

III. BAYESIAN METHODOLOGY IN ECONOMICS AND ECONOMETRICS

6. *Robert B. Litterman*, Specifying vector autoregressions for macroeconomic forecasting 79

7. *Herman K. van Dijk and Teun Kloek*, Posterior moments of the Klein–Goldberger model 95

8. *Hirotugu Akaike*, The selection of smoothness priors for distributed lag estimation 109

9. *Yoel Haitovsky*, The linear hierarchical model and its applications in econometric analysis 119

IV. SELECTED BAYESIAN TECHNIQUES

10. *Stephen J. Brown and Roger W. Klein*, Model selection in the federal courts: An application of the posterior odds ratio criterion 141

11. *Barry C. Arnold and S. James Press*, Bayesian analysis of censored or grouped data from Pareto populations 157

V. PROBABILITY ASSESSMENT

12. *James M. Dickey, A. Philip Dawid and Joseph B. Kadane*, Subjective-probability assessment methods for multivariate-t and matrix-t models 177

13. *Michael Goldstein*, Separating beliefs 197

14. *Edward E. Leamer*, Bid–ask spreads for subjective probabilities 217

15. *Arnold Zellner*, On assessing prior distributions and Bayesian regression analysis with g-prior distributions 233

VI. UTILITY AND SUBJECTIVE PROBABILITY ASSESSMENT

16. *Morris H. DeGroot and Stephen E. Fienberg*, Comparing probability forecasters: Basic binary concepts and multivariate extensions 247

17. *Robert L. Winkler*, On "good probability appraisers" 265

18. *Charles A. Holt, Jr.*, Scoring-rule procedures for eliciting subjective probability and utility functions 279

19. *Clifford Hildreth and Glen J. Knowles*, Farmers' utility functions 291

20. *Gary Klein, Herbert Moskowitz, Sathiadev Mahesh and A. Ravindran*, Simplified assessment of single and multi-attribute utility functions 319

VII. ANALYSIS OF TIME SERIES AND GENERAL LINEAR MODELS

21. *Lyle D. Broemeling and Samir Shaarawy*, A Bayesian analysis of time series 337

22. *Chan-fu Chen*, A Bayesian approach to nested missing-data problems 355

23. *K. Rao Kadiyala and Dennis Oberhelman*, Estimation of actual realizations in stochastic parameter models 363

24. *Wolfgang Polasek*, Local sensitivity analysis and Bayesian regression diagnostics 375

VIII. MODELING AND FORECASTING

25. *A. Philip Dawid*, A Bayesian view of statistical modelling 391

26. *James Q. Smith and R.B. Gathercole*, Principles of interactive forecasting 405

27. *Murray K. Clayton, Seymour Geisser and Dennis E. Jennings*, A comparison of several model selection procedures 425

IX. EXPERIMENTATION AND OPTIMAL SEARCH

28. *Wen-chen Chen and Morris H. De Groot*, Optimal search for new types 443

29. *Dennis V. Lindley*, The relationship between the number of factors and the size of an experiment 459

X. BAYESIAN AND NON-BAYESIAN INTERACTION

30. *James Berger*, Bayesian salesmanship 473

Author index 489

Subject index 493

ACKNOWLEDGEMENTS

The idea of a volume to honor Professor Bruno de Finetti was conceived at the May 1981 meeting of the NBER–NSF Seminar on Bayesian Inference in Econometrics and Statistics. The present volume has become possible because of the generous efforts of many people.

First and most important are the authors of the articles appearing in this volume, some of whose papers were presented at the semiannual meetings of the Bayesian Seminar, while others were written expressly for this volume. All the authors were most cooperative in revising the original manuscripts as suggested by the reviewers. As is usual in this series, the authors have been very generous in donating all royalties from the sale of this volume to the Savage Memorial Fund, which is quite active in promoting further research on Bayesian methodology and applications.

We must also thank the countless, not to be named, reviewers of the manuscripts. Their dedication has turned the contents of this volume into an excellent collection of Bayesian articles.

Continuous support has been provided by the National Bureau of Economic Research and the U.S. National Science Foundation for the semiannual meetings of the Bayesian Seminar. These meetings not only have had a significant impact on Bayesian research, but have also provided inspiration and material for this volume. The U.S. National Science Foundation, under grants MCS-8002336A01 at Purdue University and MCS-8400687 at The Ohio State University, has also provided partial support for the editing of this volume.

Ellen van Koten, Economics Editor, and others at North-Holland were extremely helpful in bringing this volume into existence.

Finally, Norma Lucas, Pamela Eckert (Zellner's secretary), and Mary Lee Raines (Goel's secretary) have provided invaluable assistance in the preparation of this volume using the facilities of the Department of Statistics, Purdue University; the Graduate School of Business, University of Chicago; and the Department of Statistics, Ohio State University, respectively.

We thank each and every one of them.

Arnold Zellner
Chicago, Illinois

Prem K. Goel
Columbus, Ohio

November, 1984

Chapter 1

INTRODUCTION

PREM K. GOEL

The Ohio State University

1. Background

Professor Bruno de Finetti has made some very fundamental contributions to the foundations of probability theory and neo-Bayesian statistical inference along with numerous important contributions to Bayesian methodology and its applications in many scientific disciplines, including more than 25 papers dealing with probability and statistics in economic theory. Thus, a volume to honor de Finetti in this series on *Studies in Bayesian Econometrics and Statistics* was long overdue. The decision to go ahead with the project was made at the May 1981 meeting of the NBER–NSF Seminar on Bayesian Inference in Econometrics and Statistics.

Bruno de Finetti was born on 13 June 1906 in Innsbruck, Austria. He graduated from high school at Trento and entered the Polytechnic at Milan. However, after discovering his preference for scientific rather than professional training, he entered the University of Milan, Faculty of Mathematics, and completed his degree program in four years instead of the usual five. Professor de Finetti has an illustrious record of working as a statistician at a government agency (Italian National Census Bureau) and an insurance company, and finally as a scholar of international repute. He retired as a full professor from the University of Rome in 1976.

The text of de Finetti's farewell lecture, delivered at the Istituto Mathematico G. Castelnuovo, University of Rome, which is published in Aykaç and Brumat (1977), presents his view of probability in a very clear and concise form. In fact, on reading that lecture, one can get an idea of the enormity of de Finetti's contributions to the theory of probability. Even after retirement, he has kept quite active in elucidating his ideas on probability. His recent articles, one entitled "The Proper Approach to Probability" in the Proceedings of the Conference on Exchangeability in Probability and Statistics (see Koch and Spizzichino, 1982)

Bayesian Inference and Decision Techniques, edited by P. Goel and A. Zellner

which was held in honor of de Finetti on his 75th birthday, and the other entitled "Probability and My Life" (see Gani, 1982), provide his current thoughts on important issues. Two books among several written by him, *Theory of Probability* (1970), which was translated into English (1974–75) and German (1981), and *Probability, Induction and Statistics* (1972), will always be required reading for everyone interested in learning the essence of probability. At the time these books appeared in English, Lindley [see Proceedings of the Conference on "Fondamenti dell'inferenza statistica" Firenza (1977)] aptly summarized the reaction of the scientific community by saying: "Undoubtedly, the major event since 1971 has been the appearance of the two volume work by de Finetti, in which he has presented his own strikingly original view of probability..." and suggested that there be "a moratorium on research so that we can all study it."

Professor de Finetti has been honored in various international forums. The 1981 conference on exchangeability provides us with only a glimpse of the vast impact of his thoughts on the role of probability in scientific enquiry. The present volume is a small token of our appreciation of de Finetti's contributions. Since the chapters in this volume refer time and again to de Finetti's contributions, I do not need to go any further in providing details of them, even though I had thought of writing a detailed version. Furthermore, I find myself unequal to the task. Before giving an overview of the chapters in this volume, I would like to mention that in my current investigation of matching strategies, the notion of exchangeability is playing a very important role, and I am constantly reminded of de Finetti's contributions to the field of probability and statistics.

2. Overview of volume

The primary objective of this volume is to honor de Finetti for his profound contributions to subjective probability, Bayesian statistical methodology and its application in several areas. Many of the chapters in this volume deal with topics directly in the realm of de Finetti's contributions, while others are indirectly influenced by his thoughts. Some of these chapters were presented at the Semiannual Bayesian Seminars supported by the National Bureau of Economic Research and the U.S. National Science Foundation. These seminars have been a constant source of stimulation for research on Bayesian methodology. Our secondary aim is to present some of these ongoing discussions to a wider audience in order to stimulate further research in the areas of interest to de Finetti. All the chapters included in the volume are previously unpublished. We now give an overview of the volume which has been divided into ten parts.

Part I relates most directly to de Finetti's interests. In Chapter 2, Professor Piccinato, a colleague of de Finetti's at the University of Rome, critically examines the impact of de Finetti's logic of uncertainty on statistical thinking and practice. Piccinato stresses that de Finetti's thoughts are primarily directed toward the procedures of inductive reasoning. The standard problems of inference in theoretical and applied statistics can all be put in this unified framework with a clear and conceptual meaning. However, de Finetti's theory, via the notion of *coherence*, does not provide a set of *cookbook* procedures à la frequency theory methodology. Instead he provides us with a "precise but *open* system, where the possibility of fitting real situations is adequately guaranteed". Piccinato also examines topics such as the meaning and role of subjectivity, the use of statistical models versus the predictivistic approach via the concept of exchangeability, the effect of the choice of finite additivity of probability and the relations between the neo-Bayesian approach and classical decision theory. Of course, these topics are but a small part of those considered in de Finetti's works. The chapter also provides an extremely useful set of references on a broad range of topics. In Chapter 3, Professor Jaynes discusses the extended de Finetti representation theorem for both finite and infinite exchangeable sequences. The discussion and the proof are simple and concise. Jaynes also gives an example of the profound impact of this representation theorem in statistical mechanics.

Part II deals with the implications of the assumption of finitely additive probability, an issue whose importance is emphasized by de Finetti over and over again (see appendix to de Finetti, 1977) in that one should not make an assumption for mathematical convenience only. Professors Hill and Lane discuss the concept of conglomerability in Chapter 4 and provide an alternative proof of a conjecture by de Finetti, proved earlier in Schervish et al. (1984), linking finite additivity and nonconglomerability. In Chapter 5, Professors Kadane, Schervish, and Seidenfeld discuss several interesting consequences of finitely additive probability. In view of the failure of conglomerability under finite additivity, they persuasively argue that under the circumstances, admissibility is an invalid decision criterion and marginalization and "strong inconsistency" do not lead to any paradox since they are variations of nonconglomerability.

In Part III, the four chapters discuss applications of Bayesian methodology in econometrics and economic forecasting. In Chapter 6, Dr. Litterman provides a Bayesian solution to the problem of specifying a vector autoregression for forecasting macroeconomic variables. Litterman's approach is very much in the spirit of de Finetti in that he views the specification of the "prior as a flexible means of generating a filter for the optimal extraction of information from a set of data", which requires an in-depth examination of the specific problem at hand and thus does not amount to a routine Bayesian analysis. The specification process describes the econometric modelling for short-run forecasting of monthly values of four variables of primary interest: GNP, prices, interest rates and money. It also uses three informational variables: a stock price index, the flow of total nonfinancial debt and the change in business inventories for the period 1948–1981. The example demonstrates that a carefully designed Bayesian analysis can lead to consistent improvements in out-of-sample forecast performance. In Chapter 7, Professors van Dijk and Kloek consider the revised Klein–Goldberger model in a structural form for a Bayesian analysis. The model consists of sixteen stochastic equations and four identities. The evaluation of posterior moments involves 30-dimensional numerical integration. By some standards, the computational problem may seem to be monstrous. But the authors demonstrate that because of the recent advances in computer technology and the Monte Carlo integration methods developed by them, problems of this size are computationally feasible. Another interesting outcome of their analysis is to point out that the local approximations of the covariance matrix around the posterior mode may differ considerably from the covariance matrix around the posterior mean. Professor Shiller (1973) introduced the concept of smoothness priors to derive Bayesian estimators of the lag coefficients in a stochastic linear system. These priors involve three hyperparameters. In Chapter 8, Professor Akaike suggests an empirical Bayesian approach for choosing appropriate values of the hyperparameters. The performance of these estimators is shown to be satisfactory under widely varying conditions. In Chapter 9, Professor Haitovsky provides a review of the linear hierarchical model. Then a three-stage hierarchical model is used for estimating residential demand for electricity. He also outlines some more applications of the linear hierarchical structure in econometric modeling situations.

In Part IV, we present two more applications of Bayesian methodology. In Chapter 10, Professor Brown and Dr. Klein examine the effect of the recent court rulings which have accepted hypothesis testing at a fixed 5 percent significance level, irrespective of the sample size, to evaluate evidence of discrimination cases. It is well known that the classical test of an hypothesis with a (fixed) level of significance that does not decline as the sample size increases can be very misleading since the null hypothesis can almost always be rejected by choosing a large enough sample size. Through a judicious re-analysis of data from several court cases, the authors present a convincing case for the use of the posterior odds ratio criterion, and show that outcomes could be crucially different in some of these cases. In Chapter 11, Professors Arnold and Press are concerned with socio-economic data which are modeled by a Pareto distribution that is available in grouped or censored form. They have developed Bayesian estimation techniques for censored or grouped data for the two- and three-parameter Pareto distribution and multivariate Pareto distributions.

In Part V, problems related to the assessment of prior beliefs are addressed. Two of the four chapters in this part deal with the assessment of the prior parameters in specific parametric settings, whereas the other two are concerned with foundational issues in probability assessment. In Chapter 12, Professors Dickey, Dawid and Kadane provide heuristically reasonable methods to elicit subjective probabilities based on multivariate-t and matrix-t distributions via interactive computing algorithms. The authors first show that the matrix-t distributions are uniquely determined by a pair of marginal multivariate-t vectors. This allows them to explore the assessment methods for multivariate-t models, presented earlier in a famous Kadane et al. paper in *JASA* (referenced in Chapter 12) and a new method in this paper for assessing the matrix-t models. These models encompass a large class of models in the normal theory with conjugate prior distributions. In Chapter 13, Professor Goldstein expounds on the subjectivistic theory based on de Finetti's concept of "Prévision". He contends that the formal Bayesian theory can only be viewed as a model for rational behavior and thus does not form a bedrock of subjectivistic theory in which the data consists of *everything* that has an effect on the experimenter's belief, i.e. all observations and thoughts up to the point where his beliefs are announced. Since this could mean infinitely many dimensions (parameters) for specifying one's beliefs, use of the standard subjectivist theory may be impossible in applications, Goldstein develops a fundamental notion for implementation of the subjectivistic theory. This provides the conditions under which the beliefs of an individual can be separated into subsystems which may be analyzed as separate systems. Thus, the effect of new information on one's beliefs may be determined purely by considering the effect of this information on certain limited aspects of those beliefs. He contends that these conditions can always be satisfied and provides the extent to which standard Bayesian procedures can be useful in subjectivistic theory. This chapter seems to move the de Finetti suggestions a step ahead toward application; however, researchers still need to study these aspects further before a concrete application can be made. There is a lot of food for thought in this chapter.

In Chapter 14, Professor Leamer critically examines de Finetti's demonstration of subjective probabilities being equivalent to prices of lottery tickets which are "coherent". Leamer contends that initially the notion of probability as the price of a lottery ticket is appealing to economists; however, further contemplation dims this enthusiasm since prices and valuations are ordinarily thought to differ. He emphasizes that de Finetti's subject is *obliged* to play the game, which may not be the case in actual elicitation situations unless the elicitor is willing to pay him some money in advance. Thus, the game may be very costly to the elicitor. Furthermore, the price setter is unlikely to reveal his true valuation anyway. Thus, he argues, that de Finetti's betting game is deficient as a method for elicitation of probabilities. In addition, Leamer points to the elicitation experiments by Savage and others in which the subjects from whom valuations are being elicited "*experience wavering and indecision that cannot be identified with mere indifference*. This leads to an uncomfortable gap between theorists who assume sharp, stable probabilities which can be elicited economically and accurately, and practitioners who cannot seem to find them." Of course, Bayesians will refute these experiments by saying that subjects do not understand probabilities, and they should be properly trained in probabilistic notions before elicitation games begin. Leamer suggests that most exchanges of lottery tickets involve bid–ask spreads and thus we need to elicit upper and lower probabilities (intervals of probabilities), since these can be *elicited economically* and can be more *stable* than sharp probabilities. He uses several models for intervals of probabilities to explore the accuracy and economy of various elicitation games. At least in the economic context, Leamer presents an interesting point of view which needs further investigation. In Chapter 15, Professor Zellner considers the problem of assessing prior distributions for the multiple linear regression model (MRM). Even though there are several existing methods for assessing informative prior distributions for MRM (see Chapter 12), Zellner suggests an interesting approach which is based on the rational expectations hypothesis and leads to very simple priors which he calls g-priors. These priors are shown to be a special form of a natural conjugate prior distribution for MRM, and can be viewed as reference (informative) prior distributions. Furthermore, the assessment of the hyperparameter of the g-prior is quite simple since it is one-dimensional. The chapter also gives the details of Bayesian inference methodology for the MRM based on the g-priors.

In Part VI we continue with the general area of assessment of subjective probability as well as of utility functions. However, the first two chapters concern comparison of probability forecasters rather than probability assessment per se. Professor de Finetti contributed actively to this area and feels that it is an important area. In the article entitled "Probability and My Life" (see Gani, 1982), de Finetti says that

> I am interested in (probability) prediction in appropriate fields. Football (American soccer) forecasting, although it has no truly practical applications, is a very good training ground for assessing the probability of all kinds of events.

He further adds that, "I consider it an important educational goal to estimate numerically any measure such as probability, distance, weight, temperature, and so on...".

Once probability forecasting is accepted as an important activity (and *it is* in the case of TV weather forecasting or economic forecasting), one has to develop methods for comparison of probability forecasters. Chapter 16, by Professors Feinberg and DeGroot, and Chapter 17, by Professor Winkler, provide a state of the art review of the research on comparison of probability forecasters for binary events. These papers also generalize the results to situations in which the events of interest may have more than two outcomes. Feinberg and DeGroot also provide a link between the notions in probability forecasting and those in comparison of experiments, in the area of decision theory.

In Chapter 18, Professor Holt presents a general scoring rule procedure which can be used by an elicitor *to induce* a subject to reveal his beliefs about some uncertain events. Several known probability scoring rules are shown to be special cases of this procedure, which does not involve betting schemes but pay-offs to the subject based on the ex-post facto accuracy of the subject's forecast of the realization of the random phenomena of interest. This procedure may answer some of the criticisms of elicitation of sharp probabilities by Leamer in Chapter 14. However, it does involve a cost for the elicitor to play the game. By reversing the roles of probability and utility, this procedure can also be used to induce a subject to reveal his utility function. In Chapter 19, Professors Hildreth and Knowles present the results of an actual study to elicit farmers' utility functions. This is an interesting experiment and a detailed analysis of farmers' responses has been carried out in this chapter. The estimates of the utility functions of each farmer are obtained by fitting the responses to a parametric utility function form. Some open problems for future research are also suggested. In Chapter 20,

Professors Klein et al. treat the multi-attribute utility assessment problem as an optimization problem based on individuals' responses to hypothetical gambles. They suggest using a general function for the conditional utility functions and a nonlinear programming formulation to determine the parameters of the function which best fits the qualitative and quantitative responses to hypothetical gambles. Then, assuming a general multilinear multi-attribute utility functional form, the scaling constants are calculated via linear programming by minimizing inconsistencies in expressed preferences to pairwise consequence vectors. I hope that the ease of formulation will induce researchers to carry out the viability and effectiveness of the suggested approach.

In Part VII, we present a collection of papers on Bayesian methodology for general linear models and time series analysis which are, without any doubt, the most often used tools in econometric modeling. In Chapter 21, Drs. Broemeling and Shaarawy provide an extensive review of the Bayesian contributions to time series analysis in the time domain, along with a conjugate prior Bayesian analysis of ARMA processes. This analysis is then compared with the Box–Jenkins methodology from the prediction error point of view. It is interesting to note that the two methodologies seem to provide similar estimates and forecasts. In Chapter 22, Professor Chen considers a Bayesian formulation of the nested missing data problems. An important conclusion of his investigation is that with a Normal-Wishart conjugate prior, distinct sets of parameters appearing in the factored likelihoods are independently distributed both a priori and a posteriori. This result certainly simplifies the data analysis since one can easily deal with lower dimensional numerical integrations. In Chapter 23, Professors Kadiyala and Overhelman review the problem of simultaneous estimation of parameters in a regression model and compare the Bayesian solutions to the problem with those given by the random coefficient regression model approach. A Monte Carlo study to compare the efficiency of an empirical Bayesian estimator with that of OLS provides some evidence in favor of the empirical Bayesian estimator. In Chapter 24, Professor Polasek extends regression diagnostics for the OLS regression analysis to a Bayesian linear model. The diagnostic tools are then applied to a model for an Austrian Keynesian consumption function. He also develops a sensitivity analysis for seemingly unrelated regression models with respect to the error covariance matrix and proposes overall diagnostic checks for these models.

In Part VIII, chapters relevant for modeling and forecasting are presented. In Chapter 25, Professor Dawid discusses some foundational and philosophical issues involved in defining what the term "model" means to a Bayesian. Dawid starts from a subjectivistic position, à la de Finetti, that there can be no such thing as an unknown probability. By giving several examples, he points out that for a given distribution of observables, there is no unique way of distinguishing between parameters and observables. Therefore, it is suggested that the concept of a unique "true model" may be meaningless. Some possible interpretations suggested by Dawid are: (1) a model is a purely technical device for simplifying Bayesian calculations for prediction; (2) a model represents the area of agreement among a number of different subjective opinions; (3) a model merely expresses certain natural symmetries in a problem; and (4) a model describes only the gross behavior of observables and is falsified if it does so incorrectly. Dawid also recommends a serious non-Bayesian attack on the philosophical questions: "What do we mean by a model?"; "How should we construct models?"; and "How can we falsify a model?" In Chapter 26, Professors Smith and Gaithercole seem to accept the first and fourth interpretations of a model by Dawid, given above. While adding a postulate that in business applications it is optimal to use a model which promises the greatest eventual pay-off, they propose to develop an interactive forecasting system whose associated forecast losses are as small as possible. The technique is illustrated in the context of an advertising model and compared with more conventional techniques of model selection. In Chapter 27, Professors Clayton, Geisser and Jennings seem to agree with Dawid's first interpretation and suggest that model selection criteria should be based on predictive error. The chapter provides a review of various model selection criteria and then compares these selection procedures for normal and exponential data by a simulation study. They conclude that, at least for these populations, the predictive error seems to be fairly stable for all the selection methods, even though correct selection rates may vary a little.

In Part IX, optimality considerations in two different situations are examined. In Chapter 28, Professors Chen and DeGroot discuss a more complicated version of the collector's problem in which there are K populations, and each population contains C different types of items in various proportions. The collector must try to obtain at least one item of each type by sampling from these K populations, i.e. collect a complete set of C types. They discuss optimal search procedures in order to minimize the expected number of observations necessary to obtain a complete set of C different types of items under various settings. The problem of optimal stopping in which the collector pays a cost for each observation and receives a bonus for each new type of item collected is also discussed. In Chapter 29, Professor Lindley dwells on a previously untouched topic of the effect of an increase in the number of factors in an experiment on the complexity of estimation of higher order interactions. In the usual parametric full factorial models, an increase in the number of factors does not mean the introduction of effects which are more difficult to estimate than the main effect of a single factor. However, in nonparametric formulations, Lindley describes two situations in which interactions involving m-factors get harder to estimate as m increases. The analysis does not necessarily require a Bayesian approach. Lindley's analysis suggests that experimenters should not unnecessarily increase the number of factors, even if they can afford to have more observations, and that statisticians should be more careful when analyzing data involving many factors.

Finally, we come to Part X, entitled Bayesian and Non-Bayesian Interaction. Professor Berger propounds on the effectiveness of the Conditionality–Likelihood Principle as a vehicle to convince non-Bayesians about the usefulness of the Bayesian paradigm. Mainly due to inherent difficulties in the complete and (accurate) specification of prior distributions, Berger suggests that a robust Bayesian viewpoint will be very convincing. Even though Berger agrees that in de Finetti's subjectivistic viewpoint the standard statistical setup with parameters and the objective model are deemed to be incorrect, his stress on using the Conditionality–Likelihood Principle is mainly to address the problem of selling the Bayesian viewpoint to non-Bayesians. The chapter also provides several examples in which frequentist techniques will be of value to a Bayesian. I presume we all have to learn from each other and thus believing that non-Bayesian statistics does not have much to offer to Bayesians will be suicidal for the Bayesian point of view.

This collection of essays covers a broad spectrum of relevant issues. On each re-reading of these essays I have learned something new. It is hoped that the readers of this volume will be inspired to analyze many practical and theoretical problems from the Bayesian point of view. That, in my opinion, would be the highest honor and recognition for Professor de Finetti's contributions.

References

Aykaç, A. and C. Brumat, eds. (1977) *New Developments in the Applications of Bayesian Methods* (North-Holland Publishing Company, Amsterdam).

de Finetti, B. (1977) "Probability: Beware of falsifications", text of the farewell lecture at the University of Rome, in: A. Aykaç and C. Brumat, eds., *New Developments in the Applications of Bayesian Methods* (North-Holland Publishing Company, Amsterdam).

Gani, J., ed. (1982) *The Making of Statisticians* (Springer-Verlag, New York).

Koch, G. and F. Spizzichino, eds. (1982) *Exchangeability in Probability and Statistics*, Proceedings of the International Conference, Rome, April 1981, in honor of Professor Bruno de Finetti (North-Holland Publishing Company, Amsterdam).

Schervish, M., T. Seidenfeld and J. Kadane (1984) "The extent of nonconglomerability of finitely additive probabilities", *Z. f. Wahrscheinlichkeitstheorie*, 66, 205–226.

Shiller, R.J. (1973) "A distributed lag estimator derived from smoothness priors", *Econometrika*, 41, 775–788.

PART I

ON DE FINETTI'S CONTRIBUTION

Chapter 2

DE FINETTI'S LOGIC OF UNCERTAINTY AND ITS IMPACT ON STATISTICAL THINKING AND PRACTICE

LUDOVICO PICCINATO

University of Rome

1. Logic of uncertainty and statistical inference

In his foreword to de Finetti's *Theory of Probability*, D.V. Lindley writes: "This book is about life." Indeed, the entire work of Bruno de Finetti deals with uncertainty in general and hence with the procedures for *any* inductive reasoning. Everyone is thus concerned with it; in particular the statistical community was provided with a compact and unifying framework which is fundamental for both theoretical and applied statistics.

The inferential theory of de Finetti relies on some cornerstones which have a clear conceptual meaning, but the theory itself does not imply a complete and detailed system of procedures, similar to the current *cookbooks* of the standard literature. Only in the last few years has an appreciable effort been made in research along the neo-Bayesian lines, aimed not only at deepening theoretical formulations but also at the systematization and the extension of the methods. Therefore, even if the quantity of Bayesian papers in the statistical literature is growing, the number remains, nevertheless, quite small compared to the larger set of available procedures framed in the traditional setting. Using the well-known terminology introduced by Thomas Kuhn, it can be said that we are now in a transition phase of the Bayesian paradigm, which is ceasing to be an element of *revolutionary* science and is becoming instead an element of *normal* science.

On the other hand, I think that what seems to be indeterminate in so many operational aspects of the neo-Bayesian approach, particularly in de Finetti's version, is actually closely linked to the nature of the approach itself. From a methodological standpoint, in fact, de Finetti's work is generally characterized by an effort to take into account rigorously all aspects of inductive reasoning, of

Bayesian Inference and Decision Techniques, edited by P. Goel and A. Zellner

critically evaluating standard assumptions and all the *mathematical conveniences* whenever these happen to be logically inconsistent or at least when they introduce some logically unjustifiable element. Then the goal of rigour is reached through indications which are precise but constitute on the whole an *open* system, where the possibility of fitting real situations is adequately guaranteed. Those statisticians who are inclined to reduce their work to the application of *routine* procedures – an attitude which is not particularly commendable in general – are particularly misplaced in the neo-Bayesian setting.

The fundamental bases of de Finetti's concept of probability are well known (for a concise treatment see de Finetti, 1968; the most complete reference is of course de Finetti, 1970). The notion of *coherence* is basic to all the aspects of the theory because it rules the limits of subjectivity in the evaluation of probabilities. It is almost useless to emphasize how simple the structure of this is. The coherence is in fact equivalent to a somewhat weaker version of ordinary axiomatization: the family of events to consider does not need to be a σ-algebra and the probability is required only to be *finitely* additive. The mathematics of probability remains thus largely unchanged, but a precise and meaningful interpretation is made possible in that probability becomes essentially a degree of belief in uncertain events. Given these coherence constraints, many different practical behaviours are legitimate, for instance about *how* probabilities can be evaluated, even if some current attitudes and behaviours become meaningless or need reformulating. For instance, the question "what is the *true* probability of the event E?" is senseless, in this approach, but the real problem which motivates such a question can be reformulated. In many situations distinct events are sometimes considered as repetitions of one event, say E; then the *true probability* is but the idealization of the relative frequency F_n of the "repeatable" event E in a sufficiently large number n of trials. The quantity F_n (and its asymptotic behaviour for $n \to \infty$) can plainly be discussed more clearly in the subjectivistic framework, where it has to be seen as a random quantity whose probability distribution is determined by the available and relevant information. This point will be examined in more detail below.

Even if the logical and formal bases of *frequentist* definitions of probability are not satisfactory, they constitute a genuine effort to grasp some meaningful aspects of real situations. The fundamental issue is to determine which definition is the most *adequate* when all the possibilities are seen as alternative tools for the scientific argument. The link with inductive reasoning, i.e. with the process of increase in information which occurs when new experimental data become available, seems an essential point. But only in the Bayesian framework can this problem have a satisfactory solution, because the Bayes theorem yields a synthesis of the pre-experimental information (embodied in the initial probabilities) and of the experimental information conditional on the assumed model (the likelihoods), so that the cycle can be newly activated when further information is obtained. In the Bayesian paradigm, therefore, induction can be realistically formalized as a *process*. In the more traditional statistical approaches, however, only a collection of separate inductive acts (or, rather, inductive *behaviours*; see Neyman, 1957) is available, and their mutual links are not easily formalized. It is worth remarking that even Pearson (1962) essentially acknowledged this problem, and indicated some elements which were left indeterminate in the current procedures just in order to get the possibility of recovering some pre-experimental information. Of course this solution is fragmentary in nature and essentially inadequate to assure a real dynamic structure for the inferential pattern.

The more obvious criticism of the objective approaches, in particular of the frequentist one, rely on the subjective nature of the assessment of some essential properties of the problem (e.g. that the *trials* are perfectly repeated, or that the level of imperfection is tolerable) and on the restricted applicability domain of the concept. An interesting comparative report of the discussions about these fundamental themes at an early stage of modern applied probability was given by de Finetti himself (1939).

This does not imply of course that de Finetti's subjectivistic version of statistical inference solves any problem, but allows us to look at the inferential issue only in an historical perspective. Even if a dynamic pattern is available, its application to real situations cannot be generally reduced to a trivial process of particularization; new problems in the phase of probabilistic analysis or of the modelling of examined phenomena are continuously met in practice. I cannot agree, therefore, that the neo-Bayesian approach *makes statistics dull* (a sentence that de Finetti attributes to H. Chernoff; see, for example, de Finetti, 1978b, p. 464). Quite to the contrary, using a well-known expression by Tukey (1961), which is somewhat old but still may not be out of date, I think that the neo-Bayesian approach induces us, in a natural way, to look for an approximate answer to the *right* question, even if vague, instead of the exact answer to a *wrong* question, even if precise.

2. Subjectivity and professional practice

The "subjective" character of de Finetti's approach is surely the aspect most intensively discussed and criticized; it has, in fact, been considered by a large number of critics as such an intrusion of arbitrariness so as to invalidate the power of the Bayesian arguments. It would be required, on the contrary, that the analysis of the experiments conformed to "objective" principles, and in particular that the conclusions be independent of the subject who is carrying out the analysis. Accepting the "objectivity" criterion, however, many further aspects of the analysis have to be taken into account, other than the notion of probability. Standard statistical elaborations are based for instance on suitable *statistical models*, whose adoption presupposes a certain degree of personal initiative and hence diversity. This particular level of subjectivity is usually admitted because in principle (and with some approximation) it is thinkable to repeat the experiment until it becomes clear whether the model actually fits. Thus, while in any single application the choice of the model is unavoidably subjective, it can be thought (as the "objectivistic" authors often emphasize) that at least a potentially objective justification can be granted. No kind of objectivity, however direct indirect or potential, can be seen in the process of selecting a *conclusion*. Let us consider for simplicity a statistical decision problem according to the Neyman–Pearson–Wald pattern. Under suitable conditions the "objective" analysis can at most find out a *complete* class of admissible rules (in the decision-theoretic sense), and such rules cannot be further compared without introducing "external" criteria. If, for example, we take into account the problem of testing two simple hypotheses, it is immediately seen that the optimal α-size test (in the Neyman–Pearson sense) for a random sample of n elements is mathematically coincident with a standard Bayesian elaboration (based on the minimization of the expected terminal loss) characterized by a suitable value of the ratio between the initial probabilities of the hypotheses. As simple as it is, this example is sufficient to indicate how, at least from the mathematical standpoint, the traditional non-Bayesian criteria can play the same role as the assignment of the prior probabilities. It is then difficult to defend that the level of "arbitrariness" is actually different in the two approaches, as is sometimes said, at least in polemic arguments. The main issue seems rather a different one, namely what kind of mathematical characterization allows a better and deeper awareness of the meaning and validity of any procedure? Of course the relation between Bayesian and sampling theory methods is usually much more complicated; it can be said that all types of elaborations which yield definite conclusions cannot avoid relying on additional assumptions, which cannot be seen as an "objective" outcome of the experiment. This holds in particular for the special properties of procedures that are usually suggested, e.g. unbiasedness, and, after a deeper examination of the repeated sampling approach, also for the specification of the stopping rule actually used. The latter aspect was discussed by Fisher (1956), and many authors (e.g. Bartholomew, 1971, and Lindley, 1972) suggested a kind of correspondence between the arbitrariness in the choice of the sample space and that in the choice of the prior probabilities. In practice, at least, all these choices are left to conventions rather than to a thoughtful examination of any singular instance. As de Finetti wrote (1937, p. 417 of the English translation) "when one pretends to *eliminate* the subjective factors one succeeds only in *hiding* them".

Now, given that *exogenous* assumptions are unavoidable, or, resorting to another expression by de Finetti (1962), that objectivity is a "mirage", it turns out that subjectivity is not criticizable *per se*; however, particular methods aimed at putting subjectivity into practice could be. Therefore the most correct choice is not only to critically evaluate all the assumptions, but also to behave so that such critical control can, as far as possible, be explicit and public. For more general comments about the role of subjectivity in scientific research, with special attention to statistical theory, see Scardovi (1978).

With reference to an expository paper by de Finetti (1974) it has been maintained (Hamaker, 1977) that the Bayesian approach could be a *threat* to the statistical profession, just because it relies on subjective elements. If one accepts the previous arguments it turns out clearly that the situation is actually reversed: it is the practice of hiding the relevant assumptions which makes statistical work potentially misleading, mainly when, as often happens, the statistician's role is to assist a decision-maker. The decision-maker cannot be expected to be able to penetrate the theoretical side of statistical analysis, so he must be kept adequately informed about the essential aspects of the elaboration. Therefore one task of the statistician could also be that of making this process easier; and a subject to be seriously taken into account by any Bayesian statistician, for instance, is the elicitation of probabilities (see Savage, 1971, for an in-depth survey). It cannot be seen how assumptions of the "subjective" type previously mentioned could be considered as generally irrelevant and therefore "swept under the carpet" (Good, 1976). This is rightly emphasized by Good, in my opinion, in his reply to Hamaker (Good, 1978) when he denies that statisticians should lie, because this would be strategically useful to their professional position.

Even if the neo-Bayesian approach, and de Finetti's one in particular, cannot be criticized on a *logical* basis because it resorts to subjective elements, a typical difficulty in statistical practice is that very often the *subject* is properly collective, or sometimes even imprecisely defined. To adopt conventional standards which are unjustifiable or even inconsistent can be only a superficial solution. While admitting that to overcome such a difficulty is not a simple matter, it seems important to emphasize that the real problem is rather that of the partial inadequacy of the scheme used in representing some of the actual situations. This of course would hold *a fortiori* for the non-Bayesian approaches.

A first attempt to overcome this difficulty could be based, therefore, on an explicit formulation of statistical decision problems as multi-subject problems. A step in this direction was made many years ago by de Finetti himself (1951b). A second possibility, often used, is to relate procedures of statistical analysis to the more general problems of scientific communication, stressing their instrumental relevance for such a process more than any conclusive nature. In this perspective, for instance, de Finetti and Savage (1962, p. 91) write that "it would be better to deal with every problem leaving the largest indeterminacy in the initial evaluations and showing how the final evaluations are linked to the initial ones, so that everyone is able to draw the conclusions which conform to his own opinions." For standard statistical problems this amounts to utilizing the likelihood function directly as the basic instrument for the scientific communication, at least when the main interest is centred on explicative hypotheses of a given phenomenon. Analogous arguments were also advanced, in even greater detail and depth, by Hildreth (1963) and Roberts (1974), the latter emphasizing the necessary distinction between *public* and a *private* phase of the elaboration. This way a strict connection is established with an approach originated by R.A. Fisher, the so-called *likelihood* approach (see Barnard, 1947, 1967; Birnbaum, 1962; Hacking, 1965; and many others) which is still maintained by several researchers in a way essentially distinct from any Bayesian viewpoint. These methods enjoy therefore a Bayesian "justification" because they can be thought as characterizing the first and usually interpersonal stage of a statistical analysis; this stage could even be the only one involved in the communication process.

Again in the communication perspective one can consider some of the researches about the so-called *non-informative* priors; this is particularly explicit in the works by Dickey (1973) and Bernardo (1979a), just as in the classical *principle of precise measurement* (Savage, 1962).

3. Role and limits of the statistical models

In traditional statistics, we have a *statistical model* when, a priori, the outcome is seen as a random object which takes values in a given space $\mathscr{X}$ and is given (or is *known*, according to the current terminology) a class $\mathscr{P} = \{P_\theta, \theta \in \Omega\}$ of probability distributions on $\mathscr{X}$ (which is possibly endowed with a given σ-algebra of subsets; the general advantage of using such pattern will be discussed below). It is usually said that any such distribution not known (say $P_{\theta*}$) is *true* in the sense that the data must be considered as generated according to the true *objective* probability law $P_{\theta*}$. It is clear that from a personalistic viewpoint this formulation is unacceptable; in particular, the attributes *true* and *false* are meaningless for a subjective probability. However, with suitable changes, the previous formulation could be recovered. The space Ω must then be seen as the sure event and singletons $\{\theta\}$, the *simple hypotheses*, as possible events. It is thus valid to say that one and only one of the simple hypotheses is *true*, just in the sense that any event is in principle true or false (even if, as happens typically for hypotheses, the available information makes them simply *possible*). In this sense, of course, the distributions $P_\theta, \theta \in \Omega$, are *conditional* probability measures and the *true* probability measure is but the measure conditional on the true simple hypothesis.

From a Bayesian standpoint too this pattern is remarkably useful (even if it cannot be seen as the unique way of modelling Bayesian inference), especially when the events θ are such that the specification of the corresponding conditional measures P_θ become obvious. In that case such distributions would be *intersubjective* and, once an outcome $z \in \mathscr{X}$ is obtained, they would allow us to calculate the likelihood function, i.e., a measure of the experimental evidence (conditional on the assumed model) in favour of the different hypotheses. This way we would have the obvious advantage of relying on "agreed" likelihoods (the expression *agreed probability* having been suggested by Barnard, 1980).

It must be stressed that such a direct recovery of the standard statistical model strictly depends on the previously expounded interpretation of Ω and $\mathscr{P}$. It is, however, common experience, in many applications, that the model $(\mathscr{X}, \mathscr{P})$ is actually a scheme where the hypotheses θ and the corresponding sampling probabilities P_θ are only an idealized synthesis of reality, justified at least as approximations but also chosen in such way to simplify the mathematical elaboration.

In these conditions it is clear that the model $(\mathscr{X}, \mathscr{P})$ is, in a certain sense, an hypothesis which can in turn be *true* or *false*, or at least, more realistically, *adequate* or *inadequate*. From this point of view we are led to the classical subject of the *goodness of fit* tests (or of the *conformity* to the theoretical models, using the terminology of Pompilj, 1948) with a formulation which is quite close to that originally proposed by R.A. Fisher. As many authors have remarked (e.g. Barnard, 1962; Anscombe, 1963; Cox, 1977) in this type of problem there is not a *complete* model, constituted by an exhaustive set of alternative hypotheses, but rather a limited and preliminary problem is faced, i.e. the *relation* between the given data and a particular model, usually equivalent to a composite hypothesis.

The more traditional methods try to solve this conformity problem without using any probability distribution on the values of the unknown parameters of the model and therefore rely on procedures which are identically valid (possibly with some approximation) with respect to such values. Some authors proposed instead to resort to the initial predictive distribution, which of course also carries on the initial information about the parameters, conditional on the model; see, for example, Box (1980) who distinguishes the stage of *model criticism*, whose fundamental tool is just the initial predictive distribution, and the stage of *estimation*, whose fundamental tool is instead the final distribution of the unknown parameters. Of course, the term *estimation* is used here in the original (Fisherian) sense of general inference.

These kinds of proposals, especially when they are framed in a Bayesian setting, unavoidably give rise to conceptual problems. A first objection reaffirms for the *model criticism* stage the validity of the post-experimental point of view, i.e. of a suitably formulated *likelihood principle* (see Lindley, 1980). Moreover, it could be remarked that the formal procedures for checking the goodness of fit usually introduce, *at least implicitly*, a family of alternative models, which are therefore actually compared (Smith, 1980b; an example of this type was provided by DeGroot, 1973). In this case the previous objection emerges somewhat strengthened.

On the other hand, there is no difficulty in carrying out a likelihood analysis, given a questionable model, but the assignment of prior probabilities to the simple hypotheses, when these do not constitute a partition of the sure event, again presupposes arguing conditionally on the model and a *complete* Bayesian analysis is not actually feasible without further ingredients.

Another method for overcoming these types of difficulties could be to introduce essentially non-restrictive models only, i.e. models which are *true* with practical certainty. This leads quite naturally to the topic of non-parametric models. The well-known research by Ferguson (1973, 1974) gave rise to many further Bayesian studies on that subject. It is clear that under these conditions a concrete interpretation of the simple hypotheses, as would be required for instance in de Finetti's approach, is generally rather difficult. The high generality of the model is paid for by the extreme fineness of the partition of the sure event into simple hypotheses; a deep understanding of the derived procedures, even if they are quite simple and natural from a purely mathematical viewpoint, is then made remarkably difficult. Critical advice about even the compatibility of the attributes *non-parametric* and *Bayesian* was expressed, maybe somewhat provocatively, by Smith (1980a). The criticism of approaches which are termed Bayesian only on a purely formalistic basis can be shared; an exploratory mathematical work can have a substantial importance, however, to make possible a concrete use of a given mathematical model, because its meaning and implications are then clarified. In this context a relevant example is the study of the initial and final distributions of meaningful functionals, such as the so-called *associative means*, again in the framework of a Dirichlet process (see Cifarelli and Regazzini, 1979). It is worth noting that the concept and properties of the associative means were studied originally by de Finetti himself (1931b).

A different attitude toward the statistical models, essentially closer to the spirit of de Finetti's approach, is that of introducing in the inferential scheme only events which are *strictly observable*, which implies a change of focus from the hypotheses (whose observability is at most only theoretical) to future events and then to assume the *predictive* point of view. de Finetti often recommended assigning probabilities *exclusively to facts* (see, for example, de Finetti, 1937, p. 150 of the English translation) and considering the laws and the theories only as intermediate tools for evaluations concerning concrete facts (see also de Finetti, 1970, section 11.1.4, and de Finetti, 1971). This problem will be discussed in more detail in the next section.

4. The completely predictive approach

It is well known that the predictive problems were rather neglected for a long time by the non-Bayesian schools, presumably because of the difficulty of formalizing the dependence of the future on the past. For a recent survey in the non-Bayesian framework, see, for example, Barndorff-Nielsen (1980). Here I would like only to recall that in 1985 the predictive topic is alive in all the statistical schools (see, for example, Geisser, 1980). In the Bayesian setting, even if there are no special conceptual difficulties, much more attention has been given initially to problems of the standard *hypothetical* kind (or *structural*, according to another terminology; see Kadane, 1980); this was probably due to the tendency to follow the standard approach. Anyway, a fairly complete Bayesian treatment of predictive problems was given by Roberts (1965). The predictive aspect has often been taken into account as a tool for explicating prior distributions, just because the specification of a predictive probability is a remarkably more concrete operation than the specification of a prior probability of an unobservable hypothesis. For instance, the well-known *method of imaginary results* (Good, 1957) can have this kind of application. Other relevant references (not meant to be exhaustive) are Piccinato (1977), Kadane (1980) and Winkler (1980). It is clear, however, that the treatment of predictive *problems* on its own does not necessarily require giving up the theoretical framework based on traditional models. Indeed, the text by Aitchison and Dunsmore (1975), while entirely dedicated to prediction, adopts the classical hypothetical scheme. The approach where only the strictly observable events are taken into account will be called *completely predictive*. In that case only the past and future experimental results are involved, without any further theoretical intermediation. Let us assume, for instance, that a sequence $X_1, X_2, \ldots$ of random quantities, whose joint probability distribution P is of course specified, is (at least potentially) observable, and that we are required to infer about $X_{n+1}, X_{n+2}, \ldots$ given the experience $E = (X_1 = x_1, X_2 = x_2, \ldots, X_n = x_n)$. The process of learning from experience, in this case, is represented by the calculation of the probability distribution of the process $\{X_n, n = 1, 2 \ldots\}$ *conditional on* E. Of course, essentially the same elaboration is also allowed in an hypothetical approach, where the distribution of the observable process is initially given only for each $\theta \in \Omega$. Then P corresponds to the distribution on $\Omega \times \mathscr{X}$ in the traditional case, and all that is needed is to transform the *initial* predictive probability distribution, which is marginal with respect to the hypotheses, into the *final* one, which is again marginal with respect to the hypotheses and conditional on E. What is worth noting, however, is that the recourse to the model-based pattern is not necessary either in theory or in practice, so that no problem of probability assignment to hypotheses is really needed, at least in an essential way. Besides, again within the completely predictive pattern, it is possible to recover some fundamental concepts of classical statistical theory, possibly reformulated, as happens for the concept of *sufficient statistics* (see Spizzichino, 1979; Cifarelli and Regazzini, 1980–81; Campanino and Spizzichino, 1981). Many of these concepts are essentially linked with the basic notions of *model* and *parameters*. It must be noted that in non-Bayesian statistics also there were, in these years, remarkable developments on this topic (see, for example, Lauritzen, 1974). Different developments in the same area, but in a Bayesian framework, were recently advanced by Dawid (1982).

It is well known, but worth some detailed discussion, that through the already-mentioned de Finetti representation theorem (de Finetti, 1937) the relation which exists between the completely predictive approach and the hypothetical approach can be clearly exhibited or, more precisely, that under suitable conditions the assignment of a predictive distribution implies the "existence" of a traditional model. In fact, if the infinite random quantities X_n $(n = 1, 2, \ldots)$ are exchangeable and P denotes the probability measure of the corresponding process, denoting by B an event relative to a number whatsoever of the X_n, there exists a measure μ on the space $\mathscr{F}$ of the cumulative distribution functions on R^1 such that

$$P(B) = \int P_F(B)\,\mathrm{d}\mu(F), \tag{1}$$

where $P_F(B)$ is the probability of B calculated under the assumption that all the X_n are independently distributed with a cumulative distribution function F. Apart from more technical details, for example whether or not the measure P

could be forced to be σ-additive (as in Chow and Teicher, 1978), it is important to emphasize, as in the original work by de Finetti, that roughly speaking μ represents, in the limit as $n \to \infty$, the initial opinion which is implicit in P about the behaviour of the process $\{Y_n(x), x \in R^1\}$, where

$$Y_n(x) = n^{-1} \sum_{i=1}^{n} I_{(X_i < x)}, \tag{2}$$

where I_A is the indicator of the event A. Furthermore, it can be remarked that Y_n is the *empirical distribution function* associated with the first n observations $X_1, X_2, \ldots, X_n$. The opinion embodied in μ therefore corresponds to the evaluation, which is obviously subjective, of what is expected to happen "objectively" in the long run.

Assuming again the exchangeability condition, the theorem simplifies remarkably when the X_n take only the values 0 and 1 and could then be seen as indicators of a given event in a sequence of repeated trials (for an elementary proof see Heath and Sudderth, 1976). Then $\mathscr{F}$ is the class of cumulative distribution functions with jumps of values $1-\theta$ and $\theta (0 \le \theta \le 1)$ corresponding, respectively, to the points 0 and 1, and the right-hand side of (2) can be rewritten as a Stieltjes integral. Given, for instance, that B corresponds to the result of n trials among which k are "successes" (in an arbitrary but fixed order) we then have the simple formula:

$$P(B) = \int_0^1 \theta^k (1-\theta)^{n-k} \, \mathrm{d}\Phi(\theta), \tag{3}$$

where

$$\Phi(\theta) = \lim_{n \to \infty} P\left(n^{-1} \sum X_i < \theta\right), \tag{4}$$

for every θ where Φ is continuous. In any case, coming back to formula (1), every exchangeable probability measure P on the process $\{X_n, n = 1, 2, \ldots\}$ determines essentially a traditional statistical model constituted by the *hypotheses* $F \in \mathscr{F}$ and by the so-called *sampling probabilities* $P_F(B)$.

Thus, from the viewpoint of statistical theory, the concept of *exchangeability* would allow us to tackle the problem of rigorously defining the concept of *population*, which is usually treated more or less as if it was primitive; the relevant bases have already been laid by de Finetti (1930d) when he introduced what he called *random phenomenon*. In a certain sense it can even be said that the traditional approach turns out to be justified, but the ultimate base of such a justification lies substantially in a subjective probability, initially assigned to all the uncertain events. A thorough analysis of the binomial case was made by Lindley and Phillips (1976); an interesting use of this approach was made also by Dawid (1977) who provides a Bayesian "justification" of the standard analysis of variance, different of course from that based on diffuse priors. The general validity of this kind of approach has been firmly advocated by Cifarelli and Regazzini (1982) and by Daboni and Wedlin (1982) in a recent textbook. Finally, it is relevant to recall that the same notion of exchangeability has found many other uses in statistics, for instance as a tool to obtain formalizations of prior informations closer to actual situations (see, for example, Ericson, 1969; Lindley and Smith, 1972).

5. Probabilistic framework and "mathematical convenience"

It is well known how often de Finetti polemized against the so-called "mathematical convenience", when the latter leads to logically unjustified assumptions (a synthesis is given in de Finetti, 1977). On the other hand, without a rigourous critical check, the theory of probability could not have the *normative* character which de Finetti assigns to it. In many situations of traditional studies in probability and mathematical statistics, however, it can be seen that some assumptions are taken into account mainly to simplify the mathematical treatment or to lead it back to already known and developed theories; in such cases there is therefore the risk that mathematical convenience can become a substitute for logical justification.

It is only natural that among the many issues raised by de Finetti most attention in the literature was given to those which are more connected with the general aspects of statistical theory, and in particular with the formal role and substantial meaning of the so-called "initial probabilities". On the contrary, issues having, at least seemingly, a more technical character were less actively discussed, notwithstanding their strict connection with the general framework of de Finetti's approach. The most known topic of the latter kind is the thesis about the unsuitability of assuming at a general (axiomatic) level the complete additivity of probability. This position depends also on the formulation chosen for the coherence condition; there are many possible variants, and naturally de Finetti preferred the weakest (see de Finetti, 1970, sections 3.3. and 3.11). This in turn does not imply more than simple (i.e. finite) additivity. Obviously, as we are often reminded by de Finetti himself, the adoption of simple additivity as a general norm does not prevent us from taking into account, when deemed suitable, a probability which is completely additive either in general or with respect to some special countable family of events. What is essential is that such a property be seen as a specific feature of the information embodied in *that* probability distribution and not as a characteristic of the elaboration of *any* information. For some particular inference problems (see, for example, Hill, 1980; Scozzafava, 1981a) a *not* completely additive probability has been suggested. Such situations are perfectly justifiable in de Finetti's approach, but would be inadmissible using the current axiomatization.

On the other hand, even accepting that the events can in general be represented as subsets of a given space Ω (however, de Finetti is rather critical also about this position; see de Finetti, 1970, section 2.4), it is well known that in general a σ-additive probability cannot be defined on the entire family of subsets of Ω. Therefore if f is a one-to-one non-measurable function $\Omega \rightarrow \Omega$, it can happen that the probability of the set $f(E)$ is not definable even if the probability of E is, and both represent essentially the same event. It is obvious that this is completely unnatural from a probabilistic viewpoint, although mostly it does not imply practical difficulties because the σ-algebras currently in use are sufficiently rich. It is worth noting, in this context, that recently the theory of finitely additive measures has developed enough to provide a reasonable alternative to the current approach (a systematic treatment of some of these results is provided by Scozzafava, 1982a). For instance, after the works by Dubins (1975) and others, if P is a finitely additive probability on whatever algebra $\mathscr{E}$ of events, the conditional probability $P(A \mid B)$ can be always defined for $A \in \mathscr{E}$, $B \in \mathscr{E} \setminus \{\varnothing\}$ without requiring the more restrictive condition $P(B) \neq 0$. The use of finitely additive measures therefore leads to a theory of conditional probabilities which is quite complete while still being simple and as close to the real problems as the theory of absolute probabilities. These properties are at least questionable for current σ-additive theory. It is worth remarking that a natural development of the probability theory which allows a general existence of the conditional probabilities permits us, in turn, to introduce the notion of *order of zero* of the probabilities. By means of this notion, given by de Finetti in 1936 (see de Finetti, 1972, p. 82), events with probability zero could be compared and ordered. Until the 1980s this topic received only little attention in statistical theory, but it could be useful; for instance, it has been used by Scozzafava (1981b) with the aim of providing a theoretically rigourous alternative to the improper distributions used as *non-informative* priors.

A key concept for the study of finitely additive probabilities (introduced by Dubins, 1975, but we resort here to the version by Scozzafava, 1982b) is that of *disintegrability*. Given a finite or denumerable partition $\pi = \{H_1, H_2, \ldots\}$ of the sure event Ω, a finitely additive probability on the algebra is called π-disintegrable if for every $E \in \mathscr{E}$ we have

$$P(E) = \sum_i P(H_i) P(E \mid H_i). \tag{5}$$

The role of decomposition (5) seems to be fundamental for statistical inference when the events H_i are interpreted as the possible explicative hypotheses associated with an experimental result E; for a fixed E, therefore, $P(E \mid \cdot)$ is just the ordinary likelihood function. It is evident that if π is finite, formula (5) always holds, while if π is denumerable, it can be shown that formula (5) (for *every* $E \in \mathscr{E}$) holds if and only if P is completely additive on π, that is if

$$\sum_i P(H_i) = 1. \tag{6}$$

It is worth remembering that disintegrability is connected to the known conglomerative property (which is, in substance, one of its consequences) introduced by de Finetti (1930c). The conglomerative property corresponds to the continuity of $P(A \mid B)$ with respect to B for a fixed A (de Finetti, 1972, p. 100) and is often taken into account, just because its validity is not granted, when finitely additive probabilities are used (see, for example, Hill, 1980).

Thus, using the more general approach, there can be difficulties even in the introduction of the ordinary likelihood function whose meaning, at least in a Bayesian setting, is strictly connected with formulae like (5). It seems reasonable, however, to presume that in most of the cases, relevant for applications, such difficulties will not be found. Many authors have remarked that ordinarily real problems have a *finite* nature, and only because of the (ill-famed) mathematical convenience are they formalized by means of infinite and even continuous structures (Barnard, 1967; Basu, 1975). In the finite case, however, (5) is true, and on the other hand, trivially, any finitely additive measure is countably additive too. Given the suitability of approximating finite structures by means of infinite structures, an approximation procedure would be acceptable when some elements will remain unchanged or almost unchanged in the original problem and in the new one. Then it would not be unreasonable to admit that in some cases a probability space with a finite set of elementary events would be approximated by a probability space having an infinite set of elementary events, but where the disintegrative property continues to hold. In this sense the σ-additive probability measures could be justified as approximations, at least for particular and essentially finite problems.

6. Relations with the theory of decisions

The decisional aspects are involved in de Finetti's approach at least at two distinct levels, i.e. at the foundational level and at the stage of applications. The principle of coherence, i.e. the refusal to deal with uniformly unfavourable bets (the so-called *dutch books*), assigns a sort of decisional character to the notion of subjective probability, at least in de Finetti's version. Such a decisional aspect is however *weak*, in that what is really relevant is the adoption of the coherence principle and not of the particular preference system in the subject's mind, because the essential properties of the probability do not depend on the latter one.

On this ground, it seems that the subjective probability, as advocated by de Finetti, could by full right constitute the basic tool for a theory of inductive reasoning where it should be remembered that, according to de Finetti, statistics on the whole is nothing but a particular application of the inductive method (see de Finetti, 1972, p. 155).

At the level of applications, decisional aspects are involved when a space of available acts and a utility function on the corresponding consequences are introduced. The scheme, termed *neo-Bayesian neo-Bernoullian* by de Finetti (by others, the Ramsey–de Finetti–Savage scheme) prescribes to behave in accordance with the rule of maximizing the expected utility (for general treatments see de Finetti, 1970, 1978b, and Lindley 1971; a classical relevant text is of course that by Savage, 1954).

As de Finetti himself wrote (1978a, p. 167) "the inductive [i.e. Bayesian] reasoning works in the frame of the theory of decisions as a *distinct* element in the chain of considerations which constitute in their complex the rational procedure to follow in order to reach a 'decision' in the optimal way". The reluctance to relate reasonings and behaviours to abstract situations or at least to situations which were in some way isolated from their real context, which is typical also of de Finetti's positions (see de Finetti, 1971), led many authors to give a special relevance to the decisional aspect on the whole, deeming the inductive moment as something provisional and instrumental (see, for example, Di Bacco, 1982). Apart from any precise reference to de Finetti's ideas, it is also frequently found in statistical literature that the Bayesian approach is more often taken into account for decision problems than for pure inference problems. Presumably, this depends in part on the importance given by Wald himself (1950) to *formally* Bayesian decisions. It is interesting to note that soon after de Finetti (1951a) also stressed the relevant interest of Wald's results with respect to the development of the neo-Bayesian theory. It seems, however (at least to the present writer), that the logical and methodological proposals by de Finetti would be misunderstood if they were considered as rigid schemes: the Bayesian approach has the tools to deal both with inferential and decision problems. For instance, de Finetti himself (1974, p. 26), while commenting on the well-known polemics between R.A. Fisher and J. Neyman and contrasting the opposition between *inductive reasoning* and *inductive behaviour*, criticized the accept–reject formulation of the theory of tests just because it would be misleading in comparison with a clear exhibition of the uncertainty level (for a detailed analysis of the main non-Bayesian approaches see de Finetti 1972, ch. 9).

Moreover, because the current Neyman–Pearson–Wald approach is strictly decision oriented, the decision aspect has a natural interest in that it allows a comparison to be made of the different schools. For instance, the violation of the likelihood principle, which is implicit in the use of risk functions, has particularly evident consequences just in the decision-theoretic framework (Piccinato, 1981).

On the basis of the already-mentioned results by Wald himself, it seems rather difficult to object to the ideal quality of the Bayesian approach, unless one denies even the assumption of transitive preference systems (and then, maybe, the possibility of any realistic and conclusive decision theory), and controversies can only concentrate on the concrete availability of the basic tools, i.e. essentially on the availability of a theory of probability so general to be applicable to any uncertain event.

7. Concluding remarks

Subjectivistic conceptions were virtually always present in the probabilistic literature (see Kyburg and Smokler, 1964), and in particular Ramsey (1926) outlined an approach quite close to the one de Finetti began to develop independently a few years later. In particular, Ramsey used the device of resorting to bets to get an operationally meaningful method to measure probabilities. It is remarkable that actually this idea has even older roots: de Finetti (1931a) states that he drew it from Bertrand, whose treatise was published in 1889, and Scardovi (1978 p. 74) quotes Kant stating, in 1781, that bets are the best way to ascertain the real opinion of people. Indeed, de Finetti, more than fifty years ago (1930a, 1930b, 1930d, 1931a), provided a complete and systematic theory, particularly suitable in view of the applications. It has previously been stated that de Finetti's attitude towards statistical inference follows strictly from his theory of probability and that just because of this fact his theoretical system remains open, in a certain sense, and subject to developments in different directions. The topics of statistical theory which had a treatment, or at least a hint, in de Finetti's work are of course much more numerous than we had the possibility to discuss here, and also for those that were actually taken into account it would be suitable to resort to the quoted references for further developments.

Nevertheless, we hope that this (subjective) re-reading of de Finetti's work does not hide the quality and quantity of thoughtful and constructive stimuli that de Finetti provided for the statistical community, and, as Lindley wrote, for society at large.

References

Aitchison, J. and I.R. Dunsmore (1975) *Statistical Prediction Analysis* (Cambridge University Press).

Anscombe, F.J. (1963) "Tests of goodness of fit", *J. Roy. Statist. Soc.* B, 25, 81–94.

Barnard, G.A. (1947) "Review of *Sequential Analysis* by A. Wald", *J. Amer. Statist. Assoc.*, 42, 658–664.

Barnard, G.A. (1962) "Prepared contribution" in: L.J. Savage et al., eds., *The Foundations of Statistical Inference*, (Methuen, London).

Barnard, G.A. (1967) "The use of the likelihood function in statistical practice", in: *Proc. 5th Berkeley Symp. on Math. Statistics and Probability*, vol. 1 (University of California Press).

Barnard, G.A. (1980) "Discussion of Professor Box's paper", *J. Roy. Statist. Soc.* A, 143, 404–406.

Barndorff-Nielsen, O. (1980) "Probabilistic prediction", in: Ist. Naz. Alta Matematica, *Symposia Mathematica*, vol. XXV (Academic Press, London).

Bartholomew, D.J. (1971) "A comparison of frequentist and Bayesian approaches to inference with prior knowledge", in: V.P. Godambe and D.A. Sprott, eds., *The Foundations of Statistical Inference* (Holt, Rinehart and Winston, Toronto).

Basu, D. (1975) "Statistical information and likelihood", *Sankhyā*, 37 A, 1–71.

Bernardo, J.M. (1979a) "Reference posterior distributions for Bayesian inference", *J. Roy. Statist. Soc.* B, 41, 113–147.

Birnbaum, A. (1962) "On the foundations of statistical inference", *J. Amer. Statist. Assoc.*, 57, 269–326.

Box, G.E.P. (1980) "Sampling and Bayes' inference in scientific modelling and robustness", *J. Roy. Statist. Soc.* A, 143, 383–430.

Campanino, M. and F. Spizzichino (1981) "Prediction sufficiency and representation of infinite sequences of exchangeable random variables" (preprint), *Ist. Matematico* (University of Rome).

Chow, Y.S. and H. Teicher (1978) *Probability Theory* (Springer-Verlag, New York).

Cifarelli, D.M. and E. Regazzini (1979) "Considerazioni generali sull'impostazione bayesiana di problemi non parametrici. Le medie associative nel contesto del processo aleatorio di Dirichlet", *Riv. di Matematica per le Scienze Economiche e Sociali*, 2, 39–52 and 95–112.

Cifarelli, D.M. and E. Regazzini (1980–81) "Sul ruolo dei riassunti esaustivi ai fini della previsione in contesto bayesiano", *Riv. di Matematica per le Scienze Economiche e Sociali*, 3, 109–125 and 4, 3–11.

Cifarelli, D.M. and E. Regazzini (1982) "Some considerations about mathematical statistics teaching methodology suggested by the concept of exchangeability", in: G. Koch and F. Spizzichino, eds., *Exchangeability in Probability and Statistics* (North-Holland, Amsterdam).

Cox, D.R. (1977) "The role of significance tests", *Scand. J. Statist.*, 4, 49–70.

Daboni, L. and A. Wedlin (1982) *Statistica. Un'introduzione alla impostazione neo-bayesiana* (UTET, Torino).

Dawid, A.P. (1977) "Invariant distributions and analysis of variance models", *Biometrika*, 64, 291–297.

Dawid, A.P. (1982) "Intersubjective statistical models", in: G. Koch and F. Spizzichino, eds., *Exchangeability in Probability and Statistics* (North-Holland, Amsterdam).

de Finetti, B. (1930a) "Fondamenti logici del ragionamento probabilistico", *Boll. Un. Matem. Ital.*, 9, 1–3 (also in: B. de Finetti, *Scritti scelti 1926–1930*, CEDAM, Padova, 1981).

de Finetti, B. (1930b) "Problemi determinati e indeterminati nel calcolo delle probabilità", *Rend. R. Accad. Naz. Lincei*, 12, 367–373 (also in: B. de Finetti, *Scritti scelti 1926–1930*, CEDAM, Padova, 1981).

de Finetti, B. (1930c) "Sulla proprietà conglomerativa delle probabilità subordinate", *Rend. R. Ist. Lombardo di Scienze e Lettere*, 63, 3–7, (also in: B. de Finetti, *Scritti scelti 1926–1930*, CEDAM, Padova, 1981).

de Finetti, B. (1930d) "Funzione caratteristica di un fenomeno aleatorio", *Memorie R. Accad. Naz. Lincei*, 4, 86–133 (also in: B. de Finetti, *Scritti scelti 1926–1930*, CEDAM, Padova, 1981).

de Finetti, B. (1931a) "Sul significato soggettivo della probabilità", *Fundamenta Mathematicae*, 17, 298–329.

de Finetti, B. (1931b) "Sul concetto di media", *Giorn. Ist. Ital. Attuari*, 2, 369–396.

de Finetti, B. (1937) "La prévision: Ses lois logiques, ses sources subjectives", *Ann. Inst. H. Poincaré*, 7, 1–68 (English translation in: H.E. Kyburg and H.E. Smokler, eds., *Studies in Subjective Probability*, Wiley, New York, 1964).

de Finetti, B. (1939) "Compte rendu critique du colloque de Genève sur la théorie des probabilités", *Act. Sci. Ind.*, no. 766 (Hermann, Paris).

de Finetti, B. (1951a) "L'opera di Abraham Wald e l'assestamento concettuale della statistica matematica moderna", *Statistica*, 11, 185–192.

de Finetti, B. (1951b) "Recent suggestions for the reconciliation of theories of probability, in: *Proc. 2nd Berkeley Symp. on Math. Statistics and Probability* (University of California Press).

de Finetti, B. (1962) "Obiettività e oggettività: Critica a un miraggio", *La Rivista Trimestrale*, 1, 343–367.

de Finetti, B. (1968) "Probability; interpretations", in: *International Encyclopedia of the Social Sciences*, vol. 12 (Macmillan, New York).

de Finetti, B. (1970) *Teoria della Probabilità* (Einaudi, Torino) (English translation: *Theory of Probability*, vols. 1 and 2, Wiley, New York, 1974–75).

de Finetti, B. (1971) "Probabilità di una teoria e probabilità dei fatti" in: *Studi di Probabilità Statistica e Ricerca Operativa in onore di G. Pompilj* (Ist. di Calcolo delle Probabilità, University of Rome).

de Finetti, B. (1972) *Probability, Induction and Statistics* (Wiley, New York).

de Finetti, B. (1974) "Bayesianism: Its unifying role for both the foundations and applications of statistics", *Int. Statist. Rev.*, 42, 117–130.

de Finetti, B. (1977) "Probability: Beware of falsifications!", in: A. Aykaç and C. Brumat, eds., *New Developments in the Applications of Bayesian Methods* (North-Holland, Amsterdam).

de Finetti, B. (1978a) "Inferenza statistica bayesiana", in: *I Fondamenti dell'Inferenza Statistica* (Dipart. Statistico, University of Florence).

de Finetti, B. (1978b) "Decisione", in: *Enciclopedia*, vol. 4 (Einaudi, Torino).

de Finetti, B. and L.J. Savage (1962) "Sul modo di scegliere le probabilità iniziali", in: *Sui Fondamenti della Statistica* (Ist. di Statistica (biblioteca di Metron), University of Rome) (also in: *The Writings of L.J. Savage – A Memorial Selection*, published by the Amer. Statist. Assoc. and the Inst. of Math. Statist., 1981).

DeGroot, M.H. (1973) "Doing what comes naturally: interpreting a tail area as a posterior probability or as a likelihood ratio", *J. Amer. Statist. Assoc.*, 68, 966–969.

Di Bacco, M. (1982) "On the Meaning of 'true law' in statistical inference", in: G. Koch and F. Spizzichino, eds., *Exchangeability in Probability and Statistics* (North-Holland, Amsterdam).

Dickey, J. (1973) "Scientific reporting and personal probabilities: Student's hypothesis", *J. Roy. Statist. Soc.*, B, 35, 285–305.

Dubins, L.E. (1975) "Finitely additive conditional probabilities, conglomerability and disintegrations", *Ann. Probab.*, 3, 89–99.

Ericson, W.A. (1969) "Subjective Bayesian models in sampling finite populations", *J. Roy. Statist. Soc.* B, 31, 195–233.

Ferguson, T. (1973) "A Bayesian analysis of some nonparametric problems", *Ann. Statist.*, 1, 209–230.

Ferguson, T. (1974) "Prior distributions on spaces of probability measures", *Ann. Statist.*, 2, 615–629.

Fisher, R.A. (1956) *Statistical Methods and Scientific Inference* (Oliver and Boyd, Edinburgh).

Geisser, S. (1980) "A predictivistic primer", in: A. Zellner, ed., *Bayesian Analysis in Econometrics and Statistics* (North-Holland, Amsterdam).

Good, I.J. (1957) *The Estimation of Probabilities* (M.I.T. Press, Cambridge, Mass.).

Good, I.J. (1976) "The Bayesian influence, or how to sweep subjectivism under the carpet", in: W.L. Harter and C.A. Hooker, eds., *The Foundation of Probability Theory, Statistical Inference and Statistical Theories of Science*, vol. II (Reidel, Dordrecht).

Good, I.J. (1978) "A. Alleged objectivity; A threat to the human spirit?", *Int. Statist. Rev.*, 46, 65–66.

Hacking, I. (1965) *Logic of Statistical Inference* (Cambridge University Press).

Hamaker, H.C. (1977) "Bayesianism: A threat to the statistical profession?", *Int. Statist. Rev.*, 45, 111–145.

Heath, D. and W. Sudderth (1976) "de Finetti's theorem on exchangeable variables", *The Amer. Statist.*, 30, 188–189.

Hildreth, C. (1963) "Bayesian statisticians and remote clients", *Econometrica*, 422–438.

Hill, B.M. (1980) "On finite additivity, non-conglomerability, and statistical paradoxes", in: J.M. Bernardo, M.H. DeGroot, D.V. Lindley and A.F.M. Smith, eds., *Bayesian Statistics* (University Press, Valencia).

Kadane, J.B. (1980) "Predictive and structural methods for eliciting prior distributions", in: A. Zellner, ed., *Bayesian Analysis in Econometrics and Statistics* (North-Holland, Amsterdam).

Kyburg, H.E. and H.E. Smokler, eds. (1964) *Studies in Subjective Probability* (Wiley, New York).

Lauritzen, S.L. (1974) "Sufficiency, prediction and extreme models", *Scand. J. Statist.*, 1, 128–134.

Lindley, D.V. (1971) *Making Decisions* (Wiley–Interscience, London).

Lindley, D.V. (1972) *Bayesian Statistics, A Review* (S.I.A.M., Philadelphia).

Lindley, D.V. (1980) "Discussion of Professor Box's paper", *J. Roy. Statist. Soc.* A, 143, 423.

Lindley, D.V. and L. Phillips (1976) "Inference for a Bernoulli process (a Bayesian view)", *The Amer. Statist.*, 30, 112–119.

Lindley, D.V. and A.F.M. Smith (1972) "Bayes estimates for the linear model", *J. Roy. Statist. Soc.* B, 34, 1–41.

Neyman, J. (1957) "Inductive behavior as a basic concept of philosophy of science", *Rev. Inst. Int. de Statist.*, 25, 7–22.

Pearson, E.S. (1962) "Some thoughts on statistical inference", *Ann. Math. Statist.*, 33, 394–403.

Piccinato, L. (1977) "Predictive distributions and noninformative priors", *Trans. 7th Prague Conf. on Information Theory, Statistical Decision Functions, Random Processes and 1974 Europ. Meeting of Statisticians* (Reidel, Dordrecht).

Piccinato, L. (1981) "On the orderings of decisions functions", in: Ist. Naz. Alta Matematica, *Symposia Mathematica*, vol. 25 (Academic Press, London).

Pompilj, G. (1948) "Teoria statistica della significatività e conformità dei risultati sperimentali agli schemi teorici", *Statistica*, 8, 7–42.

Ramsey, F.P. (1926) "Truth and probability", reprinted in: H.E. Kyburg and H.E. Smokler, eds., *Studies in Subjective Probability* (Wiley, New York, 1964).

Roberts, H.V. (1965) "Probabilistic prediction", *J. Amer. Statist. Assoc.*, 60, 50–62.

Roberts, H.V. (1974) "Reporting of Bayesian studies", in: S.E. Fienberg and A. Zellner, eds., *Studies in Bayesian Econometrics and Statistics* (North-Holland, Amsterdam).

Savage, L.J. (1954) *The Foundations of Statistics* (Wiley, New York).

Savage, L.J. (1962) "Subjective probability and statistical practice", in: L.J. Savage et al., eds., *The Foundations of Statistical Inference* (Methuen, London).

Savage, L.J. (1971) "Elicitation of personal probabilities and expectations", *J. Amer. Statist. Assoc.*, 66, 783–801.

Scardovi, I. (1978) "L'inferenza induttiva nella ricerca scientifica: annotazioni storiografiche", in: *I*

Fondamenti dell'Inferenza Statistica (Dipartimento Statistico, University of Florence).
Scozzafava, R. (1981a) "Non-conglomerability and the first digit problem", *Statistica*, 41, 561–565.
Scozzafava, R. (1981b) "Improper priors and finite additivity" (abstract), presented to the 14th Europ. Meeting of Statisticians, Wroclaw.
Scozzafava, R. (1982a) "Probabilità σ-additive e non", *Boll. Un. Matem. Ital.*, ser. VI, I-A, 1–33.
Scozzafava, R. (1982b) "Exchangeable events and countable disintegrations", in: G. Koch and F. Spizzichino, eds., *Exchangeability in Probability and Statistics* (North-Holland, Amsterdam).
Smith, A.F.M. (1980a) "Discussion of a paper by S.R. Dalal", in: J.M. Bernardo, M.H. DeGroot, D.V. Lindley and A.F.M. Smith, eds., *Bayesian Statistics* (University Press, Valencia).
Smith, A.F.M. (1980b) "Discussion of Professor Box's paper", *J. Roy. Statist. Soc.*, 143, 411.
Spizzichino F. (1979) "Statistiche sufficienti per famiglie di numeri aleatori scambiabili", in: *Atti del Convegno su Induzione, Probabilità e Statistica* (University of Venice).
Tukey, J.W. (1961) "The future of data analysis", *Ann. Math. Statist.*, 33, 1–67.
Wald, A. (1950) *Statistical Decisions Functions* (Wiley, New York).
Winkler, R.L. (1980) "Prior information, predictive distributions, and Bayesian model-building, in: A. Zellner, ed., *Bayesian Analysis in Econometrics and Statistics* (North-Holland, Amsterdam).

Chapter 3

SOME APPLICATIONS AND EXTENSIONS OF THE DE FINETTI REPRESENTATION THEOREM

EDWIN T. JAYNES

Washington University

1. Introduction

The contributions of Bruno de Finetti to the general philosophy of probability theory, and to the understanding of mathematical puzzles (finite additivity) and pathologies (non-conglomerability) that can arise when we try to apply it on infinite sets, have been well recognized and appreciated. To show the other side of the coin, we note some of the ways in which de Finetti's work has contributed to the technical solution of real problems.

Probability distributions that are symmetric under permutations of a finite number of variables arise constantly in applications. For example, it is known that a bank has n depositors, with total deposits T. Our joint probability distribution for the accounts will be a function $p(x_1,\ldots, x_n)$, where x_i, the size of the ith account, is presumably in $(0, T)$. Whatever other information we may possess about the distribution of wealth, banking habits, etc. in the community, in the absence of any information that could distinguish one depositor from another $p(x_1,\ldots, x_n)$ will be symmetric under permutations of the x_i.

One might then think that any such distribution could be represented in the integral form [eq. (1) below] given by de Finetti, as generalized by Hewitt and Savage (1955); however, this turns out not to be the case because of the finiteness of n. The original de Finetti representation theorem holds only for subsets of an infinitely long exchangeable sequence, and not every finite symmetric sequence is such a subset. While one may feel intuitively that the difference cannot be very serious, in principle there is still unfinished business concerning the technical

Bayesian Inference and Decision Techniques, edited by P. Goel and A. Zellner

question of applicability of the representation (1) in real problems, which always involve finite sequences.

More generally, the variables x_i might be defined on almost any set X, as Hewitt and Savage showed in far greater generality than we need. But the phenomenon we wish to show arises already for the binary set $X = \{1,0\}$ considered in the original theorem of de Finetti (Kyburg and Smokler, 1981). What we demonstrate here for the binary set will hold, mutatis mutandis, for any set X that is likely to arise in a real problem.

That there are difficulties with finite sequences was noted by Feller (1971). He proved de Finetti's representation theorem by reducing it to the Hausdorff moment problem, and then gave examples which "show that the theorem fails for finite sequences". But he offered no alternative for the finite case.

Heath and Sudderth (1976) gave such an alternative as part of a completely different proof in which the representation (1) emerged as the limit of a set of "Urn model" distributions for finite sequences. Their argument not only gives a much deeper understanding of the result, but it is so simple that it seems futile to look for a still simpler one. That such a proof ought to be possible had been suggested by de Finetti (1974), without actually carrying it out. But again, the integral representation (1) was proved only for subsets of an infinitely long sequence, and Heath and Sudderth also show by example that it can fail for finite sequences.

Of course, merely exhibiting examples of failure does not prove that (1) fails for all finite sequences; this question and also the question whether, in cases where it fails, another equally convenient integral representation may still hold, were left open by these arguments.

A closely related difficulty was discovered about twenty years ago in physics. Here a joint probability distribution for the positions of n particles, $p_n(x_1, \ldots, x_n)$, arises as the marginal distribution of a symmetric distribution function P_N for a larger number $N > n$ of particles. Since typically n was 2 or 3, while N was 10^{23}, one sought to represent p_n without bringing in the details of the larger distribution P_N. Some anomalous results were obtained before it was realized that not every symmetric n-particle function can be obtained by marginalization from a symmetric N-particle one.

The problem of finding the necessary and sufficient conditions on an n-point distribution which guarantee that it has this parentage, is called the N-representability problem in the statistical mechanics literature. That it is related to the theory of convex sets was realized at once (Coleman, 1963), but to the best of the writer's knowledge an explicit solution has not been given.

In the following we augment these discussions by showing that (a) the original de Finetti representation does indeed fail for a wide class of finite sequences, and (b) nevertheless, with a trivial modification (dropping the non-negativity condition), it is resurrected, and the integral representation (1) then holds for all finite sequences, leading to solution of the N-representability problem.

2. The generating function

Following the terminology advocated by Jimmie Savage in the 1960s, a sequence $\{x_1, x_2, \dots, x_N\}$ whose probability distribution remains symmetric as $N \to \infty$ is called *exchangeable*. The probability $p(x_1, \dots, x_N)$ can then depend only on the total number R of successes. The original theorem asserts that, given any subset of n trials from an exchangeable sequence, the probability that it contains exactly r successes can be written in the form:

$$p(r|n) = \int_0^1 \binom{n}{r} z^r (1-z)^{n-r} g(z)\,\mathrm{d}z. \tag{1}$$

where it is necessary that $g(z) \geq 0$ and that

$$\int_0^1 g(z)\,\mathrm{d}z = 1. \tag{2}$$

Thus, an exchangeable sequence is characterized completely by a single generating function $g(z)$.

Mathematically, (1) is a weighted average of binomial distributions, and on that intuitive ground the representation (1) was introduced by Laplace in 1774. He obtained many useful results – in particular the Rule of Succession – by interpreting $g(z)$ as the posterior density of a parameter z, and therefore (1) as a predictive distribution for future observations. But for over a century Laplace's method was in disrepute because of conceptual difficulties [$g(z)\mathrm{d}z$ seemed to be a "probability of a probability"].

In 1937 de Finetti obtained the representation (1) by a completely different argument, which showed that Laplace's fault lay not in the alleged metaphysics of his method, but only in his failure to establish its essential assumptions and generality. The results Laplace obtained are now reinstated as important parts of Bayesian statistics (Lindley, 1976). Henceforth we shall call (1) the Laplace–de Finetti (LdF) representation; its apparent failure for finite sequences remains as a puzzle to resolve.

3. An example

As soon as we know what to look for, it is easy to produce almost trivial-looking examples of this failure and its relation to the N-representability problem; it arises already in the comparison of two and three tosses of a coin. A two-point symmetric probability distribution $p(x_1, x_2) = p(x_2, x_1)$ on the set $X = \{0, 1\}$ is determined by three numbers:

$$\begin{aligned} p(11) &= A, \\ p(01) = p(10) &= B, \\ p(00) &= C, \end{aligned} \tag{3}$$

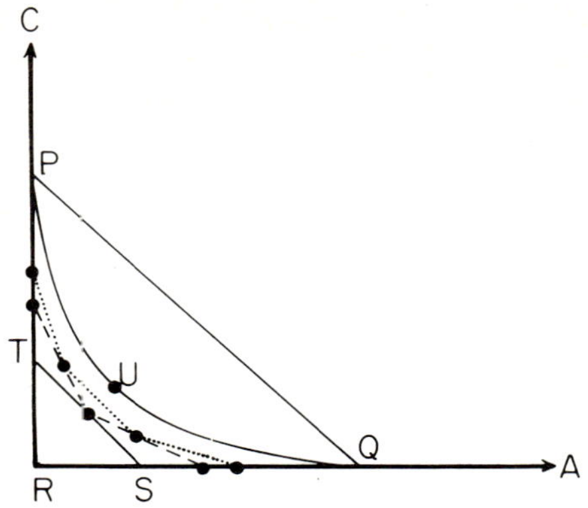

Fig. 3.1. Exchangeable two-point distributions derivable from N-point ones. $A = P(2 \text{ heads})$, $C = P(2 \text{ tails})$. PQR = all possible distributions. Derivability from a three-point function truncates this to $PQST$. Four-point and five-point parentage further truncate $PQST$, as indicated by the dashed and dotted lines. As $N \to \infty$ these nested polygons tend to the parabola, representing independent Bernoulli trials. The original Laplace–de Finetti representation yielded only distributions on or above the parabola, where negative correlations are impossible.

and the possible distributions are represented by all non-negative A, B, C satisfying

$$A + 2B + C = 1. \tag{4}$$

Thus, any point on or within the triangle PQR of fig. 3.1 represents a possible two-point symmetric probability distribution.

But now suppose our two-point function is to be derived from a symmetric three-point one. For this we may write:

$$\begin{aligned} p(111) &= a, \\ p(011) = p(101) = p(110) &= b, \\ p(100) = p(010) = p(001) &= c, \\ p(000) &= d. \end{aligned} \tag{5}$$

The possible three-point symmetric functions are given by all non-negative (a, b, c, d) satisfying

$$a + 3b + 3c + d = 1. \tag{6}$$

But these distributions are related by marginalization:

$$A = a + b,$$
$$B = b + c,$$
$$C = c + d,$$

and now the normalization (6) imposes the constraint:

$$1/3 \leq (A + C) \leq 1, \tag{7}$$

since $(b+c)$ cannot exceed $1/3$. Therefore for a two-point symmetric function to be derivable from a symmetric three-point one, its representative point must lie on or within the quadrilateral *PQST* in fig. 3.1. For it to be derivable from a symmetric four-point function, we find a further restriction to a five-sided polygon inside *PQST*, etc. These polygons were found independently by Diaconis (1977); we generalize them to arbitrary n, N below.

What two-point functions are representable in the original LdF form? This requires

$$A=\int_0^1 z^2 g(z)\,\mathrm{d}z,$$

$$B=\int_0^1 z(1-z)g(z)\,\mathrm{d}z,$$

$$C=\int_0^1 (1-z)^2 g(z)\,\mathrm{d}z. \tag{8}$$

After a little algebra, we find that this implies:

$$2(A+C)-(A-C)^2-1=2\int_0^1 \mathrm{d}y\int_0^1 \mathrm{d}z(y-z)^2 g(y)g(z), \tag{9}$$

and so if $g(z)\geq 0$ we can generate only those two-point functions for which (9) is non-negative. This limits us to points on or above the tilted parabola in fig. 3.1. The two-point distributions representable in LdF form are therefore limited to the parabolic slice *PQU*.

We can understand this restriction intuitively by noting that the right-hand side of (9) is just four times the covariance of (x_1, x_2). The limitation to the parabolic slice therefore signifies that the LdF representation is incapable of giving negative correlations. Yet in our opening example of the bank depositors, knowledge of the total deposits T clearly does impose a negative correlation in $p(x_1,\dots,x_n)$. Thus, the bank distribution has no generalized LdF representation.

Of course, to represent a symmetric distribution of finite length, the de Finetti–Heath–Sudderth Urn representation is available. But while their proof was very simple, their result (a probability mixture of hypergeometric distributions) is not. As an analytical tool, the integral form (1) would be much easier to use. Also, a simpler Urn model may be defined, in which we put in (1):

$$g(z)=\sum_{k=0}^{n} U_k\delta\left(z-\frac{k}{n}\right). \tag{10}$$

(Here the kth Urn, containing a fraction k/n of red balls, is chosen with probability, U_k, and n balls are drawn from it with replacement.) However, an argument like (9) applies; if $U_k\geq 0$ not every symmetric distribution $p(x_1,\dots,x_m)$, $m<n$, can be generated from (10).

As this example shows, the general N-representability problem is non-trivial and threatens to become very complicated. Marginalization from an N-point function to an n-point one involves a summation,

$$p(x_1,\ldots,x_n) = \sum p(x_1,\ldots,x_N), \tag{11}$$

over $\{x_{n+1},\ldots,x_N\}$. The resulting intricate details suggest that we are using an awkward set of variables for, abstractly, marginalization is nothing but a projection operation. All its properties would be retained, and the problem should become easy, if we could find a "natural" coordinate system in which this marginalization reduces to an elementary geometrical projection, parallel to the axes, of a point rather than a distributed mass.

The LdF representation comes close to accomplishing this. If we view the marginalization (11) in the LdF representation, what is happening is that the part $\hat{g}(z)$ of $g(z)$ that affects only details of sequences longer than n drops out from the expression for $p(r|n)$. Therefore, $\hat{g}(z)$ must be orthogonal to the coefficients of $g(z)$ in (1). These coefficients are Bernstein polynomials of degree n; for $r=(0,1,\ldots,n)$ they form a complete set for the expansion of any polynomial of degree N. The projection property we seek is therefore one of orthogonality to all polynomials of a given degree.

4. Orthogonal expansions

The polynomials that are orthogonal with respect to uniform weight over a finite interval are not the Bernstein polynomials, but the Legendre polynomials $P_n(x)$ scaled to the interval

$$\int_0^1 P_n(2z-1)P_m(2z-1)\,\mathrm{d}z = (2n+1)^{-1}\delta_{nm}. \tag{12}$$

Since $P_n(2z-1)$ is itself a polynomial in z of degree n, $P_k(2z-1)$ is orthogonal to every polynomial of degree $n<k$:

$$\int_0^1 P_k(2z-1)z^n\,\mathrm{d}z = 0, \quad 0 \le n < k. \tag{13}$$

Therefore, if we expand the Laplace–de Finetti generating function in the form

$$g(z) = \sum_{k=0}^{\infty} (2k+1)a_k P_k(2z-1), \tag{14}$$

where the factor $(2k+1)$ is inserted to simplify later formulae, the LdF representation reduces to a finite sum:

$$p(r|n) = \sum_{k=0}^{n} A_{rk}^{(n)} a_k, \quad 0 \le r \le n, \tag{15}$$

where the coefficients $A_{rk}^{(n)}$, evaluated explicitly in the appendix, represent the transformation from Bernstein to Legendre polynomials:

$$\binom{n}{r} z^r (1-z)^{n-r} = \sum_{k=0}^{n} A_{rk}^{(n)} P_k(2z-1), \quad 0 \le r \le n. \tag{16}$$

Now the crucial point is the observation that for $r = (0, 1, \dots, n)$ the Bernstein polynomials, although not orthogonal, are linearly independent; therefore the $(n+1)\times(n+1)$ matrix $A^{(n)}$ is non-singular and (15) is uniquely invertible:

$$a_k = \sum_{r=0}^{n} B_{kr}^{(n)} p(r|n), \quad 0 \le k \le n, \tag{17}$$

where $B^{(n)}$, the inverse matrix to $A^{(n)}$, is also given explicitly in the appendix.

The expansion coefficients $\{a_0, a_1, a_2, \dots\}$ are therefore the "natural coordinates" that we sought. For any n, the probabilities $f_r \equiv p(r|n)$, $0 \le r \le n$, defining an n-point symmetric probability distribution $p(x_1, \dots, x_n)$, are determined uniquely by the first $(n+1)$ expansion coefficients $\{a_0, \dots, a_n\}$; but these coefficients are in turn determined uniquely by $\{f_0, \dots, f_n\}$. This makes both the finiteness puzzle and the N-representability problem simple.

5. Finite sequences

Comparing (15) and (9), we see that the LdF representation was unable to give all finite symmetric sequences because the allowable expansion coefficients a_i had been restricted by the non-negativity condition $g(z) \ge 0$. But this was in fact the only reason, for (14) and (17) constitute a proof by construction that for any finite symmetric distribution $p(x_1, \dots, x_n)$ we can find a set of expansion coefficients, and therefore a generating function $g(z)$ for which the representation (1) will hold; with the only difference that if there are negative correlations, then $g(z)$ must be allowed to become negative.

In our example of $n = 2$, we find from the value of $B^{(2)}$ given in the appendix that any point in the triangle *PQR* of fig. 3.1 can now be reached by the generating function:

$$g(z) = 1 + 3(A - C)(2z - 1) + 5(A - 4B + C)(6z^2 - 6z + 1). \tag{18}$$

But when the point (A, C) is outside a weird region bounded by two straight lines and the arc of an ellipse, $g(z)$ becomes negative somewhere in $(0, 1)$.

With hindsight it is easy to see heuristically how non-negativity got into the original theorem. When $n \to \infty$ the binomial distribution in (1) goes into a delta-function, $(n+1)^{-1}\delta(z - r/n)$, and so the probability that the frequency (r/n) lies in $(z < r/n < z + \mathrm{d}z)$ tends to $g(z)\mathrm{d}z \ge 0$, in agreement with Laplace's intuitive meaning of the generating function $g(z)$. Thus, it was only the desire to represent infinitely long sequences that brought in non-negativity in the first place. The properties of Legendre polynomials give us a proof of the integral representation that applies equally well to the finite or infinite case, making this obvious. However, the important new content of de Finetti's theorem did not concern the length of the sequence, but rather the demonstration of something that Laplace probably never suspected, namely that every exchangeable sequence can be so represented.

Our conclusion is that we have only to drop the non-negativity condition, and then every finite exchangeable sequence has a de Finetti integral representation. This gives us the means to solve many real problems, including N-representability.

We find similarly that the simpler Urn model (10) can represent any finite exchangeable sequence if we allow some of the U_k to become negative.

6. *N*-representability

Given an exchangeable N-point distribution $p(x_1,\dots,x_N)$, the n-point distribution $n<N$, obtained from it by marginalization, is the one whose first $(n+1)$ expansion coefficients $\{a_0, a_1,\dots,a_n\}$ are the same (but $a_0=1$ is the normalization condition always satisfied). Conversely, given an n-point distribution, what is the condition that it be derivable from some N-point one? We cannot just assign new expansion coefficients $\{a_{n+1},\dots,a_N\}$ arbitrarily, because in general this would not lead to non-negative probabilities for the N-point distribution:

$$p(R|N)=\sum_{k=0}^{N} A_{Rk}^{(N)}a_k \geq 0, \quad 0\leq R\leq N. \tag{19}$$

So our expansion coefficients have been freed from one non-negativity condition $g(z)\geq 0$ only to become entangled in a new one (19).

But because of the invertibility (17), this new condition can be dealt with at once. Think of the quantities $\{f_R=p(R|N), 0\leq R\leq N\}$ as cartesian coordinates of a point in an $(N+1)$-dimensional space, restricted by $\sum f_R=1,\ f_R\geq 0$, to an N-dimensional convex set F, a generalization of the familiar simplex triangle in three dimensions. The $N+1$ vertices of F (the "vertices of the triangle") are determined by $\{f_R=1, f_m=0, m\neq R;\ 0\leq R\leq N\}$. Therefore, in the inversion,

$$a_K=\sum_{R=0}^{N} B_{KR}^{(N)}f_R, \quad 0\leq K\leq N, \tag{20}$$

the point $\{a_0,\dots,a_N\}$ is restricted to a convex set (polyhedron) S_N whose Rth vertex has coordinates:

$$\{a_0,\dots,a_N\}_R=\{B_{0R}^{(N)},\dots,B_{NR}^{(N)}\}, \quad 0\leq R\leq N. \tag{21}$$

From any point $\{a_0,\dots,a_N\}$ of S_N, marginalization to a lower sequence of length n consists simply of parallel projection, throwing away the coordinates $\{a_{n+1},\dots,a_N\}$ and retaining $\{a_0,\dots,a_n\}$.

The solution of the N-representability problem is therefore: the necessary and sufficient condition that a symmetric n-point distribution $p(x_1, \ldots, x_n)$ can be obtained by marginalization from a symmetric N-point one is that its representative point $\{a_0, \ldots, a_n\}$ must lie on the projection of S_N. This is a polyhedron S_n whose vertices are contained in the set of projected vertices of S_N (some of these may be interior points of S_n).

In our example of fig. 3.1, for the two-point distribution to be derivable from an N-point one, $\{a_1, a_2\}$ must lie on the $(N+1)$-sided polygon whose Rth vertex is

$$a_1 = B_{1R}^{(N)} = \frac{2R - N}{N},$$

$$a_2 = B_{2R}^{(N)} = \frac{6R^2 - 6NR + N^2 - N}{N(N-1)}. \tag{22}$$

Four of these polygons are imaged in fig. 3.1, which is an oblique projection of the (a_1, a_2)-plane. $N = 2,3,4,5$ gives the regions PQR, $PQST$, and the further truncations of the dashed and dotted lines. As $N \to \infty$ they approach the parabola.

Likewise, for a symmetric three-point distribution to be derivable from some n-point one, it is necessary and sufficient that its expansion coefficients lie on or in the polyhedron whose vertices are

$$\{a_1, a_2, a_3\} = \{B_{1r}^{(n)}, B_{2r}^{(n)}, B_{3r}^{(n)}\}, \quad 0 \le r \le n. \tag{23}$$

But as $n \to \infty$, we state without proof that $B^{(n)}$ goes asymptotically into

$$B_{kr}^{(n)} \sim P_k\left(\frac{2r}{n} - 1\right) + O(n^{-1}), \tag{24}$$

and (23) goes into a twisted smooth curve (the generalization of the parabola) whose parametric equations are

$$\{a_1, a_2, a_3\} = \{P_1(x), P_2(x), P_3(x)\}, \quad -1 \le x \le 1. \tag{25}$$

The possible three-point distributions are then defined by the convex hull of the curve (25); in this limit we regain the result of the original LdF representation.

For many applications we need to generalize these results, in the manner of Hewitt–Savage, to larger sets than the binary one $X = \{0, 1\}$. One can, of course, find new functions which generalize the properties of Legendre functions to higher dimensions. A more powerful and abstract approach, which does not require us to go into all that detail, was discovered by Dr. Eric Mjolsness while he was a student of the present author. We hope that, with its publication, the useful results of this representation will become more readily obtainable.

7. Conclusion

With the clearing up of one small technical detail, de Finetti's famous representation theorem is valid for all exchangeable sequences, finite or infinite. This enables us to make a significant reduction in many real problems; we note briefly how the appearance of kinetic theory is changed by this result.

A great deal of work has been done on what is called the Bogoliubov–Born–Green–Kirkwood–Yvon (BBGKY) hierarchy, a set of distribution functions for the positions x_i (in classical theory, also the momenta g_i) of one particle, two particles, etc. on up to perhaps 10^{23} particles. The n-particle distribution $p_n = p(x_1, \ldots, x_n; t)$ is not only symmetric and derived by marginalization from all higher distributions; it depends explicitly on the time t because of Schrödinger's (or Newton's) equations of motion.

Now it turns out that almost all observable macroscopic properties of a system – reversible or irreversible in the sense of thermodynamics – could be predicted if we could calculate how the two-particle function evolves in time. But in this endeavor we have been frustrated for decades by the fact that the recursion relations go the wrong way, i.e. the time evolution of p_2 depends on p_3, which is unknown. The evolution of p_3 depends in turn on p_4, which is even more unknown – and so on. In other words, the evolution of p_2 depends on all the correlations in all the higher order distributions. The problem appears hopeless without making drastic simplifying assumptions that nobody believes.

But now, let us note the de Finetti representation. The distribution function for any order n is determined by a single generating functional $g[f(x)]$:

$$p(x_1, \ldots, x_n) = \int f(x_1) f(x_2) \cdots f(x_n) g[f(x)] \, \mathrm{d}f. \tag{26}$$

This is a functional integral over functions $f(x)$, generalizing eq. (1) above. Furthermore, any physical prediction we wish to make can be extracted directly from $g[f(x)]$ by another functional integral like (26).

Therefore, we can reformulate kinetic theory in a way that sidesteps the recursion problem by never introducing the hierarchy of distributions at all. The object of our study becomes the generating functional $g[f(x)]$, which contains full information about all correlation effects of arbitrarily high order. We seek the time evolution of g:

$$\mathrm{d}g/\mathrm{d}t = Kg, \tag{27}$$

where K is an operator that we need not write down here, but that physicists know how to find.

Although this functional integral was pointed out by de Finetti almost fifty years ago, the idea is still new in physics and not well explored. But its superior precision and simplicity make us predict that, fifty years from now, kinetic theory will be based on the idea of the de Finetti functional, and will be an incomparably more powerful tool for prediction than our present one. As de Finetti has noted, it requires an incredibly long time for new ideas to be appreciated in probability theory.

Appendix

From (16), the transformation matrix $A^{(n)}$ is given by

$$A_{rk}^{(n)} = (2k+1)\int_0^1 \binom{n}{r} z^r (1-z)^{n-r} P_k(2z-1)\,\mathrm{d}z. \tag{A1}$$

The integral is a ${}_3F_2$ hypergeometric function whose power series terminates, giving the result

$$A_{rk}^{(n)} = (2k+1)\sum_{m=0}^{k} (-1)^m \frac{n!(k+m)!(n-r+m)!}{(n-r)!(k-m)!(n+m+1)!(m!)^2}. \tag{A2}$$

The columns of $A^{(n)}$ are orthogonal vectors, but not normalized; one can show by induction that

$$\sum_{r=0}^{n} A_{rk}^{(n)} A_{rs}^{(n)} = \frac{(2k+1)(n!)^2}{(n-k)!(n+k+1)!}\delta_{ks}, \quad 0 \le k, s \le n. \tag{A3}$$

Therefore, the inverse of $A^{(n)}$ is

$$B_{kr}^{(n)} = \frac{(n-k)!(n+k+1)!}{(2k+1)(n!)^2} A_{rk}^{(n)}, \quad 0 \le k, r \le n. \tag{A4}$$

The first few of these matrices are:

$$B^{(1)} = \begin{pmatrix} 1 & 1 \\ -1 & 1 \end{pmatrix},$$

$$B^{(2)} = \begin{pmatrix} 1 & 1 & 1 \\ -1 & 0 & 1 \\ 1 & -2 & 1 \end{pmatrix},$$

$$B^{(3)} = \begin{pmatrix} 1 & 1 & 1 & 1 \\ -1 & -1/3 & 1/3 & 1 \\ 1 & -1 & -1 & 1 \\ -1 & 3 & -3 & 1 \end{pmatrix}.$$

References

Coleman, A.J. (1963) "Structure of fermion density matrices", *Rev. Mod. Phys.*, 35, 668–687.

Diaconis, P. (1977) "Finite forms of de Finetti's theorem on exchangeability", *Synthese*, 36, 271–281.

Feller, W. (1971) *An Introduction to Probability Theory and Its Applications*, Vol. II (John Wiley & Sons, Inc., New York), pp. 228–230.

de Finetti, B. (1974) *Theory of Probability*, 2 vols. (John Wiley & Sons, Inc., New York).

Heath, D. and W. Sudderth (1976) "de Finetti's theorem on exchangeable variables", *Amer. Stat.*, 30, 188–189.

Hewitt, E. and L.J. Savage (1955) "Symmetric measures on cartesian products", *Trans. Amer. Math. Soc.*, 80, 470–501.

Kendall, D.G. (1967) "On finite and infinite sequences of exchangeable events", *Studia Scv. Math. Hungar.*, 2.

Kyburg, H.E. and H.E. Smokler (1981) *Studies in Subjective Probability*, 2nd edn. (John Wiley & Sons, Inc., New York).

Lindley, D.V. (1976) "Inference for a Bernoulli process (a Bayesian view)", *Amer. Stat.*, 30, 112–119.

PART II

FINITE ADDITIVITY

Chapter 4

CONGLOMERABILITY AND COUNTABLE ADDITIVITY

BRUCE M. HILL* and DAVID LANE

University of Michigan and University of Minnesota

1. Introduction

Over a half century ago, de Finetti initiated a bold and imaginative approach to the theory of probability. He started with operational definitions of probability numbers and criteria for coherence, and with rare and admirable obstinacy he refused to compromise his study with restrictions imposed solely "for mathematical convenience". In particular, he insisted that, when probability numbers are assigned to subsets of a fixed set Ω, they be represented by finitely, but not necessarily countably, additive measures. Dispensing with the axiom of countable additivity has, we think, great value from a foundational and inferential point of view, but of course it introduces some new mathematical puzzles for those schooled in the Kolmogorov system. In this chapter we discuss an interesting, perhaps surprising, implication of the axiom of countable additivity, first conjectured by de Finetti (1972, p. 99). This implication has to do with the relation between conditional and unconditional probability.

Suppose Ω is a set. Let $\mathscr{P}$ be the set of subsets of Ω and $\mathscr{P}^0$ the nonempty elements of $\mathscr{P}$. A *probability* γ on Ω is a function on $\mathscr{P}$ satisfying

(a) for A in $\mathscr{P}$,

$$0 \leq \gamma(A) \leq 1 = \gamma(\Omega); \tag{1.1}$$

(b) if A and B are disjoint elements of $\mathscr{P}$,

$$\gamma(A \cup B) = \gamma(A) + \gamma(B).$$

*Research partially sponsored by the National Science Foundation under Grant MCS-8026531.

Bayesian Inference and Decision Techniques, edited by P. Goel and A. Zellner

A *full conditional probability* P on Ω is a function on $\mathscr{P} \times \mathscr{P}^0$ satisfying

(a) for each A in $\mathscr{P}^0$, $P(\circ|A)$ is a probability on Ω such that

$P(A|A)=1$;

(b) for $A \subset B \subset C \subset \Omega$, $B \neq \varnothing$, (1.2)

$P(A|C) = P(A|B)P(B|C)$.

Define the probability π by

$$\pi(A) = P(A|\Omega); \tag{1.3}$$

then π is called the *unconditional probability* for P. If π and P satisfy (1.3), we sometimes say P *extends* π. Objects satisfying (1.1) and (1.2) coincide in an obvious way with objects obeying axioms 1–3 of de Finetti (1975, pp. 335–340).

Now suppose $\mathscr{H}$ is a partition of Ω. A full conditional probability P is called *conglomerable with respect to* $\mathscr{H}$ if

for each A in $\mathscr{P}$,

$$\sup_{H \text{ in } \mathscr{H}} P(A|H) \geq \pi(A) \geq \inf_{H \text{ in } \mathscr{H}} P(A|H). \tag{1.4}$$

That is to say, if the conditional probabilities of an event, given partition elements, are restricted to a certain range of values, then the unconditional probability of the event must be in that range also. A P which is conglomerable with respect to every partition of Ω is called *conglomerable*.

At first sight, conglomerability appears to be almost a logical requirement for the theory of probability. Indeed, arguments deriving such a requirement, at least for certain special partitions, from criteria of coherence are presented, for example in Heath and Sudderth (1978) and Lane and Sudderth (1982). Yet numerous examples of de Finetti and others show that it is not really a necessary requirement in general; see de Finetti (1972, p. 102 and p. 205) and (1974, p. 178 and p. 275). Other examples have been introduced to show that the interpretation of some types of data support the need to violate conglomerability, as in Hill (1980).

The subject of this chapter is the connection between conglomerability of a full conditional probability and countable additivity of its unconditional probability. The connection can be seen most vividly when Ω is a countable set (and so, in particular, all partitions of Ω are countable), and so we shall henceforth suppose that Ω is countable. If π is countably additive on Ω, and P extends π, then, from (1.2b), if H has positive π-measure and A is any event,

$$P(A|H) = \frac{\pi(A \cap H)}{\pi(H)}.$$

Thus, if $\mathscr{H} = \{H_1, H_2, \ldots\}$ is a partition of Ω, for any event A,

$$\pi(A) = \sum_{i=1}^{\infty} \pi(A \cap H_i)$$

$$= \sum_{i=1}^{\infty} \pi(H_i) P(A|H_i). \tag{1.5}$$

Consequently, P is conglomerable with respect to every partition of Ω.

Krauss (1968) and Dubins (1975) have shown that every probability π can be extended to a full conditional probability. If π is not countably additive, there are many possible extensions of π. Are *any* of these extensions conglomerable? According to de Finetti's conjecture, the answer to this question is *no*: assuming the axiom of countable additivity is equivalent to requiring full conditional probabilities to be conglomerable with respect to every partition of Ω. (Note that section 3 below gives a more careful statement of this equivalence.)

In the next section of this chapter we verify de Finetti's conjecture. This result has already been given, along with other interesting material, in Schervish et al. (1984), in a more general setting and using different arguments. The purpose of our exposition is to present what is an important, but essentially simple, result in as elementary and self-contained a manner as possible. Consequently, our discussion, unlike that in Schervish et al. avoids references to general results in the theories of finitely additive probabilities and ultrafilters.

2. Theorems

The main result of this section is theorem 2.1, which shows by how much a full conditional probability whose unconditional probability is not countably additive fails to be conglomerable. Theorem 2.2, which characterizes the structure of diffuse probabilities, may be of independent interest.

Here is an outline of this section. In subsection 2.1 we state theorem 2.1. Its proof is divided into four cases, of which only one, case 3, is at all complicated. The argument for case 3 is presented in subsection 2.4; subsection 2.5 provides the proof of the essential step in this argument. The proofs for cases 1, 2, and 4 occupy subsections 2.2, 2.3, and 2.5, respectively.

2.1. Statement of the theorem

How nonconglomerable a full conditional probability P, with unconditional probability π, is can be measured by its *extent of nonconglomerability*, which is the supremum, taken over all subsets A and partitions $\mathscr{H}$ of Ω of the quantities:

$$\pi(A) - \sup_{H \text{ in } \mathscr{H}} P(A|H).$$

Define $\mathscr{A}$ and α as follows:

$$\mathscr{A} = \{\omega \text{ in } \Omega : \pi\{\omega\} > 0\}, \tag{2.1}$$

$$\alpha = \sum_{\omega} \pi\{\omega\}. \tag{2.2}$$

If more than one probability is under discussion, denote the $\mathscr{A}$ and α associated with γ as $\mathscr{A}(\gamma)$ and $\alpha(\gamma)$. Note that if γ is not countably additive, $\alpha(\gamma)$ is less than one.

Theorem 2.1. Suppose P is a full conditional probability on Ω, whose unconditional probability function π is not countably additive. Then P is not conglomerable, and the extent of its nonconglomerability is $1-\alpha$, where α is given by (2.2).

In order to prove theorem 2.1, suppose $\varepsilon > 0$ is fixed. We shall exhibit a subset A and a partition $\mathscr{H}$ of Ω satisfying

$$\pi(A) > 1 - \alpha - \varepsilon,$$

$$\sup_{H \text{ in } \mathscr{H}} P(A|H) < \varepsilon. \tag{2.3}$$

2.2. *Case 1: $\mathscr{A}$ is infinite and $\pi(\mathscr{A}) = 1$*

The idea in this case is simple. Since $\pi(\mathscr{A}) = 1$, essentially every singleton has π-mass attached to it. Thus, if H is a finite set and B is a subset of H, $P(B|H)$ is just the ratio of the masses attached to elements in B to masses attached to all the elements in H. But π is not countably additive, so it is not determined by the masses attached to singletons. We try to find a set A whose π-measure is much greater than the sum of the masses attached to its elements, and to form a partition whose elements are finite sets, each consisting of some elements in A combined with elements *not* in A which, collectively, have much greater mass attached to them.

To start, partition $\mathscr{A}$ into finite sets $L_1, L_2, \ldots$, satisfying, for each k,

$$\pi(L_k) > \varepsilon^{-1} \sum_{j>k} \pi(L_j). \tag{2.4}$$

(One way to do this is to arrange the elements of $\mathscr{A}$ in a sequence along which the π-mass is nonincreasing; take for L_1 a large enough initial segment of the sequence to satisfy (2.4); take a large enough initial segment of $\mathscr{A} - L_1$ for L_2; and so on). Next, choose an integer N larger than one, such that $\sum_N^\infty \pi(L_j) < \varepsilon$, and set $L = \bigcup_N^\infty L_j$. Then, since $\mathscr{A} - L$ is finite, $\pi(L) > 1 - \alpha$. For n an integer larger than ε^{-1}, partition L into $B_0, \ldots, B_{n-1}$ where, for $0 \le i \le n-1$,

$$B_i = \bigcup_{r=0}^{\infty} L_{N+i+rn}$$

(that is, B_i is the union of every nth L_j, starting with L_{N+i}).

Since the B_i's are disjoint, for some index, say r,

$$\pi(B_r) \le 1/n < \varepsilon;$$

set $A = L - B_r$. Thus,

$$\pi(A) > 1 - \alpha - \varepsilon. \tag{2.5}$$

To construct the partition $\mathscr{H}$, first define, for each non-negative integer k,

$$C_k = \bigcup_{i=0}^{n-1} L_{N+i+kn} - L_{N+r+kn},$$

so each C_k is a finite set, and the C_k's partition A. Now define

$$H_0 = L^c \cup C_0,$$

and for $k \ge 1$,

$$H_k = L_{N+r+(k-1)n} \cup C_k.$$

Then,

$$\begin{aligned} P(A \mid H_0) &= P(C_0 \mid H_0) \\ &= \frac{\pi(C_0)}{\pi(L^c) + \pi(C_0)} \\ &< \frac{\sum_N^{\infty} \pi(L_j)}{\pi(L_1)} \\ &< \varepsilon. \end{aligned} \tag{2.6}$$

The final inequality is from (2.4). A similar calculation gives, for $k \ge 1$,

$$\begin{aligned} P(A \mid H_k) &< \frac{\pi(C_k)}{\pi(L_{N+r+(k-1)n})} \\ &< \frac{\sum_{j > N+r+(k-1)n} \pi(L_j)}{\pi(L_{N+r+(k-1)n})} \\ &< \varepsilon. \end{aligned} \tag{2.7}$$

Together, (2.5), (2.6), and (2.7) are (2.3).

2.3. Case 2: $\mathscr{A}$ is infinite and $\pi(\mathscr{A}) < 1$

First, suppose that $\alpha = \pi(\mathscr{A})$. Let A be $\mathscr{A}^c$, write $A = \{a_1, a_2, \dots\}$ and $\mathscr{A} = \{b_1, b_2, \dots\}$, and set, for each positive integer k, $H_k = \{a_k\} \cup \{b_k\}$. Then,

$$\pi(A) = 1 - \alpha,$$

while for each k,

$$P(A \mid H_k) = \frac{\pi\{a_k\}}{\pi\{a_k\} + \pi\{b_k\}} = 0,$$

since b_k is in $\mathscr{A}$, while a_k is not. This verifies (2.3).

Now suppose that α is less than $\pi(\mathscr{A})$. The argument in the above paragraph still shows that P is not conglomerable, but it does not give the extent of its nonconglomerability. To do this, we combine it with one additional idea; we relativize P to $\mathscr{A}$, and apply the result for case 1.

Definition 2.1. Suppose P is a full conditional probability on Ω with unconditional probability π. Suppose F is a subset of Ω with positive π-probability. The *relativization* of P to F is the full conditional probability P_F on F given by

for C and B subsets of F, B nonempty,

$$P_F(C \mid B) = P(C \mid B).$$

The unconditional probability π_F for P_F is then given by

for B a subset of F,

$$\pi_F(B) = \frac{\pi(B)}{\pi(F)}.$$

Relativize P to $\mathscr{A}$ and notice that $\mathscr{A}(\pi_{\mathscr{A}}) = \mathscr{A}$, and $\alpha(\pi_{\mathrm{A}}) = \alpha(\pi)/\pi(\mathscr{A}) < 1$. Thus, $\pi_{\mathscr{A}}$ is not countably additive. Since $\pi_{\mathscr{A}}(\mathscr{A}) = 1$, the result of case 1 applies to $P_{\mathscr{A}}$. We may thus find a subset $\tilde{A}$ in $\mathscr{A}$ and a partition $\tilde{H}_1, \tilde{H}_2, \dots$ of $\mathscr{A}$ such that

$$\pi_{\mathscr{A}}(\tilde{A}) > 1 - \frac{\alpha}{\pi(\mathscr{A})} - \frac{\varepsilon}{\pi(\mathscr{A})},$$

and for each positive integer k,

$$P_{\mathscr{A}}(\tilde{A} \mid \tilde{H}_k) < \varepsilon.$$

Equivalently,

$$\pi(\tilde{A}) > \pi(\mathscr{A}) - \alpha - \varepsilon, \tag{2.8}$$

and for each positive integer k,

$$P(\tilde{A} \mid \tilde{H}_k) < \varepsilon. \tag{2.9}$$

Now write $\mathscr{A}^c = \{b_1, b_2, \ldots\}$, and set $A = \tilde{A} \cup \mathscr{A}^c$. For each positive integer k, set $H_k = \tilde{H}_k \cup \{b_k\}$, so the H_k's form a partition of Ω. We have [from (2.8)]:

$$\begin{aligned}\pi(A) &= \pi(\tilde{A}) + \pi(\mathscr{A}^c)\\ &> 1 - \alpha - \varepsilon.\end{aligned} \tag{2.10}$$

For each k,

$$P(A \mid H_k) = P(\tilde{A} \mid H_k) + P(\mathscr{A}^c \mid H_k). \tag{2.11}$$

Since b_k is not in $\mathscr{A}$, but $\tilde{H}_k$ is, H_k has positive π-probability and $\{b_k\}$ does not. So $\pi(H_k) = \pi(\tilde{H}_k) > 0$. Hence,

$$\begin{aligned}P(\tilde{A} \mid H_k) &= \frac{\pi(\tilde{A}\tilde{H}_k)}{\pi(\tilde{H}_k)}\\ &= P(\tilde{A} \mid \tilde{H}_k)\\ &< \varepsilon.\end{aligned} \tag{2.12}$$

The last inequality follows from (2.9).

Also,

$$P(\mathscr{A}^c \mid H_k) = \frac{\pi\{b_k\}}{\pi(H_k)} = 0. \tag{2.13}$$

Taking (2.11)–(2.13) together gives, for each positive integer k,

$$P(A \mid H_k) < \varepsilon. \tag{2.14}$$

Now (2.10) and (2.14) are (2.3).

2.4. Case 3: $\mathscr{A}$ is empty

If $\mathscr{A}$ is empty, π is *diffuse*: it gives zero probability to every finite set. Our strategy for proving theorem 2.1 in the case when π is diffuse is based upon theorem 2.2 below. Call a set B *σ-decomposable* with respect to a measure γ if B can be partitioned into an infinite number of subsets each with positive γ-measure. Call a set B *indecomposable* with respect to γ if B has positive γ-measure, and for each subset C of B, either C or $B - C$ has γ-measure zero.

Theorem 2.2. Suppose γ is a probability measure on Ω. Then either:

(1) γ admits a σ-decomposable set, or

(2) there is a finite partition of Ω, each of whose elements is indecomposable with respect to γ.

Proof. Let R denote the range of γ. Suppose R is infinite. Then there exists a nested infinite sequence of subsets of Ω, $B_1 \supset B_2 \supset \ldots$, such that, for $i = 1, 2, \ldots$, γ restricted to B_i has infinite range and $\gamma(B_{i+1}) < \gamma(B_i)$. Set $C_\infty = \bigcap_{i \geq 1} B_i$ and, for

$n=1,2,\ldots, C_n = B_n - B_{n+1}$. Then B_1 is σ-decomposable with respect to γ, since the C_i's are disjoint, $B_1 = \bigcup_{i=1}^{\infty} C_i$, and each C_i, except possibly C_∞, has positive probability.

Next, suppose R is finite. Let $\alpha_1 = \min\{\alpha : \alpha \text{ in } R, \text{ and } \alpha > 0\}$ and pick A_1 such that $\gamma(A_1) = \alpha_1$. Clearly, A_1 is indecomposable with respect to γ. Next, having selected $A_1, \ldots, A_j$, if $\gamma(\bigcup_{i=1}^{j} A_i)^c = 0$, adjoin $(\bigcup_{i=1}^{j} A_i)^c$ to A_j and terminate; otherwise, select A_{j+1} such that $\gamma(A_{j+1}) = \alpha_{j+1}$, where α_{j+1} is the minimum of the range of γ restricted to $(\bigcup_{i=1}^{j} A_i)^c$.

Suppose that the unconditional probability π admits a σ-decomposable set. It is easy to exhibit A and $\mathscr{H}$ verifying (2.3) in this case. Here is the argument. Suppose B is σ-decomposable and $\pi(B) < \varepsilon$. [To see that there is such a B, suppose that C is σ-decomposable, with $C = \bigcup_{i=1}^{\infty} C_i$ and $\pi(C_i)$ positive for each i; then choose n larger than ε^{-1} and, for $1 \le k \le n$, set $D_k = \bigcup_{j=0}^{\infty} C_{k+jn}$; now note that each D_k is σ-decomposable, and one of them has π-probability less than ε.] Partition B into sets $B_1, B_2, \ldots$, each with positive π-probability. Let A be the complement of B, and write $A = \{a_1, a_2, \ldots\}$. For $k \ge 1$, set $H_k = \{a_k\} \cup B_k$. Then A and $\mathscr{H}$ verify (2.3): in fact, since π is diffuse,

$$P(A \mid H_k) = \frac{\pi\{a_k\}}{\pi\{B_k\} + \pi\{a_k\}} = 0.$$

So, to finish proving theorem 2.1 when π is diffuse, we turn to unconditional probabilities which satisfy condition (2) of theorem 2.2. There are two steps to this part of the argument. First, we consider the special case when π is *remote*: that is, Ω itself is indecomposable with respect to π. The proof of theorem 2.1 when π is remote is given in subsection 2.5. Next, we note that condition (2) of theorem 2.2 just asserts that there is a finite partition of Ω such that the relativization of P to each element of the partition has a remote unconditional probability. Thus, the following lemma shows that theorem 2.1 is true for all π's which satisfy condition (2) of theorem 2.2, if it is true for remote π's.

Lemma 2.1. Suppose P is a full conditional probability on Ω with unconditional probability π. Suppose $B_1, \ldots, B_k$ partition Ω, each B_i has positive π-probability, and the relativization of P to B_i has extent of nonconglomerability 1 (for $1 \le i \le k$). Then P has extent of nonconglomerability 1.

Proof. Fix $\varepsilon > 0$. For each i between 1 and k, select a subset A_i and a partition $\mathscr{H}^i = \{H_1^i, H_2^i, \ldots\}$ of B_i satisfying

$$\pi_{B_i}(A_i) - \sup_{H \text{ in } \mathscr{H}^i} P_{B_i}(A_i \mid H) > 1 - \frac{\varepsilon}{2} \cdot \pi(B_i).$$

In particular, note that

$$\pi_{B_i}(A_i) > 1 - \frac{\varepsilon}{2} \cdot \pi(B_i)$$

and

$$\sup_{H \text{ in } \mathscr{H}^i} P_{B_i}(A_i|H) < \frac{\varepsilon}{2} \cdot \pi(B_i) < \frac{\varepsilon}{2}.$$

Set $A = \bigcup_{i=1}^k A_i$ and $\mathscr{H} = \bigcup_{i=1}^k \mathscr{H}^i$. Then $\mathscr{H}$ is a partition of Ω. Moreover

$$\begin{aligned}\pi(A) &= \sum_{i=1}^{k} \pi(A_i)\\ &= \sum_{i=1}^{k} \pi(B_i)\pi_{B_i}(A_i)\\ &> 1 - \frac{\varepsilon}{2}.\end{aligned}$$

Also, for H in $\mathscr{H}$, H is in some $\mathscr{H}^i$ for some index i, say r, and so

$$\begin{aligned}P(A|H) &= \sum_{i=1}^{k} P(A_i|H)\\ &= P(A_r|H)\\ &< \varepsilon/2.\end{aligned}$$

Hence

$$\pi(A) - \sup_{H \text{ in } \mathscr{H}} P(A|H) > 1 - \varepsilon.$$

2.5. *Remote* π

To prove theorem 2.1 when the unconditional probability π is remote, we introduce an ordering relation $<$ on the elements of Ω. For a and b in Ω, $a(\leq)b$ if and only if $P[\{a\}|\{a\}\cup\{b\}](\leq)\frac{1}{2}$. That this relation is transitive is an easy consequence of (1.2b). Arranging singletons according to this relation lets us exhibit the nonconglomerability of P. For example, suppose the elements of Ω were well-ordered by $<$:

$$\Omega = \{a_1, a_2, \ldots\} \text{ and } a_1 < a_2 < \ldots.$$

Let $O = \{a_1, a_3, \ldots\}$ and $E = \{a_2, a_4, \ldots\}$. Since $\Omega = O \cup E$, one or the other of these sets, say O, has π-probability one. Now partition Ω as follows: $H_1 = \{a_1, a_2\}$, $H_2 = \{a_3, a_4\}, \ldots$. Then, $P[O|H_j] = P[\{a_{2j-1}\}|\{a_{2j-1}, a_{2j}\}] < \frac{1}{2}$ since $a_{2j-1} < a_{2j}$. Thus, P is nonconglomerable.

Of course, $<$ need not well-order Ω. However, a relatively simple extension of this idea allows us to find a *usefully* ordered subset of Ω which gives the result of theorem 2.1 when π is remote.

To start, well-order Ω and call this ordering the w-order. The relation $<$ is as defined above; we refer to it as the $\not p$-order.

Arrange the elements of Ω in a doubly-infinite array as follows, where a_{ij} refers to the element in the ith row and jth column of the array, where i and j are positive integers.

(1) a_{11} is the w-smallest element in Ω. Complete the first column by selecting a_{k1} subject to (a) a_{k1} has not been previously selected, and (b) $a_{k1} \geq a_{(k-1)1}$. This process either generates an infinite column, or for some k it terminates. If the latter, a_{n1}, the last element, is strictly $\wp$-larger than any unselected element of Ω.

(2) a_{12} is the w-smallest unselected element of Ω. Proceed as in step (1) to construct the second column.

(3) Continue as in (a) and (b), constructing columns 3, 4,... (terminating, of course, if every element in Ω has been selected).

Note that since the mth w-smallest element of Ω appears in the first m columns of the array, every element of Ω appears in the array.

Let I be the union of all elements which appear in some *infinite* column of the array. Each element not in I appears in a column of finite length. Let c_i be the *last element* in the ith finite column. Then

$$c_1 > c_2 > \ldots .$$

We now construct a subset of I^c as follows. Set $d_{1,0} = c_1$. The only elements in I^c which are at least as $\wp$-large as c_2 must appear in the first two finite columns. Suppose there are m_1 such elements. Label them, in decreasing $\wp$-order, as $d_{1,1}, \ldots, d_{1,m_1}$. Then

$$d_{1,0} = c_1 \geq d_{1,1} \geq \ldots \geq d_{1,m_1} \geq c_2 .$$

Next, set $d_{2,0} = c_2$. The only elements in I^c which are at least as large as c_3 appear in the first three finite columns. Suppose there are m_2 such elements which are not yet included in the d-sequence. Label them, in decreasing $\wp$-order, as $d_{2,1}, \ldots, d_{2,m_2}$.

Continuing in this way yields a sequence $d_{i,j}$ which is lexicographically ordered. Let D consist of all the d_{ij}'s. For each positive integer i, there are only a finite number of elements in D with first index i. Thus, we may relabel the elements in D using the positive integers as index set, such that $D = \{d_1, d_2, \ldots\}$ and

$$d_1 \geq d_2 \geq \ldots . \tag{2.15}$$

Let $B = I^c - D$. Then, for each b in B and d in D, b is $\wp$-smaller than d by the construction of D. Since I, B, and D partition Ω, and π is remote, one of the three sets has π-probability one. We now consider each possibility in turn.

First, suppose $\pi(I) = 1$. Relabel the elements in I so that a_{ij} represents the element in the ith row of the jth *infinite* column in the array as constructed above. For each j, then

$$a_{1j} \leq a_{2j} \leq a_{3j} \leq \ldots . \tag{2.16}$$

Now pick an integer n larger than ε^{-1}, and for $1 \le i \le n$ set

$$C_i = \bigcup_{j=1}^{\infty} \bigcup_{k=0}^{\infty} \{a_{(kn+i)j}\}$$

(that is, having deleted all finite columns, take the union of every nth row, starting with the ith).

Since I is the union of the C_i's, one of them – say C_k – has π-probability equal to one. Let $A = C_k$.

To form the partition $\mathscr{H}$, for each pair of positive integers i and j, set

$$H_{ij} = \{a_{(i-1)n+kj}, a_{(i-1)n+k+1j}, \ldots, a_{in+k-1j}\}.$$

That is, H_{ij} consists of one element in $A(a_{(i-1)n+kj})$ and $n-1$ elements not in A, each of which is at least as $\not{p}$-large as the element in A, by (2.16). It is an easy consequence of (1.2b) that

$$P(A \mid H_{ij}) \le 1/n < \varepsilon.$$

Finally, let H_0 consist of all elements of Ω not in some H_{ij} (that is, B, D, and the first $k-1$ rows of the array). Since H_0 is disjoint from A, $P(A \mid H_0) = 0$. Thus, A and $\mathscr{H} = H_0 \bigcup_{i,j} H_{ij}$ verify (2.3), proving theorem 2.1 if $\pi(I) = 1$.

Next, suppose $\pi(D) = 1$. With n as above, let

$$C_0 = \{d_1, \ldots, d_n\},$$

and for $i = 1, \ldots, n$, let

$$C_i = \bigcup_{j=1}^{\infty} \{d_{jn+i}\};$$

that is, C_i consists of every nth element in D, starting with d_{n+i}. Now C_0 is finite, so $\pi(C_0) = 0$; since the C_i's partition D, and π is remote, for some integer between 1 and n, say k, $\pi(C_k) = 1$. Let A be C_k.

Construct the partition $\mathscr{H}$:

$$H_0 = I \cup B \cup \{d_k\},$$

and for $i \ge 1$,

$$H_i = \left[\bigcup_{j=1}^{n} \{d_{(i-1)n+j}\} - \{d_{(i-1)n+k}\}\right] \cup \{d_{in+k}\}.$$

That is to say, H_0 is disjoint from A, and, for $i \ge 1$, H_i consists of one element in $A(d_{in+k})$ and $(n-1)$ elements not in A, each at least as $\not{p}$-large as the element in A. The argument from here proceeds as in the previous $[\pi(I) = 1]$ case.

Finally, suppose $\pi(B) = 1$. Since B is infinite, D must be infinite also. Partition D into disjoint sets $D_1, D_2, \ldots$, each consisting of $(n-1)$ elements. For this case, A is the set $B = \{b_1, b_2, \ldots\}$. To form the partition $\mathscr{H}$, let H_0 be I, and for each positive integer i, set

$$H_i = \{b_i\} \cup D_i.$$

Again, H_0 is disjoint from A, and the other partition elements each contain one member of A and $(n-1)$ $\not{p}$-larger elements not in A.

2.6. Case 4: $\mathscr{A}$ is finite

Since $\mathscr{A}$ is finite, $\alpha = \pi(\mathscr{A})$. Relativize P to $\mathscr{A}^c$ and, since $\pi_{\mathscr{A}^c}$ is diffuse, apply case 3.

3. On the equivalence of conglomerability and countable additivity

3.1. Call a full conditional probability P countably additive if $P(\cdot \mid B)$ is countably additive for each nonempty subset B of Ω. If Ω is countable and P is conglomerable, the unconditional probability π must be countably additive, by theorem 2.1. More is true. If B has positive unconditional probability, then $P(\cdot \mid B)$ is obtained from π by relativization to B, and hence is countably additive also. So it is reasonable to ask: if P is conglomerable must it be countably additive also? The answer is no. By standard extension techniques, as in Dubins (1975), it is possible to have π countably additive, while $P(\cdot \mid B)$ is diffuse for some B with $\pi(B) = 0$. Notice that the relativization (in the obvious generalized sense) of P to B is then nonconglomerable, even though P itself is conglomerable – and the relativization P_B and P agree on all subsets of B. This is possible, of course, because, if $\pi(B) = 0$, the unconditional probability associated with P_B is not determined by π; and conglomerability has to do with the relation between a full conditional probability and the *relevant* unconditional probability. Thus, in the sense indicated by this paragraph, with Ω countable, requiring that *all* probabilities be countably additive is equivalent to requiring that *all* full conditional probabilities be conglomerable; even though, without such a requirement, conglomerable, noncountably additive full conditional probabilities exist (even abound).

What if Ω is not countable? It is easy to show that a full conditional probability P is conglomerable with respect to *countable* partitions if and only if its unconditional probability is countably additive; all you need to show is that, given a noncountably additive probability π on Ω, you can find a countable partition of Ω such that π restricted to the σ-algebra consisting of all unions of partition elements is not countably additive, and then use theorem 2.1. There is a little more to be done to get the extent of nonconglomerability; see Schervish et al. (1984). On the other hand, if you require conglomerability with respect to *all* partitions – even uncountable ones – full conditional probabilities with countably additive unconditional probabilities can fail to be conglomerable. It is clear from the context that de Finetti's conjecture had to do only with countable partitions.

References

de Finetti, Bruno (1972) *Probability, Induction and Statistics* (John Wiley & Sons, Inc.).

de Finetti, Bruno (1974) *Theory of Probability*, Vol. 1 (John Wiley & Sons, Inc.).

de Finetti, Bruno (1975) *Theory of Probability*, Vol. 2 (John Wiley & Sons, Inc.).

Dubins, Lester (1975) "Finitely additive conditional probabilities, conglomerability, and disintegrations", *Annals of Probability*, 3, 89–99.

Heath, David and William Sudderth (1978) "On finitely additive priors, coherence, and extended admissibility", *Annals of Statistics*, 6, 333–345.

Hill, Bruce M. (1980) "On some statistical paradoxes and nonconglomerability", in: J. Bernardo, M. De Groot, D. Lindley, A. Smith, eds., *Bayesian Statistics: Proceedings of the First International Meeting in Bayesian Statistics* (University Press, Valencia, Spain).

Krauss, P.H. (1968) "Representation of conditional probability measures on Boolean algebras", *Acta Math. Sci. Hungar. Tomus*, 19 (3–4), 229–241.

Lane, David and William Sudderth (1982) "Coherent predictive inference", University of Minnesota Technical Report No. 399.

Schervish, M., T. Seidenfeld and J. Kadane, (1984), "The extent of nonconglomerability of finitely additive probabilities", *Z. f. Wahrscheinlichkeits theorie*, 66, 205–226.

Chapter 5

STATISTICAL IMPLICATIONS OF FINITELY ADDITIVE PROBABILITY

JOSEPH B. KADANE, MARK J. SCHERVISH and TEDDY SEIDENFELD*

Carnegie-Mellon University

1. Introduction

In his classic monograph on the foundations of the theory of probability, Kolmogorov (1956) introduces a postulate [P-6], equivalent to the principle of countable additivity, which he justifies as a mathematical expedient for infinite probability structures. Countable additivity requires that the denumerable union of pairwise disjoint events has probability equal to the sum of the individual probabilities, i.e. if $A = \bigcup_{i=1}^{\infty} A_i$ $(A_i \cap A_j = \varnothing, i \neq j)$, then $P(A) = \sum_{i=1}^{\infty} P(A_i)$. His theory less P-6 is hereafter described as the theory of finitely additive probability.

Our dispute with Kolmogorov's characterization of countable additivity as an "expedient" does not stem from the fear of non-measurability arising in routine problems of statistical inference. For our purposes it is enough that probability be defined over a σ-field. The questions we ask are:

(1) Does countable additivity build in unexpected statistical consequences?

(2) Does Bayesian statistics mandate the added restrictions countable additivity imposes?

We argue that the answers are Yes and No (respectively).

Without countable additivity, finitely additive probability undergoes a failure of "conglomerability" (discussed in section 2). Without conglomerability, familiar

* We thank Morris DeGroot, Isaac Levi and Herman Rubin for their suggestions and criticisms.

Bayesian Inference and Decision Techniques, edited by P. Goel and A. Zellner

decision theoretic principles, e.g. admissibility (discussed in section 3) and other forms of dominance rules (discussed in section 4) are invalid. What reason is there for taking seriously the finitely additive theory if dropping countable additivity undermines what has become accustomed statistical decision theory?

First, if one adopts the standards of coherence defended by de Finetti (1974), Savage (1974), and Lindley (1981), one's beliefs are modeled by a probability that is finitely but not necessarily countably additive. (See Seidenfeld and Schervish, 1983, for discussion of this view.) Second, as we investigate in section 5, there are important connections between finite additivity and Bayesian reconstructions of standard (orthodox) statistical inference – reconstructions using "improper" priors in the fashion of Jeffreys (1961) and Lindley (1970). By interpreting improper distributions as finitely and not countably additive probabilities, the way is paved to resolve a host of anomalies thought by some to be evidence of inconsistency in Bayesian theory. In this we agree with Heath and Sudderth (1978), Hill (1980), and Levi (1980).

2. Non-conglomerability of merely finitely additive probability

Define P to be *conglomerable in* π (see de Finetti, 1972, p. 99) when for every event E such that $P(E|h_i)$ is defined for all i, and for all constants k_1, k_2, if $k_1 \le P(E \mid h_i) \le k_2$ for all $h_i \in \pi$, then $k_1 \le P(E) \le k_2$. That is to say, conglomerability asserts that, for each event E, if all the conditional probabilities over a partition π are bounded by two quantities, k_1 and k_2, then the unconditional probability for that event is likewise bounded by these two quantities. de Finetti draws attention to non-conglomerability of finitely additive probability in denumerable partitions.

Example 2.1 (due to P. Lévy, 1930; see de Finetti, 1972, p. 102). Let $P(\cdot)$ be a finitely additive probability defined over the field of all subsets of the denumerable set of points $\{\langle i, j\rangle : i, j \text{ are positive integers}\}$. Thus, one may think of $P(\cdot)$ defined over subsets of integer coordinate points $\langle i, j\rangle$ in the first quadrant. Constrain $P(\cdot)$ so that for any point $\langle i, j\rangle$, $P(\langle i, j\rangle) = 0$. Hence, $P(\cdot)$ is not countably additive. Also, for any pair $\langle i, j\rangle$, let $P(\langle i, j\rangle | B) = 0$ if B is an infinite set. Hence, conditional on an infinite set B, $P(\cdot | B)$ is again not-countably additive.

Now, using the finitely additive probability $P(\cdot)$ (given above), we see that conglomerability must fail with respect to the event $A = \{\langle i, j\rangle : j \ge i\}$. That is, let A be the region whose lower boundary is the diagonal "$i = j$". Consider the partition $\pi_1 = \{h_i : h_i = \{\langle i, j\rangle : j < \omega\}$ and $i < \omega\}$. That is, π_1 is the partition into "vertical" sections. Since, for each $i < \omega$, h_i is an infinite set, $P(A | h_i) = 1$ (since there are only finitely many points below the point $\langle i, i\rangle$, for each $i < \omega$). Let $\pi_2 = \{h'_j : h'_j = \{\langle i, j\rangle : i < \omega\}$ and $j < \omega\}$. That is, π_2 is the partition into "horizontal" sections. Since, for each $j < \omega$, h'_j is an infinite set, $P(A | h_j) = 0$

(since there are only finitely many points to the left of the point $\langle j, j \rangle$, for each $j < \omega$). Thus, conglomerability must fail for at least one of π_1, π_2 as $P(A)$ must differ from at least one of the two values 0, 1.

In a recent paper (Schervish, Seidenfeld and Kadane, 1984), we investigate several general questions about the existence and magnitude of failures of de Finetti's conglomerability principle. Using a result due to Dubins (1975) (that conglomerability in a partition is equivalent to "disintegrability" in that partition), we provide least upper bounds on the failures of conglomerability with respect to *denumerable* partitions. We also show that for finitely additive probabilities, non-conglomerability in *denumerable* partitions characterizes those which fail to be countably additive, confirming a statement of de Finetti (1972, p. 99).

3. Admissibility

Non-conglomerability of finitely additive probabilities, i.e. the phenomenon that an unconditional probability may lie outside the range of values of conditional probabilities over an exhaustive partition, quickly leads to a violation of a familiar decision-theoretic principle, admissibility.

Let O_1 and O_2 be two options in a decision and let π be a partition into states (independent of the options) such that for each state the same option, say O_1, is strictly preferred to the other option, O_2. Then O_2 is *inadmissible*, i.e. O_1 is strictly preferred to O_2 unconditionally.

However, if we consider a choice between O_1: bet on the event A (as defined in example 2.1) at even odds, and O_2: bet on the complementary event $\bar{A}$ at even odds, then for each $h_i \in \pi_1, O_1$ is strictly preferred (in expectation) to O_2; while for each $h'_j \in \pi_2$, O_2 is strictly preferred (in expectation) to O_1. It cannot be both that O_1 is strictly preferred to O_2 *and* that O_2 is strictly preferred to O_1. Therefore, failures of conglomerability entail failures of admissibility. This observation permits a simplification of Arrow's (1972) axiom system for preferences which include both a principle of Montone Continuity (Villegas, 1964) and an admissibility rule, called Dominance. Since the Dominance principle implies countable additivity, Monotone Continuity is redundant.

The fundamental property of admissible procedures, namely that they alone are Bayesian procedures or limits of them, fails for finitely additive probabilities. There are reasonable acts, Bayesian with respect to a finitely additive opinion, which are inadmissible.

Historically, admissibility took on its greatest importance for statistics when Stein (1955) showed that $\bar{x}$ is inadmissible as an estimate of the mean μ of a normal vector with dimension greater than two. Stein showed that drawing in the components of $\bar{x}$ toward an arbitrary origin was a strict improvement over $\bar{x}$ in the admissibility sense. Lindley (1962) later showed in a simple Bayesian model in which μ itself is considered to be normally distributed with mean μ_0, that $\bar{x}$

should be drawn toward μ_0 by an amount determined by the prior variance of μ around μ_0, and the sampling variance of $\bar{x}$ around μ. These observations in turn led to an increased interest in empirical Bayes methods, using the data to estimate hyperparameters like μ_C. In general, the move away from the automatic use of $\bar{x}$, occasioned by Stein's and Lindley's work, has been a healthy development for statistics, we think. Statisticians have been encouraged by these results to consult their prior beliefs more systematically than they had been before.

A consequence of accepting the coherence of merely finitely additive options is to reduce the importance of the concept of admissibility. In particular, there are finitely, but not countably additive, prior distributions for which $\bar{x}$ is an optimal Bayesian act regardless of the dimension of $\bar{x}$. We do not interpret this to mean that now people should go back to using $\bar{x}$ (if they ever stopped) without concern for their prior opinion. Rather, we think that our results re-emphasize the importance of using your opinion within a Bayesian paradigm that does not insist on countably additive "priors".

4. Coherence and "strong inconsistency" with finitely additive probability

Associated with a failure of conglomerability are instances of "strong inconsistencies", as Stone (1976) calls them. These are colorfully illustrated by the following game (Stone, 1981), which is a rewording of Stone's (1976) "Flatland" example.

A regular tetrahedral die is rolled a very large number of times. The faces of the die are labelled: e^+ (positron); e^- (electron); μ^+ (muon); and μ^- (antimuon). A record is kept of the outcomes subject to the constraint that if complementary events occur successively in the record, they "annihilate" each other and the record contracts, without trace of the "annihilation". Thus, the sequences $\ldots e^+e^- \ldots$, $\ldots e^-e^+ \ldots$, $\ldots \mu^+\mu^- \ldots$, and $\ldots \mu^-\mu^+ \ldots$, cannot occur in the record. At some (arbitrary) point in the sequence of rolls the player, who is ignorant of the outcome to date, calls for one last roll after which he is shown the final record. He is then asked to gamble on the outcome of the last toss of the die.

Consider a finitely additive prior over the countable set S of all possible states of the record prior to the final toss defined as follows. First arrange all elements of S into a single sequence, and let B_n be the set consisting of the first n of them. For $B \subseteq S$, let $\lambda_n(B) = \#(B \cap B_n)/n$. Take any limit point of the sequence λ_n as a prior λ. Since every finite subset of S has zero probability under λ, define for any finite set C, $\lambda(A \mid C) = \#(A \cap C)/\#(C)$. This set of conditional probabilities is consistent in the sense that if A and B are any subsets of S, and C is any finite subset such that $A \cap C \neq \varnothing$, $\lambda(B \mid A \cap C)\ \lambda(A \mid C) = \lambda(B \cap A \mid C)$. If the player is a Bayesian who adopts this prior distribution, then upon seeing the record he will assign equal probability to the four possible outcomes compatible with the record. For example, if the record ends $\ldots e^+\mu^+$, he will assign probability 1/4 to the four possible outcomes of the final toss. But then the player assigns probability

1/4 to each of the following four states of the record as it existed immediately before the final roll of the die:

$\ldots e^{+}$ (corresponding to a final toss landing μ^{+}),

$\ldots e^{+}\mu^{+}e^{+}$ (corresponding to a final toss landing e^{-}),

$\ldots e^{+}\mu^{+}e^{-}$ (corresponding to a final toss landing e^{+}) and

$\ldots e^{+}\mu^{+}\mu^{+}$ (corresponding to a final toss landing μ^{-}).

Thus, he assigns probability 3/4 to the event that the final toss resulted in an "annihilation" in the record for the final entry. Moreover, with the prior described above, the player assigns probability 3/4 to the event "annihilation" in the record for the final entry (call this event A) for each observation of a non-vacuous record. If he were to observe a blank record, then for certain the last roll resulted in A. Hence,

$$3/4 \leq P(A|x) \leq 1, \quad \text{for each observation } x. \tag{1}$$

Let us call the state of the record just prior to the final toss, θ, the parameter. Then, on the assumption that the die is fair,

$$0 \leq P(A|\theta) \leq 1/4, \quad \text{for each } \theta. \tag{2}$$

Note that $P(A|\theta) = 0$ only when $\theta = \theta_0$, corresponding to a blank record just prior to the final toss. $P(A|\theta) = 1/4$ for all other parameter states.

If conglomerability applies, then incoherence results as $3/4 \leq P(A)$ by (1) and $P(A) \leq 1/4$ by (2). It is trivial to "make book" in such circumstances, by betting at, say 1:3 odds both against and for the event A. The player who (incoherently) adopts conglomerability is led to accept these gambles, as shown by the two (inconsistent) inequalities for $P(A)$.

What if the player adopts conglomerability in only one of the two partitions? For instance, Levi (1980, pp. 284–287) requires conglomerability in the $\pi_\theta = \{\theta_i\}$ partition, as a consequence of his theory of "direct inference". It appears that most other writers assume conglomerability in this, the margin of the "parameter". The result is a finitely additive distribution which is coherent, in the sense that no finite collection of bets suffice to "make book". However, on pain of a sure loss, the player with such a coherent, finitely additive distribution will decline the offer of an infinite class of "fair" bets, where he is agreeable to every finite subclass of wagers. This is, according to Stone, a "strong inconsistency".

For example, let conglomerability apply in $\pi_\theta = \{\theta_i, i = 0, \ldots\}$. Then prior to the game the player has odds of 1:3 on A (an "annihilation" on the final roll), as $P(A) = 1/4$ (since $P(\theta_0) = 0$). Prior to the game, the player also holds the infinite collection of conditional odds, given $x = x_i (i = 0, \ldots)$, of (at least) 3:1 on A, for each possible observation x_i. But the player will not accept all the denumerably many called-off bets (called off in case $x = x_i$ fails to occur) at the conditional odds of 3:1 on A, while also agreeing to wager on A at the unconditional odds of 1:3. To do so would expose him to a sure loss. Of course, the player is willing to accept any finite subset of this infinite set of wagers. No finite subset is sufficient to fix a sure loss. Hence, the player is coherent in de Finetti's sense (1974, ch. 3).

A similar argument applies in case the player holds conglomerability only in $\pi_x = \{x_i\}$, the margin of the "observable". Then $P(A) = 3/4$. The requisite infinite class of "fair" gambles is constructed by considering the conditional odds on the event A, given the state of the record immediately prior to the final toss (θ_i). The player is agreeable to each of the denumerably many called-off bets on A, called-off in case $\theta = \theta_i$ fails, at odds of 1:3, as $P(A|\theta_i) \leq 1/4$ ($i = 0, \ldots$). Once again, the player will not accept all such bets at once. Instances of "strong inconsistencies" are thus not violations of coherence.

Additional cases of "strong inconsistency" (without failure of coherence in de Finetti's sense) can arise when a posterior probability is calculated by Bayes' Theorem for densities with an "improper" prior, subject to the assumption of conglomerability in the margin of the parameter (the unobserved quantity). Heath and Sudderth (1978) illustrate this with their example 5.2. We discuss their theory in the next section.

5. "Improper" priors, coherent posteriors, and attempts at curtailing non-conglomerability

The use of improper distributions to represent ignorance has a long history (see, for example, Jeffreys, 1961 [first edition 1939], Lindley, 1970 [first edition 1965], and Hartigan, 1964). Improper distributions are not probabilities because they assign infinite mass to the universal set, e.g. Lebesgue measure on the real line. An improper distribution may be translated directly into a set of finitely additive probabilities, as does Levi (1980, pp. 125–131) who treats improper distributions as σ-finite representations of finitely additive measures. Alternatively, an improper distribution may arise as a limit of finite measures which can be normalized to be probabilities (see Renyi, 1955).

The limits of the normalized measures often do not exist, however. For example, if the universal set is the real numbers, let λ_n be Lebesgue measure restricted to the interval $[-n, n]$. Each λ_n can be normalized to $\mu_n = (2n)^{-1}\lambda_n$ which is a probability. Whereas λ_n converges to Lebesgue measure on the entire line, the sequence $\{\mu_n\}_{n=1}^{\infty}$ does not converge to a countably additive measure. There exist subsequences of $\{\mu_n\}$, however, which converge to finitely additive probabilities. This can be seen as follows. Each probability μ_n is a function from a field of events $\mathscr{F}$ into the interval [0, 1]. The collection of all such functions is a compact space (in the product topology). Hence, every sequence $\{\mu_n\}$ in this space has a limit point. Since limits preserve finite additivity, each such limit point μ will be a finitely additive probability.

The same reasoning can be applied to any improper prior to produce a collection of finitely additive probabilities associated with it, as follows. For each improper prior λ there exist finitely additive probabilities μ, each of which is a limit point of a sequence of probabilities $\{\mu_n\}$, and each μ_n is λ restricted to a set of finite measures and normalized to be a probability. This establishes a connection between improper priors and finitely additive probabilities.

Having established the connection, we can ask if the inferences made using improper distributions remain valid in the finitely additive theory. For example, suppose the distribution of $X = (\bar{X}, S)$ given $\theta = (\mu, \sigma)$ is that $\bar{X}$ and S are independent, $\bar{X}$ has a normal $N(\mu, \sigma^2/n)$ distribution and nS^2/σ^2 has a chi-squared distribution with $n-1$ degrees of freedom. If we pretend that the improper prior $1/\sigma$ is a density for (μ, σ), then use of Bayes' Theorem for densities leads to the conclusion that the posterior distribution of μ given X is such that $(n-1)^{1/2}(\mu - \bar{X})/S$ has a t distribution with $n-1$ degrees of freedom, a proper distribution.

In their important paper, Heath and Sudderth (1978) take the following approach to inferences involving improper distributions. They define the posterior distribution of a parameter θ given the data X as the conditional distribution necessary to make the joint distribution of (X, θ) conglomerable in *both* the X and θ margins if such a posterior distribution exists. That is to say, if the conditional distribution of X given θ (i.e. the likelihood) is $p(\mathrm{d}x|\theta)$ and the prior for θ is $\pi(\mathrm{d}\theta)$, then the marginal for X is $m(\mathrm{d}x) = \int p(\mathrm{d}x|\theta)\pi(\mathrm{d}\theta)$ and the posterior $q(\mathrm{d}\theta|x)$, if it exists, will satisfy

$$\int\int \phi(x, \theta) q(\mathrm{d}\theta|x) m(\mathrm{d}x) = \int\int \phi(x, \theta) p(\mathrm{d}x|\theta)\pi(\mathrm{d}\theta), \tag{3}$$

for all bounded measurable ϕ. In the above example concerning (μ, σ), Heath and Sudderth find a class of measures, to which $\mathrm{d}\lambda = \mathrm{d}\mu\, \mathrm{d}\sigma/\sigma$ belongs, which have the property, among others, that there exists a sequence $\{B_n\}_{n=1}^{\infty}$ of sets with $0 < \lambda(B_n) < \infty$, and $\bigcup_{n=1}^{\infty} B_n$ equal to the space of all (μ, σ) pairs. They then form the sequence of probabilities $\lambda_n(\cdot) = \lambda(\cdot \cap B_n)/\lambda(B_n)$. Each limit point π of this sequence is a distinct finitely additive prior for (μ, σ) which has a "posterior" distribution for μ given X agreeing with the one obtained above by use of Bayes' Theorem for densities applied to the improper prior $\mathrm{d}\lambda$.

However, Heath and Sudderth (H–S) use the term "posterior" in an overly restricted sense, in our opinion. The implication of their definition is that if no such q exists, there is no "posterior". The work of Schervish, Seidenfeld and Kadane (1984) shows that under mild conditions, each merely finitely additive probability, even one that satisfies (3), will have countable partitions (margins) in which it is not conglomerable. The requirement that a probability be conglomerable in a pair of given (albeit uncountable) partitions before admitting the existence of a posterior is, perhaps, too harsh. After all, the prior $\pi(\mathrm{d}\theta)$, if it is merely finitely additive, will not be conglomerable in all countable partitions which result from coarsening the parameter space. Consider the following example:

Example 5.1. Assume $\theta \in \{0,1\}$, $X \in \{1,2,3,\ldots\}$ with a distribution satisfying $p(x|\theta) = 2^{-(x+\theta)}$ and $\pi(\theta) = \frac{1}{2}$. There are many such finitely additive distributions. Each has some probability adherent along the sequence of (x, θ) values $\{(n,1)\}_{n=1}^{\infty}$ in the sense of de Finetti (1974, p. 240). It follows that the marginal for X is $m(x) = (3)2^{-(x+2)}$ and Bayes' Theorem gives the posterior of θ given X

to be $q(\theta|x)=(2-\theta)/3$ for all x. Let $\phi(x,\theta)$ equal 1 if $\theta=1$ and zero otherwise. Then

$$\int\int\phi(x,\theta)p(\mathrm{d}x|\theta)\pi(\mathrm{d}\theta)=\tfrac{1}{2}$$

and

$$\int\int\phi(x,\theta)q(\mathrm{d}\theta|x)m(\mathrm{d}x)=\tfrac{1}{3},$$

where the integrals are defined as in Dunford and Schwartz (1958). Yet it is reasonable to claim that q is the posterior for θ given X once finite additivity is accepted. This example makes clear the need for a less restrictive definition of posterior distribution that will allow inference even when a probability cannot be made conglomerable in a specific partition.

Another problem with requiring probabilities to be conglomerable in a specific partition is illustrated by this example. Schervish, Seidenfeld and Kadane (1984) show that there are finitely additive probabilities on the space of all (x,θ) pairs which assign probability one to the points $\{(x,1)\}_{x=1}^{\infty}$, are conglomerable in the x margin, and assign zero probability to each individual point. Let P_1 be such a probability. There is a countably additive probability P_0 which assigns probability 2^{-x} to the point $(x,0)$ for each x. Of course, this probability is conglomerable in the x margin. But $P=\frac{1}{2}P_0+\frac{1}{2}P_1$ has exactly the form of the probability described in example 2.1, where we showed that P is not conglomerable in the x margin. Hence, the collection of probabilities conglomerable in a specified partition is not closed under convex combination, and H–S coherence is not preserved under mixtures of H–S coherent distributions.

Finitely additive distributions that are coherent in the usual sense but not H–S coherent reflect "strong inconconstencies" similar to the anomaly of Stone's example (section 4). Likewise, coherent but H–S incoherent distributions fail Robinson's (1979) criterion that there be no "super-relevant" betting procedures.

In contrast to H–S coherence, where "posteriors" are free of strong inconsistency in the two canonical margins (π_x and π_θ), Levi's (1980) theory requires little more than coherence in de Finetti's sense, and suggests one view of how to calculate posterior probability with an "improper" prior. Levi's (1980, p. 129) analysis supports the familiar manuever with the Bayes' formula if the likelihood function is countably additive.

Whether or not one is prepared to insist on H–S coherence (we are not), it seems reasonable to investigate "conditional properties" of coherent distributions. That is, from a Bayesian point of view the distinction between "absolute" and "conditional" probability is tenuous, at best. What is the effect of strengthening conglomerability to fix conditional probability values?

Conditional conglomerability. Let F be an event for which $P(\cdot|F)$ is defined, and let $\pi=\{h_i\}$ be a partition. Let $h_{iF}=h_i\cap F$, and assume $P(\cdot|h_{iF})$ is defined for all i. Then P is *conditionally conglomerable in* π *with respect to* (*given*) F if for each $E \subseteq F$, and for all constants k_1 and k_2 such that

$$k_1 \le P(E|h_{iF}) \le k_2, \quad \text{for all } i; \quad k_1 \le P(E|F) \le k_2.$$

Trivially, if conditional conglomerability with respect to all events is satisfied in a specified partition, then conglomerability also holds in that margin. The converse is false, however, as we show using the example constructed by Buehler and Feddersen (1963), which was a rebuttal to arguments of Fisher (1956). Thus, H–S coherence admits distributions which are conditionally H–S incoherent.

Let (x_1, x_2) be i.i.d. $N(\mu, \sigma^2)$, with both parameters "unknown". It is obvious that

$$P(x_{\min} \le \mu \le x_{\max}|(\mu, \sigma^2)) = 0.5, \quad \text{for each pair } (\mu, \sigma^2). \tag{4}$$

Let $t = (x_1 + x_2)/(x_1 - x_2)$. Buehler and Feddersen (1963) show that

$$P(x_{\min} \le \mu \le x_{\max}|(\mu, \sigma^2), \quad |t| \le 1.5) > 0.518, \tag{5}$$

for each pair (μ, σ^2), despite (4).

Moreover, Heath and Sudderth (1978) show there is a class of finitely additive "prior" probabilities over pairs (μ, σ^2) such that for each "prior" conglomerability is satisfied in *both* partitions $\pi_{(x_1, x_2)}$ and $\pi_{(\mu, \sigma^2)}$. (Note that each partition has cardinality of the continuum.) Also, they show that each finitely additive "prior" probability induces a familiar countably additive "posterior" distribution, where

$$P(x_{\min} \le \mu \le x_{\max}; (x_1, x_2)) = 0.5, \quad \text{for each pair } (x_1, x_2). \tag{6}$$

Now, since t is a function of the (x_1, x_2) pairs, we can partition the event $|t| \le 1.5$ by the set of pairs (x_1, x_2) for which this inequality holds. Thus, by conditional conglomerability applied to this partition of $|t| \le 1.5$, (7) is a consequence of (6):

$$p(x_{\min} \le \mu \le x_{\max} |\, |t| \le 1.5) = 0.5. \tag{7}$$

But, also we may partition the event $|t| \le 1.5$ into the continuum of states $(|t| \le 1.5, \mu, \sigma^2)$, so that with conditional conglomerability applied to this partition of $|t| \le 1.5$, (8) is a consequence of (5):

$$P(x_{\min} \le \mu \le x_{\max} |\, |t| \le 1.5) > 0.518. \tag{8}$$

At least one of (7) and (8) must fail, hence conditional conglomerability fails in at least one partition where conglomerability holds (see Seidenfeld, 1979).

The failure of conditional conglomerability in a partition that admits conglomerability is restricted to events that have zero probability. We show this as follows. Let $P(F) \neq 0$, and argue that

$$\begin{aligned}\int_\theta P(E|F,\theta)\,\mathrm{d}P(\theta|F) &= \int_\theta P(E|F,\theta)\frac{P(F|\theta)\,\mathrm{d}P(\theta)}{P(F)}\\ &= \int_\theta \frac{P(E\cap F|\theta)\,\mathrm{d}P(\theta)}{P(F)}\\ &= P(E\cap F)/P(F)\\ &= P(E|F),\end{aligned}$$

where the first equality is by conglomerability in π_θ and Bayes' Theorem, the second and fourth equalities are by the multiplication theorem for conditional probability, and the third equality is again by conglomerability in π_θ. Hence, in the Fisher–Buehler–Feddersen problem, it must be that $P(|t| \leq 1.5) = 0$.

Examples such as the Fisher–Buehler–Feddersen paradox, where conditional conglomerability fails in a margin for which conglomerability holds, illustrate the violation of Robinson's (1979) criterion that there be no "relevant" betting procedures. Thus, H–S coherence is insufficient to preclude "relevant" betting schemes, in Robinson's sense. Also, subject to conditional conglomerability, advantage may switch repeatedly between two gamblers by an iteration of who has last say in determining when a bet is "on". This is illustrated by Fraser's (1977) "balanced procedure".

In contrast to the anomaly of "strong inconsistency" that is a consequence of non-conglomerability, we can identify a "weak inconsistency" of finitely additive distributions that arises from failure of a simple dominance (admissibility) rule even where conglomerability applies. Consider the choice between "no-bet" and a gamble that pays off $1/i$ if θ_i occurs ($i = 1,\ldots$). Clearly, for each value of θ the gamble is strictly preferred to "no-bet". However, if one imposes a merely finitely additive probability distribution over θ, such that $P(\theta_i) = 0$ (all i), E(no-bet) = E(gamble) = 0. That is, one becomes indifferent between the two options even though simple dominance obtains.

We can express this failure of dominance as a violation of a strengthened version of conglomerability. That is, "weak inconsistency" results from violating the following:

$$\text{If there exists a constant } k, \text{ such that } P(E|\theta_i) \gtrless k \text{ for all } i, \text{ then } P(E) \gtrless k. \tag{9}$$

In terms of the above example, if $P(E|\theta_i) = 1/i$ and we receive \$1 if E occurs when we choose to gamble, then for each θ_i we gain (in expectation) $1/i$ if θ_i occurs. However, though $P(E|\theta_i) > 0$ (all i), $P(E) = 0$ under the merely finitely additive prior described above, in violation of (9).

A somewhat different example is as follows. Let $x \sim N(\mu, 1)$. As Heath and Sudderth (1978) show, suitably invariant "priors" over μ exist for which conglomerability is satisfied simultaneously in π_μ and π_x, and where (for each x)

$P(\mu|x)$ is $N(x,1)$. Hence, for each x, $P(x \leq \mu|x) = 0.5$. Assume that the prior chosen assigns positive probability to $\{x : x < 0\} = X^-$, i.e. $P(X^-) \neq 0$. (We make this assumption to avoid unnecessary questions about probability conditional upon an event of zero probability.) Since we know conglomerability holds in π_x (and since $P(X^-) \neq 0$) we have that $P(x \leq \mu|X^-) = 0.5$. However, we note that, for each μ, $P(x \leq \mu|X^-, \mu) > 0.5$. (In fact, $P(x \leq \mu|X^-, \mu) = 1$ for all $\mu \geq 0$.) Thus, $P(\cdot\,|X^-)$ fails (9) in π_μ, though conditional conglomerability, given X^-, holds.

Finally, we come to those "marginalization paradoxes" involving the transformation of continuous random variables. These include all but the first of those of (Dawid, Stone and Zidek, 1973) and (Stone and Dawid, 1972). Here, by transformation, e.g. changing from $N(\mu, \sigma)$ to (τ, σ), where $\tau = \mu/\sigma$, we create the situation that:

(i) $P(\tau|x) = P(\tau|t)$, for t a function of $x = \{x_1, \ldots, x_n\}$, $x_i \sim N(\mu, \sigma)$;

(ii) $P(t|(\mu, \sigma)) = P(t|\tau)$, i.e. t depends solely on τ;

(iii) conglomerability holds simultaneously in $\pi_{(\mu,\sigma)}$ and π_x;

(iv) yet the probability density $p(t|\tau)$ [from (ii)] does not factor the density $p(\tau|t)$ [from (i)] as a function of τ.

As is argued by Seidenfeld (1981) and Sudderth (1980), no violation of conglomerability is present here. Only (9) is violated. But, above we saw that (9) can fail in partitions where conglomerability holds. Where conditional conglomerability obtains, but (9) does not, we are faced with betting schemes that violate Robinson's (1979) criterion of "semi-relevance", though satisfying his requirement that there be no "super-relevant" or "relevant" betting policies.

In conclusion, we have noted several strengthened versions of conglomerability and have described statistical anomalies that reflect violation of each. Since each coherent finitely additive distributions must fail conglomerability in some denumerable partition (unless the distribution is countably additive), our reaction to these anomalies is to see them as further evidence that the attempt to modulate non-conglomerability (as in Robinson's criteria prohibiting betting schemes with "super-relevant" or "relevant" selections) is misguided. If conglomerability is necessary for "consistency", then nothing less than countably additive distributions suffice with denumerable partitions, and even "proper" priors may suffer non-conglomerability in non-denumerable partition, as we discuss in the next section.

6. Countable additivity and non-conglomerability in non-denumerable partitions

Since non-conglomerability in denumerable partitions characterizes merely finitely additive probability, it is a mistake to advocate *both* a finitely additive theory of probability *and* standards of coherence entailing dominance or disintegrability in particular partitions (margins). One may cite non-conglomerability as reason enough to reinstate countable additivity. Non-conglomerability may be too high a price to pay for the convenience of "uniform" distributions.

However, example 6.1 shows that non-conglomerability, which is characteristic of merely finitely additive probability in denumerable partitions, can occur with countably additive measures in non-denumerable partitions.

Example 6.1. Suppose X_1 and X_2 are independent standard normal random variables except that $X_2 = 0$ is impossible (this does not change the joint distribution function). Since X_1 and X_2 are independent, the distribution of X_2 given $A = \{X_1 = 0\}$ is standard normal. Using the usual transformation of variables technique, the conditional distribution of X_2 given $B = \{X_1/X_2 = 0\}$ is that of $(-1)^Y Z$, where $P[Y=1] = P[Y=2] = \frac{1}{2}$, Z has chi-squared distribution with two degrees of freedom, and Y and Z are independent. The two events A and B are identical, however.

The fact that the conditional distribution of X_2 given A differs from that given B is an example of the well-known Borel paradox (Kolmogorov, 1956). This paradox was also discussed by Hill (1980, p. 44). The seeming contradiction is often resolved by claiming that the transformation of variables only yields conditional probability given the sigma field of events determined by the random variable X_1/X_2, not given individual events in the sigma field. This approach is unacceptable from the point of view of the statistician who, when given the information that $A = B$ has occurred, must determine the conditional distribution of X_2. A more reasonable approach is to consider the theory of conditional probability spaces as defined by Dubins (1975), among others. In such a theory $P[X_2 \in E | A]$ means the conditional probability that X_2 is in the set E given the event A and is not relative to a sigma field. This, then, is the meaning of conditional probability one assumes when one conditions on the occurrence of a particular event.

To conclude example 6.1 suppose one must determine conditional probabilities given all events of the form $A_{a,c} = \{X_1 - cX_2 = a\}$. The distribution of X_2 given $A_{a,c}$ can be calculated from the transformation of variables technique, if one wishes, as

$$\text{Normal } N\left(\frac{-ac}{1+c^2}, \frac{1}{1+c^2}\right).$$

In particular, if $a = 0$ the distribution of X_2 is normal with mean zero and variance $(1+c^2)^{-1}$. Consider the uncountable partition of the sample space via the events $\{A_{0,c} | c \text{ a real number}\} = \theta$. Let Φ denote the standard normal distribution function and $E = \{X_2 > 1, X_1 \neq 0\}$. Then $P[E | A_{0,c}] = 1 - \Phi((1+c^2)^{1/2}) < 1 - \Phi(1) = P(E)$, for all c. Hence, P is not conglomerable in the partition θ. Theorem A1 in the appendix shows that non-conglomerability in uncountable margins is quite common for countably additive probabilities with continuous distributions, just as non-conglomerability in countable margins is quite common for merely finitely additive probabilities. In fact, the theorem implies that no matter how one defines conditional probability given $A_{a,c}$ in the above example, conglomerability will fail in some partition.

Hence, it appears that finite additivity cannot be rejected merely on the grounds that it allows failures of conglomerability, unless countable additivity is also to be rejected. Since few, if any, statisticians are willing to reject countable additivity, we suggest that finite additivity be judged on issues other than conglomerability.

7. Conclusion

What attitude does our research lead us to with respect to finite additivity? We are comforted to know that various seeming paradoxes in statistical theory are understood once the lack of conglomerability of merely finitely additive probabilities is accounted for. Thus, marginalization, the Fisher–Buehler–Feddersen paradox, etc. no longer are troublesome as problems illustrating a failure in the foundations of probability and (subjective Bayesian) statistical theory.

Should we then advocate the use of merely finitely additive probabilities as likelihoods and/or prior distributions? So much is unknown, or known only to a very few, about the probability theory of merely finitely additive probabilities that, as of 1985, it is very difficult to make such a judgement. On the one hand, merely finitely additive distributions do allow certain kinds of invariant distributions not allowed under the requirement of countable additivity, such as uniform distributions on countable sets, and translation invariant measures on the real line. For certain purposes these may be useful objects (Jeffreys, 1961; Zellner, 1971; Box and Tiao, 1973). Yet countably additive distributions do offer a wide range of expression of opinion as it is. For the moment, then, we propose the dual strategy of encouraging the development of the probability theory of finite additivity, while leaving open the matter of whether the extra generality permitted by going beyond countably additive opinions is worth the cost.

We conjecture that, ultimately, the combined force of de Finetti's argument that only finitely additive probabilities are required for coherence, and the kind of considerations that led Dubins and Savage (1965) to write their book in a finitely additive framework, will lead to a recognition that finitely additive probability is the proper setting for subjective Bayesian inference. The pace at which this innovation occurs will be governed largely, we think, by the development of the necessary probability theory.

Appendix: Non-conglomerability of countably additive probabilities

To continue example 6.1, consider two independent standard normal random variables X_1 and X_2 with $X_2=0$ impossible. Define $Y_c = X_1 - cX_2$ and

$Z_a = (X_1 - a)/X_2$. Then $A_{a,c} = \{Y_c = a\} = \{Z_a = c\}$. The conditional density of X_2 given $Y_c = a$, $f_c(x_2|a)$, is proportional to

$$\exp\left\{-\tfrac{1}{2}(1+c^2)\left[x_2 + ca(1+c^2)^{-1}\right]^2\right\},$$

and the conditional density of X_2 given $Z_a = c$, $f^a(x_2|c)$, is proportional to

$$|x_2|\exp\left\{-\tfrac{1}{2}(1+c^2)\left[x_2 + ca(1+c^2)^{-1}\right]^2\right\}.$$

The set $\{(a,c,x): f^a(x|c) = f_c(x|a)\}$ has zero three-dimensional Lebesgue measure. If we assume that the conditional density of X_2 given $A_{a,c}$, $f(x_2|a,c)$, is defined for all a and c and is a measurable function of (x_2, a, c), then the conditions of theorem A1 below are satisfied. The conclusion is that conglomerability fails in some uncountable partition. In fact, the partition will be the one determined by one of the random variables Y_c or one of the Z_a.

Theorem A1. Suppose two random variables (X_1, X_2) have a joint density $f(x_1, x_2)$ which is strictly positive over some measurable set G of positive Lebesgue measure. Let $\mathbb{R}$ denote the real numbers. Suppose there exist two sets of random variables $\{Y_c : c \in \mathbb{R}\}$ and $\{Z_a : a \in \mathbb{R}\}$, all measurable functions of (X_1, X_2), such that $Y_c = a$ if and only if $Z_a = c$, and (Y_c, X_2) and (Z_a, X_2) are each one-to-one functions of (X_1, X_2) for all a and c. Suppose the conditional density of X_2, given $A_{a,c} = \{Y_c = a\} = \{Z_a = c\}$, is $f(x_2|a,c)$ and is a measurable function of (x_2, a, c). Suppose that the transformation of variables technique gives the conditional density of X_2, given $Y_c = a$, as $f_c(x_2|a)$, and given $Z_a = c$, as $f^a(x_2|c)$, with the three-dimensional Lebesgue measure of $\{(a,c,x): f^a(x|c) = f_c(x|a)\}$ equal to zero. Then there exists $\delta > 0$, an event E, and a partition $\pi = \{h_b : b \in \mathbb{R}\}$ such that

$$P(E) - \int_\pi P(E|h_b)\,\mathrm{d}P(h_b) > \delta \text{(or } < -\delta) \tag{A1}$$

and conglomerability fails in the partition π.

Before proving theorem A1, we state and prove the following lemma.

Lemma A2. Under the conditions of theorem A1, there exists $\varepsilon > 0$, such that for some c the intersection of one of the following two sets with G has positive Lebesgue measure:

$$G_1(c,\varepsilon) = g_c^{-1}\{(a, x_2): f(x_2|a,c) - f_c(x_2|a) > \varepsilon\},$$

$$G_2(c,\varepsilon) = g_c^{-1}\{(a, x_2): f(x_2|a,c) - f_c(x_2|a) < -\varepsilon\}.$$

or for some a, the intersection of one of the following two sets with G has positive

Lebesgue measure:

$$G_3(a, \varepsilon) = (g^a)^{-1}\{(c, x_2): f(x_2|a, c) - f^a(x_2|c) > \varepsilon\},$$

$$G_4(a, \varepsilon) = (g^a)^{-1}\{(c, x_2): f(x_2|a, c) - f^a(x_2|c) < -\varepsilon\}.$$

Proof of lemma A2. Let $(Y_c, X_2) = g_c(X_1, X_2)$ and $(Z_a, X_2) = g^a(X_1, X_2)$ be the one-to-one functions guaranteed by the hypothesis of theorem A1. Since g^a and g_c are both one-to-one measurable functions, if D is any subset of G, then D has positive Lebesgue measure if and only if $g_c(D)$ has positive measure for all c and $g^a(D)$ has positive measure for all a. Assume that for all $\varepsilon > 0$, and a the intersections of $G_3(a, \varepsilon)$ and $G_4(a, \varepsilon)$ with G have zero Lebesgue measure. It follows that the intersection of G with

$$(g^a)^{-1}\{(c, x_2): f(x_2|a, c) \neq f^a(x_2|c)\}$$

has zero Lebesgue measure for all a. It follows that the Lebesgue measure of

$$g^a(G) \cap \{(c, x_2): f(x_2|a, c) \neq f^a(x_2|c)\}$$

is zero for all a. Integrating the measure of this set with respect to one-dimensional Lebesgue measure $\mathrm{d}a$ gives that the three-dimensional measure of

$$\left[\bigcup_a \{(a, c, x_2): (c, x_2) \in g^a(G)\}\right] \cap \{(a, c, x_2): f(x_2|a, c) \neq f^a(x_2|c)\} \tag{A2}$$

is zero. Call the set on the left of the intersect symbol in eq. (A2) A and call the set on the right B. Next, consider the set

$$A \cap \{(a, c, x_2): f(x_2|a, c) = f_c(x_2|a)\}. \tag{A3}$$

Call the set on the right of the intersect symbol in eq. (A3) C. It follows from the hypotheses of theorem A1 that $B^c \cap C$ has zero measure, where B^c is the complement of B. Write

$$A \cap C = (A \cap C \cap B) \cup (A \cap C \cap B^c).$$

Since $A \cap B$ and $C \cap B^c$ both have zero measure, so does $A \cap C$. Since $g_c(x_1, x_2) = (a, x_2)$ if and only if $g^a(x_1, x_2) = (c, x_2)$, it follows that we can also write A as

$$\bigcup_c \{(a, c, x_2): (a, x_2) \in g_c(G)\}.$$

The three-dimensional measure of $A \cap C$ equals the integral over $\mathbb{R}$ with respect to $\mathrm{d}c$ of the two-dimensional measure of

$$g_c(G) \cap \{(a, x_2): f(x_2|a, c) = f_c(x_2|a)\},$$

which must be zero for almost all c by Tonelli's Theorem. Since $g_c(G)$ has positive measure, the intersection of $g_c(G)$ with either

$$\{(a, x_2): f(x_2|a, c) > f_c(x_2|a)\}$$

or

$$\{(a, x_2) : f(x_2|a, c) < f_c(x_2|a)\}$$

has positive measure for some c. Hence, there exists $\varepsilon > 0$ and c such that the intersection of G with one of the sets $G_1(c, \varepsilon)$ or $G_2(c, \varepsilon)$ has positive measure. ■

Proof of theorem A1. In the notation of lemma A2, suppose that the intersection of $G_1 = G_1(c_0, \varepsilon)$ with G has positive Lebesgue measure. The proof is similar for the others. Let $f_{c_0}(a)$ denote the marginal density of Y_{c_0}. Consider the partition $\pi = \{h_a : a \in \mathbb{R}\}$, where $h_a = A_{a,c_0}$. This is the partition determined by Y_{c_0}, hence $\mathrm{d}P(h_a) = f_{c_0}(a)\mathrm{d}a$. If E is any event of the form $\{(X_1, X_2) \in D\}$, then

$$P(E) = \int\int_D f(x_1, x_2)\,\mathrm{d}x_1\,\mathrm{d}x_2 = \int\int_{g_{c_0}(D)} f_{c_0}(x_2|a) f_{c_0}(a)\,\mathrm{d}x_2\,\mathrm{d}a. \tag{A4}$$

Let $D = G_1$ and $E = \{(X_1, X_2) \in D\}$. Note that $P(E)$ is positive since the intersection of G and D has positive Lebesgue measure. Let $G_0 = g_{c_0}(G_1)$. Since g_{c_0} is one-to-one, eq. (A4) implies:

$$P(E) = \int\int_{G_0} f_{c_0}(x_2|a) f_{c_0}(a)\,\mathrm{d}x_2\,\mathrm{d}a. \tag{A5}$$

From the definitions of G_0 and G_1 and eq. (A5), it follows that

$$P(E) < \int\int_{G_0} f(x_2|a, c0) f_{c_0}(a)\,\mathrm{d}x_2\,\mathrm{d}a - \varepsilon\int\int_{G_0} f_{c_0}(a)\,\mathrm{d}x_2\,\mathrm{d}a. \tag{A6}$$

Since $P(E)$ is positive, it follows from eq. (A6) that f_{c_0} cannot be zero almost everywhere in G_0, hence the second integral on the right-hand side of eq. (A6) is positive. Let $\delta = \varepsilon\int\int_{G_0} f_{c_0}(a)\mathrm{d}x_2\,\mathrm{d}a > 0$. If we let $G(a) = \{x_2 : (a, x_2) \in G_0\}$, then $E \cap A_{a,c_0}$ equals $\{a\} \times G(a)$ and

$$P(E|A_{a,c_0}) = \int_{G(a)} f(x_2|a, c_0)\,\mathrm{d}x_2, \tag{A7}$$

by the definition of $f(x_2\ a, c_0)$. The first integral on the right-hand side of (A6) can be rewritten, using eq. (A7), as

$$\int_{\mathbb{R}}\int_{G(a)} f(x_2|a, c_0) f_{c_0}(a)\,\mathrm{d}x_2\,\mathrm{d}a$$

$$= \int_{\mathbb{R}} P(E|A_{a,c_0}) f_{c_0}(a)\,\mathrm{d}a = \int_{\pi} P(E|h_a)\,\mathrm{d}P(h_a). \tag{A8}$$

Together eq. (A6), eq. (A8), and the definition of δ imply eq. (A1) (with $< -\delta$). In the terminology of Dubins (1975), we have proven that the joint distribution of X_1 and X_2 is not disintegrable in π. Dubins proves that this implies that conglomerability fails in π also. ■

To conclude example 6.1, assume that the conditional density of X_2 given $A_{a,c}$ is given by $f(x_2|a,c)$. As long as this is a measurable function of (x_2,a,c), Theorem A1 shows that conglomerability fails in some partition determined by one of the Y_c or one of the Z_a.

References

Arrow, K.J. (1972) "Exposition of the theory of choice under uncertainty", in: C.B. McGuire and R. Radner, eds., *Decision and Organization* (North Holland, Amsterdam).

Box, G.E.P. and Tiao G. (1973) *Bayesian Inference in Statistical Analysis* (Addison-Wesley Publishing Co., Reading).

Buehler, R.J. and A.P. Feddersen (1963) "Note on a conditional property of Student's t", *Ann. Math. Stat.* 34, 1098–1100.

Dawid, A.P., M. Stone and J.V. Zidek (1973) "Marginalization paradoxes in Bayesian and structural inference", *J.R.S.S.*, Ser. B 35, 189–233 (with discussion).

de Finetti, B. (1972) *Probability, Induction and Statistics* (Wiley, New York).

de Finetti, B. (1974) *Theory of Probability*, Vol. 1 (Wiley, New York).

Dubins, L. (1975) "Finitely additive conditional probabilities, conglomerability and disintegrations", *Ann. Probability*, 3, 89–99.

Dubins, L. and L.J. Savage (1965) *How to Gamble if You Must: Inequalities for Stochastic Processes* (McGraw-Hill, New York).

Dunford, N. and J.T. Schwartz (1958) *Linear Operations, Part I: General Theory* (Interscience Publishers).

Fisher, R.A. (1956) "On a test of significance in Pearson's Biometrika Tables (No. 11)", *J. Roy. Statist. Soc.*, Ser. B 18, 56–60.

Fraser, D.A.S. (1977) "Confidence, posterior probability, and the Buehler example", *Ann. Stat.*, 5, 892–898.

Hartigan, J. (1964) "Invariant prior distributions", *Ann. Math. Statist.*, 35, 836–845.

Heath, J.D. and W. Sudderth (1978) "On finitely additive priors, coherence, and extended admissibility", *Ann. Stat.*, 6, 333–345.

Hill, B. (1980) "On some statistical paradoxes and non-conglomerability", in: Bernardo, DeGroot, Lindley and Smith, eds., *Bayesian Statistics* (University Press, Valencia, Spain), pp. 39–49 (with discussion).

Jeffreys, H. (1961) *Theory of Probability* (Oxford University Press).

Kolmogorov, A. (1956) *Foundations of the Theory of Probability* (Chelsea, New York).

Levi, I. (1980) *The Enterprise of Knowledge* (The MIT Press, Boston).

Lindley, D.V. (1962) "Discussion of C. Stein's paper, 'Confidence sets for the mean of a multivariate normal distribution'", *J. Roy. Statist.Soc.*, Ser. B 24, 285–287.

Lindley, D.V. (1970) *Introduction to Probability and Statistics* (Cambridge University Press).

Lindley, D.V. (1981) "Scoring rules and the inevitability of probability", unpublished report ORC 81-1, Operations Research Center, University of California at Berkeley.

Renyi, A. (1955) "On a new axiomatic theory of probability", *Acta Math. Acad. Sci. Hung.*, 6, 285–335.

Robinson, G.K. (1979) "Conditional properties of statistical procedures", *Ann. Stat.*, 7, 742–755.

Savage, L.J. (1974) *Foundations of Statistics* (Wiley, New York).

Schervish, M., T. Seidenfeld and J. Kadane (1984) "The extent of non-conglomerability of finitely additive probabilities", Z. f. Wahrscheinlichkeitstheorie, 66, 205–226.

Seidenfeld, T. (1979) *Philosophical Problems of Statistical Inference* (D. Reidel, Dordrecht, Holland).

Seidenfeld, T. (1981) "Paradoxes of conglomerability and fiducial inference", in: J. Los and H. Pfeiffer, eds., *Proceedings of the 6th International Congress on Logic, Methodology and Philosophy of Science* Amsterdam: (North Holland, Amsterdam).

Seidenfeld, T. and M.J. Schervish (1983) "A conflict between finite additivity and avoiding dutch-book", *Phil. Sci.*, 50, 398–412.

Stein, C. (1955) "Inadmissibility of the usual estimator for the mean of a multivariate normal distribution", in: *Third Berkeley Symposium* (University of California Press, Berkeley), pp. 197–206.

Stone, M. (1976) "Strong inconsistency from uniform priors", *J.A.S.A.*, 71, 114–116 (with discussion. 117–125).

Stone, M. (1981) "Review and analysis of some inconsistencies related to improper priors and finite additivity", in: J. Los and H. Pfeiffer, eds., *Proceedings of the 6th International Congress on Logic, Methodology and Philosophy of Science* (North Holland, Amsterdam).

Stone, M. and A.P. Dawid (1972) "Un-Bayesian implications of improper Bayes inference in routine statistical problems", *Biometrika*, 59, 369–375.

Sudderth, W. (1980) "Finitely additive priors, coherence and the marginalization paradox", *J. Roy. Statist. Soc.*, Ser. B 42, 339–341.

Villegas, C. (1964) "On quantitative probability σ-algebras", *Ann. Math. Statist.*, 35, 1787–1796.

Zellner, A. (1971) *An Introduction to Bayesian Inference in Econometrics* (Wiley, New York).

PART III

BAYESIAN METHODOLOGY IN ECONOMICS AND ECONOMETRICS

Chapter 6

SPECIFYING VECTOR AUTOREGRESSIONS FOR MACROECONOMIC FORECASTING

ROBERT B. LITTERMAN*

Federal Reserve Bank of Minneapolis

1. Introduction

This chapter describes a Bayesian solution to the problem of specifying a vector autoregression for forecasting macroeconomic variables. Although the specification process described here is unorthodox, it does have antecedents in two areas: the work of Hoerl and Kennard (1970), Stein (1974), Shiller (1973) and Leamer (1972, 1978) on shrinkage estimation and its Bayesian interpretation; and the work of Sims (1980, 1981, 1982) and Litterman (1980, 1983) on specifying loosely parameterized vector autoregressions. The motivation for this approach is described first, followed by an example that demonstrates how it can be used to improve macroeconomic forecasting.

2. The specification problem

Although it does not usually get much attention, the specification process is a critical stage in empirical analysis. In specifying a model, a researcher constructs the instrument that will be used to filter information from the data. Bayesian statistics focuses on the process of combining information in a prior with information from a set of data, and thus a Bayesian approach to the specification process naturally gives an important role to the construction of an informative

*The views expressed herein are those of the author and not necessarily those of the Federal Reserve Bank of Minneapolis or the Federal Reserve System.

Bayesian Inference and Decision Techniques, edited by P. Goel and A. Zellner

prior. A Bayesian specification search is described here in which the construction of a prior is viewed as a flexible means of generating a filter for the optimal extraction of information from a set of data.

The particular specification described here generates a set of loosely parameterized equations designed for forecasting macroeconomic variables. It differs significantly from the usual approaches to the specification of macroeconomic models, which adopt a much narrower focus. A specific behavior, often the solution to a particular optimization problem, is the guide to finding the right specification for an equation in most macroeconometric applications. This is a reasonable approach to explain many individual economic phenomena; but it seems likely to fail to explain aggregate economic behavior, which reflects the solutions of many different agents to a multitude of different problems. With respect to the cause and propagation of business cycles, for example, numerous explanations have been suggested, most probably containing a degree of truth and many probably having had empirical relevance at one time or another. It is doubtful, however, that any one paradigm can explain more than a very small part of the behavior of any macroeconomic time series.

For forecasting purposes, it seems quite reasonable to suppose that small bits of useful information concerning the macroeconomy are scattered throughout the data, and a narrowly focused approach is thus unlikely to find much useful information. The problem of econometric models is to filter out as much of the information as possible from the data and to give each little bit an appropriate weight. The proposed solution for the problem of filtering information useful for forecasting is motivated by two assertions, which I take to be self-evident, concerning the location and amount of information available for macroeconometric forecasting. The first assertion is that there is a very low signal-to-noise ratio in macroeconomic data, by which I mean that the predictable movement in such variables based on their own past is only a small fraction of their total variation. The second assertion is that the current state of macroeconomic theory leaves a great deal of uncertainty concerning which economic structures are useful for macro-forecasting, particularly at the level of detail that is necessary for short-run prediction.

From a Bayesian perspective, the latter assertion is that economic theory gives small positive probability to a large number of economic structures, each of which can be represented as an equation with a flat prior distribution over a wide range of parameter values. If that is true, then the true nature of prior information may be badly misrepresented by the usual approach of focusing on a particular economic theory and (unless it is rejected by the data) imposing the restrictions it suggests. Nonetheless, such restrictions are used because they are considered necessary to reduce the number of estimated parameters.

Unfortunately, testing economic hypotheses is especially difficult given the sample sizes of macroeconomic time series and the low signal-to-noise ratio in such data. When this is the case, the data alone cannot reject many restrictions, nor can they provide useful estimates, for forecasting purposes, of coefficients in an unrestricted specification. Faced with these problems, the specification process boils down to a set of decisions concerning what functional form to adopt and which explanatory variables to include. The choices that are made in specifying an equation determine, in effect, a filter that is used to extract information from the data. When an important variable is left out of the equation, a potential channel for extracting information in the data is closed. When too many variables are included in the equation, the filter is too wide and noise in the data will obscure the relatively weak signal. Thus, the choice of which variables to include involves a crucial tradeoff between oversimplification and overparameterization.

Neither of the two strategies commonly used in specifying macroeconomic forecasting equations addresses this tradeoff adequately. The structural approach relies on the restrictions suggested by one or more economic theories. The specification in this case includes a few of the many possible explanatory variables suggested by different economic theories. The alternative approach is to use a time series representation, which usually means inclusion of a few autoregressive and moving-average parameters. When such specifications are univariate, as is almost always the case, they cannot capture the interactions among variables. In this case, the focus on generating a parsimonious representation will lead to oversimplification of multivariate processes. In the standard approach to either structural or time series modeling, the tradeoff is crudely struck, essentially by restricting oneself to only a few parameters.

A Bayesian approach allows one to generate a class of estimators that highlight the tradeoffs between oversimplification and overparameterization. Moreover, for forecasting purposes, the use of out-of-sample prediction errors provides a natural measure for use in picking a specification that balances these tradeoffs.

Consider the following type of question, which typically arises in the context of a specification search: Should past values of variable x be included in the equation for another variable, y? Such a question is asked when one or more theories suggest that variations in x should contain information about future movements in y. There are probably many such x variables, however, only a few of which can reasonably be included in the equation for y. Economic theory does not provide enough information to make the decision. The usual approach is to choose the variables to include on the basis of statistics generated from classical tests of the hypotheses that the coefficients on variables in question are zero.

A Bayesian interpretation of the above procedure suggests that it is an inadequate way to combine prior information with evidence in the data. The procedure leads to a specification that can be viewed as reflecting very strong priors that coefficients are zero on all the excluded variables and as reflecting flat priors for the coefficients on the variables that are included in the equation. This result does not reflect the original symmetrical nature of the prior information.

As an alternative, I suggest a procedure in which the number of included explanatory variables is large, determined by computational expense considerations rather than the limited information content of the data; and in which the relatively symmetrical prior information on all of the included variables is used to balance the tradeoff between oversimplification and overparameterization. The basic idea is to specify a relatively unrestricted vector autoregression and a prior that can be varied along one or more dimensions affecting this tradeoff. Consideration of out-of-sample forecast errors as a function of movement in these dimensions is then used to find the optimal balance for forecasting purposes. In effect, these parameters are used to fine-tune the prior, which then acts as a filter to extract as much information from the data as possible. Implementation of this specification process is described in the context of the following example.

3. The model

The specification search described here is designed to generate a model for short-run forecasting of monthly values of a set of macroeconomic variables. Included are four variables of primary interest: measures of output, prices, interest rates, and money; along with three informational variables: a stock price index, the flow of total nonfinancial debt, and the change in business inventories. The data are described fully in the appendix. Observations begin in 1948 : 1 and end as of 1981 : 12. All variables, except changes in business inventories and the interest rate, are logged.

As a first step in the specification process, a set of benchmark univariate autoregressive forecasting equations is estimated by least squares. This benchmark, and the measures of fit that are used throughout, are based on out-of-sample forecast errors. Because I search over many forecasting models and choose among them based on this measure, it is true that the ultimate specification is, in a sense, fit "in-sample". It is important to recognize, however, that what is being fit is a balance between oversimplification and overparameterization based on the out-of-sample performance of estimators derived from a particular prior.

For each of the model specifications described in this chapter, the following procedure is used to generate the out-of-sample forecast errors. Estimation is carried out one observation at a time using a Kalman filter algorithm. The posterior distribution from a given set of observations is used as the prior distribution to be combined with the next observation. An approximation to the posterior mean is then used to generate a set of forecasts for 1 through 12 steps ahead. In making forecasts at each point in time, for computational reasons I followed the usual procedure of treating the coefficient estimates as fixed, even though that procedure does not generate the Bayesian posterior mean predictions that minimize mean square error. The updating and forecasting procedure is continued for each observation in the sample. The computations were carried out at the University of Minnesota using the Regression Analysis of Time Series (RATS) package, which is available from the author. Each cycle through the estimation and forecasting procedure required 20 seconds of computer time on a Cray 1 and cost about $10.

The out-of-sample forecast errors form the raw data for the measures of fit. The overall measure of fit is the log of the determinant of the sample covariance matrix of the one-step-ahead forecast errors. This criterion is suggested by Sims (1982) when the desire is to study a procedure's predictive accuracy and one does not "believe literally in the fine detail of a model's nominal probability structure", as is the case here. In addition to the log-det measure, I also calculate the standard errors of the forecasts for each variable. Both the log-det and the standard errors are calculated for forecast errors 1, 3, 6, and 12 steps ahead and for each of three subperiods as well as the full forecast sample. The purpose in considering these different measures is to be able to observe any systematic changes in the predictive accuracy of different models over time or with respect to different variables. The forecasts are begun in 1951 : 1, with the subperiods defined as follows:

(1) 1952 : 1 through 1961 : 12,
(2) 1962 : 1 through 1971 : 12,
(3) 1972 : 1 through 1981 : 12.

The full sample consists of the sum of the three subperiods and thus contains 360 forecast errors for each variable and for each horizon, a total of more than 10,000 forecast errors. The overall pattern of results was quite stable across subperiods, variables, and forecast horizons. To save space, few of these results are reported here; the complete set of results is available from the author on request.

The main purpose in estimating a set of univariate autoregressive models is to provide a benchmark against which to compare other specifications. Another natural choice would be to use ARIMA models as a benchmark. I rejected this approach primarily because of the computational expense it would have involved. Because they are nonlinear, even parsimoniously parameterized ARIMA models are expensive relative to the models estimated here. Moreover, the judgmental process involved in identifying ARIMA models makes them less suitable as benchmarks and raises difficult issues during implementation. For example, does one use the sample specification of (p, d, q) throughout, or how often does one identify the equation? Finally, as shown in Litterman (1983), low-order autoregressive models often perform better out of sample without the addition of moving-average terms.

These univariate autoregressive models also show how important the tradeoff is between oversimplification and overparameterization. For these models the tradeoff arises in the context of choosing a lag length, and the best forecasting performances were generated with a surprisingly small number of lags. There is a large literature addressing the problem of choosing lag length by associating penalties with longer lags and balancing these against the improvements in fit obtained as lags are increased. (See, for example, Geweke and Meese, 1981.) The best one-step-ahead forecasts, using the overall log-det measure, were generated with only four lags. These results are also not sensitive to the subperiod or forecast horizon. One might suppose, for example, when starting with more than

250 monthly observations and wanting to forecast a year ahead, that it would help to include at least a year's worth of lags; however, in general that is not the case. For all 12 of the combinations of subperiods and horizons, the four-lag specification by this measure performs better than 12 lags. In fact, for 10 of these 12 measures, even the one-lag specification performs better than 12 lags (see table 6.1).

Given a set of univariate benchmark models, it is natural to seek improvement in forecast performance by allowing multivariate interaction. This is the motivation for the second step in the specification process, which is to define a class of estimators that are derived as approximations to the posterior means associated with prior distributions. The priors are allowed to vary along a dimension that determines how likely coefficients on lags of other variables are to deviate from a

Table 6.1. Univariate autoregression forecast performance as measured by log determinants of forecast errors.

Number of lags in model	Overall	Subperiods 1	2	3
		One-step horizon results		
1	−42.312	−44.804	−45.936	−41.481
2	−42.643	−45.077	−46.489	−41.701
3	−42.615	−44.966	−46.543	−41.718
4	−42.702	−45.095	−46.746	−41.783
5	−42.676	−45.043	−46.693	−41.783
6	−42.626	−44.940	−46.673	−41.772
9	−42.481	−44.405	−46.584	−41.738
12	−42.190	−44.017	−46.307	−41.458
		Twelve-step horizon results		
1	−23.395	−23.781	−28.593	−25.519
2	−24.084	−24.878	−28.506	−25.171
3	−24.295	−24.696	−28.433	−25.459
4	−24.564	−24.730	−28.631	−25.967
5	−24.078	−24.016	−28.620	−25.863
6	−24.001	−23.637	−28.668	−25.756
9	−23.994	−22.939	−28.462	−25.876
12	−22.960	−21.168	−28.042	−25.809

One-step standard error results
(All units are percentages, except change in business inventories (Inven.), which is expressed in billions of 1972 dollars.)

Number of lags in model	GNP	M1	Stocks	Tbills	Debt	Prices	Inven.
1	0.8673	0.3817	3.330	0.5367	14.46	0.2073	5.582
2	0.8550	0.3808	3.275	0.5237	14.39	0.1862	5.493
3	0.8552	0.3823	3.290	0.5239	14.37	0.1856	5.488
4	0.8450	0.3818	3.291	0.5274	13.67	0.1855	5.499
5	0.8435	0.3800	3.295	0.5313	13.69	0.1862	5.516
6	0.8457	0.3812	3.306	0.5381	13.73	0.1866	5.506
9	0.8549	0.3847	3.300	0.5261	13.99	0.1893	5.617
12	0.8687	0.3891	3.322	0.5432	14.07	0.1940	5.715

prior mean of zero. Movement along this dimension includes the univariate specification at one end and an unrestricted vector autoregression at the other.

The representation to which this prior applies is a sixth-order autoregression of the vector of current observations, X_t, given by

$$X_t = \sum_{s=1}^{6} A_s X_{t-s} + C + \varepsilon_t, \tag{1}$$

where A_s are 7×7 coefficient matrices and C is a vector of constant terms. The choice of six lags is motivated by computational considerations, as is the limitation of the estimation procedures to single-equation techniques. It would be prohibitively expensive to form the more accurate approximation of the posterior mean generated by stacking the coefficient vectors of the multiequation system. A viable alternative suggested by Sims (1982) would be to form a recursive system for which single-equation methods are appropriate. Although we would expect some improvement when applying such a system to the exercises considered here, the results would probably not be very different because the forecast errors are not based on contemporaneous observations. In a practical forecasting problem, however, some variables are observed before others, and when that is the case this alternative approach may have significant benefits. On the other hand, such an approach is not simply a different estimation technique; it also forces one to make substantive decisions about an ordering of the variables.

At this stage in the specification process the prior distribution is flat with respect to own-lag coefficients and has a mean of zero on lags of other variables in each equation. In the prior, the distributions for each coefficient are treated as independent and normal. The variances of the distributions for coefficients on lags vary as a function of the lag number, being tighter around lags further back in the distribution. The variance of the prior distribution for the coefficient on lag l of variable j in equation i is given by the formula:

$$\frac{\pi_1 \sigma_i^2}{l \sigma_j^2}. \tag{2}$$

For this prior, and all others described below, the constant term in each equation is given a prior with a mean of zero and a variance 10^6 times π_1 to represent lack of knowledge about the means of the variables. The forecast results were not significantly affected by order-of-magnitude changes in this variance.

The prior distribution is parameterized by π_1, which determines the tightness of the prior around zero for each of the coefficients on variables other than own lags in each equation. The σ's are the standard errors of errors in six-lag univariate regressions over the full sample; they are included to scale the prior to make it independent of the units of the variables. The computation of the posterior mean is an application of Theil's (1971) mixed estimation technique, as described in Litterman (1980). The basic idea is to formulate a prior as a set of dummy observations that is added to the data. Ordinary least squares is then applied, producing an approximation of the posterior mean. The updating of estimates is accomplished using the Kalman filter algorithm, a description of which can be found in Bertsekas (1976).

By varying π_1 and following the out-of-sample forecasting procedure described above, it is possible to map different prior distributions, which vary according to how much multivariate interaction is allowed, into a measure of predictive accuracy. The results of such an experiment are shown in table 6.2. This mapping is the first example illustrating the assertion made above that a Bayesian procedure could generate a class of models in which a balance could be struck between oversimplification and overparameterization.

For this mapping, the results show that such a balance requires tight priors around zero on coefficients of other variables in each equation. In the model that forecasts best, overall, the variance around the first lag (aside from a scale factor) is only 0.0001. Although the improvement is not large in absolute magnitude, it is quite consistent across variables, subperiods, and forecast horizons. Notice also that the largest improvement is apparent in subperiod 1, followed by subperiod 2 and then subperiod 3. This result is consistent with the expectation that prior information is most helpful when there is the least amount of data.

Table 6.2. Forecast results from the search along a dimension of the prior allowing varying amounts of multivariate interaction.

One-step log-det results

Values of π_1	Overall	Subperiods 1	2	3
Univariate	−42.626	−44.940	−46.673	−41.772
0.00005	−42.692	−45.091	−46.809	−41.849
0.0001	−42.696	−45.079	−46.787	−41.904
0.0005	−42.689	−45.005	−46.702	−42.019
0.001	−42.678	−44.982	−46.646	−42.022
0.005	−42.615	−44.891	−46.417	−41.927
0.01	−42.535	−44.736	−46.269	−41.832
0.05	−42.140	−43.989	−45.834	−41.455
0.1	−41.885	−43.524	−45.632	−41.250
1.0	−40.881	−41.612	−45.103	−40.712

One-step standard error results

Values of π_1	GNP	M1	Stocks	Tbills	Debt	Prices	Inven.
Univariate	0.8457	0.3812	3.305	0.5380	13.72	0.1866	5.506
0.00005	0.8394	0.3790	3.328	0.5344	13.45	0.1837	5.513
0.0001	0.8357	0.3780	3.329	0.5339	13.40	0.1843	5.518
0.0005	0.8272	0.3740	3.336	0.5298	13.35	0.1885	5.536
0.001	0.8257	0.3713	3.343	0.5264	13.36	0.1911	5.545
0.005	0.8325	0.3644	3.378	0.5148	13.55	0.1986	5.590
0.01	0.8423	0.3626	3.402	0.5092	13.71	0.2029	5.628
0.05	0.8844	0.3640	3.494	0.4995	14.32	0.2176	5.790
0.1	0.9111	0.3671	3.552	0.4974	14.71	0.2258	5.894
1.0	1.0361	0.3831	3.779	0.4966	16.22	0.2589	6.376

To get a rough guide to the interpretation of the magnitude of changes in the log-det statistics, notice that (aside from the covariance terms) changes in the log-det approximate a sum of the percent changes in the variance of forecast errors from each equation. Thus, to get a rough estimate of the average percentage of reduction in forecast standard errors in each equation, multiply the change in the log-det by $100/(7\times 2)$, approximately 7. By this guide the reduction in the one-step log-det, from -42.626 in the univariate case to -42.696 for the best value of π_1, represents an average reduction of about 0.5 percent. The actual average reduction in standard errors in this case was 0.78 percent.

Starting from the $\pi_1 = 0.001$ model of the previous exercise, it is natural to ask what possible improvement could be obtained by restricting the own-lag coefficients in some manner. This value of π_1 was chosen despite the fact that the overall log-det was minimized with $\pi_1 = 0.0001$. There are several reasons for this choice. First, notice that this value of π_1 is best in the most recent subperiod. Also, the differences in log-dets between the values of π_1 in this range of values are not large. Finally, these results were obtained with a flat prior on own-lag coefficients and no differentiation among coefficients of variables, except own versus other. When, in later steps in the specification process, such additional prior information is added, one might expect more room for other variables to enter in each equation. In effect, this search over π_1 is serving several roles, one of which is to describe the tradeoff along this dimension and another to serve as a rough guide to the location in a larger dimensional space where the best forecasting performances can be found. If the computational expense were not excessive, one would prefer to search directly over the larger space. The prior mean for the own lags is taken to be a random walk specification; that is, all coefficients are given a prior mean of zero except the own first lag, which is given a prior mean of one. Another parameter, π_2, determines a variance of the prior distributions around this mean. The variance of own lag l is given by the formula:

$$\pi_2/l. \tag{3}$$

Again, by varying π_2 a class of estimators is generated that balances tradeoff between oversimplification and overparameterization, and again there is an interior minimum with respect to movements along this dimension. The results of this search are given in table 6.3.

The specification of π_1 and π_2 has generated a prior-shown schematically in fig. 6.1. Although this prior contains no economic theory and treats each equation identically, it can lead to improvements in forecasting. The symmetry of the prior is limiting and unnecessary, however, and a reasonable next step is to ask whether further improvements can be obtained by specifying prior information that is more specific to each equation. One might want to include in a prior, for example, one's knowledge that stock prices are more likely to respond like a random walk than are changes in business inventories. A way to include this information is to vary the variances of distributions for coefficients on different variables in each

Table 6.3. Forecast results from the search along a dimension of the prior that imposes a random walk specification for own lags.

One-step log-det results

Values of π_2	Overall	Subperiods 1	2	3
Infinite	−42.678	−44.982	−46.646	−42.022
1.0	−42.717	−45.150	−46.668	−42.034
0.1	−42.790	−45.311	−46.694	−42.094
0.05	−42.808	−45.344	−46.683	−42.127
0.01	−42.772	−45.261	−46.547	−42.182

One-step standard error results

Values of π_2	GNP	M1	Stocks	Tbills	Debt	Prices	Inven.
Infinite	0.8257	0.3713	3.343	0.5264	13.37	0.1912	5.546
1.0	0.8251	0.3686	3.340	0.5245	13.34	0.1909	5.539
0.1	0.8225	0.3672	3.328	0.5201	13.30	0.1901	5.519
0.05	0.8221	0.3672	3.324	0.5185	13.32	0.1902	5.519
0.01	0.8270	0.3671	3.325	0.5166	13.50	0.1932	5.588

equation. For an n-variable vector autoregression, there are n decisions to be made for each equation. Another way to differentiate different equations is to change the prior mean; in particular, for variables that are clearly stationary, it may be appropriate to use a prior mean on the first own-lag coefficient that is somewhat less than 1.

Although we could use intuition guided by economic theory to suggest more or less weight for each variable in a particular equation, it is much more challenging to translate those qualitative feelings into a quantitative prior. One way to proceed is to specify a set of relative weights for the purpose of scaling the variances for each variable in each equation, and then to search over a dimension that determines how much effect these weights have. Such a search is demonstrated here by specifying a set of weights, w_{ij}, and defining the variance of the distribution for lag l of variable i in equation j to be

$$\frac{\pi_1\sigma_i^2\exp(-\pi_3 w_{ij})}{l\sigma_j^2}. \tag{4}$$

The search over π_3 defines a dimension of more or less differentiation among variables. When π_3 is zero, all variables are treated symmetrically; as π_3 increases, the limiting specification has zero restrictions on all coefficients for which w_{ij} is positive. The weights used in this search are given in table 6.4. Note that the larger the weight, the faster the decrease of the variance on that coefficient as π_3 increases. The positive diagonal elements for stock prices and interest rates, for example, reflect a prior that those variables are more likely to follow a random walk process.

At the same time that these asymmetries in the prior variances are introduced, the mean values for the first lag of two variables – changes in business inventories

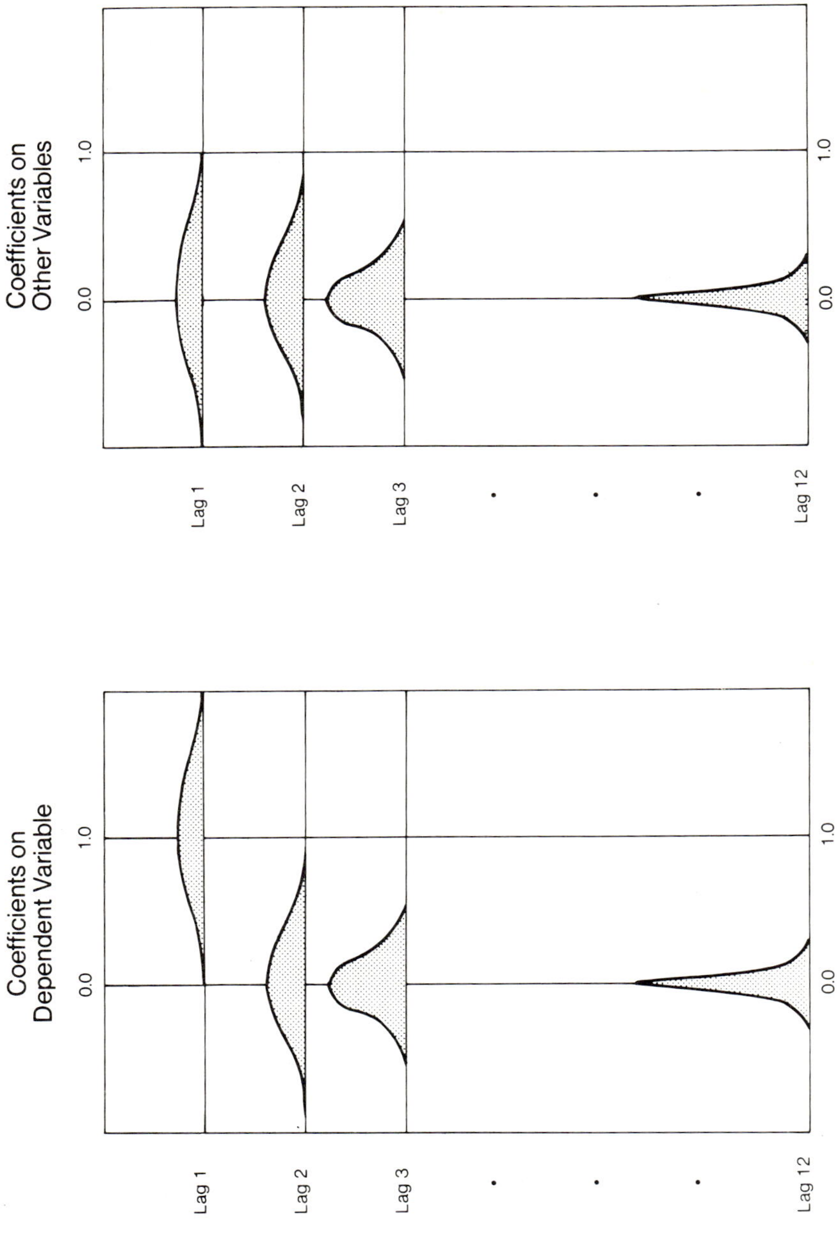

Fig. 6.1. A schematic representation of the prior.

Table 6.4. Relative weights of variables in different equations.

Equation	GNP	M1	Stocks	Tbills	Debt	Prices	Inven.
GNP	0.0	1.0	1.0	0.0	1.0	1.0	1.0
M1	1.0	0.0	1.0	0.0	1.0	1.0	1.0
Stocks	5.0	5.0	1.0	3.0	5.0	5.0	5.0
Tbills	2.0	1.0	2.0	1.0	2.0	2.0	2.0
Debt	1.0	1.0	1.0	0.0	0.0	1.0	1.0
Prices	1.5	1.0	1.5	1.0	1.5	0.0	1.5
Inven.	1.5	3.0	3.0	1.5	4.0	3.0	0.0

and GNP – are changed in their own equation from 1 to 0.8 and 0.95, respectively. The prior means of 1 seemed reasonable for the other variables. The value of π_2 was set at 0.05 for this search, and π_1 remained at 0.001. The results of the search are shown in table 6.5. The effects of the changes in the prior means alone can be seen by comparing the line $\pi_3 = 0.0$, in table 6.5, with the line $\pi_2 = 0.05$, in table 6.3.

The final step in this specification search is to estimate a time-varying parameter representation. Up to this point, the underlying model has been one in which it is assumed that the true coefficients are constant through time. At best, this is thought to be a reasonably good approximation, which is necessary given the limitations in the information content of the data. We could, however, proceed in a manner similar to the three searches described above and generate a class of estimators that include constant-coefficient specifications as one special case and that allow time-varying parameters with various amounts of freedom. Again, by varying a parameter, π_4, a balance can be obtained between the oversimplification of a constant-coefficient specification and the overparameterization of a time-varying coefficients model. Sims (1982) reports an informal search of exactly this

Table 6.5. Forecast results from the search along a dimension of the prior that allows asymmetric treatment of variables in different equations.

One-step log-det results

		Subperiods		
Values of π_3	Overall	1	2	3
0.0	−42.806	−45.346	−46.685	−42.122
1.0	−42.854	−45.319	−46.755	−42.169
1.5	−42.868	−45.326	−46.793	−42.164
2.0	−42.875	−45.338	−46.810	−42.159
3.0	−42.871	−45.334	−46.819	−42.122

One-step standard error results

Values of π_3	GNP	M1	Stocks	Tbills	Debt	Prices	Inven.
0.0	0.8220	0.3672	3.324	0.5185	13.32	0.1902	5.510
1.0	0.8248	0.3686	3.313	0.5227	13.29	0.1847	5.501
1.5	0.8261	0.3689	3.305	0.5246	13.28	0.1829	5.499
2.0	0.8273	0.3690	3.299	0.5258	13.27	0.1818	5.492
3.0	0.8292	0.3687	3.305	0.5259	13.26	0.1813	5.488

Table 6.6. Forecast results from the search along a dimension of the prior that allows time-varying coefficients.

One-step log-det results

		Subperiods		
Values of π_4	Overall	1	2	3
0.0	−42.876	−45.338	−46.810	−42.159
0.1×10^{-8}	−42.884	−45.338	−46.825	−42.170
0.5×10^{-8}	−42.900	−45.337	−46.854	−42.192
0.1×10^{-7}	−42.910	−45.336	−46.870	−42.205
0.5×10^{-7}	−42.920	−45.324	−46.887	−42.220
0.1×10^{-6}	−42.913	−45.308	−46.881	−42.207
0.1×10^{-5}	−42.785	−45.118	−46.766	−42.031

One-step standard error results

Values of π_4	GNP	M1	Stocks	Tbills	Debt	Prices	Inven.
0.0	0.8273	0.3690	3.299	0.5258	13.27	0.1818	5.492
0.1×10^{-8}	0.8264	0.3680	3.299	0.5258	13.27	0.1817	5.492
0.5×10^{-8}	0.8239	0.3661	3.300	0.5257	13.27	0.1819	5.493
0.1×10^{-7}	0.8222	0.3651	3.300	0.5257	13.27	0.1821	5.493
0.5×10^{-7}	0.8202	0.3644	3.301	0.5254	13.27	0.1824	5.493
0.1×10^{-6}	0.8216	0.3654	3.302	0.5250	13.27	0.1825	5.494
0.1×10^{-5}	0.8385	0.3756	3.307	0.5219	13.32	0.1842	5.509

type. Estimation of a time-varying parameters representation requires a specification of the nature of this variation and of the variances of the disturbances that cause the parameters to change each period. The process considered here follows Sims' suggestion that the parameters follow a random walk and that the variances of the disturbances for each coefficient be proportional to their variances in the prior distribution. The factor of proportionality is π_4; thus, when $\pi_4=0.0$ a constant-coefficients result is obtained. As π_4 increases, more parameter variation is allowed. The search takes values of π_1 and π_2 as before, and π_3 is set at 2.0. The results of this search, displayed in table 6.6 again indicate that an improvement in forecast performance is possible. A comparison of the forecast results between the best univariate benchmark model and those generated with the best value of π_4 shows an average reduction of 1.74 percent in the standard errors.

4. Conclusion

This chapter has attempted to motivate and illustrate a specification search for a vector autoregressive representation to be used in short-run macroeconomic forecasting. In designing a model for use in forecasting, we encounter a particularly crucial tradeoff between oversimplification and overparameterization. A Bayesian approach to the specification process suggests several ways to balance this tradeoff. An example demonstrates that such a procedure can lead to consistent improvements in out-of-sample forecast performance.

Appendix

The seven variables used in this study are monthly measures of real GNP, the GNP deflator, three-month Treasury bill yields, the M1 measure of the money supply, the flow of total nonfinancial debt, the Standard and Poor's stock price index, and change in business inventories. Only three of the series are published on a monthly basis; the other four are constructed by interpolating the quarterly series using related monthly series. All series are seasonally adjusted, except the interest rate and stock price index. The money series is extended back from 1948 : 1 through 1958 : 12 using the old M1 series scaled by a constant factor.

The interpolation is accomplished using the Chow and Lin (1971) first-order Markov (M) procedure, which in a few cases is modified to be a random walk Markov (RWM) procedure, as suggested in Litterman (1982). Monthly real GNP is constructed as the sum of eight real components, each of which is either available monthly or is interpolated. For each component and for the other interpolated series, the following list shows which procedure – M or RWM – was used and which monthly series were used as explanatory variables. In several cases, only nominal monthly series were available for interpolating real quarterly data. In these cases, flagged by the letters DFL, the monthly interpolated GNP deflator was used to deflate the monthly series. In one case, flagged SA, the monthly series required seasonal adjustment, which was accomplished using a regression of the logged data on monthly seasonal dummies. All equations included a constant and trend. In some cases where monthly series were not available for the entire sample, additional constant and trend dummies were added for those periods.

Real Consumption (available on a monthly basis)

Real Residential Fixed Investment (RWM)
- Personal income DFL
- New private construction DFL
- Expenditures on private construction DFL
- Total private construction put in place SA

Real Nonresidential Fixed Investment (M)
- Contracts and orders for plant and equipment
- Capital investment commitments 1967 = 100
- New orders, nondefense capital goods
- Three-month Treasury bill yields
- Industrial production index
- Commercial and industrial loans DFL

Real Change in Business Inventories, Durable Goods (M)
  Net change in inventories on hand and on order
  Change in wholesale inventories, durable goods DFL
  Change in inventories, durable goods DFL
  Change in finished goods inventories, durable goods DFL

Real Change in Inventories, Nondurable Goods (M)
  Net change in inventories on hand and on order
  Change in wholesale inventories, nondurable goods DFL
  Change in inventories, nondurable goods DFL
  Change in finished goods inventories, nondurable goods DFL

Real Government Purchases of Goods and Services (RWM)

Real Exports
  Exports, excluding military aid shipments DFL

Real Imports
  General imports, F.A.S. basis DFL

GNP Price Deflator (RWM)
  Consumer price index
  Producer price index

Flow of Total Nonfinancial Debt (M)

Change in the Level of Consumer Installment Debt
  Standard and Poor's stock price index
  Three-month Treasury bill yield
  Consumer price index
  Commercial and industrial loans

References

Bertsekas, D.P. (1976) *Dynamic Programming and Stochastic Control* (Academic Press, New York).

Chow, G. and An-loh Lin (1971) "Best linear unbiased interpolation, distribution and extrapolation of time series by related series", *Review of Economics and Statistics*, 53, 372–375.

Geweke, J. and R. Meese (1981) "Estimating regression models of finite but unknown order", *International Economic Review*, 22, 50–70.

Hoerl, A.E. and R. W. Kennard (1970) "Ridge regression: Biased estimation for nonorthogonal problems", *Technometrics*, 12, 55–67.

Leamer, E.E. (1972) "A class of informative priors and distributed lag analysis", *Econometrica*, 40, 1059–1081.

Leamer, E.E. (1978) *Specification Searches, Ad Hoc Inference with Nonexperimental Data* (Wiley, New York).

Litterman, R.B. (1980) "Techniques for forecasting with vector autoregressions", Doctoral thesis, University of Minnesota.
Litterman, R.B. (1982) "A random walk, Markov model for the interpolation of time series", *Journal of Business and Economic Studies*, 1, 169–173.
Litterman, R.B. (1983) "A use of index models in macroeconomic forecasting", in: A. Zellner, ed., *Applied Time Series Analysis of Economic Data* (National Bureau of the Census).
Shiller, R.J. (1973) "A distributed lag estimator derived from smoothness priors", *Econometrica*, 41, 775–788.
Sims, C.A. (1980) "Macroeconomics and reality", *Econometrica*, 48, 1–48.
Sims, C.A. (1981) "Scientific standards in econometric modeling", Working paper, University of Minnesota; forthcoming in Proceedings of the Conference in honor of the Rotterdam Econometrics Institute.
Sims, C.A. (1982) "Policy analysis with econometric models", *Brookings Papers on Economic Activity*, 1, 107–154.
Stein, C.M. (1974) "Multiple regressions", in: I. Olkin, ed., *Contributions to Probability and Statistics: Essays in Honor of Harold Hotelling* (Stanford University Press), Ch. 37.
Theil, H. (1971) *Principles of Econometrics* (Wiley, New York).

Chapter 7

POSTERIOR MOMENTS OF THE KLEIN–GOLDBERGER MODEL

HERMAN K. VAN DIJK and TEUN KLOEK*

Erasmus University, Rotterdam

1. Introduction

It has been suggested in the literature that Bayesian analysis in systems which contain more than two or three stochastic equations is a computational challenge; see Rothenberg (1973, p. 153) and Richard (1973, p. 10). This chapter intends to demonstrate that, due to the advances in modern computer technology and by using a recently developed Monte Carlo integration method (van Dijk and Kloek, 1983a), it has become computationally feasible to analyse medium-size econometric models. As an example we take the structural form of the revised Klein–Goldberger model; see Klein (1969). This model consists of 16 stochastic equations and four identities. The model is nonlinear in the endogenous variables since most stochastic equations are linear in real variables, while the accounting identities are linear in nominal variables. Since the absolute price level is an endogenous variable, the system can only be solved in terms of the endogenous variables by an iterative method.

We follow Klein (1969) in the use of exact prior information with respect to the zero constraints on a subset of the structural parameters. In particular, it is assumed that part of the system is recursive. Our prior on the nuisance parameters (constant terms and covariance matrix of the disturbances) is such that these can be integrated out of the joint posterior by an analytical integration procedure. The same holds for the parameters of the three recursive equations. One ends up

*An earlier version of this paper was presented at the European Meeting of the Econometric Society, Dublin, 1982. The authors thank A.S. Louter for his assistance in preparing the necessary computer programs and the referees for several suggestions to improve the presentation.

Bayesian Inference and Decision Techniques, edited by P. Goel and A. Zellner

with 30 structural parameters, on 29 of which we specify a uniform prior on a finite region. The marginal posterior density of these thirty structural parameters has to be computed by numerical integration methods. We shall make use of the method of mixed integration (see van Dijk and Kloek, 1983a) for the computation of posterior moments. Although an integration procedure has been specified that is capable of computing the required integrals, several problems remain for further research. Some comments on these are given in the conclusions.

The organization of the chapter is as follows. The next section contains the model and the marginal posterior density. Mixed integration is summarized in section 3. Numerical results and additional details with respect to the prior specification of the thirty structural parameters are presented in section 4. Finally, some conclusions are given in section 5.

2. Model and posterior density

We start with the discussion of some modifications of the standard simultaneous equations model (SEM). (See Drèze and Richard, 1983, and Zellner, 1979, for a discussion of SEM.) These are due to nonlinearity and recursiveness. Let the equations of the model be given as

$$F(y_t, x_t, \theta, \alpha) = u_t'; \quad t = 1, \ldots, T,$$

where y_t is a G-vector of endogenous variables and x_t a K-vector of predetermined variables. The vector α contains the constant terms and the vector θ contains the remaining parameters. Only the latter are supposed to be of interest. Note that identifying zero restrictions and normalization restrictions are directly substituted in the model. The G-vectors u_t', $t = 1, \ldots, T$, are assumed to be independently, normally distributed with mean zero and covariance matrix:

$$\begin{bmatrix} \Sigma & 0 \\ 0 & 0 \end{bmatrix},$$

where Σ is a $G_1 \times G_1$ nonsingular matrix corresponding to the stochastic equations. The set of functional operators F is well known for the revised Klein–Goldberger model: F is linear with respect to the structural parameters θ and α (and the predetermined variables x_t), but it is nonlinear with respect to a subset of the endogenous variables y_t. (For a more detailed discussion of nonlinear simultaneous equation models, see Amemiya, 1983.)

In order to compute the likelihood of the parameters, one has to apply a transformation of variables, namely from the disturbances to the endogenous variables. The Jacobian of this transformation $|J_t|$ is not constant (as in linear models) but different from period to period due to the nonlinearity of the system. The absolute value of this Jacobian is denoted by $||J_t||$. Details with respect to the treatment of identities are given by Dagenais (1978, p. 1354). We assume that the model can be solved in terms of the endogenous variables y_t for all t, by means of an iterative method, i.e. we assume that the transformation from u_t to y_t is locally one-to-one for every t. The likelihood function can then be written as

$$L(\theta, \alpha, \Sigma \mid Y, X) \propto \prod_{t=1}^{T} \|J_t\| \, |\Sigma|^{-(1/2)T} \exp\left[-\tfrac{1}{2}\operatorname{tr}(Q\Sigma^{-1})\right],$$

where $Q = U_1'U_1$ and U_1 is a $T \times G_1$ matrix of disturbances corresponding to the stochastic equations.

The data do not satisfy the sample size requirement $T \geq G_1 + K$. (For more details, see Brown, 1981.) There are 31 observations, 16 stochastic equations and 19 predetermined variables. Clearly, there exists a degrees of freedom problem in this case. We follow Klein (1969) in assuming that part of the system is recursive. There are three equations in which all explanatory variables are predetermined. A reordering such that these are the last three equations yields

$$\|J_t\| = \begin{Vmatrix} J_{1t} & \cdots \\ 0 & I_3 \end{Vmatrix} = \|J_{1t}\|.$$

With respect to Σ it is assumed that

$$\Sigma = \begin{bmatrix} \Sigma_s & 0 \\ 0 & \Sigma_r \end{bmatrix},$$

where Σ_r is a 3×3 diagonal matrix corresponding to the last three stochastic equations. The assumption of a recursive subsystem of three equations enables one to reduce the number of stochastic simultaneous equations to 13 and the number of predetermined variables to 12. Given 31 observations, the number of degrees of freedom is small but positive. The other exact restrictions (zero and normalizing constraints) were also taken from Klein (1969). They are presented in tables 7.1 and 7.2 and are discussed below. Of course, the posterior results may indicate that the exact restrictions have been imposed incorrectly. Then one faces the problem of misspecification in a Bayesian framework; see Box (1980). This subject, although of considerable interest, it outside the scope of the present chapter.

Before we state the stochastic prior information on the remaining structural parameters we partition these parameters θ and α into two parts. Let θ_s and α_s contain the parameters of the thirteen stochastic simultaneous equations and θ_r and α_r contain the parameters of the three stochastic recursive equations. We shall take a noninformative approach with respect to the prior specification of the nuisance parameters (the constant terms α_s and α_r and the covariance matrix Σ_s and the diagonal elements of Σ_r). That is, the prior on α_s and α_r is uniform on a large region. The prior on the diagonal elements of Σ_r is proportional to σ_j^{-1}, $j = 14, 15, 16$. (See, for example, Zellner, 1971.) The noninformative prior on Σ_s is one suggested by Malinvaud (1980, p. 242). The prior on θ_r is uniform on a large region. Assuming prior independence one can write the prior density as

$$p(\theta_s, \alpha_s, \theta_r, \alpha_r, \Sigma_s, \Sigma_r) \propto p(\theta_s)|\Sigma_s|^{-(1/2)h}\sigma_{14}^{-1}\sigma_{15}^{-1}\sigma_{16}^{-1},$$

where $p(\theta_s)$ is specified in section 4 and $h = 2G_2 = 2 \times 13$. A similar comment applies here as has been made after the specification of the exact restrictions. The posterior results may be sensitive for the specification of the prior information of the nuisance parameters in an undesired way. This is again an indication of misspecification. (In van Dijk and Kloek, 1977b, we have illustrated the sensitivity of the posterior results for the choice of the prior parameter h within the context of Klein's model I.) One may, of course, use more informative prior information with respect to the nuisance parameters. A difficulty is, in our opinion, the specification of the parameters of such informative priors. Furthermore, a noninformative approach has the advantage that possible misspecification of the model can be studied separately from possible misspecification due to incorrect informative prior information.

Next, we combine the likelihood and the prior by means of Bayes rule and obtain the joint posterior density of all unknown structural parameters. In order to obtain the marginal posterior density of θ_s we make use of a multi-step analytical integration procedure with respect to $(\alpha_s, \theta_r, \alpha_r, \Sigma_s, \Sigma_r)$. After some tedious algebra using several integral transformations involving the inverted-gamma-2 density, the Student density and the inverted-Wishart density, one obtains:

$$p(\theta_s | Y, X) \propto p(\theta_s) \prod_{t=1}^{T} \|J_{1t}\| \, |Q_2|^{-(1/2)(T+G_2-2)},$$

where in our case θ_s is a 30-dimensional parameter. In van Dijk and Kloek (1977a) we have spelled out the details of a derivation for the case of the linear SEM. The extensions to the nonlinear Klein–Goldberger model with a recursive subsystem are trivial. (The Jacobian factor does not enter the derivation and one can separate the last three stochastic recursive equations from the first thirteen stochastic simultaneous equations.) Note that

$$Q_2 = U_2' N U_2,$$

where $N = I - (1/T)\iota\iota'$, and ι is a vector of unit elements. U_2 relates to the first $G_2 = 13$ equations which remain after the identities and the last three stochastic equations have been removed.

Next, we list the exact restrictions by stating the variables and the equations of the so-called revised Klein–Goldberger model in tables 7.1 and 7.2. The model is to a large extent a Keynesian-type model that explains the components of aggregate demand (given in eqs. (1)–(5) and (14)). The main components are consumption expenditures disaggregated into two groups and investment expenditures disaggregated into three groups. The linearized production function (6) is a supply equation. Gross national product is distributed according to eq. (19).

Table 7.1. Variables.[a]

$*C_d$	Consumption of durables
$*C_n$	Consumption of nondurables and services
$*D$	Capital consumption allowances, billions of current dollars
D_u	Dummy variable, 0 for 1929–1941, 1 for 1947–1964
E	Exports
G	Government expenditures
$*h$	Index of hours worked per week, 1954 = 1.00
$*H$	Stock of inventories
$*I$	Investment in plant and equipment
$*I_m$	Imports
N_g	Government employees, millions
N_l	Total labor force, millions
N_s	Self-employed workers, millions
$*N_w$	Wage and salary workers, millions
$*p$	Implicit GNP deflator, 1954 = 1.00
$*P_c$	Corporate profits including inventory valuation adjustment
$*\Pi - P_c$	Proprietors' income
$*\Pi_r$	Rental income and net interest
p_m	Implicit price deflator for imports, 1954 = 1.00
$*r$	Average yield on corporate bonds (Moody's), percent
$*R$	Residential construction
r_d	Average discount rate at all Federal Reserve Banks, percent
R_e	Year end ratio of member banks' excess to required reserves
$*r_s$	Yield on prime commercial paper, 4–6 months, percent
$*S_c$	Corporate saving, including inventory valuation adjustment
T	Personal taxes plus contributions for social insurance less government and business transfer payments less interest on government debt, billions of current dollars
T_c	Corporate profit taxes, billions of current dollars
T_i	Reconciling item between net national product and national income, billions of current dollars
$*w$	Annual earnings, thousands of dollars
$*W$	Wages and salaries and supplements to wages and salaries
W_g	Government wages and salaries
$*X$	Gross national product
$*Y$	Personal disposable income

[a]An asterisk (*) denotes an endogenous variable. Variables are in billions of 1954 dollars (per year) unless otherwise specified.

The factor income equations are the private wage income equation (8), the proprietors income equation (12) and the rental income equation (13). The depreciation equation (15) and the corporate savings equation (11) show that the private sector determines these variables endogenously. The wage rate equation (9) reflects a Phillips curve effect. Together with the hours worked equation (7), eq. (9) relates to the determination of the money wage rate on the labor market. Two interest rate equations are given in (10) and (16). Identities (17), (18) and (20) close the model.

The short-run structural dynamics is in 12 out of 16 stochastic equations represented as a simple (mostly partial) adjustment scheme.

We note that eqs. (14), (15) and (16) have been dropped out of the full information statistical analysis because of our assumption that they form a recursive subsystem.

Table 7.2. Klein–Goldberger model (revised).[a]

$$C_d - 0.7C_{d,-1} = \theta_1(Y - 0.7Y_{-1}) - \theta_2 C_{d,-1} + \alpha_1 + u_1 \quad (1)$$

$$C_n = \theta_3 Y + \theta_4 C_{n,-1} + \alpha_2 + u_2 \quad (2)$$

$$R = \theta_5 Y - \theta_6 r_{-1} + \theta_7 R_{-1} + \alpha_3 + u_3 \quad (3)$$

$$H = \theta_8(X - \Delta H) + \theta_9 H_{-1} + \alpha_4 + u_4 \quad (4)$$

$$I_m = \theta_{10} X - \theta_{11}(p_m - p) + \theta_{12} I_{m,-1} + \alpha_5 + u_5 \quad (5)$$

$$X - W_g - 0.95(X - W_g)_{-1} = \theta_{13}(I + R) + \theta_{14}[N_w - N_g + N_s - 0.95(N_w - N_g + N_s)_{-1}] + \theta_{15}(h - 0.95h_{-1}) + \alpha_6 + u_6 \quad (6)$$

$$h = -\theta_{16}\Delta w - \theta_{17}(N_l - N_w - N_s) + \alpha_7 + u_7 \quad (7)$$

$$W - W_g = \theta_{18}(X - W_g) + \theta_{19}(W - W_g)_{-1} + \alpha_8 + u_8 \quad (8)$$

$$\Delta w = -\theta_{20}(N_l - N_w - N_s) + \theta_{21}\Delta p_{-1} + \alpha_9 + u_9 \quad (9)$$

$$r = \theta_{22} r_s + \theta_{23} r_{-1} + \alpha_{10} + u_{10} \quad (10)$$

$$pS_c = \theta_{24}(pP_c - T_c) - \theta_{25}(pP_c - T_c - pS_c)_{-1} + \alpha_{11} + u_{11} \quad (11)$$

$$p(\Pi - P_c) = \theta_{26} pX + \theta_{27}[p(\Pi - P_c)]_{-1} + \alpha_{12} + u_{12} \quad (12)$$

$$p\Pi_r = \theta_{28} p(I + R) + \theta_{29}\Delta r + \theta_{30}(p\Pi_r)_{-1} + \alpha_{13} + u_{13} \quad (13)$$

$$I - 0.95I_{-1} = \theta_{31}(X - W_g)_{-1} - \theta_{32} r_{-1} - \theta_{33} I_{-1} + \alpha_{14} + u_{14} \quad (14)$$

$$D = \theta_{34} \sum_{i=1}^{20} [p(I + R)]_{-i} + \theta_{35} D_u + \alpha_{15} + u_{15} \quad (15)$$

$$r_s = \theta_{36} r_d - \theta_{37} R_{e,-1} + \theta_{38} D_u + \alpha_{16} + u_{16} \quad (16)$$

$$X = C_d + C_n + I + R + \Delta H + G + E - I_m \quad (17)$$

$$pY = pX - D - T_i - pS_c - T_c - T \quad (18)$$

$$p\Pi = pX - D - T_i - pW - p\Pi_r \quad (19)$$

$$pW = whN_w \quad (20)$$

[a] The indexes r and s of the vector θ have been deleted for notational convenience.

3. Mixed integration

In this section we give a brief description of the mixed integration method, which is a combination of classical numerical integration and Monte Carlo integration. For a more detailed analysis we refer to van Dijk and Kloek (1983a). Mixed integration is intended for the computation of moments of a unimodal multivariate posterior density with different tail properties in different directions.[1]

Let $g(\theta)$ be a measurable function of the parameters of interest θ, where g may be a scalar, a vector or a matrix. Frequent choices for g are $g(\theta) = \theta$ and $g(\theta) = \theta\theta'$. Our purpose is the efficient computation of

$$Eg(\theta) = \frac{\int g(\theta)p(\theta)\,\mathrm{d}\theta}{\int p(\theta)\,\mathrm{d}\theta},$$

[1] The arguments for choosing Monte Carlo as an approach to solving the numerical integration problems that have to be faced when computing posterior moments and densities of parameters of econometric models are discussed in Kloek and van Dijk (1978).

where $p(\theta)$ is the kernel of the marginal posterior density specified in section 2. (The index s of the vector θ and the conditioning arguments (Y, X) have been deleted for notational convenience.)

Mixed integration is based on a transformation of random variables. The method proposed transforms the s-dimensional parameter space of vectors θ into another s-dimensional parameter space of pairs (η, ρ), where η is an $(s-1)$-dimensional vector and ρ a scalar. For ρ one takes $\pm d$, where d is a measure of the distance between a point $\theta^{(i)}$, generated at random, and a point θ^0, a location estimate of θ. As a location estimate one can take, for example, the posterior mode or a preliminary estimate of the posterior mean. The vector η is taken as the direction $(\theta^{(i)} - \theta^0)/\rho^{(i)}$ with one coordinate deleted in order to avoid degeneracy. The sign of this coordinate is preserved in the sign of $\rho = \pm d$ in order to obtain a one-to-one transformation.

Let the inverse transformation formula, which carries (η, ρ) into θ, be denoted by $\theta = T^{-1}(\eta, \rho)$. One can write the numerator of $Eg(\theta)$ in terms of η and ρ as

$$
\begin{aligned}
A &= \int g(\theta)p(\theta)\,\mathrm{d}\theta \\
&= \int\int g[T^{-1}(\eta,\rho)]\,p[T^{-1}(\eta,\rho)]|J|\,\mathrm{d}\rho\,\mathrm{d}\eta \\
&= \int\left[\int h(\eta,\rho)\,\mathrm{d}\rho\right]|J_2(\eta)|\,\mathrm{d}\eta,
\end{aligned}
$$

with

$$
\begin{aligned}
h(\eta,\rho) &= g[T^{-1}(\eta,\rho)]\,p[T^{-1}(\eta,\rho)]|\rho^{s-1}|, \\
|J| &= |\rho^{s-1}|\,|J_2(\eta)|.
\end{aligned}
$$

The proper expressions for the transformation formulae and for the Jacobian factor are given in van Dijk and Kloek (1983a, section 2).

After having performed the transformation one generates an $(s-1)$-dimensional vector $\eta^{(i)}$ by means of Monte Carlo from a density that is proportional to $|J_2(\eta)|$ and one applies classical numerical integration with respect to ρ given $\eta^{(i)}$. The Monte Carlo estimator for A can be written as

$$
\hat{A} = \frac{1}{2N}\sum_{i=1}^{N}\int h(\eta^{(i)},\rho)\,\mathrm{d}\rho.
$$

In van Dijk and Kloek (1983a) it is shown that the random vectors $\eta^{(i)}$ may be generated as follows. One generates a random vector $\theta^{(i)}$ from a multivariate normal density function with mean θ^0 and covariance matrix V and one applies the transformation that carries $\theta^{(i)}$ into $(\eta^{(i)}, \rho^{(i)})$. Additional details with respect to the regions of integration and with respect to the value of the denominator ($2N$ instead of N) are given in the reference cited above.

We summarize the main steps of an algorithm for the mixed integration method as follows.

Step 1. Generate a random vector $\theta^{(i)}$ from a multivariate normal density function with mean θ^0 and covariance matrix V. Preliminary parameter estimates for θ^0 and V are, for instance, the posterior mode as an estimate for μ and minus the inverse of the Hessian of the log posterior density as estimate for V.

Step 2. Apply the transformation of random variables from $\theta^{(i)}$ into $(\eta^{(i)}, \rho^{(i)})$.

Step 3. Given $\eta^{(i)}$, perform a one-dimensional classical numerical integration with respect to ρ.

Step 4. After repeating steps 1–3 a sufficient number of times one computes the posterior moments of $g(\theta)$ by means of standard Monte Carlo procedures.

One may use this algorithm in two stages in the sense that steps 1–4 can be repeated with the posterior mean of θ as an estimate for θ^0 and with the posterior covariance matrix as an estimate for V.

Given our experience, it appears that the mixed integration method has several advantages. First, it is *flexible* in the sense that one perfectly follows the shape of the posterior density when the integration with ρ is performed. Second, it is *parsimonious* in the sense that only preliminary estimates for θ^0 and V are required. Third, it is *robust* in the sense that it can be applied in cases where the tail behavior of the posterior varies strongly in different directions. Finally, we mention that mixed integration may not be always *efficient*. In van Dijk and Kloek (1983b) we have compared mixed integration with another Monte Carlo integration method (importance sampling), where an importance function is used that consists of a finite mixture of multivariate normal densities. The finite mixture approach was more efficient than the mixed integration approach for the cases studied. We emphasize that the results are based on a limited number of experiments on a small model. Clearly, there exists a need for further experiments in this area with different models and methods.

4. Results[2]

In this section we report estimates of the posterior means, standard deviations and correlations of the 30 structural parameters θ_s and compare these estimates with the posterior mode and the full information maximum likelihood estimates.

The prior density on 29 elements of the vector θ_s is uniform on a bounded region (a Cartesian product of bounded intervals). The exact values of the bounds are given in table 7.3. It appears that one can use some information from economic theory for the specification of these bounds. For instance, the parame-

[2]We are indebted to L.R. Klein for private communication on a few problems regarding the data and the model. One should take Δp_{-1} in his eq. (3.9) instead of Δp. The value of T_c should be 1.3 in 1928 instead of zero. Initial values of p and H are lacking. We estimated these given Klein's regression results. As a check we re-estimated all equations with ordinary least squares yielding results which coincide with Klein's results except for a minor difference with his eq. (3.11a).

Table 7.3. Posterior results for structural parameters θ_s.

Equation	θ_s	Prior bounds	FIML	Posterior mode	Posterior mean	Error
(1) C_d	1	0,1	0.11(0.06)	0.19(0.05)	0.21(0.02)	0.0017
	2	0,0.25	−0.15(0.12)	0.008(0.108)[a]	0.056(0.037)	0.0026
(2) C_n	3	0,0.5	0.23(0.04)	0.24(0.3)	0.24(0.03)	0.0022
	4	0,1	0.74(0.05)	0.72(0.05)	0.72(0.04)	0.0027
	5	0,0.25	0.05(0.007)	0.055(0.007)	0.054(0.006)	0.0003
(3) R	6	0,1	0.03(0.21)	0.03(0.18)	0.16(0.10)	0.0077
	7	0,1	0.34(0.10)	0.32(0.09)	0.34(0.08)	0.0050
(4) H	8	0,1	0.19(0.02)	0.17(0.02)	0.19(0.02)	0.0023
	9	0,1	0.14(0.11)	0.23(0.10)	0.15(0.11)	0.0107
	10	0,1	0.03(0.007)	0.025(0.006)	0.026(0.007)	0.0005
(5) I_m	11	0,20	7.11(2.75)	6.65(2.36)	6.35(2.31)	0.2110
	12	0,1	0.52(0.16)	0.56(0.13)	0.52(0.13)	0.0106
	13	0,1	0.21(0.10)	0.27(0.06)	0.23(0.07)	0.0058
(6) $X - W_g$	14	0,10	6.17(1.88)	4.78(0.88)	5.82(0.91)	0.0637
	15	0,150	96.90(41.4)	71.90(30.2)	86.60(30.5)	2.5800
(7) h	16	0,2	1.00(−)	1.15(0.61)	1.11(0.24)	0.0205
	17	0,0.2	0.03(0.003)	0.037(0.015)	0.035(0.006)	0.0004
(8) $W - W_g$	18	0,1	0.36(0.03)	0.38(0.02)	0.38(0.03)	0.0023
	19	0,1	0.39(0.06)	0.34(0.04)	0.34(0.05)	0.0042
(9) Δw	20	0,1	0.02(0.003)	0.021(0.003)	0.019(0.002)	0.0002
	21	0,1	0.43(0.14)	0.35(0.23)	0.39(0.14)	0.0091
(10) r	22	0,1	0.13(0.04)	0.14(0.03)	0.15(0.05)	0.0062
	23	0,1	0.80(0.06)	0.83(0.05)	0.80(0.07)	0.0085
(11) pS_c	24	0,1	0.75(0.06)	0.77(0.06)	0.76(0.05)	0.0024
	25	0,1	0.55(0.13)	0.60(0.12)	0.54(0.10)	0.0051
(12) $p\Pi - pP_c$	26	0,0.1	0.003(0.011)	0.002(0.009)[a]	0.008(0.004)	0.0002
	27	0,1	0.98(0.12)	0.99(0.10)[a]	0.93(0.04)	0.0019
	28	0,0.5	−0.01(0.02)	0.024(0.015)[a]	0.029(0.006)	0.0004
(13) $p\Pi_r$	29	0,7	3.02(1.04)	2.23(0.83)	1.72(0.65)	0.0650
	30	0,1	1.07(0.05)	0.995(0.34)[a]	0.989(0.010)[a]	0.0006

[a]A penalty function restricted these elements of the posterior mode.

ter θ_1, which reflects the short-run structural effect of disposable income on consumption expenditures, is supposed to be restricted to the interval between 0 and 1. The same range has been specified for many so-called adjustment parameters, e.g. $\theta_4, \theta_7, \theta_9$. Other values of the bounds, for instance the upper bound of θ_2, have been taken from a preliminary diagnostic analysis, which indicated that the posterior density is almost zero beyond these bounds. One may question whether a uniform prior on a bounded interval is an adequate description of one's prior knowledge. In van Dijk and Kloek (1980) we commented upon this question as follows:

> The reason for choosing such a uniform prior is that it is easy to understand and to specify. Suppose, for example, that the most important theoretical information on some parameter α is that it is positive, but that one has no idea whether it will be close to zero or closer to 0.5, say. So one wants to specify a prior with the following properties: (i) the implied prior probability on the

interval $(-0.5,0)$ should be zero or small; (ii) the implied prior probability on $(0,0.5)$ should be substantial; (iii) the prior density on $(0,0.5)$ should be constant or at least approximately constant. It is impossible to specify a normal or Student prior which simultaneously has these three properties. The simplest way to avoid this problem is to specify a uniform prior on $(0, b)$, where b is given some appropriate value. In that interval the likelihood determines the posterior. It goes without saying that such a prior need not reflect in all detail the betting odds one might be willing to accept.

Note that for θ_{16} a normal prior with mean 1.0 and standard deviation 0.4 has been chosen, truncated at the prior bounds given in the first column of table 7.3. In addition, the prior of θ_{16} is assumed to be independent of the joint prior of the other parameters. The prior specification of θ_{16} resulted from a preliminary diagnostic analysis on which we comment below.

Before discussing results, we comment briefly on the computational procedures. Minus the log of the concentrated likelihood function of θ_s and minus the log of the posterior density were minimized numerically be means of a variable metric method; see Fletcher (1980, p. 44). We made use of an analytical gradient. This increases the numerical precision and the rate of convergence, in particular around the optimum of a function where the gradient is almost zero and where the rounding errors of a numerically approximated gradient may cause serious convergence problems. (A similar experience has been reported by Belsley, 1979.) Furthermore, the evaluation of minus the inverse of the Hessian of the log of the concentrated likelihood around its maximum (or, alternatively, of the log posterior around its mode) is simplified by the use of an analytical gradient since this gradient can be applied effectively in the evaluation of the numerically approximated second order derivatives; see Goldfeld and Quandt (1972, p. 19).

We made use of the method of mixed integration for the computation of the posterior first- and second-order moments. One needs estimates of the location parameter vector θ^0 and the covariance matrix V of a 30-dimensional multivariate normal importance function in order to apply mixed integration. The full information maximum likelihood estimates are not suitable since they are outside the a priori plausible region for parameters 2, 28 and 30. A brief comment on the interpretation of this phenomenon will be given in section 5, below. Furthermore, parameters 6, 26 and 27 are too close to the boundary of the a priori interval. For these reasons we minimized minus the log posterior augmented with a penalty function. The penalty function is a fifth-degree polynomial in each of the restricted parameters. This choice implies that the objective function remains twice continuously differentiable. We emphasize that this penalty function is used only to obtain preliminary estimates of θ^0 and V by means of an unconstrained minimization method. Our original prior of $p(\theta_s)$ is maintained for the posterior analysis; see van Dijk and Kloek (1980, p. 317).

As an estimate for V we computed minus the inverse of the Hessian of the log posterior around the restricted mode. The square roots of the diagonal elements are listed in table 7.3. We applied mixed integration in two steps. In the first step (based on 2000 random drawings) preliminary estimates of the posterior first- and second-order moments were obtained. These estimates were used as estimates of the importance function parameters θ^0 and V in the second step, which was based on 10 000 random drawings. The numerical precision of the posterior results showed a considerable improvement in the second round of mixed integration.

Posterior results are reported together with the FIML estimates in table 7.3. Selected values of the parameter correlations are reported in table 7.4. The complete correlation matrices are deleted in order to save space. They are available from the authors upon request. The numbers in parentheses in table 7.3 are asymptotic standard errors for FIML and the square roots of the diagonal elements of minus the inverse of the Hessian of the log posterior for the posterior mode. The posterior standard deviations are reported in parentheses next to the posterior means. In the final column of the table we have given the standard deviation of the Monte Carlo estimate of the posterior mean as a measure of absolute error; for details, see Kloek and van Dijk (1978, section 6). We make the following remarks about the results.

Table 7.4. Selected values of parameter correlations.

Parameter combination	Equation(s)	Importance function correlations	Posterior correlations
(a) Large values of correlations			
(1, 2)	1	0.980	0.893
(3, 4)	2	−0.991	−0.986
(5, 7)	3	−0.955	−0.943
(8, 9)	4	−0.992	−0.991
(10, 12)	5	−0.983	−0.982
(18, 19)	8	−0.994	−0.994
(26, 27)	12	−0.978	−0.838
(28, 30)	13	−0.947	−0.559
(16, 17)[a]	7	0.983	0.920
(b) Difference between importance function correlation and posterior correlation greater than |0.4|			
(5, 15)	3 and 6	−0.251	0.175
(6, 27)[b]	3 and 12	0.372	−0.152
(8, 28)	4 and 13	−0.169	0.276
(9, 28)	4 and 13	0.156	−0.291
(12, 22)	5 and 10	−0.026	0.388
(13, 14)	6	−0.524	−0.089
(18, 22)	8 and 10	0.008	−0.455
(19, 22)	8 and 10	−0.019	0.427

[a] The only parameter combination of (a) that does not refer to a simple dynamic adjustment scheme.
[b] Difference is larger than |0.5|.

First, several marginal posterior densities are very skew due to truncation implied by our specification of prior information. That is, in particular, the case for parameters 2, 6, 26 and 27.

Second, the standard deviations in the local approximation case (posterior mode) are in 18 cases larger than in the global approximation case (posterior means). In nine cases they are smaller and in three cases they are approximately equal.

Third, the selected values of the posterior correlations of table 7.4 indicate that there are five correlations between the lagged dependent variable and a dependent variable at the right-hand side of the equation larger than 0.9. This number increases to eight when one considers the importance function correlations. This suggests that it may be difficult to estimate the precise lag structure. Furthermore, there are only eight cases where the difference between an importance function correlation and a posterior correlation is larger than 0.4 in absolute value. This suggests that a small subset of the parameters causes nonlinearities in the surface of the posterior density.

Fourth, we have an almost singular posterior covariance matrix. This might be due to our Monte Carlo integration procedure. In particular, a poor estimate of the covariance matrix V of the importance function around the posterior mode may affect the numerical precision strongly. To verify this conjecture we multipled the diagonal elements of the covariance matrix of the normal importance function by a scalar k for several values of k in the range $[0.01, 1.10]$. Even for $k = 1.01$ the numerical accuracy of the posterior means decreased because the number of important drawings dropped. This supports the hypothesis that the collinearity of the posterior means is caused by the specification of the model and the small sample size.

Fifth, the absolute numerical errors are small; compare the results in the last column of table 7.3. But if one considers relative precision, in the sense of the ratio ($\times 100$) of the standard deviation of the Monte Carlo estimate of the posterior mean and the posterior standard deviation, it is seen that these errors are between 5 and 13 percent. This is not extremely precise, but it seems tolerable for practical purposes.

Sixth, our FIML estimates differ from Klein's estimates. In preliminary work we obtained a nonrestricted stationary point of θ, but we did not succeed in determining the Hessian in a numerically stable way. It turns out that parameter 16 is extremely collinear with parameters 17, 20 and 21. The correlations are larger than 0.995 in absolute value. Therefore we restricted parameter 16 at 1.0. This restriction has a chi-square value of 0.44 (in a likelihood ratio test), while the restriction $\theta_{16} = 0$ yields a significant chi-square value of 5.32. (Note that such test statistics are convenient diagnostics to check whether prior and sample information are contradictory.) In the Bayesian analysis we used a normal prior for θ_{16} with mean 1.0 and standard deviation 0.4 and we truncated this density at the prior boundaries. In a sense one can say that the results from the numerical optimization of the concentrated likelihood function and the posterior density of

θ_s have provided diagnostic information for the Bayesian analysis. The posterior results depend, of course, on the value of the informative prior of θ_{16}. We repeat the statements of section 2 that there exists a need for assessing the correctness of the prior information in a Bayesian framework.

Finally, we investigated the sensitivity of the results for the choice of the boundaries. As was to be expected from the results of the numerical optimization, several parameters are sensitive, in particular, the starred ones of table 7.3. Also, the problem of collinearity becomes more serious in the large region that it was in the small region.

5. Conclusions

We have demonstrated the computational feasibility of posterior analysis of the structural form of a nonlinear econometric model in 30 dimensions. This requires high-speed modern computers and flexible Monte Carlo integration procedures. The method of mixed integration appears to be such a method. A next step will be to make a standard program available.

The results of this exercise point also to several problems of research. When the models are highly nonlinear in both the endogenous variables and the parameters the method of mixed integration may be too restrictive. Then other Monte Carlo integration procedures will have to be devised. Preliminary experiments in small models with a composition technique which is based on a finite mixture of multivariate densities appear promising. Another problem for research is to devise diagnostic tests of misspecification in a Bayesian context. We conjecture that, along with the small sample problem, there is a misspecification problem with the Klein–Goldberger model. At least, this is our interpretation of the implausible values of several FIML estimates reported in section 4. Investigation of this problem is outside the scope of the present chapter. Finally, we remark that we did not tackle the reduced form analysis. Computationally it is very time consuming to have an iterative method, for solving a nonlinear model, built into a Monte Carlo integration procedure.

References

Amemiya, T. (1983) "Nonlinear regression models", in: Z. Griliches and M. Intriligator, eds., *Handbook of Econometrics*, Vol. 1 (North-Holland, Amsterdam).

Belsley, D.A. (1979) "On the computational competitiveness of full-information maximum likelihood and three-stage least squares in the estimation of nonlinear, simultaneous equations models", *Journal of Econometrics* 9, 315–342.

Box, G.E.P. (1980) "Sampling and Bayes inference in scientific modelling and robustness" (with discussion), *Journal of the Royal Statistical Society*, A, 143, 383–430.

Brown, B.W. (1981) "Sample size requirements in full information maximum likelihood estimation", *International Economic Review*, 22, 443–459.

Dagenais, M.G. (1978) "The computation of FIML estimates as iterative generalized least squares estimates in linear and nonlinear simultaneous equations models", *Econometrica*, 46, 1351–1362.

Drèze, J.H. and J.F. Richard (1983) "Bayesian analysis of simultaneous equation systems", in: Z. Griliches and M. Intriligator, eds., *Handbook of Econometrics*, Vol. 1 (North-Holland, Amsterdam).

Fletcher, R. (1980) *Practical Methods of Optimization*, Vol. 1 (Wiley, New York).

Goldfeld, S.M. and R.E. Quandt (1972) *Nonlinear Methods in Econometrics* (North-Holland, Amsterdam).

Klein, L.R. (1969) "Estimation of interdependent systems in macroeconometrics", *Econometrica*, 37, 171–192.

Kloek, T. and H.K. van Dijk (1978) "Bayesian estimates of equation system parameters; an application of integration by Monte Carlo", *Econometrica*, 46, 1–19; reprinted in: A. Zellner, ed. (1980) *Bayesian Analysis in Econometrics and Statistics, Essays in Honor of Harold Jeffreys* (North-Holland, Amsterdam).

Malinvaud, E. (1980) *Statistical Methods of Econometrics*, 3rd edn. (North-Holland, Amsterdam).

Richard, J.F. (1973) *Posterior and Predictive Densities for Simultaneous Equation Models* (Springer-Verlag, Berlin).

Rothenberg, T.J. (1973) *Efficient Estimation With A Priori Information* (Yale University Press, New Haven).

van Dijk, H.K. and T. Kloek (1977a) "Predictive moments of simultaneous econometric models: A Bayesian approach", in: A. Aykac and C. Brumat, eds., *New Developments in the Applications of Bayesian Methods* (North-Holland, Amsterdam).

van Dijk, H.K. and T. Kloek (1977b) "Likelihood diagnostics and posterior analysis of Klein's model I", Working paper, Econometric Institute, Erasmus University, Rotterdam.

van Dijk, H.K. and T. Kloek (1980) "Further experience in Bayesian analysis using Monte Carlo integration", *Journal of Econometrics*, 14, 307–328.

van Dijk, H.K. and T. Kloek (1983a) "Posterior moments computed by mixed integration", Report 8318/E, Econometric Institute, Erasmus University, Rotterdam, Rotterdam.

van Dijk, H.K. and T. Kloek (1983b) "Experiments with some alternatives for simple importance sampling in Monte Carlo integration", Report 8326/E, Econometric Institute, Erasmus University, Rotterdam; to appear in: Proceedings of the Second International Meeting on Bayesian Statistics, 6–10, September 1983 (Valencia, Spain).

Zellner, A. (1971) *An Introduction to Bayesian Inference in Econometrics* (Wiley, New York).

Zellner, A. (1979) "Statistical analysis of econometric models", *Journal of the American Statistical Association*, 74, 628–643.

Chapter 8

THE SELECTION OF SMOOTHNESS PRIORS FOR DISTRIBUTED LAG ESTIMATION

HIROTUGU AKAIKE

The Institute of Statistical Mathematics, Tokyo

1. Introduction

Shiller (1973) introduced the concept of smoothness priors to define a Bayesian estimator of the lag coefficients, or the impulse response function, of a linear system. In this approach the prior preference of the smoothness of the lag function is expressed by a spherical normal distribution of the fixed order differences of the lag coefficients.

Outstanding problems in the application of Shiller's estimator are the choices of the lag length, the order of differencing and the variance of the spherical normal prior distribution. In this chapter we propose a practical solution to this problem obtained by maximizing a properly defined likelihood of the Bayesian model.

The performance of the estimator is checked by numerical examples. The results show that the estimator performs satisfactorily under widely varying conditions. They also show that an arbitrary subjective choice of the prior distribution can produce an unreasonable estimate.

2. Smoothness prior for distributed lag estimation

Consider the stochastic linear system defined by

$$y_n = \sum_{m=0}^{M} a_m x_{n-m} + w_n, \tag{1}$$

Bayesian Inference and Decision Techniques, edited by P. Goel and A. Zellner

where y_n, x_n and w_n denote the output, input and error term of the system, respectively. Here $\{w_n\}$ is assumed to be a sequence of random variables which are independent of $\{x_n\}$ and are independently identically distributed as normal with mean zero and variance σ^2.

Shiller introduced the smoothness prior defined by a spherical normal distribution of the dth differences of the distributed lag coefficients $\{a_m\}$. This prior distribution is given by

$$p(a|\sigma,\lambda)=\left(\frac{1}{2\pi\sigma^2}\right)^{(M+1)/2}|\lambda^2\boldsymbol{R}_d\boldsymbol{R}'_d|^{1/2}\exp\left(-\frac{\lambda^2}{2\sigma^2}\boldsymbol{a}'\boldsymbol{R}'_d\boldsymbol{R}_d\boldsymbol{a}\right),$$

where $|\cdot|$ denotes the determinant, $\boldsymbol{a}'=(a_0,a_1,\dots,a_M)$, λ is the ratio of the standard deviation σ of w_n to that of the dth difference of a_m, and $\boldsymbol{R}_d=\boldsymbol{R}_1^d$, where $\boldsymbol{R}_1$ is an $(M+1)\times(M+1)$ matrix defined by

$$\boldsymbol{R}_1=\begin{bmatrix}1 & -1 & 0 & \cdot & \cdot & 0\\ 0 & 1 & -1 & \cdot & \cdot & 0\\ \vdots & \vdots & \vdots & \cdot & \cdot & \vdots\\ 0 & 0 & 0 & \cdot & \cdot & 1\end{bmatrix}.$$

It should be noticed here that the last d rows of $\boldsymbol{R}_d$ define incomplete differences. However, as was mentioned by Shiller, this is equivalent to connecting the final coefficients to $a_m=0$ for $m>M$, which will be a reasonable assumption for practical application if the lag length M is taken sufficiently large.

By combining the present prior distribution with the data distribution

$$p(y|\boldsymbol{a},\sigma)=\left(\frac{1}{2\pi\sigma^2}\right)^{N/2}\exp\left(-\frac{1}{2\sigma^2}\|\boldsymbol{y}-\boldsymbol{X}\boldsymbol{a}\|^2\right),$$

where $\boldsymbol{y}=(y_1,y_2,\dots,y_N)'$, $\|\cdot\|$ denotes the Euclidean norm, and

$$\boldsymbol{X}=\begin{bmatrix}x_1 & x_0 & \cdot & \cdot & \cdot & x_{1-M}\\ x_2 & x_1 & \cdot & \cdot & \cdot & x_{2-M}\\ \vdots & \vdots & \cdot & \cdot & \cdot & \vdots\\ x_N & x_{N-1} & \cdot & \cdot & \cdot & x_{N-M}\end{bmatrix},$$

the estimate $\boldsymbol{a}^*$ of the lag coefficients is defined as the posterior mean of $\boldsymbol{a}$ and is given by

$$\boldsymbol{a}_*=\left(\boldsymbol{X}'\boldsymbol{X}+\lambda^2\boldsymbol{R}'_d\boldsymbol{R}_d\right)^{-1}\boldsymbol{X}'\boldsymbol{y}.$$

Shiller demonstrated by numerical examples the superiority of this type of estimator to both the ordinary least squares estimator and the Almon lag estimator. This is one of the earliest examples of successful Bayesian modeling of practical importance.

The crucial point in applying Shiller's estimator is the choice of the hyperparameter λ. Shiller suggested some rule of thumb for the choice of λ. However, this is deeply concerned with the basic problem of the selection of a Bayesian model which is vital in implementing a Bayesian procedure.

3. Selection of a prior distribution by likelihood

When there are a finite number of Bayesian models defined by the data distributions $f_k(\cdot|\theta_k)$ and corresponding prior distributions $\pi_k(\theta_k)$ $(k=1,2,\dots,K)$ the posterior probability of each model is given by

$$p(k|y)=\frac{f(y|k)C_k}{\sum_{k=1}^{K} f(y|k)C_k},$$

where y denotes the observation, C_k denotes the prior probability of the kth model and $f(y|k)$ is defined by

$$f(y|k)=\int f_k(y|\theta_k)\pi_k(\theta_k)\,\mathrm{d}\theta_k.$$

The above formula for $p(k|y)$ shows that it is natural to call $f(y|k)$ the likelihood of the Bayesian model specified by $f_k(\cdot|\theta_k)$ and $\pi_k(\theta_k)$.

In practical applications of the smoothness prior for the distributed lag estimation we usually do not know the value of σ. Accordingly, we have to specify a prior distribution of σ. To make the resulting estimator applicable to observations in arbitrary scale units we consider the use of Jeffrey's ignorance prior, $\sigma^{-1}\mathrm{d}\sigma$. For each particular application we may consider a proper prior distribution $C(u,v)\sigma^{-1}$ obtained by restricting the range of σ to a finite interval (u,v) of positive numbers. However, since the integral of $p(y|\boldsymbol{a},\sigma)p(\boldsymbol{a}|\sigma,\lambda)\sigma^{-1}$ with respect to $\mathrm{d}\boldsymbol{a}\,\mathrm{d}\sigma$ is finite, we may use this integral as the likelihood of the Bayesian model specified by $p(y|\boldsymbol{a},\sigma)$ and $p(\boldsymbol{a}|\sigma,\lambda)C(u,v)\sigma^{-1}$ with sufficiently small u and sufficiently large v. Since only ratios of the likelihoods are of interest we define the likelihood of the Bayesian model, specified by the data distribution $p(y|\boldsymbol{a},\sigma)$ and the (improper) prior distribution $p(\boldsymbol{a},\sigma|\lambda)=p(\boldsymbol{a}|\sigma,\lambda)\sigma^{-1}$, by

$$p(y|\lambda)=\int\int p(y|\boldsymbol{a},\sigma)p(\boldsymbol{a},\sigma|\lambda)\,\mathrm{d}\boldsymbol{a}\,\mathrm{d}\sigma.$$

Since the hyperparameter λ has a clearly defined technical meaning as the ratio of the standard deviation of an error term to that of a difference of lag coefficients, it is not difficult to find a reasonable selection of finite number of possible values of λ. The model selection is realized by maximizing the likelihood over this set of λ.

For our present model we have

$$p(y|a,\sigma)p(a,\sigma|\lambda) = \left(\frac{1}{2\pi\sigma^2}\right)^{(N+M+1)/2} |\lambda^2 R_d' R_d|^{1/2} \frac{1}{\sigma}$$
$$\times \exp\left[-\frac{1}{2\sigma^2}(a - a_*)'(X'X + \lambda^2 R_d' R_d)(a - a_*)\right]$$
$$\times \exp\left[-\frac{1}{2\sigma^2} S(\lambda)\right],$$

where $a_* = (X'X + \lambda^2 R_d' R_d)^{-1} X'y$ and $S(\lambda) = \| y\|^2 - a_*'(X'X + \lambda^2 R_d' R_d) a_*$. The likelihood $p(y|\lambda)$ of this model is then given by

$$p(y|\lambda) = \frac{1}{2}\left(\frac{1}{\pi}\right)^{N/2} \Gamma\left(\frac{N}{2}\right) S(\lambda)^{-N/2} |X'X + \lambda^2 R_d' R_d|^{-1/2} |\lambda^2 R_d' R_d|^{1/2}.$$

Obviously the estimator a_* is given as the solution of the least squares problem that minimizes $\| y^* - X^* a\|^2$, where

$$X^* = \begin{bmatrix} X \\ \lambda R_d \end{bmatrix}, \qquad y^* = \begin{bmatrix} y \\ 0 \end{bmatrix}.$$

Here it holds that $S(\lambda) = \| y^* - X^* a_*\|^2$. By successively applying the Householder transformation to $[X^* y^*]$, first to transformation λR_d into upper triangular form and then to transform the whole matrix into upper triangular form, the necessary quantities for the likelihood computation can easily be obtained during the process of the least squares computation. For the purpose of the comparison of models we may ignore the constant factor $2^{-1}\pi^{-N/2}\Gamma(N/2)$ of the likelihood.

It should be noted here that the likelihood $p(y|\lambda)$ is also a function of d, the order of differencing, and M, the lag length. Thus, our search for the best model is realized by maximizing $p(y|\lambda)$ over some finite number of possible combinations of λ, d and M. The practical utility of this procedure will be demonstrated by numerical examples in the next section. The feasibility of this type of procedure was first discussed in Akaike (1980a) and its application to seasonal adjustment was discussed by Akaike (1980b).

4. Numerical investigation

In this section we discuss the practical utility of the distributed lag estimation realized by the maximum likelihood selection of the smoothness prior.

First, we consider an artificial data set generated by the model discussed by Shiller as his second example. In this data set the input series was the series of four- to six-month commercial paper rate which gave a typical nearly collinear matrix X. The output series y was generated by the relation $y = Xa + w$, with $w = 0.05e$, where e is a vector of standard normal random numbers. The lag coefficients were defined by $a_m = \phi((1/2)(m-9))$, with $M = 19$, where $\phi(t)$

denotes the density of the standard normal distribution. The dimension, or the length, of y was given by $N = 40$. One data set generated by this model was used in the subsequent analysis.

The search for the smoothness prior was extended over the models defined with $d = 1,2,3$, $M = -1,(1),19$, where $M = -1$ denotes zero regression, and $\lambda = 5 \times 2^k$ $(k = -10,(1),10)$. Since we are accustomed to the use of the log-likelihood ratio test we used $(-2)\log p(y|\lambda)$ as our criterion. By ignoring the additive constant, the criterion is given by

$$\text{ABIC} = N \log S(\lambda) + \log|\boldsymbol{X}'\boldsymbol{X} + \lambda^2 \boldsymbol{R}'_d \boldsymbol{R}_d| - \log|\lambda^2 \boldsymbol{R}'_d \boldsymbol{R}_d|,$$

where N is the dimension of the vector y, log denotes natural logarithm and ABIC stands for a Bayesian information criterion. Our search for the model was realized by finding the minimum of ABIC. ABIC was first introduced in Akaike (1980a) where the criterion was derived by using the maximum likelihood estimate of σ^2 instead of the improper prior distribution.

The best choice in terms of the criterion was given by $d = 2$, $M = 13$ and $\lambda = 5 \times 2^{-3}$. The resulting estimate is shown in table 8.1 along with the true values and the ordinary least squares estimate for $M = 19$, denoted by LS. By table 8.1 it is almost obvious that the present estimation procedure produces a significantly improved estimate over the ordinary least squares estimate. Although the present result is not directly comparable with Shiller's result due to the different realiza-

Table 8.1. Results of an experiment of Shiller's second example (LS denotes least squares estimate).

m	0	1	2	3	4
True	0.000	0.000	0.001	0.004	0.018
$\boldsymbol{a}^*$	−0.011	−0.001	0.008	0.014	0.015
LS	−0.010	0.021	−0.045	0.037	0.078
m	5	6	7	8	9
True	0.054	0.130	0.242	0.352	0.399
$\boldsymbol{a}^*$	0.045	0.136	0.240	0.347	0.400
LS	−0.074	0.255	0.113	0.462	0.334
m	10	11	12	13	14
True	0.352	0.242	0.130	0.054	0.018
$\boldsymbol{a}^*$	0.370	0.250	0.125	0.060	0
LS	0.359	0.329	0.046	0.072	0.042
m	15	16	17	18	19
True	0.004	0.001	0.000	0.000	0
$\boldsymbol{a}^*$	0	0	0	0	0
LS	−0.018	−0.050	0.065	−0.008	−0.008

$\text{SSEC}^{a}(\boldsymbol{a}^*) = 0.0019$ $\quad$ $\text{SSER}^{b}(\boldsymbol{a}^*) = 0.2578$

SSEC (LS) = 0.0954 $\quad$ SSDR (LS) = 0.0430

[a] $\text{SSEC}(\boldsymbol{a}^*) = \|\boldsymbol{a}^* - \boldsymbol{a}_{\text{true}}\|^2$.

[b] $\text{SSER}(\boldsymbol{a}^*) = \|\boldsymbol{X}(\boldsymbol{a}^* - \boldsymbol{a}_{\text{true}})\|^2$.

tions of the error term, the shape of the estimated distributed lag coefficients is quite similar to some of the best ones given by Shiller.

The values of the sum of squared errors of the coefficients (SSEC) given in table 8.1 show the superiority of $\boldsymbol{a}^*$ to LS. However, the sum of squared errors of the regression (SSER) is smaller for LS than for $\boldsymbol{a}^*$. For $\boldsymbol{a}^*$ defined with $M=14$ these figures were both very small with SSEC = 0.0009 and SSER = 0.0186. This point will be discussed further in the final chapter.

One might be concerned with the possibility of non-smooth behavior of distributed lag coefficients. To check the performance of our procedure under such circumstances we considered a system defined by

$$y_n = 1.2x_n - 0.6x_{n-1} + 0.4x_{n-2} + w_n,$$

where the input x_n was the same as that in the preceding example and thus $N=40$, and w_n was also normal with mean zero and $\sigma=0.05$. The range of the search for the parameters was the same as in the preceding example. The minimum of ABIC was attained at $d=1$, $M=2$ and $\lambda=5\times2^{-7}$. The resulting estimate, $\boldsymbol{a}^*$, and the estimate $\boldsymbol{a}^{**}$, which was obtained by increasing the value of λ from 5×2^{-7} to 5×2^{-3}, the value from the preceding example, are given in table 8.2 along with the true values and LS, the ordinary least squares estimate for $M=19$. To see the effect of increasing M, another estimate, $\boldsymbol{a}^{***}$, obtained with $d=1$, $M=17$ and $\lambda=5\times2^{-3}$, is also included in table 8.2.

Taking into account the poor modeling of the situation and the wide ambiguity of d, M and λ, it is remarkable that our procedure could produce a very good result. Another estimate, $\boldsymbol{a}^{**}$, which was obtained by assuming the exact order but with λ from the former example, produced a very poor result, much worse than LS with $M=19$. The numerical output also showed that with the proper choice of $\lambda=5\times2^{-7}$ the increase of M from 2 to 19 produced an estimate which was close to LS of table 8.2. These results demonstrate the danger of applying a Bayesian model based on an arbitrary choice of the prior distribution. Clearly, a proper adaptive procedure is necessary for the practical application of the smoothness prior.

Having confirmed the performance of our procedure at the two extreme situations, most favorable and unfavorable, we now turn to the example of real data, handled by Shiller as his first example. In this example Shiller analyzed the response of the Federal Reserve Board Aaa new issue yield series to the four- to six-month prime commercial paper rate. Due to the unavailability of the Federal Reserve Board Aaa new issue yield series we used the corresponding series of Moody's AAA bond yield as the output series. The similarity of our result to Shiller's confirms that the substitution did not change the essential aspect of the problem.

In this example it is already noticed by Shiller that the coefficient a_0 has different characteristics from other coefficients and should be freed from the rest. The validity of this hypothesis can easily be checked by our present approach by considering models obtained by multiplying the first row of the matrix R_d,

Table 8.2. Test result for non-smooth distributed lag coefficients.

m	0	1	2	3	4
True	1.2	−0.6	0.4	0	0
a^*	1.206	−0.601	0.394	0	0
a^{**}	0.912	−0.047	0.133	0	0
LS	1.193	−0.587	0.390	−0.044	0.045
a^{***}	0.879	0.007	−0.33	0.068	0.066
m	5	6	7	8	9
True	0	0	0	0	0
a^*	0	0	0	0	0
a^{**}	0	0	0	0	0
LS	0.018	−0.018	0.037	−0.037	0.026
a^{***}	0.047	−0.023	−0.019	0.037	−0.019
m	10	11	12	13	14
True	0	0	0	0	0
a^*	0	0	0	0	0
a^{**}	0	0	0	0	0
LS	−0.029	0.030	−0.027	0.059	−0.067
a^{***}	−0.007	0.048	−0.001	−0.023	−0.24
m	15	16	17	18	19
True	0	0	0	0	0
a^*	0	0	0	0	0
a^{**}	0	0	0	0	0
LS	0.028	−0.015	−0.052	0.092	−0.043
a^{***}	0.015	−0.011	−0.028	0.043	−0.019

$\mathrm{SSEC}(a^*) = 0.0001$	$\mathrm{SSER}(a^*) = 0.0007$
$\mathrm{SSEC}(a^{**}) = 0.4611$	$\mathrm{SSER}(a^{**}) = 0.3819$
$\mathrm{SSEC(LS)} = 0.0327$	$\mathrm{SSER(LS)} = 0.0281$
$\mathrm{SSEC}(a^{***}) = 0.6790$	$\mathrm{SSER}(a^{***}) = 0.2957$

a^{**} denotes the estimate obtained with an increased value of λ.

a^{***} denotes the estimate obtained by further increasing M to 19.

defined in section 2, by a small positive number δ. For the present example ABIC kept decreasing when δ was decreased from 1 to 0.0001 in several steps. This confirmed the validity of Shiller's argument. The following result was obtained with $\delta = 0.0001$ and the same range of the parameters as in the preceding examples.

The minimum of ABIC was attained at $d = 3$, $M = 19$, and $\lambda = 5 \times 2^5$. The amount of reduction of the minimum ABIC obtained by reducing δ from 1 to 0.0001 was 20.7. The estimate $\boldsymbol{a}_*$ is given in table 8.3 along with the estimate $\boldsymbol{a}_*^{1.0}$ obtained by putting $\delta = 1.0$. In this example the constant term a_{-1} was included to represent the effect of non-zero average values of the input and output series. The estimate denoted by $\boldsymbol{a}_*^d$ was obtained from the series of first-order differences of the input and output series. This was to check the possible effect of the trends of both series. The estimate $\boldsymbol{a}_*^d$ was obtained with $d = 3$, $M = 19$ and $\lambda = 5 \times 2^4$.

Table 8.3. Result of application to real economic data.

m	0	1	2	3	4
$\boldsymbol{a}^*$	0.212	0.029	0.045	0.058	0.068
$\boldsymbol{a}_*^{1,0}$	0.104	0.088	0.076	0.070	0.068
$\boldsymbol{a}_*^d$	0.281	0.005	0.031	0.052	0.068
LS	0.352	−0.132	0.081	0.127	−0.009
m	5	6	7	8	9
$\boldsymbol{a}^*$	0.074	0.079	0.078	0.076	0.072
$\boldsymbol{a}_*^{1,0}$	0.069	0.072	0.075	0.076	0.075
$\boldsymbol{a}_*^d$	0.078	0.083	0.084	0.081	0.076
LS	0.106	0.106	0.084	0.054	0.144
m	10	11	12	13	14
$\boldsymbol{a}^*$	0.066	0.058	0.049	0.040	0.031
$\boldsymbol{a}_*^{1,0}$	0.072	0.065	0.056	0.046	0.035
$\boldsymbol{a}_*^d$	0.068	0.059	0.049	0.039	0.030
LS	0.077	0.058	0.057	0.025	0.038
m	15	16	17	18	19
$\boldsymbol{a}^*$	0.022	0.014	0.008	0.004	0.001
$\boldsymbol{a}_*^{1,0}$	0.024	0.014	0.007	0.002	0
$\boldsymbol{a}_*^d$	0.022	0.014	0.008	0.004	0.001
LS	0.079	0.048	0.063	−0.042	0.037

$\boldsymbol{a}_*^{1,0}$ denotes the result obtained under the assumption of smooth connection to a_0.

$\boldsymbol{a}_*^d$ denotes the result obtained from differenced series.

The corresponding ordinary least squares estimate for the differenced series is given by LS.

The similarity between $\boldsymbol{a}_*$ and $\boldsymbol{a}_*^d$ is remarkable and confirms that the smooth behavior of $\boldsymbol{a}_*$ does not represent the spurious response due to the trend components. Also, the distortion caused by the inclusion of a_0 can be seen clearly from $\boldsymbol{a}_*^{1,0}$. The usual erratic pattern of the least squares estimate persists in this example.

6. Discussion

The purpose of the present chapter has been to show the feasibility of an objectively defined procedure for the selection of the smoothness prior for the estimation of distributed lag coefficients. It was confirmed that the maximum likelihood procedure proposed in this chapter can produce results comparable to the results reported by Shiller in his original paper. Since Shiller's results may be considered as typical examples produced by an expert, this shows that the present procedure produces a good approximation to the judgement procedure of an expert.

The idea of using the method of maximum likelihood for the Bayesian model selection might look rather disturbing. One might even think that the use is contrary to Bayesian philosophy. However, any statistician with some experience of handling real problems will admit that there is always enormous uncertainty in choosing a probabilistic structure for a specific application. The result of table 8.2 demonstrated the danger inherent in a purely subjective selection of a prior distribution.

The uncertainty in choosing a prior distribution is typically represented by the existence of various possibilities of the choice. Due to the limitation of the available time we are often unable to collect sufficient information to uniquely specify a prior distribution over these possibilities. If the reality of this type of situation is accepted, then there is room for the maximum likelihood method. The idea is not new. Good (1965) considered the use of type II maximum likelihood, the maximization of the likelihood of a Bayesian model with respect to the hyperparameter, as one example of the compromise between Bayesian and non-Bayesian method.

There is a more positive aspect in the use of the likelihood. The very basic element in statistics is the concept of the *true* distribution. For conventional statisticians the true distribution is always unknown. For practical Bayesians who admit the uncertainty in the choice of a probabilistic structure, the true distribution is also unknown. Once this unknowness is accepted, the situation must be handled by a logic outside the Bayesian framework. This is the situation considered by the empirical Bayes approach; see, for example, Morris (1983).

It is known that the expected log likelihood with respect to the true distribution provides a reasonable measure of the goodness of a statistical model; see, for example, Akaike (1983a). When a set of data is obtained the log likelihood provides a natural unbiased estimate of the expected log likelihood, *irrespective of the form of the true distribution*. Thus, at least in terms of expectation, the log likelihood provides a reasonable observable measure of goodness of fit of a statistical model, be it Bayesian or non-Bayesian.

To clarify the distinction between the use of the likelihood in the model evaluation and Bayesian modeling it is necessary to differentiate the log likelihood from the likelihood. The use of the log likelihood characterizes the phase of model evaluation and the likelihood is directly used in Bayesian modeling. The distinction between the log likelihood and the likelihood might look subtle, but it is of great practical significance. In particular, since the expected log likelihood admits a specific interpretation originated in Boltzmann's probabilistic treatment of entropy (Akaike, 1983a), *the difference of two log likelihoods has a natural unit for its scaling*, i.e. we have an objective measure that allows us to communicate our experience of statistical model evaluation to others.

Coming back to the discussion of the smoothness prior, it is now obvious that the next step of the modeling is the specification of a prior distribution for the lag length M. We tried experimentally the simple assumption of equal prior probabilities for M's. In the case of the example of table 8.1 the model produced an estimate with smaller SSEC and SSER than the present $\boldsymbol{a}^*$. A good result was obtained also for the example of table 8.2. Obviously it is necessary to develop a systematic analysis of the construction of a prior distribution over finite alternatives. An example of such a study is reported in Akaike (1983b).

Although the result reported in this chapter has demonstrated the feasibility of the smoothness prior selection for distributed lag estimation, the practical applicability of the single-input–single-output model to the analysis of economic data is rather limited. This is due to the common existence of feedback between the input and output in econometric applications. In such a case multivariate time series modeling is required. Whether the smoothness prior can find a useful application in this case is a subject for further study.

Acknowledgements

This work is a continuation of the former work "Smoothness priors and the distributed lag estimator", Technical Report No. 40, Stanford University, 1979, which was supported in part by the Office of Naval Research Contract, N00014-75-C-0442. By the introduction of the model selection procedure discussed in this chapter some of the conclusions of the original work have been completely revised. The present work was partly supported by the United States Army under Contract No. DAAG29-80-C-0041 at the Mathematics Research Center, University of Wisconsin-Madison. The author is grateful to the editor, referees and W. Polasek for very helpful comments.

References

Akaike, H. (1980a) "Likelihood and the Bayes procedure", in: J.M. Bernardo, M.H. DeGroot, D.V. Lindley and A.F.M. Smith, eds., *Bayesian Statistics* (University Press, Valencia, Spain), pp. 143–166.

Akaike, H. (1980b) "Seasonal adjustment by a Bayesian modeling", *Journal of Time Series Analysis*, 1, 1–13.

Akaike, H. (1983a) "Statistical inference and measurement of entropy", in: G.E.P. Box, T. Leonard and C.F. Wu, eds., *Scientific Inference, Data Analysis, and Robustness* (Academic Press, New York), pp. 165–189.

Akaike, H. (1983b) "On minimum information prior distributions", *Annals of the Institute of Statistical Mathematics*, 35A, 139–149.

Good, I.J. (1965) *The Estimation of Probabilities* (M.I.T. Press, Cambridge, Mass.).

Morris, C.N. (1983) "Parametric empirical Bayes confidence intervals", in: G.E.P. Box, T. Leonard and C.F. Wu, eds., *Scientific Inference, Data Analysis, and Robustness* (Academic Press, New York), pp. 25–50.

Shiller, R.J. (1973) "A distributed lag estimator derived from smoothness priors", *Econometrica*, 41, 775–788.

Chapter 9

THE LINEAR HIERARCHICAL MODEL AND ITS APPLICATIONS IN ECONOMETRIC ANALYSIS

YOEL HAITOVSKY*

The Hebrew University of Jerusalem

1. Introduction

Many econometric models involve different *levels* of relationships among variables and/or observation units. For example, a random coefficients specification using panel data essentially utilizes time-series data on the micro units to estimate regression parameters for each micro model. Then, the micro units are tied together by the assumption that the respective micro coefficients average to yield a set of common coefficients. In another version of this model similar specifications are postulated on the micro units and on the aggregates, while each of the micro parameters is allowed to deviate from the macro parameters by a random vector with specified distribution. As an example consider the case where the micro units are firms comprising a particular industry, each with a production function with coefficients that might be either random or fixed. Then, a second stage is postulated, representing the econometrician's concept of the variations of the individual firm's production patterns across the industry. Surely, if such a variational pattern is modelled, it should be taken into account in the estimation of the individual micro functions, each borrowing strength from the others. See Dempster, Rubin and Tsutakawa (1981).

*This work was supported in part by NSF Grant SOC-7909674 to the University of Southern California and completed at the National Research Institute for Mathematical Sciences of the CSIR where I held visiting appointments. I wish to thank Dennis V. Lindley for arousing my interest in this model and to Prem K. Goel and two anonymous referees for their comments, which led to improvements in the exposition.

Bayesian Inference and Decision Techniques, edited by P. Goel and A. Zellner

In this chapter we utilize the Linear Hierarchical Model, originally devised by Lindley and Smith (1972) as a Bayesian alternative to least-squares estimates of the linear model. In the Lindley–Smith approach two phases of the model are distinguished. In the first a linear model is postulated to "explain" a vector of observations. In the second, prior information expressible as a *linear* probability structure in terms of a vector of hyperparameters, is imposed. Each of the two phases may comprise a linear hierarchical structure, namely its vector of parameters can be expressed as a linear model in terms of other vectors of parameters, but of smaller dimensions. Thus, while the behavioral model is designed to express different levels of relationships, the linear prior on the highest of such level parameters can take a sequence of decreasing dimensions until either a proper Bayesian, or at least an admissible (with respect to a quadratic loss) estimate, is obtained.

It is implicit in the present chapter that the distinction between the two phases is, in many cases, artificial and the whole model, except possibly at its two extreme ends θ_0 and θ_k (to be defined below), could be more usefully looked upon as a modelling device; hence all the different stage parameters should in principle be potentially estimable.

In section 2 we present the k-stage linear hierarchical model and examine its properties. Special emphasis will be given to the case in which the modelling process will end with a non-informative prior on the final stage parameter vector. Section 3 presents the estimators of the parameters of the model with known covariances and outlines approaches to the estimation with unknown covariances. In section 4 we apply the three-stage model to estimating residential demand for electricity. Three approaches to the estimation of the model parameters are presented in section 5, where the three vary only regarding the estimation of the model covariances. In section 6 we suggest a further elaboration of the model to take into account possible homogeneous household groups via the use of exchangeable priors in these groups. Finally, in section 7 we outline a few new possible applications of the linear hierarchical structure to econometric models.

2. The linear hierarchical model

We start by presenting the Linear Hierarchical Model (LHM). Following Lindley and Smith (1972) (hereafter L–S), Smith (1973), Haitovsky (1979a) and Goel and DeGroot (1981), the k-stage LHM is defined by the set of equations:

$$\theta_{i-1} = A_i\theta_i + \varepsilon_i, \quad i=1,\dots,k, \tag{1}$$

where θ_0 is an observable p_0-dimensional random vector; θ_i is a p_i-dimensional vector of parameters or hyperparameters, $i=1,\dots,k$; A_i is $p_{i-1}\times p_i$ known design matrix, and ε_i is a $p_{i-1}\times 1$ error vector, independent of θ_i, and of the other $\varepsilon_j (j\neq i)$, with mean 0, and dispersion matrix C_i, a $p_{i-1}\times p_{i-1}$ positive definite symmetric matrix. It will be convenient for later use to impose a distributional assumption on the vector θ_{i-1}, $i=1,\dots,k$, and in particular, that conditional on θ_i it has a multivariate normal distribution, i.e.

$$\theta_{i-1}|\theta_i \sim N(A_i\theta_i, C_i), \quad i=1,\ldots,k. \tag{2}$$

This formulation was originally devised by L–S to yield a Bayesian alternative to least-squares (or Maximum Likelihood when normality is assumed) estimates of the linear model. They regarded the first-stage equations, i.e. $i=1$, as a description of, or an approximation to, some physical phenomenon, namely the process generating the vector of observations θ_0. By allowing specification of *a prior* structure on the model parameters in the form of a linear model, where the vector of coefficients θ_1 plays the role of the dependent variable in the ordinary linear model, the second stage is obtained. Linear probability structures of the same form can be further imposed on the previously imposed priors, forming the higher stages with $\theta_2, \theta_3, \ldots, \theta_k$ as hyperparameters.

Smith (1973) uses the first two stages for a specification of regression of the second kind, in sampling-theory terminology, while the third stage provides the vehicle for imposing prior structures on the second stage parameters.

Haitovsky (1979a) offered a frequentist interpretation to the general form of the Bayesian linear model, justifying a Maximum Joint Probability estimation when normality is assumed, or an extended least-squares estimate when normality is not assumed. A special form of the two-stage model has become known as the Random Coefficient Regression (RCR) model (see Swamy, 1970), or the mixed regression model.

In the L–S set-up there are two important features that will be assumed throughout.

(a) The *hierarchical feature* of the model, namely that at each stage, coefficients are expressed as a linear function of a *smaller* number of coefficients. Or, symbolically, $p_0 > p_1 > \cdots > p_k > 0$. In practice quite often $p_i \gg p_{i+1}$ for some $i = 0, 1, \ldots, k-1$. It will also be assumed here for convenience that $p_0 > \sum_{i=1}^{k} p_i$.

(b) The *Markov property* of the model, namely that

$$\begin{aligned} E(\varepsilon_i \varepsilon_j^{\mathrm{T}}) &= C_i, \quad \text{whenever } i = j, \\ &= 0, \quad \text{whenever } i \neq j. \end{aligned}$$

The significance of these features is evident from the two well-known propositions:

Proposition 1. For $h=1,2,\ldots,k$, the distribution of θ_h, conditional on the *lower* hierarchical vectors $\theta_{h-1},\ldots,\theta_0$ is equal to the distribution of θ_h conditional on θ_{h-1}. Likewise, the distribution of θ_h conditional on the *higher* hierarchical vectors $\theta_{h+1},\ldots,\theta_k$ is equal to the distribution of θ_h conditional on θ_{h+1}, i.e.

$$\left.\begin{aligned} &\theta_h|\theta_{h-1},\ldots,\theta_1,\theta_0 \sim \theta_h|\theta_{h-1} \\ &\text{and} \\ &\theta_h|\theta_{h+1},\ldots,\theta_k \quad\ \sim \theta_h|\theta_{h+1} \end{aligned}\right\} h=1,\ldots,k.$$

Proposition 2. Conditional on $\{C_i; i=1,\dots,k\}$ and θ_k, every k-stage model is reducible, without loss of generality, to an h-stage model, $h<k$, where the intermediate stages can be consolidated into any one or more stage equations.

Proposition 2 is very useful in practical applications. The LHM provides us with a modelling device which is frequently very powerful. One of the main objects of the LHM is to assist the researcher in his modelling process. However, at the end of this process he might be interested in estimating only a particular subvector from the whole hierarchical structure. Furthermore, in the typical application the prior mean vectors, except possibly the last one, θ_k, do not assume particular values and one might want to eliminate all these intermediate vectors. Proposition 2 allows us to focus on the conditional distribution of a particular vector in the hierarchical structure, since the reduction of the model has the effect of integrating out $k-h$ vectors of parameters, or of condensing all the prior mean vectors into one such vector, or both. For instance, by successive substitution one can condense all the first $i(<k)$ stages into the first stage and all the remaining $k-i$ stages into the last stage, obtaining a two-stage LHM:

$$\theta_0 = B_{0i}\theta_i + \sum_{j=0}^{i-1} B_{0j}\varepsilon_{j+1}, \qquad \theta_i = B_{ik}\theta_k + \sum_{j=i}^{k-1} B_{ij}\varepsilon_{j+1}, \quad i=1,\dots,k-1, \tag{3}$$

with the corresponding covariance matrices:

$$V(\theta_0|\theta_i) = \sum_{j=0}^{i-1} B_{0j}C_{j+1}B_{0j}^{\mathrm{T}} \quad \text{and} \quad V(\theta_i|\theta_k) = \sum_{j=i}^{k-1} B_{ij}C_{j+1}B_{ij}^{\mathrm{T}}, \qquad i=1,\dots,k-1, \tag{4}$$

where

$$B_{ij} = A_{i+1}\dots A_j, \quad \text{whenever i} < j,$$
$$= I, \quad \text{whenever } i \geq j.$$

Until now we have assumed that the last stage vector of parameters θ_k is known. However, in its more general formulation three possible termination regimes can be distinguished:

(a) θ_k is known with certainty ("completely known");

(b) no a priori knowledge whatsoever about θ_k is available ("completely unknown"); and

(c) prior knowledge about θ_k is expressible in the form of a known distribution.

Here we shall adopt the following convention:

(a) When θ_k is "completely known" and hence is not subject to inference, it will be denoted by μ with $C_j=0$ for $j>k$. The joint probability distribution of the model is then written as $p(\theta_0,\Theta|\mu)$, where $\Theta^{\mathrm{T}}=(\theta_1^{\mathrm{T}},\dots,\theta_{k-1}^{\mathrm{T}})$ and hence the posterior distribution will be conditional on μ as well.

(b) We interpret the statement that θ_k is "completely unknown" to mean that it has a flat prior. It can then be formally modelled by carrying the hierarchy one step further and letting the precision matrix, associated with θ_k, $C_{k+1}^{-1}=0$. Consequently, for the sake of uniformity we will instead redefine k so that the last stage is $\theta_{k-1}=A_k\theta_k+\varepsilon_k$ with $C_k^{-1}=0$.

(c) In the case when θ_k is neither "completely known" nor "completely unknown", the hierarchy can again be stepped up with either (a) or (b) as the terminating regime, depending on the nature of the postulated prior, but again we will rename the last stage, as the case may be, the kth stage.

The advantage of this convention is that in all three cases there exists the vector Θ of unknown parameters that one may wish to make inference about, and $C_{k+j}=0$, $j=1,2,\ldots$. Hence, a unified estimation theory can be offered for all three regimes, as will become evident in the next sections.

3. Estimation

3.1. Covariance matrices are known

Bayes estimates under quadratic loss of θ_i are the means of the posterior distribution of θ_i conditional on θ_0 and μ (when it exists). This posterior distribution will be obtained by applying the reducibility proposition of the previous section and by recognizing that under the normality assumption the posterior mean is the matrix weighted average of the least-squares estimate applied to the first stage of the condensed model and the prior mean of the remaining stages, with the associated precision matrices serving as weights. That is to say, using eqs. (3) and (4), one obtains the posterior distribution of θ_i conditional on θ_0 and μ:

$$\theta_i|\theta_0,\mu\sim N(D_i d_i, D_i), \tag{5}$$

with

$$d_i=B_{0i}^{\mathrm{T}}\left\{\sum_{j=0}^{i-1}B_{0j}C_{j+1}B_{0j}^{\mathrm{T}}\right\}^{-1}\theta_0+\left\{\sum_{j=1}^{k-1}B_{ij}C_{j+1}B_{ij}^{\mathrm{T}}\right\}^{-1}B_{ik}\theta_k, \tag{6}$$

and

$$D_i^{-1}=B_{0i}^{\mathrm{T}}\left\{\sum_{j=0}^{i-1}B_{0j}C_{j+1}B_{0j}^{\mathrm{T}}\right\}^{-1}B_{0i}+\left\{\sum_{j=1}^{k-1}B_{ij}C_{i+1}B_{ij}^{\mathrm{T}}\right\}^{-1}, \tag{7}$$

for $i=1,\ldots,k-1$.

Alternatively, define the recurrence relations:

$$G_j = A_j G_{j+1} A_j^{\mathrm{T}} + C_j, \quad \text{whenever } j = 1, 2, \ldots, i-1, i+1, \ldots, k-1$$

$$\text{and} \tag{8}$$

$$G_j = C_j, \quad \text{whenever } j = i, k,$$

then

$$\left.\begin{aligned} d_i &= B_{0i}^{\mathrm{T}} G_1^{-1} \theta_0 + G_{i+1}^{-1} B_{ik} \theta_k \\ D_i^{-1} &= B_{0i}^{\mathrm{T}} G_1^{-1} B_{0i} + G_{i+1}^{-1} \end{aligned}\right\}, \quad i = 1, \ldots, k-1. \quad \begin{aligned}(6\text{a})\\(7\text{a})\end{aligned}$$

Notice that the first term on the left-hand side of (6) is the least-squares estimate of θ_i, weighted by its precision matrix, which is the first term on the right-hand side of (7). These are generalizations of Smith (1973, eq. 10), derived there for $k = 2$.[1] Lindley and Smith's (1972) *matrix lemma* can now be applied to eq. (8), obtaining for computational convenience:

$$G_j^{-1} = C_j^{-1} - C_j^{-1} A_j \left(A_j^{\mathrm{T}} C_j^{-1} A_j + G_{j+1}^{-1} \right)^{-1} A_j^{\mathrm{T}} C_j^{-1},$$
$$j = 1, \ldots, i-1, i+1, \ldots, k-1. \tag{8a}$$

The computation convenience arises from the fact that normally C_i, often a diagonal or block diagonal matrix, is relatively easy to invert, while matrices of the form $A_j^{\mathrm{T}} C_j^{-1} A_j$ are of order $p_{i-2} \times p_{i-2}$, which are typically of much smaller order than $A_j G_{j+1} A_j^{\mathrm{T}}$, a $p_{i-1} \times p_{i-1}$ matrix, which we would have had to invert, had we used eqs. (6a) and (7a).

There are situations where more than one θ_i parameter vectors need to be estimated. A case in point is model II, or random effects, in which both θ_1 and θ_2 need to be estimated. See, for example, Smith (1973). A more general class of problems, in which the whole $\Theta = (\theta_1^{\mathrm{T}}, \ldots, \theta_{k-1}^{\mathrm{T}})^{\mathrm{T}}$ vector is of interest and needs to be estimated, is where the LHM is given a frequentist interpretation. That is, the LHM constitutes a description of the generating mechanism of θ_0, and hence a physical model. In this case, the scientist may wish to estimate the whole vector Θ. An obvious estimator is that which maximizes, with respect to Θ, the *joint probability density* of $\theta_0, \theta_1, \ldots, \theta_{k-1}$ for a given μ. This is a natural extension of the maximum likelihood principle applied to the extended *physical* model (2).

An information theoretic justification to ML was given by Akaike (1973). He proved that ML can be considered as a method of asymptotic realization of an estimate with respect to a very general information theoretic criterion. Akaike's argument can be easily extended to the estimation of the LHM. From the Bayesian point of view, maximum likelihood estimates are Bayes estimates under 0–1 loss functions.

[1] The least-squares precision matrix can also be expressed in the recurrence relations $\hat{D}_i^{-1} = A_i^{\mathrm{T}}(C_i + \hat{D}_{i-1})^{-1} A_i$ until $\hat{D}_0 = 0$, generalizing Smith (1973, eq. 11).

Thus, we maximize

$$P(\theta_0, \Theta|\mu) \propto \exp\left\{-\tfrac{1}{2}\sum_{i=1}^{k}(\theta_{i-1}-A_i\theta_i)^{\mathrm{T}}C_i^{-1}(\theta_{i-1}-A_i\theta_i)\right\} \tag{10}$$

jointly with respect to Θ by minimizing the exponent, yielding the point estimates for θ_i,

$$\hat{\theta}_i = \left(A_i^{\mathrm{T}}C_i^{-1}A_i + C_{i+1}^{-1}\right)^{-1}\left(A_i^{\mathrm{T}}C_i^{-1}\hat{\theta}_{i-1} + C_{i+1}^{-1}A_{i+1}\hat{\theta}_{i+1}\right), \quad i=1,\ldots,k-1, \tag{11}$$

with the associated covariance matrix:

$$V(\hat{\boldsymbol{\theta}}) = \left((h_{i,j})\right)^{-1},$$

where

$$\begin{aligned} h_{i,i} &= A_i^{\mathrm{T}}C_i^{-1}A_i + C_{i+1}^{-1}, && i=1,2,\ldots,k-1, \\ h_{i,i+1} &= -C_{i+1}^{-1}A_{i+1},\ h_{i+1,i} = -A_{i+1}^{\mathrm{T}}C_{i+1}^{-1}, && i=1,\ldots,k-2 \\ h_{i,j} &= 0, && \text{otherwise.} \end{aligned} \tag{12}$$

Note that due to the Markov property of the model, $\hat{\theta}_i$ is a function of the two neighboring vector estimates $\hat{\theta}_{i-1}$ and $\hat{\theta}_{i+1}$. Likewise, its precision matrix is a tridiagonal block matrix.

The minimization of the exponent in eq. (10) leads to a least-squares-like estimator of the extended model (1). To this end we write the model as

$$\begin{bmatrix} \theta_0 \\ 0 \\ 0 \\ \vdots \\ 0 \\ A_k\mu \end{bmatrix} = \begin{bmatrix} A_1 & 0 & 0 & \ldots & 0 & 0 \\ -I & A_2 & 0 & \ldots & 0 & 0 \\ 0 & -I & A_3 & \ldots & 0 & 0 \\ \vdots & \vdots & \vdots & \ddots & \vdots & \vdots \\ 0 & 0 & 0 & \ldots & -I & A_{k-1} \\ 0 & 0 & 0 & \ldots & 0 & I \end{bmatrix} \begin{bmatrix} \theta_1 \\ \theta_2 \\ \theta_3 \\ \vdots \\ \theta_{k-1} \end{bmatrix} + \begin{bmatrix} \varepsilon_1 \\ \varepsilon_2 \\ \varepsilon_3 \\ \vdots \\ \varepsilon_{k-1} \\ -\varepsilon_k \end{bmatrix}, \tag{13}$$

or, more compactly, as

$$Y = A\Theta + \varepsilon. \tag{13a}$$

Then the least-squares estimates are

$$\tilde{\Theta} = (A^{\mathrm{T}}C^{-1}A)^{-1}A^{\mathrm{T}}C^{-1}Y; \qquad V(\tilde{\Theta}) = (A^{\mathrm{T}}C^{-1}A)^{-1}, \tag{14}$$

where

$$C^{-1} = \begin{bmatrix} C_1^{-1} & & & 0 \\ & C_2^{-1} & & \\ & & \ddots & \\ 0 & & & C_k^{-1} \end{bmatrix}$$

because of the Markov property.

Notice that, $A^{\mathrm{T}}C^{-1}A$ is identical to the right-hand side of (12) before inversion and that

$$A^{\mathrm{T}}C^{-1}Y=\begin{bmatrix} A_1^{\mathrm{T}}C_1^{-1}\theta_0 \\ 0 \\ \vdots \\ 0 \\ C_k^{-1}A_k\mu \end{bmatrix}.$$

Remark. Using well-known results of inversion of partitioned matrices, $\tilde{\theta}_i$ and its variance, given in eqs. (14), can be shown to be identical to the posterior mean and variance (4), and to $\hat{\theta}_i$, and $V(\hat{\theta}_i)$ in (11) and (12). These identities can also be obtained as a direct consequence of the optimization criteria underlying the Bayesian and ML estimation principles. See Haitovsky (1979a) for the details.

In actual econometric applications, as in many other fields, special attention should be given to the "completely unknown" termination stage, i.e. where the final stage hyperparameters assume a vague prior. In this case, we have

$$\left.\begin{aligned} &\text{(i)} \quad d_i = B_{0i}^{\mathrm{T}}G_1^{-1}\theta_0,\\ &\text{(ii)} \quad G_{k-1}=C_{k-1}^{-1}-C_{k-1}^{-1}A_{k-1}\left(A_{k-1}^{\mathrm{T}}C_{k-1}^{-1}A_{k-1}\right)^{-1}A_{k-1}^{\mathrm{T}}C_{k-1}^{-1},\\ &\text{(iii)} \quad \theta_{k-1}|\theta_0 \sim \left(A_{k-1}^{\mathrm{T}}C_{k-1}^{-1}A_{k-1}\right)^{-1}A_{k-1}C_{k-1}^{-1}\theta_{k-2}|\theta_0, \end{aligned}\right\} \tag{15}$$

and

(iv) estimates derived from the hierarchical structure using the posterior mean are admissible, with respect to quadratic loss, provided that any impropriety introduced in the final stage is confined to at most two dimensions, that is, θ_k is such that $p_k \leq 2$.

Statements (i)–(iii) are obtained by recalling that in the vague final stage case $C_k^{-1}=G_k^{-1}=0$ and by inserting it in (6a), (8a) and (11). A discussion of (iv) is contained in Smith (1973).

Property (iv) provides one of the major incentives for using the LHM as a potentially admissible alternative to least-squares estimates. Property (iii) can be very profitably employed to decrease the dimensionality of the estimation problem, at times substantially, using the following argument. In the extended least-squares representation (13a) all data are included in Y. In the present case $\tilde{\theta}_{k-1}$ can also be considered known by virtue of property (iii). Thus, the last row and column of submatrices in A in eq. (13a) will be deleted and conformably the last subvectors in Y, Θ and ε. The last p_{k-2} zeroes in Y will be replaced by $A_{k-1}\theta_{k-1}$, with the consequent reduction of dimentions of $A^{\mathrm{T}}C^{-1}A$ and $A^{\mathrm{T}}C^{-1}Y$.

3.2. *Covariance matrices unknown*

In actual research the covariances are unknown and need to be estimated. In order to obtain Bayesian estimates one proceeds as follows. One first assigns prior distributions to all the parameters in the model, i.e. to all the θ_i's and C_i's $(i = j, \ldots, k)$ and combines them with the likelihood function, to obtain the posterior distribution. One then proceeds by integrating out all the θ_i's which are of no direct interest (the nuisance θ_i-parameters). The resulting marginal joint posterior distribution of the θ_i's of interest (with all the C_i's) is then used to estimate C_i's and the remaining θ_i's. The estimation of the C_i's is more problematic and thus we defer its discussion until later. Given an initial estimate of the C_i's, the non-nuisance θ_i's are estimated, under quadratic loss, by the means of their posterior joint distribution, conditional on the estimates of the C_i's. Obviously, these conditional means are identical in form to the posterior means given by (4) and (15), with the C_i's replaced by their estimates. Having obtained these (conditional) estimates of the θ_i's, one then returns to the marginal likelihood of the C_i's conditional on these estimates of the θ_i's, which are used in turn to provide updated estimates of the C_i's. The first step is then repeated – the non-nuisance θ_i's are estimated conditional on the new estimates of the C_i's – and the process is repeated iteratively.

In actual application we will use modal estimates for the covariances. This will be demonstrated in the application discussed in section 4. There, Wishart priors are assigned to the unknown covariances. A justification for this procedure is given by L–S, who use modal estimates for both θ's and C's as an approximation to the means required for Bayesian estimates. In spite of theoretical pitfalls of modal estimates (see, for example, O'Hagan, 1976), they are often found to approximate the posterior means of the *regression coefficients* well (see, for instance, the example produced in O'Hagan, 1976), the reason being that in regression models the main discrepancies between the means and the modes stems from the fact that in the latter it is not clear what is the appropriate number of degrees of freedom (the divisor) in the estimation of the variance and covariance matrices. However, these estimates enter only indirectly into the calculation of the regression coefficients, and often appear identically in both the numerator and denominator (as is the case in eq. (27) below), so that the final estimates are often insensitive to the number of degrees of freedom.

Our procedure is akin to the EM algorithm (see Dempster, Rubin and Tsutakawa, 1981), where the mean of the marginal joint distribution, conditional on the estimated variances and covariances, are iteratively computed, and then using the provisional estimates of the posterior means, the variances and the covariances are re-estimated.

For the non-Bayesian, two versions are proposed for which no prior distributions on the variances and covariances are required, at the cost of a more restrictive structure of the variance matrix. These two versions are developed in the application of the LHM to the residential demand for electricity in section 5 and are dubbed LHM and Empirical Bayes estimates, respectively.

4. The residential demand for electricity

It will now be shown how the three-stage model, with a vague final stage, might be usefully applied to estimation of residential demand for electricity. The dependent variable is the household "load" on which data are typically available in many studies every 15 minutes over a long period of time. At the first level of the analysis, there is the problem of data reduction: the characterization of each household's daily demand pattern. At the second level, the first-level estimated (random) parameters are linked to household socioeconomic and demographic data through the specification of a direct functional relationship, complete with error terms and assumptions about their moments. This general approach to residential demand for electricity was used by Granger et al. (1979) who fitted hourly aggregates of household electricity usages to a common set of explanatory variables depicting time of day, working days vs. weekend or public holidays, peak-load vs. normal pricing, and a few temperature measures. Similarly, Hendricks et al. (1979) fitted a spline function to daily demand, producing a smooth characterization of each household's daily "load curve". In both, the first-stage households' time-series parameters were regressed on the households' sociodemographic characteristics and their stock of appliances (i.e. cross-sectional data), allowing the causes determining the shapes of the daily demand curve to interact with the demographics. The traditional econometric approach to estimating such models has been to treat each level of analysis separately. However, efficiency can be gained by using the complete model structure to jointly estimate both the first- and second-stage parameters.

As part of the estimation process one needs to estimate the unknown covariance matrices, and hence methodologies of covariance estimation will be demonstrated in the course of deriving the model estimates.

4.1. The model and its distributional assumptions

We first characterize the residential household demand for electricity as an electricity load curve. Thus, the ith household load curve is given by

$$y_i = X_i\beta_i + \zeta_i \quad (i=1,\ldots,N),$$

where y_i is a $T\times 1$ vector of observations on the household demand for electricity in prespecified time periods $t=1,\ldots,T$; X_i is a $T\times K$ matrix of observations on K physical characteristics such as time of day and weather variables, which determine the use of electricity by the household; β_i is a $K\times 1$ vector of parameters; and ζ_i is the corresponding $T \times 1$ vector of disturbances satisfying $E(\zeta_i) = 0$ and $E(\zeta_i \zeta_k^{\mathrm{T}}) = \omega_{ik} I_T$, for all $i, k = 1, \ldots, N$.

Next we relate the coefficients of the load curve to the household's socioeconomic and demographic determinants by expressing the jth component of β_i for all households as a linear function of L such determinants. Thus, we construct an $N \times K$ matrix of parameters B whose ith row is β_i^{T}, to give

$$B^{\mathrm{T}} = (\beta_1, \beta_2, \ldots, \beta_N)$$

and define

$$B = (b_1, b_2, \ldots, b_K),$$

then

$$b_j = Z\gamma_j + \eta_j, \quad j = 1, \ldots, K,$$

where Z is an $N \times L$ matrix of observations on the socioeconomic and demographic determinants, γ_j is an $L \times 1$ vector of parameters, and η_j is an $N \times 1$ vector of second-stage disturbances.

As before we assume that $E(\eta_j) = 0$, but that

$$E\left(\eta_j \eta_l^{\mathrm{T}}\right) = g_{jl} D, \quad j, l = 1, \ldots, K,$$

where D is an $N \times N$ covariance matrix and both D and $G = \{g_{jl}\}$ are positive definite matrices.

We can now combine the two stages into a multistage model as follows.

Stage 1:

$$y = X\beta + \zeta, \qquad E(\zeta) = 0, \qquad V(\zeta) = \Omega \otimes I_T, \tag{20}$$

where $y = (y_1^{\mathrm{T}}, y_2^{\mathrm{T}}, \ldots, y_N^{\mathrm{T}})^{\mathrm{T}}$ and $\zeta = (\zeta_1^{\mathrm{T}}, \zeta_2^{\mathrm{T}}, \ldots, \zeta_N^{\mathrm{T}})^{\mathrm{T}}$ are the stacked $NT \times 1$ vectors of observations on all N households' electricity usage and equation disturbances, respectively; $X = \mathrm{diag}\{X_1, X_2, \ldots, X_N\}$ is an $NT \times NK$ block diagonal matrix of observations on the first-stage explanatory variables; and $\beta = (\beta_1^{\mathrm{T}}, \beta_2^{\mathrm{T}}, \ldots, \beta_N^{\mathrm{T}})^{\mathrm{T}}$ is the stacked vector of all N households' parameters, as before. This stage can be recognized as a system of seemingly unrelated regressions, see Zellner (1962).

Stage 2:

$$B = Z\Gamma + H, \qquad E(H) = 0, \qquad V(H) = \Psi, \tag{21}$$

where $\Gamma = (\gamma_1, \gamma_2, \ldots, \gamma_K)$ is an $L \times K$ matrix of second-stage parameters and $H = (\eta_1, \eta_2, \ldots, \eta_K)$ is an $N \times K$ matrix of disturbances. This is a multivariate regression formulation, where the jth regression coefficients ($j = 1, \ldots, K$) for all N households in the sample play the role of the dependent variable. Eq. (21) can be rewritten as

$$\beta = (Z \otimes I_K)\gamma + h, \qquad E(h) = 0, \qquad V(h) = D \otimes G, \tag{21a}$$

where $\beta = \text{vec}\{B^{\mathrm{T}}\}, \gamma = \text{vec}\{\Gamma^{\mathrm{T}}\}, h = \text{vec}\{H^{\mathrm{T}}\}$, $\otimes$ denotes the usual Kronecker product of matrices, and the operator vec stacks its argument column by column into a column vector.

Note that although the researcher might prefer to have an unrestricted covariance matrix associated with h, in practical situations where the covariances are unknown and need to be estimated, a more restrictive structure must be imposed on the second-stage error covariances. The covariance matrix given in eq. (21a) is the least restrictive identifiable matrix that maintains the properties of symmetry and positive definitiveness. We shall see below, however, that in some actual applications it will need to be further restricted.

Stage 3: requires specification of Γ. We have no structure to offer for it at this point, and hence we assume it *non-informative*.

It will be useful at this point in the development to assume normality of the two vectors of disturbances, and hence the model can be compactly written as

$$y \sim N(X\beta; \Omega\otimes I_T), \tag{20a}$$

$$\beta \sim N[(Z\otimes I_K)\gamma; D\otimes G], \tag{21b}$$

$$\gamma \sim (\cdot\,; \infty I_{KL}). \tag{22}$$

Identifying now y, β and γ with $\theta_0, \theta_1, \theta_2$, respectively; X and $Z\otimes I_K$ with A_1 and A_2; and $\Omega\otimes I_T$ and $D\otimes G$ with C_1 and C_2 in eq. (2), the theory developed in the previous section is applicable.

5. Estimation

If all covariances in the model (20)–(21) were known, we could have used the posterior means as our parameter estimates. However, since they are not known in actual research, we offer here three approaches to their estimation.

5.1. Bayesian estimates

In the Bayesian approach, the elements of Γ are considered hyperparameters, and hence will have to be removed by integrating them out of the joint probability function which, for our model is proportional to

$$\begin{aligned} |\Omega\otimes I_T|^{-1/2}|D\otimes G|^{-1/2}\exp\{&-\tfrac{1}{2}(y-X\beta)^{\mathrm{T}}(\Omega^{-1}\otimes I_T)(y-X\beta) \\ &-\tfrac{1}{2}[\beta-(Z\otimes I_K)\gamma]^{\mathrm{T}}(D\otimes G)^{-1}[\beta-(Z\otimes I_K)\gamma]\}. \end{aligned} \tag{23}$$

In addition, it is necessary to specify prior distributions for the unknown covariances. We use here the appropriate conjugate family, which is the Wishart distribution. We assume that Ω^{-1}, D^{-1} and G^{-1} are independently Wishart distributed with v, r and s degrees of freedom and V, R and S matrices of proper dimensions, with the respective densities given by

$$p(\Omega^{-1}) \propto |\Omega|^{-(1/2)(v-N-1)}\exp\left(\mathrm{tr}\{-\tfrac{1}{2}\Omega^{-1}V\}\right),$$
$$p(D^{-1}) \propto |D|^{-(1/2)(r-N-1)}\exp\left(\mathrm{tr}\{-\tfrac{1}{2}D^{-1}R\}\right)$$

and

$$p(G^{-1}) \propto |G|^{-(1/2)(s-K-1)}\exp\left(\mathrm{tr}\{-\tfrac{1}{2}G^{-1}S\}\right).$$

Combining (23) with these densities and integrating out γ, one obtains the marginal posterior distribution of the remaining parameters, which is proportional to

$$|Z^{\mathrm{T}}D^{-1}Z|^{-K/2}|\Omega|^{-(1/2)(T+v-N-1)}|D|^{-(1/2)(K+r-N-1)}|G|^{-(1/2)(N-L-K+s-1)}$$
$$\exp\Big(\mathrm{tr}\Big\{-\tfrac{1}{2}\big[(y-X\beta)^{\mathrm{T}}(\Omega^{-1}\otimes I_T)(y-X\beta)+V\Omega^{-1}\big]$$
$$-\tfrac{1}{2}[\beta-(Z\otimes I_K)\bar{\gamma}]^{\mathrm{T}}(D\otimes G)^{-1}[\beta-(Z\otimes I_K)\bar{\gamma}]-\tfrac{1}{2}D^{-1}R-\tfrac{1}{2}G^{-1}S\Big\}\Big), \tag{24}$$

where

$$\bar{\gamma}=\big[(Z^{\mathrm{T}}D^{-1}Z)^{-1}Z^{\mathrm{T}}D^{-1}\otimes I\big]\beta. \tag{25}$$

To obtain estimates for β and γ, denoted by $\tilde{\beta}$ and $\tilde{\gamma}$, one uses the posterior means given by (4) and (9) with the corresponding prior covariances. Thus,

$$\tilde{\beta}=\big[X^{\mathrm{T}}(\tilde{\Omega}^{-1}\otimes I_T)X+M_{Z\tilde{D}}\otimes\tilde{G}^{-1}\big]^{-1}X^{\mathrm{T}}(\tilde{\Omega}^{-1}\otimes I_T)y \tag{26}$$

and

$$\tilde{\gamma}=\big[(Z^{\mathrm{T}}\tilde{D}^{-1}Z)^{-1}Z^{\mathrm{T}}\tilde{D}^{-1}\otimes I\big]\tilde{\beta}, \tag{27}$$

which agrees with (25), with

$$\tilde{\Omega}=\{\tilde{\omega}_{ik};\, i,k=1,\dots,N\}, \tag{28}$$
$$\tilde{\omega}_{ik}=(T+v-N-1)^{-1}\big[v_{ik}+(y_i-X_i\tilde{\beta}_i)^{\mathrm{T}}(y_k-X_k\tilde{\beta}_k)\big],\quad i,k=1,\dots,N, \tag{29}$$

$$\tilde{D}=(K+r-N-1)^{-1}\big[R+H\tilde{G}^{-1}H^{\mathrm{T}}+KZ(Z^{\mathrm{T}}\tilde{D}^{-1}Z)^{-1}Z^{\mathrm{T}}\big],[2] \tag{30}$$

$$\tilde{G}=(N-L-K+s-1)^{-1}[S+H^{\mathrm{T}}\tilde{D}^{-1}H], \tag{31}$$

[2] This equation can be solved for $\tilde{D}$:

$$\tilde{D}=(K+r-N-1)^{-1}$$
$$\times\Big\{R+H\tilde{G}^{-1}H^{\mathrm{T}}+K(r-N-1)^{-1}Z\big(Z^{\mathrm{T}}[R+H\tilde{G}^{-1}H^{\mathrm{T}}]^{-1}Z\big)^{-1}Z^{\mathrm{T}}\Big\}.$$

obtained by maximizing (23) with respect to Ω^{-1}, D^{-1} and G^{-1}, where

$$H = \tilde{B} - Z\tilde{\Gamma},$$
$$M_{Z\tilde{D}} = (\tilde{D}^{-1/2})^{\mathrm{T}}\left[I_N - W(W^{\mathrm{T}}W)^{-1}W^{\mathrm{T}}\right]\tilde{D}^{-1/2},$$
$$W = \tilde{D}^{-1/2}Z \quad \text{and} \quad \tilde{D}^{-(1/2)\mathrm{T}}\tilde{D}^{-1/2} = \tilde{D}^{-1},$$

where $\tilde{D}^{-1/2}$ is any $N \times N$ square root of $\tilde{D}^{-1}$, which is assured by the positive definiteness of $\tilde{D}$.

Finally, the covariance matrices of $\tilde{\beta}$ and $\tilde{\gamma} = \mathrm{vec}\{\tilde{\Gamma}^{\mathrm{T}}\}$ can be estimated, respectively, by

$$V(\tilde{\beta}) = \left[X^{\mathrm{T}}(\tilde{\Omega}^{-1}\otimes I_T)X + M_{Z\tilde{D}}\otimes\tilde{G}^{-1}\right]^{-1} \tag{32}$$

and

$$V(\tilde{\gamma}) = \left[(Z^{\mathrm{T}}\otimes I_K)\left\{\tilde{D}\otimes\tilde{G} + \left[X^{\mathrm{T}}(\tilde{\Omega}^{-1}\otimes I_T)X\right]^{-1}\right\}^{-1}(Z\otimes I_K)\right]^{-1}. \tag{33}$$

That is, the estimates of β and γ are given by their posterior means with the covariances replaced by their estimates.

5.2. *The linear hierarchical model estimates*

In this approach one maximizes the joint probability function (23) with respect to all parameters one wishes to estimate, i.e. β, γ, Ω, D and G.

The resulting modal estimates are

$$\hat{\omega}_{ik} = \frac{1}{T}(y_i - X_i\hat{\beta}_i)^{\mathrm{T}}(y_k - X_k\hat{\beta}_k), \quad i,k = 1,\ldots,N, \tag{34}$$

$$\hat{D} = \frac{1}{K}(\hat{B} - Z\hat{\Gamma})\hat{G}^{-1}(\hat{B} - Z\hat{\Gamma})^{\mathrm{T}}, \tag{35}$$

$$\hat{G} = \frac{1}{N}(\hat{B} - Z\hat{\Gamma})^{\mathrm{T}}\hat{D}^{-1}(\hat{B} - Z\hat{\Gamma}), \tag{36}$$

with the estimates for β, Γ and their associated covariance matrices as before – eqs. (26), (27), (32) and (33) – with Ω, D and G estimated by eqs. (34)–(36). The latter estimates are of the same form as those obtained in Haitovsky (1979b, 1980) and they come naturally in view of the standard maximum likelihood results, but also suffer from the same shortcomings, namely either $\hat{G}$ or $\hat{D}$ must be singular.[3]

To get around the singularity problem we introduce an additional assumption regarding the structure of the second-stage covariances:

[3] The rank of $\hat{B} - Z\hat{\Gamma} = (I - M_{Z\hat{D}})\hat{B} \leq \min\{N - L, K\}$, while its order is $N \times K$.

Assumption 5.2. D is a diagonal matrix.

Since normally $N \gg K, L$, as long as all elements of any single row in $\hat{B} - Z\hat{\Gamma}$ are not all zeros, D^*, the diagonalized matrix in the left-hand side of eq. (35), will now be positive definite.[4]

5.3. *Empirical Bayes estimates*

Following the approach as outlined in Maritz (1979), we start with the posterior mean of β, eq. (26), and then estimate the unknown covariances from the data. Since the actual estimates depend on specification of the covariances, and since the Granger et al. (1979) estimates come very close to being of the empirical Bayes type, we use their covariance specification here. They assume that Ω and G are diagonal matrices and that $D = I_N$. Their application is further simplified by the fact that observations on the X-matrix for all households are identical. Thus, the posterior means of β and γ become, respectively,

$$\left[I + \Omega M_Z \otimes (G\bar{X}^{\mathrm{T}}\bar{X})^{-1}\right]^{-1} y \quad \text{and} \quad \left[(Z^{\mathrm{T}}Z)^{-1}Z^{\mathrm{T}} \otimes I_K\right]\beta^*, \tag{37}$$

where $M_Z = I_N - Z(Z^{\mathrm{T}}Z)^{-1}Z^{\mathrm{T}}$ and $\bar{X}$ denotes the identical X matrix for all households.[5] Granger et al. estimate Ω by S, where $\{s_{ii}\} = S$ is the diagonal matrix of ordinary least-squares estimates of the regression error variances for each of the households, and $G = \{g_{ii}\}$, consisting of the least-squares residual sums-of-squares in the regression of the columns of the *estimated* B matrix on Z. That is, let $\tilde{H} = B^* - Z(Z^{\mathrm{T}}Z)^{-1}Z^{\mathrm{T}}B^* = M_Z B^*$, where B^* is the matrix of least-squares estimates, and examine the estimator $\operatorname{diag}\{\tilde{H}^{\mathrm{T}}\tilde{H}\}$ for G. It can be shown (see Granger et al., 1979) that

$$E\left(\operatorname{diag}\{\tilde{H}^{\mathrm{T}}\tilde{H}\}\right) = (N-L)G + \operatorname{tr}\left[\Omega - (Z^{\mathrm{T}}Z)^{-1}Z^{\mathrm{T}}\Omega Z\right]\operatorname{diag}(\bar{X}^{\mathrm{T}}\bar{X})^{-1},$$

and hence an unbiased estimator for G is

$$G^* = (N-L)^{-1}\operatorname{diag}\left\{\tilde{H}^{\mathrm{T}}\tilde{H} - (\bar{X}^{\mathrm{T}}\bar{X})^{-1}\operatorname{tr}\left[S - (Z^{\mathrm{T}}Z)^{-1}Z^{\mathrm{T}}SZ\right]\right\}.$$

Note that $(Z^{\mathrm{T}}Z)^{-1}Z^{\mathrm{T}}\beta^*$, the estimate for Γ in the computation of $\tilde{H}$, is unbiased.

Finally, S and G^* are substituted back into eq. (37) to yield the Empirical Bayes estimates.

[4] Our solution to the singularity problem is similar to Theil's (1973).

[5] Note that the unconditional distribution of the OLS estimate of $\operatorname{vec}\{B\}$ is $\operatorname{vec}\{B^*\} \sim N[(I_K \otimes Z)\operatorname{vec}\{\Gamma\}; G \otimes I_N + (\bar{X}^{\mathrm{T}}\bar{X})^{-1} \otimes \Omega]$ and for a single column in B^*, $b_j^* \sim N[Z\gamma_j; g_{jj}I_N + (\bar{X}^{\mathrm{T}}\bar{X})^{jj}\Omega]$, where $(\bar{X}^{\mathrm{T}}\bar{X})^{jj}$ is the jjth element of $(\bar{X}^{\mathrm{T}}\bar{X})^{-1}$.

5.4. Computation

A typical study of the residential demand for electricity might include a few hundred households with between six and twenty first-stage explanatory variables. Estimating β using eq. (26) as it is, is impractical because of the scale of the matrix inversion involved. The computations can be simplified substantially if one is willing to assume that D and Ω are both diagonal. Then eq. (26) can be written as

$$\tilde{\beta}=\left\{Q^{-1}+Q^{-1}(\tilde{D}^{-1}Z\otimes\tilde{G}^{-1})\left[(Z^{\mathrm{T}}\tilde{D}^{-1}Z\otimes\tilde{G}^{-1})-(Z^{\mathrm{T}}\tilde{D}^{-1}\otimes\tilde{G}^{-1})Q^{-1}\right.\right.$$
$$\left.\left.\times(\tilde{D}^{-1}Z\otimes\tilde{G}^{-1})\right]^{-1}(Z^{\mathrm{T}}\tilde{D}^{-1}\otimes\tilde{G}^{-1})Q^{-1}\right\}X^{\mathrm{T}}(\tilde{\Omega}^{-1}\otimes I_T)y, \tag{38}$$

where Q^{-1} is a block diagonal matrix with a typical submatrix

$$Q_{ii}^{-1}=\left(\tilde{\omega}_{ii}^{-1}X_i^{\mathrm{T}}X_i+\tilde{d}_{ii}^{-1}G^{-1}\right)^{-1}=\left(\tilde{d}_{ii}\tilde{G}\left(\frac{\tilde{d}_{ii}}{\tilde{\omega}_{ii}}X_i^{\mathrm{T}}X_i\tilde{G}+I_K\right)^{-1}\right).$$

Whereas before we had to invert an $NK\times NK$ matrix, the largest one we need to invert now is of dimension $LK\times LK$. Similarly for $V(\tilde{\beta})$ and $V(\tilde{\gamma})$. The former is given by the expression within the curly brackets, while the latter is the inverse of the expression within the square brackets in (38).[6]

6. Further elaboration

Individual households' data are very noisy, since electricity usage depends very much not only on the life-style of the family, but on what its members happened to be doing on the particular days of the experiment. Evidently, additional structure imposed on the individual household is a potential source of improvement. One way is to divide all households into m homogeneous groups, $A, B, C,\ldots$, each of which is comprised of $a, b, c,\ldots$ households, respectively; hence $a+b+c+\cdots=N$. For instance, Hendricks et al. (1979) divide their sample into three groups: "high use", "moderate use" and "low use" according to the electrical appliance in the households' possession. Granger et al. (1979) divide their sample into five groups according to the previous year's usage level.

We model the homogeneity within each group by assuming exchangeability of the first-stage parameter vectors within each group, i.e.

$$\beta_i=\beta_a+\xi_i,\quad \xi_i \text{ i.i.d.} \sim (0,\Sigma_a)\quad \text{for all } i\in A,$$
$$\beta_i=\beta_b+\xi_i,\quad \xi_i \text{ i.i.d.} \sim (0,\Sigma_b)\quad \text{for all } i\in B, \text{etc.}$$

where β_i, β_a, $\beta_b,\ldots$, and ξ_i are all $K\times 1$ vectors.

[6] Programs to implement the theory discussed here were developed by W. Polasek and D.G. Fiebig using the SAS/Matrix language.

If, in addition, the researcher believes that the groups are "similar" in the parametric characterizations of the load curves, a "between groups" exchangeability can be imposed by letting

$$\beta_\alpha = \bar{\beta} + \eta_\alpha, \quad \alpha = a, b, \dots, m,$$

where η_α i.i.d. $(0, \Sigma_\eta)$.

Combining the hierarchical structure with the previous one we obtain for the LHM first-stage parameters estimates formulae similar to eq. (38), but with Q redefined to be

$$Q = X^{\mathrm{T}}(\Omega^{-1} \otimes I_T) X + D^{-1} \otimes G^{-1} + \begin{bmatrix} I_a \otimes \Sigma_a^{-1} & 0 & \cdots & 0 \\ 0 & I_b \otimes \Sigma_b^{-1} & \cdots & 0 \\ 0 & 0 & \ddots & \\ \vdots & \vdots & & \ddots \\ 0 & 0 & & \end{bmatrix}$$

$$- \begin{bmatrix} 1_a \otimes \Sigma_a^{-1} & 0 & \cdots & 0 \\ 0 & 1_b \otimes \Sigma_b^{-1} & \cdots & 0 \\ 0 & 0 & \ddots & \\ \vdots & \vdots & & \ddots \\ 0 & 0 & & \end{bmatrix}$$

$$\times \left[\begin{pmatrix} a\Sigma_a^{-1} & & \cdots & 0 \\ 0 & b\Sigma_b^{-1} & \cdots & 0 \\ 0 & 0 & \ddots & \\ \vdots & \vdots & & \ddots \\ 0 & 0 & & \end{pmatrix} + U_m \otimes \Sigma_\eta^{-1} \right]^{-1}$$

$$\times \begin{bmatrix} 1_a^{\mathrm{T}} \otimes \Sigma_a^{-1} & & \cdots & 0 \\ 0 & 1_b^{\mathrm{T}} \otimes \Sigma_b^{-1} & \cdots & 0 \\ 0 & 0 & \ddots & \\ \vdots & \vdots & & \ddots \\ 0 & 0 & & \end{bmatrix}, \tag{39}$$

where

$$U_m = I_m - \frac{1}{m} 1_m 1_m^{\mathrm{T}}.$$

In section 5, Q comprises the two first matrices of (39). The two additional terms result from the exchangeability assumptions. Furthermore, if, as in section 5.4, Ω and D are taken to be diagonal, the first three terms in (39) are block diagonal matrices with a typical submatrix along the main diagonal:

$$\omega_{ii}^{-1}X_i^{\mathrm{T}}X_i+\Sigma_\alpha^{-1}+d_{ii}^{-1}G^{-1},\quad i=1,\dots,N;\quad \alpha=a,b,\dots,$$

and hence can easily be inverted. Special inversion methods are available for the remaining terms.

After β is estimated with its variance (eq. (32)), eqs. (27) and (33) can be used to estimate γ and its corresponding covariance matrix. Then, the estimation of β_α $(\alpha=a,b,\dots,m)$ and $\bar{\beta}$ are given by

$$\beta_\alpha^*=W_\alpha\bar{\beta}_\alpha^*+(I-W_\alpha)\left(\sum_{j=a}^{m}W_j\right)^{-1}\sum_{j=a}^{m}W_j\bar{\beta}_j^*,\quad \alpha=a,b,\dots,m,$$

and its corresponding covariance matrix:

$$V(\beta_\alpha^*)=\left\{W_\alpha+(I-W_\alpha)\left(\sum_{j=a}^{m}W_j\right)^{-1}W_\alpha\right\}\alpha^{-1}\Sigma_\alpha,\quad \alpha=a,b,\dots,m$$

where

$$W_\alpha=\left(\alpha\Sigma_\alpha^{-1}+\Sigma_\eta^{-1}\right)^{-1}\alpha\Sigma_\alpha^{-1},\quad \alpha=a,b,\dots,m,$$

$$\bar{\beta}_a^*=\frac{1}{a}\sum_{j=1}^{a}\beta_j^*,\qquad \bar{\beta}_b^*=\frac{1}{b}\sum_{j=a+1}^{a+b}\beta_j^*,\ \text{etc.}$$

are the componentwise means of the β_j^*'s, and similarly

$$\bar{\beta}^*=\frac{1}{m}\sum_{j=a}^{m}\beta_j^*,$$

with

$$V(\bar{\beta}^*)=\left\{\sum_{j=a}^{m}\left(\alpha^{-1}\Sigma_\alpha+\Sigma_\eta\right)^{-1}\right\}^{-1}.$$

Cf. Lindley and Smith (1972, p. 10) and Fearn (1975, p. 91).

Finally, the estimators of Σ_α and Σ_η are obtained in a straight-forward manner by differentiation of the joint probability function. The resulting estimates are of the modal L–S type.

Thus, let

$$B_a=\{\beta_1,\dots,\beta_i,\dots\},\quad \text{for } i\in A,$$

then

$$\Sigma_a^*=\frac{1}{a}\left(B_a^*-1_a^{\mathrm{T}}\otimes\beta_a^*\right)\left(B_a^*-1_a^{\mathrm{T}}\otimes\beta_a^*\right)^{\mathrm{T}}=\frac{1}{a}\sum_{i\in A}\left(\beta_i^*-\beta_a^*\right)\left(\beta_i^*-\beta_a^*\right)^{\mathrm{T}},$$

where B_α^* and β_α^* are the LHM estimates of B_α and β_α $(\alpha=a,b,c,\dots)$, respectively. Similarly for Σ_b, etc.

Σ_η is estimated by

$$\Sigma_\eta^* = \frac{1}{m}(\bar{B}^* - 1^{\mathrm{T}} \otimes \bar{\beta}^*)(\bar{B}^* - 1^{\mathrm{T}} \otimes \bar{\beta}^*)^{\mathrm{T}} = \frac{1}{m} \sum_{i=a}^{m} (\beta_i^* - \bar{\beta}^*)(\beta_i^* - \bar{\beta}^*)^{\mathrm{T}},$$

where $\bar{B} = (\beta_a, \beta_b, \ldots)$ with the corresponding $\bar{B}^*$ estimates.

Note that for the Σ_α the number of first-stage coefficients must not be smaller than the corresponding $\alpha = a, b, c, \ldots$. Likewise m, the number of groups, must at least be as large as K. If any of these conditions is not satisfied, priors need to be imposed.

7. Additional econometric applications

We have used here the three-stage LHM to construct households' "load curve" in the first stage and then to "explain" the reaction of the different households to the determinants of the load curve by another set of explanatory variables. This is the second stage. It will be useful to outline other econometric applications of this sort. Then a special case of the model of section 4 is also outlined.

7.1. Returns to schooling

Expected economic returns to schooling are commonly derived by means of a regression equation. It is constructed as the regression coefficients associated with, say, number of years of school attendance. Then a related model is specified to explain the "returns to schooling" coefficient by appropriately specified explanatory variables, as in Welch (1966). For obvious reasons, economic returns to schooling vary among persons by occupation and region. This model is a typical two-stage model (with a possible third vague prior stage) and should be estimated as such. The construction of measures by means of regression coefficients introduce heteroscedasticity that should not be ignored in the estimation. LHM estimation takes care of this problem.

7.2. Unemployment coefficients in wage equations

A related problem is an attempt to explain an industry's relative response to unemployment rates by non-competitiveness of the industry's product and labor markets. At the first step, relative wages in the various industries are regressed on aggregate unemployment rates, on value added, on prices and on some dummy variables. Then the unemployment coefficients from the various industries' wage equations are regressed on some measures of concentration and on the percent of unionized workers in the industry, making the second stage. See Wachter (1970) and Amemiya (1978).

7.3. *Seemingly unrelated regressions applied to cross-sections of time series*

A simpler version of the three-stage LHM of section 4 was used by the author to estimate the seemingly unrelated regressions model (see Haitovsky, 1980). It can easily be derived from section 4 by letting $Z = 1_N$, an N-dimensional vector of ones. This model was applied to Grunfeld's (1958) data. The resulting estimates made more sense than Smith's (1973) estimates obtained with a more restrictive covariance matrix.

References

Akaike, H. (1973) "Information theory and an extension of the maximum likelihood principle", in: B.N. Petrov and F. Csáki, eds. *2nd International Symposium on Information Theory* (Akademiai Kiado, Budapest), pp. 267–281.

Amemiya, T. (1978) "A note on a random coefficient model", *International Economic Review*, 19, 793–796.

Dempster, A.P., D.B. Rubin and R.K. Tsutakawa (1981) "Estimation in covariance components models", *Journal of the American Statistical Association*, 76, 341–353.

Fearn, T. (1975) "A Bayesian approach to growth curves", *Biometrika*, 62, 89–100.

Goel, P.K. and M.H. DeGroot (1981) "Information about hyperparameters in hierarchical models", *Journal of the American Statistical Association*, 76, 140–147.

Granger, Clive W.J., Robert Engle, Ramu Ramanathan and Allan Anderson (1979) "Residential load curves and time-of-day princing", *Journal of Econometrics*, 9, 13–32.

Grunfeld, Y. (1958) "The determinants of corporate investment", Unpublished Ph.D. Thesis, University of Chicago.

Haitovsky, Yoel (1979a) "The linear hierarchical model I: Estimation", Modelling Research Group Working Paper 7902, Department of Economics, University of Southern California.

Haitovsky, Yoel (1979b) "On multivariate ridge regression", Modelling Research Group Working Paper 7915, Department of Economics, University of Southern California.

Haitovsky, Yoel (1980) "Seemingly unrelated regressions with linear hierarchical priors", Modelling Research Group Paper 8018, Department of Economics, University of Southern California.

Hendricks, Wallace, Roger Koenker and Dale J. Poirier (1979) "Residential demand for electricity", *Journal of Econometrics*, 9, 33–57.

Lindley, D.V. and A.F.M. Smith (1972) "Bayes estimates for the linear model", *Journal of the Royal Statistical Society*, Ser. B, 34, 1–41.

Maritz, J.S. (1979) *Empirical Bayes Methods* (Methuen & Co. Ltd., London).

O'Hagan, A. (1976) "On posterior joint and marginal modes", *Biometrika*, 63, 329–333.

Smith, A.F.M. (1973) "A general Bayesian linear model", *Journal of the Royal Statistical Society*, Ser. B, 35, 67–75.

Swamy, P.A.V.B. (1970) "Efficient inference in a random coefficient regression model", *Econometrica*, 38, 311–323.

Theil, Henri (1973) "A simple modification of the two-stage least-squares procedure for undersized samples", in: *Structural Equation Models in the Social Sciences* (Seminar Press, Inc.), pp. 113–129.

Wachter, Michael L. (1970) "Relative wage equations for U.S. manufacturing industries 1947–1967", *Review of Economics and Statistics*, 52, 405–410.

Welch, F. (1966) "Measurement of the quality of schooling", *American Economic Review*, 56, 379–392.

Zellner, Arnold (1962) "An efficient method of estimating seemingly unrelated regressions and tests for aggregation bias", *Journal of the American Statistical Association*, 57, 348–368.

PART IV

SELECTED BAYESIAN TECHNIQUES

Chapter 10

MODEL SELECTION IN THE FEDERAL COURTS: AN APPLICATION OF THE POSTERIOR ODDS RATIO CRITERION

STEPHEN J. BROWN and ROGER W. KLEIN*

Yale University and Bell Communications Research, Inc.

1. Introduction

As a result of two recent Supreme Court rulings, federal courts have begun to accept hypothesis tests at fixed significance levels of 5 and 1 percent. Such tests have been employed as an appropriate standard of proof to establish a prima facie case of discrimination in disparate treatment actions brought under Title VII of the Civil Rights Act. We will argue that this standard is subject to abuse as both plaintiffs and defendants vary the size of the sample on which such a test is based. In this chapter we advocate a posterior odds ratio standard and illustrate its application in the context of several recent disparate treatment cases. In section 2 we review the concept of disparate treatment and outline the statistical methods used to establish it in two simple cases. Section 3 re-examines these cases using a posterior odds ratio criterion, while section 4 extends the analysis to address the comparison of regression models used to establish disparate treatment. Section 5 presents some concluding remarks.

2. Disparate treatment and fixed significance levels

In 1979 Andra Capaci filed an employment discrimination action against her former employer, Katz & Besthoff Inc. (K&B), a Louisiana-based drugstore

*The views expressed are those of the authors and do not necessarily reflect those of Bell laboratories or of the Bell system. We wish to acknowledge the comments of Catherine Paulo and references supplied by John Abowd. Remaining errors are entirely our own.

Bayesian Inference and Decision Techniques, edited by P. Goel and A. Zellner

Table 10.1. Evidence of disparate treatment: *Capaci* and *Teamsters*.

1. Hire and promotion Data of Katz & Besthoff Inc.[a]

	7-12/65	1966	1967	1968	1969	1970	1971	1972	1973	1974	1975	1976	1977	1978
M	12	20	28	31	32	42	47	54	71	93	55	67	85	1
F	0	0	0	0	0	0	0	0	0	1	3	5	8	11
T	12	20	28	31	32	42	47	54	71	94	58	72	93	12

Referent groups (percent male)

1. Civilian labor force	63.24	5. Department and sales managers: retail trade	71.91
2. All managers	83.76	6. EEO-1 Report by State: officials and managers, New Orleans	86.80
3. All retail trade managers	81.84	7. EEO-1 Report by State: officials and managers, Louisiana	88.60
4. All general merchandise retail store managers	75.60		

2. Membership of Road Bargaining Unit, T.I.M.E. D.C.[b]

A. Atlanta:	W	57	Referent groups:	SMSA	77.41
	B	0	(percent white)	City	48.40
	T	57			
B. Dallas:	W	20	Referent groups:	SMSA	83.23
	B	0	(percent white)	City	74.15
	T	20			
C. Los Angeles:	W	322	Referent groups:	SMSA	85.42
	B	2	(percent white)	City	77.19
	T	324			

[a]*Capaci v. Katz & Besthoff* 525 F. Supp. 317 (1981), 326.
[b]*United States v. T.I.M.E.* D.C. 517 F. 2d 299, 315n 29 (1975).

chain. She alleged among other things that the drugstore chain discriminated against her by refusing her promotion to a managerial position, subjecting her to disparate terms and conditions of employment, and retailiating against her after she filed sex discrimination charges through EEOC in October 1973.

In her U.S. District Court action (*Capaci v. Katz & Besthoff, Inc.*, 525 F. Supp. 317 (1981)), Capaci sought to establish a prima facie case of employment discrimination. Her expert noted that there were three women in the group of 201 officials and managers at K&B. He observed that the chance of choosing three or fewer women at random from various management groups in Louisiana would be "less than one in one billion". When K&B countered that many in the management group had been hired prior to the Civil Rights Act of 1964, the expert noted that of 265 individuals hired in the interval 1965–1973 all were male. Again, given a variety of labor force referent groups the chance that this would occur through random selection was "less than one in one billion". While K&B questioned the validity of the referent groups, their expert also suggested that the data should be broken down and tested as to each locality in which applications were received. To show what impact this would have, the hire and promotion data were broken down by year (table 10.1). In comparing these data with male proportions in several referent groups, the discrimination effect was not statistically significant in many instances. For this reason the court found that the plaintiff had not established a prima facie case and decided against her.[1]

Capaci v. Katz & Besthoff Inc. is typical of a large number of so-called "disparate treatment" cases[2] that have come before the federal courts as a result of the passage of the Civil Rights Act of 1964. Disparate treatment is said to occur where a defendant covertly treated some people less favorably than others because of their group status.[3]

For a case in which the plaintiffs were found to establish a prima facie case, consider *Teamster v. United States* 431 U.S. 324 (1977). In *Teamsters*, plaintiffs alleged that a multistate trucking firm, T.I.M.E. D.C. inter alia had systematically excluded blacks from the more desirable long distance haul Road Bargaining Group. While the facts were more complex then in *Capaci,* similar statistics were used in part to establish the prima facie case in the lower court (see *United States v. T.I.M.E. D.C.* 517 F2d 299, 315n 2,9 (1975)). A portion of these data was reported in the trail court decision and is reported in table 10.1. In particular, if we compare the Dallas data and SMSA referent group cited in *Teamsters* with the 1966 data and "All Managers" referent group of *Capaci,* we find identical numbers: of a group of 20 hires or promotions, all belong to the majority group. Yet, the majority group comprises only 84 percent of the reference population. While the *Capaci* court found this particular datum "not significant", the appeals court in the *Teamsters'* case cited "the inexorable zero" (i.e. the fact that there were simply no blacks in the Road Bargaining Groups of Atlanta or Dallas) in its determination that the defense had failed to rebut these statistics (*United States v. T.I.M.E. D.C.* 517 F2d 299, 315). How do commonly accepted motions of "statistical significance" accord with the "inexorable zero" standard used in Teamsters?

The Supreme Court first laid down a rule on statistical significance in *Casteneda v. Partida*, 430 U.S. 482 (1977), a jury selection case. This rule was followed in *Hazelwood School District v. United States,* 433 U.S. 299 (1977), a teacher hiring

[1]The Court found unpersuasive arguments that the statistical insignificance of the year by year numbers was an artifact of the small sample size (despite *Friend v. Leidinger* 588 F2d 61 (1978), *Harper v. Trans World Airlines* 525 F2d 412 (1975), *Turner v. Texas Instrument* 555 F2d 1251 (1977) and *Morita v. So. Cal. Permanente Medical Group* 541 F2d 217 (1976) and other cases that specifically warn against this), and rejected plaintiff's attempt to pool data from 1965 to 1973. For other recent cases where subsampling by the defendant is used to reduce the statistical significance of the plaintiff's case, see *Little v. Master Bilt Products, Inc.* 506 F. Supp. 319 (1980) and *Pouncy v. Prudential Insurance Co. of America* 499 F. Supp. 427 (1980).

[2]The Courts draw distinctions between "disparate treatment", "differential impact", and other theories of discrimination. While these distinctions affect the required proof (Baldus and Cole, 1980, pp. 24ff), they are not central to the discussion that follows.

[3]An excellent discussion of these cases and the statistical issues raised by them can be found in Baldus and Cole (1980), which has itself been cited in a number of subsequent cases, for example *Villafane v. Manson* 504 F. Supp. 78 (1980), *Vuyanich v. Republic Nat. Bank of Dallas* 505 F. Supp. 224 (1980), and *Reynolds v. Sheet Metal Workers Local 102* 498 F. Supp. 952 (1980).

case where the Court stated:

> "[a]s a general rule for large samples if the difference between the expected value and the observed number is greater than two or three standard deviations" then the hypothesis that teachers were hired without regard to race would be suspect (433 U.S. 299, 308).

This criterion is generally understood to sanction hypothesis tests at the 5 or 1 percent level (e.g. *Pouncy v. Prudential Insurance Co. of America* 499 F. Supp. 427 (1980)) and as such is now cited whenever tests of statistical significance are introduced into evidence in discrimination cases.

Capaci represents a valid application of this standard of statistical significance. However, there is something inherently unreasonable about this application. For any result that is statistically significant at a fixed significance level, there is always a sample size sufficiently small obtained through subsampling for which the result is insignificant. This property of the standard was exploited in *Capaci*. In fact the large sample qualification of the Supreme Court standard could be interpreted as an effort by the court to block the reductio ad absurdam of *Capaci*.[4]

There is another possible standard of statistical significance which avoids the *Capaci* problem without relying on the ill-defined large sample qualifications. The two or three standard deviation rule, even if interpreted in a hypothesis testing framework, does not directly address the preponderance of the evidence standard used in these disparable treatment cases. A Bayesian approach that examined the probability of disparate treatment relative to the probability of no disparate treatment – a posterior odds criterion – would address this evidence standard. This point is argued persuasively in Finkelstein and Fairley (1970), where preponderance of evidence is specified in terms of a posterior probability.[5] Baldus and Cole (1980, pp. 304–305) concede this argument, but note that such methods are difficult to apply. In the next section we take issue with this presumption, and apply the posterior odds analysis to the *Capaci/Teamsters* example.

[4] Baldus and Cole (1980, p. 297) interpret this qualification either as a degrees of freedom correction or as allowing for asymptotic approximations. This is certainly valid if we accept the fixed significance level as an appropriate standard. However, the large sample qualification is interpreted differently in at least one court. *Friend v. Leidinger* 588 F 2d 61 (1978) cites the qualification and explains "the larger the sample the greater the reliability of the findings and samples of under 30 should be avoided" (at 66).

[5] Finkelstein and Fairley (1970, p. 508) observe that for the preponderence of evidence rule:

> The choice of fifty percent probability as a criterion may seem to be an overly literal application of the preponderence of evidence rule. However, it can easily be shown that this rule of decision would result in less error than under any other posterior probability criterion.

Although this odds ratio of 1 : 1 may be reasonable in civil cases where the preponderence of evidence rule is applied, in criminal cases the standard of reasonable doubt is quite different. For the civil discrimination cases discussed here, we take the critical odds value of 1 : 1. However, the purpose of this chapter is to demonstrate the advantages to the posterior odds criterion, and not to advocate any given critical odds level. The arguments presented here are valid for any such level.

3. Disparate treatment: The binomial model

In the simplest models of disparate treatment, a prima facie case is made out when the proportion of majority group members hired or promoted differs significantly from the referent population from which they are drawn. In the *Capaci* and *Teamsters* examples, the proportion of the majority group in the referent population was approximately 0.84. If we denote by θ the true proportion of the majority group hired or promoted, then the discrimination model M_{D} and non-discrimination model M_{ND} can be represented:[6]

$$M_{\mathrm{D}}: \theta \neq 0.84; \qquad M_{\mathrm{ND}}: \theta = 0.84.$$

Assuming a binomial model,[7] an exact test used in *Capaci* finds the non-discrimination model M_{ND} cannot be rejected at the 5 percent level.

In contrast, suppose that we compute the posterior probabilities of these models. Under the binomial i.i.d. assumption, the data are generated according to the likelihood:

$$L = \theta^{m}(1-\theta)^{n-m}, \tag{1}$$

where m is the number of majority group members out of a sample of n promotions or hires. Then, the posterior odds favoring model M_{D} are given by:

$$P(M_{\mathrm{D}})/P(M_{\mathrm{ND}}) = [P_0(M_{\mathrm{D}})/P_0(M_{\mathrm{ND}})] \cdot \\ \mathrm{E}_{\theta}\theta^{m}(1-\theta)^{n-m}/\theta_{\mathrm{ND}}^{m}(1-\theta_{\mathrm{ND}})^{n-m}, \tag{2}$$

where $P_0(\cdot)$ is the prior probability of the indicated model, E_{θ} is the expectation taken over the prior distribution of θ, and θ_{ND} is the value of θ under model M_{ND}.

To explicitly compute the odds in (2), for illustrative purposes, let the prior distribution for θ in model M_{D} be the conjugate beta prior:

$$P_0(\theta/M_{\mathrm{D}}) \propto \theta^{\gamma\theta_0 - 1}(1-\theta)^{\gamma(1-\theta_0)-1}, \tag{3}$$

where θ_0 is the prior mean and γ is a parameter whose value will, determine the weight of the prior evidence. Since the variance of θ for the prior is (3) is $\theta_0(1-\theta_0)/(\gamma+1)$, as γ increases the prior will center more closely on the prior mean, θ_0. If we further assume that the prior distribution has the null value of θ

[6]As is common in many of the disparate treatment cases, in particular *Capaci*, M_{D} reflects discrimination against *either* minority or majority groups, despite the strong presumption that discrimination favors the majority group. Baldus and Cole (1980, pp. 307–308) defend this practice.

[7]For a justification of the binomial model in this context and a discussion of the conditions under which it may or may not be appropriate, see Finkelstein (1966) and Baldus and Cole (1980, p. 325).

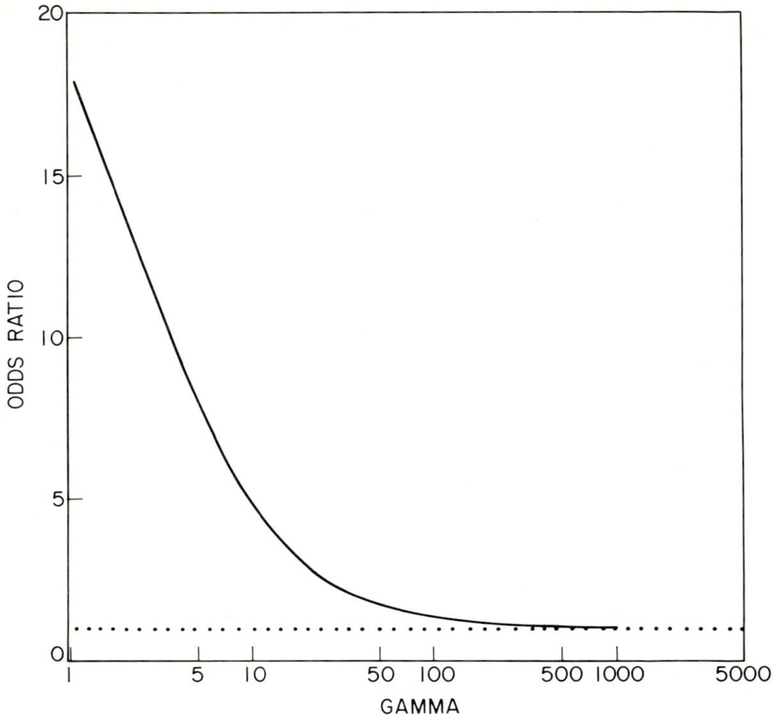

Fig. 10.1. Posterior odds as a function of prior information.

as its mean: $\theta_0 = \theta_{ND}$, and that the prior odds are 1.0, then one can derive the odds in favor of discrimination.[8]

For the data introduced in *Capaci* and *Teamsters* ($M = n = 20$, $\theta_{ND} = 0.84$), fig. 10.1 shows the odds as a function of the prior evidence parameter, γ from fig. 10.1, when $\gamma = 1.0$, the posterior odds are about eighteen to one in support of discrimination (model M_D). In this case, the posterior mean of θ is about 1.0, which is the data-based estimate of θ. Consequently, in this dimension the data "dominates" the prior. Even with $\gamma = 1000$, the posterior odds still slightly favor model M_D. In this case, the prior evidence is sufficiently strong so as to move the posterior mean, now at 0.843, away from the data-based estimate (1.0) and close

[8] $$P(M_D)/P(M_{ND}) = [\Gamma(m + \gamma\theta_{ND})\Gamma(n - m + \gamma(1 - \theta_{ND}))/\Gamma(n + \gamma)] \times \{\Gamma(\gamma)/[\Gamma(\gamma\theta_{ND})\Gamma(\gamma(1 - \theta_{ND}))]\}/L_{ND},$$

where $\Gamma(r)$ is the gamma function ($\Gamma(r) = (r-1)\Gamma(r-1)$) and L_{ND} is the likelihood under model M_{ND}:

$$L_{ND} = \theta_{ND}^m (1 - \theta_{ND})^{n-m}.$$

to the prior mean (0.840).[9] Since the odds still favor discrimination, it would appear that in terms of the preponderance of the evidence standard the inexorable zero of *Teamsters* is very persuasive indeed!

To explain these conflicting interpretations of the same sample evidence, we note that as discussed in the literature (e.g. Leamer, 1978, pp. 93–107; Lindley, 1957; and Zellner, 1971), it is misleading to employ a significance level fixed independently of the sample size. As the sample size increases, we are afforded the opportunity to purchase less of type I (significance level) and type II errors. By fixing the significance level, the entire increase in sample size is expended to reduce the type II error. It seems intuitive that if we treat these errors in this asymmetric manner, we will all too easily reject the restrictive null hypothesis as the sample size increases. This is precisely the issue raised by Lindley's paradox (see Lindley, 1957, and Zellner, 1971, pp. 303–305; see also Shafer, 1982, and Discussion for a debate on the Bayesian approach and the meaning of Lindley's paradox). Just the reverse is the case for small samples. As the sample size declines, in general it will be appropriate to "accept" more of each error. In this instance, as *Capaci* demonstrates, a fixed significance level will result in the null hypothesis not being rejected as often as it should.

In the next section we examine a regression model of disparate treatment, in which we find the sample size effect is also pronounced.

4. Disparate treatment: Regression models

In a disparate treatment context, regression models can be used to measure differences due to membership in a particular group controlling for factors such as education and productivity.[10] This application is interesting because the plaintiff not only attempts to establish his prima facie case, but also anticipates the defendant's rebuttal by controlling for all potential non-discriminatory reasons for observed differences. Typically the defendant responds with another regression model, which led one court to remark (*Vuyanich v. Republic Nat. Bank of Dallas* 521 F. Supp. 656, 664 (1981), that compared to the traditional model of "thrust parry and engage this was statistical blitzkrieg versus jihad". Unlike more traditional hypothesis testing procedures the posterior odds criterion is uniquely suited to this regression model comparison context.

We shall illustrate the posterior odds criterion with the *Vuyanich v. Republic National Bank of Dallas* 505 F. Supp. 224 (1980). For this illustration we shall

[9] In general, with m males in a sample of size n and a prior mean of θ_0, the posterior mean will be:

$$\bar{\theta} = [m/\gamma + \theta_0]/[n/\gamma + 1].$$

[10] For examples of the use of such models in discrimination cases, see *Pouncy v. Prudential Insurance Co. of America* 499 F. Supp. 427 (1980), *Valentino v. U.S. Postal Service* 511 F. Supp. 917 (1981), *Vuyanich v. Republic National Bank of Dallas* 505 F. Supp. 224 (1980), 521 F. Supp. 656 (1981). Finkelstein (1973) provides an excellent survey of the possible application of regression models in the law.

assume information in the prior is minimal, in the sense defined by Klein and Brown (1984).[11] Furthermore, unlike *Capaci* and other cases, Vuyanich presumes that discrimination can occur in only one direction. Therefore, we extend recent results of Klein and Brown (1984) to derive in the appendix a simple expression for the posterior odds by minimizing prior information subject to this condition.

Vuyanich was a class action brought against Republic National Bank of Dallas charging that the bank discriminated on the basis of sex and race in hiring, promotion and compensation. We shall examine the compensation model.

The litigants utilized a human capital theory which specified that absent discrimination, wages should depend on individual productivity. The court accepted this theory and was led to compare models of the following type

$$\begin{aligned} M_{\mathrm{D}}&: y = X\beta_1 + d\beta_{\mathrm{D}} + u, \quad \beta_{\mathrm{D}} < 0, \\ M_{\mathrm{ND}}&: y = X\beta_1 + u, \end{aligned} \tag{4}$$

where y represents wages, X represents those variables affecting individual productivity, u is a zero mean normally distributed variable and

$$d = \begin{cases} 1, & \text{if the individual belongs to the affected group,} \\ 0, & \text{otherwise.} \end{cases}$$

We had insufficient data in the trial court decision to examine the plaintiff's models, but did have the data to re-examine the defendant's findings. For each of two job types (professional and non-professional) and at two different time periods (at hire and year end), discrimination against females (F) and blacks (B) was examined. In the case of race discrimination for non-professionals, the defendant also split the sample (B and B′) and analyzed each separately.

From the appendix, the posterior odds ratio that minimizes the constrained information in the prior is both intuitive and relatively easy to compute.[12] Table 10.2 shows the defendant's results and the associated posterior odds for discrimination. Here, N refers to the sample size and K_{D} to the dimension of the discrimination model, M_{D}. The dimension of the non-discrimination model, as illustrated in (4), is $K_{\mathrm{D}} - 1$.

From table 10.2, *given* defendant's models M_{D} and M_{ND}, the odds "agree" with defendant's results for females in that the odds favor non-discrimination ($O_{\mathrm{D,ND}} \equiv P(M_{\mathrm{D}}/P(M_{\mathrm{ND}}) < 1$). As to racial discrimination, in general the odds

[11] This may answer in part some of Professor Tribe's objections (Tribe, 1971a) to the use of Bayesian analysis in the courtroom.

[12] The posterior odds in favor of discrimination are given by:

$$[P(M_{\mathrm{D}})/PM_{\mathrm{ND}})] = \{N^{-[K(1)-K(2)]/2}\}\{\max \hat{L}_{\mathrm{D}}/\max \hat{L}_{\mathrm{ND}}\}\{2\cdot G_N(-\hat{t})\},$$

where N is the sample size, $K(i)$ is the parameter space dimension in model $M_i (M_1 \equiv M_{\mathrm{D}}, M_2 \equiv M_{\mathrm{ND}})$, max $\hat{L}_i$ is the maximized likelihood in model M_i, G_N is the cumulative t-distribution with N degrees of freedom, and $\hat{t}$ is the t-ratio (the least-squares estimate of β_{D} divided by its estimated standard error). Intuitively, the first and second terms reflect respectively a tradeoff between parsimony and data-fit. The presence of the latter term is due to the a priori direction of discrimination in model M_{D}.

Table 10.2. The defendant's models in Vuyanich v. Republic National Bank of Dallas.

Model comparison	At hire			Year end		
	β_d	$\hat{t}$	$O_{D,ND}$	β_d	$\hat{t}$	$O_{D,ND}$
F: Professional	−6.99	−1.24	0.36	−2.53	−0.46	0.13
Sample size dimension M_D		$N=119$; $K_D=8$			$N=135$; $K_D=8$	
F: Non-professional	−3.52	−1.03	.25	−0.00	−0.00	0.09
Sample size dimension M_D		$N=148$; $K_D=10$			$N=131$; $K_D=10$	
B: Professional	−6.26	−0.72	0.18	−6.14	−0.80	0.19
Sample size dimension M_D		$N=119$; $K_D=8$			$N=135$; $K_D=8$	
B: Non-professional	−3.45	−1.01	0.30	−3.44	−1.06	0.32
Sample size dimension M_D		$N=98$; $K_D=9$			$N=95$; $K_D=9$	
B′: Non-professional	−9.56	−1.63	1.29	−4.11	−0.92	0.48
Sample size dimension M_D		$N=50$; $K_D=9$			$N=36$; $K_D=9$	

do not support discrimination, *given* defendant's models. There is, however, one striking divergence in the results. For non-professionals, with $N=50$ observations, $K_D=9$, and the period taken at time of hire, the defendant found a t-ratio of -1.63. At the 5 percent significance level, the non-discrimination model could not be rejected. Yet, the posterior odds are 1.29 to 1 in favor of discrimination.

The plaintiff examined an entirely different set of models. We had insufficient information in the record to evaluate these models or to compare them with the defendant's. This latter non-nested comparison would have been possible, because the posterior odds criterion is valid whether or not the models are nested. We simply note here that the plaintiff employed much larger sample sizes than the defendant (cf. *Capaci*). Perhaps, not surprisingly, the plaintiff rejected the non-discrimination model for both females and blacks.[13]

As is evident in the case above, when the significance level is fixed independent of the sample size, statistical tests of a model can be misleading. Clearly, to balance type I and type II errors, the appropriate significance level should decline as the sample size increases. To demonstrate that the posterior odds criterion here exhibits this property, note that the odds (see footnote 12 above) depend on the t-value. Accordingly, as a function of the sample size, we can determine negative "critical t-values" corresponding to a critical value of the odds, which for

[13] The court found the evidence for sex discrimination to be unconvincing, but upheld the plaintiff's evidence for racial discrimination.

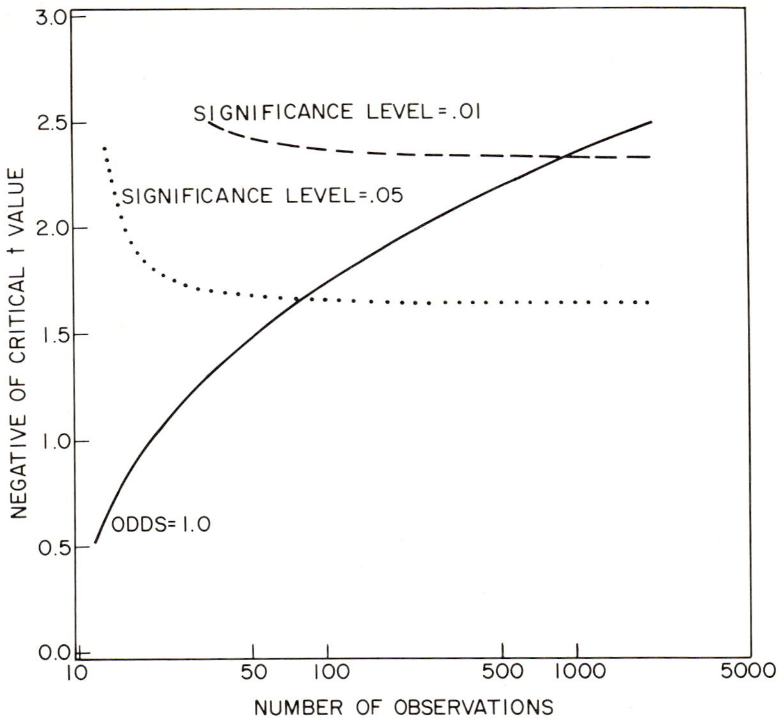

Fig. 10.2. Critical t-values corresponding to equal posterior model probabilities.

illustrative purposes we take as 1.0.[14] With $K_D = 10$ and $K_{ND} = 9$, these critical t-values, $\hat{t}_1^*, \hat{t}_2^*, \ldots,$ are shown in fig. 10.2. If the observed t-value is below the negative critical value for the relevant sample size, then the odds will favor the less restrictive discrimination model, M_D. From fig. 10.2 it is clear that as the sample size increases, the critical t-value becomes more negative. Accordingly, this posterior odds criterion can be viewed as a standard hypothesis test where the significance level decreases as the sample size increases.

In fig. 10.2 we have also shown the locus of t-values, as a function of the sample size, that correspond to a given significance level. According to the posterior odds criterion, notice that with $N < 50$ observations, the 5 percent

[14] The "indifference" level for the odds need not be 1.0 where the models are equally likely. In general, the level of the odds such that we would just be indifferent between the models (choosing one model if we are below this indifference level and the other model otherwise) depends on the underlying loss structure. Fig. 10.2 could be constructed for any indifference level (see footnote 5 above).

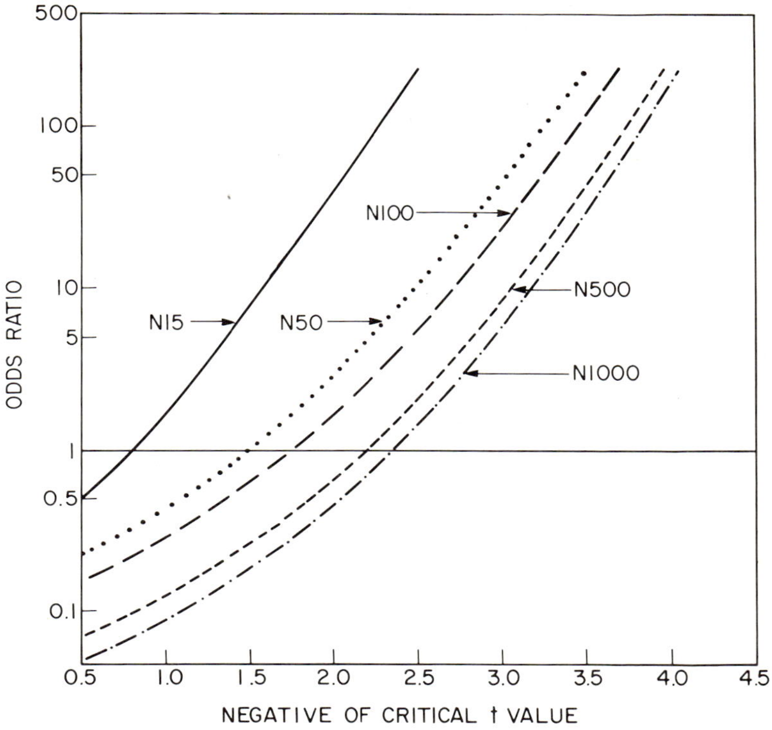

Fig. 10.3. The posterior odds.

significance level is "too low" (i.e. the t-values are "too negative"). With $N > 1000$ observations, the 1 percent significance level is "too high".

By employing a significance level not appropriate to the sample size, the results of classical hypothesis testing can be misleading. To focus on this issue, fig. 10.3 shows for each of several sample sizes the odds as a function of the t-value. Again, as in fig. 10.2, notice that with the odds set at 1.0, the corresponding negative t-values decrease with the sample size (only negative t-values are considered). More importantly, fig. 10.3 demonstrates explicitly the problem with a fixed significance level. For example, with $N = 15$ observations, the 5 percent t-value (in a one-tailed test) is $t_{0.05}(5) = -2.02$. With an observed t-value of $\hat{t} = -1.40$, we would not reject the restrictive model (M_{ND}) at the 5 percent level. Yet, the odds are 5 to 1 in favor of the alternative model, M_{D}. On the other hand, suppose that we have $N = 1000$ observations. Now, the 5 percent t-value is $t_{0.05}(990) = -1.65$. With an observed t-value of $\hat{t} = -2$, we would reject the more restrictive model, M_{ND}, at the 5 percent level. Yet, the odds are $1:0.5 = 2:1$ in favor of this model. Consequently, by failing to adjust the significance level for the sample size, in both small and large samples we can easily reject the more probable model.

5. Conclusion

In this chapter we have examined standards of statistical significance used in the federal courts to determine whether people are treated less favorably because of their particular group status. The Supreme Court advocates a standard that translates into a fixed significance level regardless of sample size. This standard is subject to some abuse in practice, and we advocate an alternative posterior odds ratio standard. This latter standard is easy to apply, and we show how it would affect the conclusions of several recent federal cases. In the regression model framework with a minimal information prior, this standard translates into a significance level that varies with the sample size.

Appendix

The models that we will be comparing have the following structure:

$$M_1: y_j = X_j(1)\beta(1) + u_j \equiv X_j\beta_{\mathrm{ND}}(1) + d_j\beta_{\mathrm{D}} + u_j(1), B_{\mathrm{D}} < 0$$

$$M_2: y_j = X_j(2)\beta(2) + u_j \equiv X_j\beta_{\mathrm{ND}}(2) + u_j(2), \beta_{\mathrm{D}} = 0, \tag{A1}$$

where with $\{u_j(i)\}$ i.i.d. $N(0, \sigma_i^2)$, $j = 1, \ldots, N$, M_1 is the discrimination model and M_2 is the non-discrimination model. With y denoting wages (or, as usually the case, ln wages) for observation j, $j = 1, \ldots, N$, interpret X_j as the jth observation on a collection of non-discriminatory variables and d_j as a dummy variable that is zero if the individual is in the potential discrimination group and one otherwise. Therefore, for example, with females as the potential discrimination group, under model M_1 females are paid less than males for equal ability, etc. (X). In this context, it is natural to have the same non-discriminatory variables in the models, in which case $X_j(1) = X_j(2)$ and model M_2 is nested within model M_1. We will, however, retain the notation $X_j(i)$ to emphasize that the criterion derived here applies whether or not the models are nested, and define $X(i): N \times K_i$ as the matrix of N observations on the K_i exogenous variables in model M_i.

To compare these models in a partial information setting, we require a measure of information:

Definition 1. Let $C(i)$ be the $K_i \times K_i$ matrix: $C'(i)C(i) \equiv X'(i)X(i)/T$. Let f_{Ri} be the normal density: $N[C(i)\beta(i), \sigma_i^2 \cdot I]$, where I is a $K_i \times K_i$ identity matrix. Then, with $m(i) \equiv [C(i)\beta(i), \ln \sigma_i]$ and $p(\cdot)$ as the prior density for $m(i)$, prior information in model M_i is defined as the relative measure:

$$I[M_i] \equiv I_s[p(m(i))] - E_p[I_s(f_{Ri})],$$

where E_p denotes an expectation under p and $I_s(g)$ is the Shannon (1948) measure of information in the probability density g: $I_s(g) \equiv E_g ln(g)$.

This measure is a variant of one proposed in a fundamental paper by Zellner (1977) and is discussed in detail in Klein and Brown (1984). It consists of the difference between the Shannon information in the prior, $p(m(i))$, and in the data-generating process, f_{Ri}.

As to the properties of this measure, note first that information is defined in terms of a metric, $m(i) \equiv [C(i)\beta(i), \ln \sigma_i]$, for which the likelihood is "data-translated." Box and Tiao (1973, pp. 20–60) argue that such a metric is appropriate for representing and then minimizing prior information. Second, this measure is invariant to "model-preserving" transformations in the exogenous variable matrix. Transform X to $X^* \equiv XA$ and β to $\beta^* \equiv A^{-1}\beta$, with A any non-singular matrix. The measure in definition 1 is invariant to such transformation, because in the metric $C(i)\beta(i)$ we "keep track" of the transformed exogenous variable matrix. Third, as the other "model-preserving" transformation of interest, this measure is invariant to renormalizing y to y^*/s, β to β^*/s and σ_i to σ_i/s, where s is an arbitrary scalar. This invariance follows from the prior metric employed and because this measure is defined in terms of the data-generating process, f_{Ri}.

Fourth, for purposes of defining relative prior model information, we have defined the data-generating process in K_i dimensions in terms of the prior metric $C(i)\beta(i)(f_{Ri} = N[C(i)\beta(i), \sigma_i^2 \cdot I_{K_i}])$. It can be argued that such a formulation is "representative" of the data in K_i dimensions (see Klein and Brown, 1984). In so doing, we have insured that this measure depends on the characteristics of the exogenous variable matrix, but not on the sample size. It should also be noted that in this formulation information does not depend on realizations of the endogenous variable or on the distribution for the exogenous variables.

The measure in definition 1 should be viewed as a conditional measure, because it takes the model under consideration as given. For our purposes, we must take into account not only uncertainty on the parameter values of a given model, but also uncertainty as to the model itself. Accordingly, we seek an overall information measure that incorporates information on the prior model probabilities. To this end, we have from Klein and Brown (1984):

Definition 2. For models M_1 and M_2, with prior model probabilities P_1 and P_2, and $I(M_i)$ as given in definition 1, prior information in the set of models is given by

$$IM \equiv \{P_1 \ln P_1 + P_2 \ln P_2\} + \{P_1 I[M_1] + P_2 I[M_2]\}.$$

The two terms in this definition can be interpreted respectively as the Shannon information in the prior model probabilities and, given the prior model probabilities, expected model information.

To minimize this measure, first prior model probabilities must satisfy:

$$\ln P_1 - \ln P_2 = I(M_2) - I(M_1), \tag{A2}$$

which is obtained by differentiating $I(M)$ with respect to $P_i(P_1 + P_2 = 1)$. Second, to minimize $I(M)$, prior distributions must be chosen to minimize $I(M_i)$. For a given mean and covariance matrix for $\beta(i)$ and a given support for the disturbance variance, it can be shown that $I(M_i)$ is minimized with the prior of the form:[15]

$$\bar{p}_i^*[\beta(i), \ln\sigma_i] \,\alpha N[\bar{\mu}_i^*, \bar{V}_i^*], \beta(1) \equiv [\beta_{\mathrm{ND}}(1), \beta_{\mathrm{D}}], \beta(2) \equiv \beta_{\mathrm{ND}}(2),$$
$$-\infty < \beta_{\mathrm{ND}}(i) < \infty, -\infty < \beta_{\mathrm{D}} < 0, \delta_{1i} < \ln\sigma_i < \delta_{2i}. \tag{A3}$$

To determine the limiting expression for the odds given in footnote 6 above, we consider a sequence of prior specifications along which information is reduced. Then we derive the corresponding odds sequence and its limit. To specify this sequence, assume that $\bar{V}_i \equiv \bar{V}_i^*/\sigma_i$ and $\bar{\mu}_i \equiv \bar{\mu}_i^*/\sigma_i$ do not depend on σ_i (see Klein and Brown, 1984, Theorems 1–2). Let $\bar{V}_i$ increase in a positive definite sense without bound (i.e. for every vector z of given length, $z'\bar{V}_i z \to \infty$ along our sequence). Next, let the support for the disturbance variance approach $(0, \infty)$. We can now obtain the limiting odds expression.

Formally, to obtain this limit, from (A2) with EL_i as the expected likelihood for model M_i and O_{12} as the odds in favor of model M_1:

$$O_{12} = (P_1/P_2)(EL_1/EL_2) = \mathrm{e}^{-IM_1}EL_1/(\mathrm{e}^{-IM_2}EL_2)$$
$$= \{|\bar{V}_1|^{1/2}\alpha \mathrm{e}^{[(\gamma-1)/2]}|C_1|[\delta_{21}-\delta_{11}]/|\bar{V}_2|^{1/2}|C_2|[\delta_{22}-\delta_{12}]\}EL_1/EL_2,$$
$$\alpha = \Phi(\bar{u}_{\mathrm{D}}/\bar{\sigma}_{\mathrm{D}}), \gamma = E\{[(\beta_{\mathrm{D}} - \bar{u}_{\mathrm{D}})/\bar{\sigma}_d]^2\}, \tag{A4}$$

where $\Phi(\cdot)$ is the cumulative normal, the expectation E is taken over the prior in model M_1, and $\bar{u}_{\mathrm{D}}$ and $\bar{\sigma}_d$ are the appropriate elements of $\bar{\mu}$ and $\bar{V}_1$, respectively. With $EL_1/EL_2 \equiv A_1/A_2$ and $\beta^*(1) \equiv \beta(1)/\sigma_1$, we have:

$$A = \left(\frac{1}{\alpha}\right)|\bar{V}_1|^{-1/2}[\delta_{21}-\delta_{11}]^{-1}|\bar{V}_1^{-1} + X_1'X_1|^{-1/2}$$
$$\times \int_{e^{\delta_{11}}}^{e^{\delta_{21}}} (1/\sigma_1)^{N+1} \exp[(-\tfrac{1}{2}\sigma_1^2)(S_1^2 + e_1^2)]$$
$$\times \int_{-\infty}^{0}\int_{-\infty}^{\infty}\cdots\int_{-\infty}^{\infty} |\bar{V}_1^{-1} + X_1'X_1|^{1/2}/(2\pi)^{K_1/2}$$
$$\times \exp\{-\tfrac{1}{2}[\beta^*(1) - \bar{\beta}(1)/\sigma_1]'$$
$$\times(\bar{V}_1^{-1} + X_1'X_1)[\beta^*(1) - \bar{\beta}(1)/\sigma_1]\}\,\mathrm{d}\beta^*(1)\,\mathrm{d}\sigma_1, \tag{A5}$$

[15] Notice that the marginal prior for β_{D} is a truncated normal. The same arguments in Klein and Brown (1984) readily yield this result.

where along our sequence, e_1^2 is a term that tends to zero and $\bar{\beta}(1)$ is a posterior-type mean tending to the least squares estimate, $\hat{\beta}(1)$. A similar expression holds for A_2. Next, note that a density in normal form can be written as the product of a conditional normal ($\beta_{nd}(1)$) and a marginal normal "form" (β_D). Furthermore, note that since $z'V_1z/z'z \to \infty$, every characteristic root of V_1^{-1} approaches zero. Therefore, $e^{-IM_1}A_1$ approaches the limit:

$$|C_1||X_1'X_1|^{-1/2}\int_0^\infty \frac{\hat{h}_D^{1/2}}{\sqrt{2\pi}}\int_{-\infty}^0 (1/\sigma_1)^{N+1} \times \exp\left\{-\left(\tfrac{1}{2}\sigma_1\right)\left[S_1^2+\hat{h}_d\left(\beta_d-\hat{\beta}_d\right)^2\right]\right\}\mathrm{d}\beta_D\,\mathrm{d}\sigma_1, \tag{A6}$$

where $\hat{h}_D$ is the diagonal element of X'_1X_1 corresponding to β_d and we have made a change of variable from β_D/σ_1 to β_D. Finally, note that $|C_1||X_1'X_1|^{-1/2} = N^{-K_1/2}$ and that the integrand is in the inverted gamma form. Therefore, integrating first over σ_1 will yield a single dimensional integral with integrand in the Student t-form. Then, with $t \equiv \beta_D\hat{h}_D^{1/2}/(S_1^2/N)^{1/2}$ and $\hat{t} = \hat{\beta}_D/[\hat{h}_D/S_1^2/N]^{1/2}$, we have $e^{-IM_1}A_1$ converging to:

$$\Gamma(N/2)N^{-K_1/2}\left(S_1^2\right)^{-n/2}G_N(-\hat{t})\left(\tfrac{1}{2}\right)^{-N/2}, \tag{A7}$$

where G_N is a cumulative Student-t distribution with N degrees of freedom. An analogous argument yields:

$$e^{-IM_1}A_2 \to \Gamma(N/2)T^{-K_2/2}\left(S_2^2\right)^{-N/2}\left(\tfrac{1}{2}\right)\left(\tfrac{1}{2}\right)^{-N/2}. \tag{A8}$$

The limiting value of O_{12} is then the ratio of the expression in (A7) to that in (A8), as claimed:

$$O_{12} = \left\{N^{-[K_1-K_2]/2}\right\}\left\{\max \hat{L}_1/\max \hat{L}_2\right\}\left\{2G_N(-\hat{t})\right\}. \tag{A9}$$

References

Baldus, David and James Cole (1980) *Statistical Proof of Discrimination* (Shepards/McGraw-Hill, Colorado Springs).

Box, G.E. and G.C. Tiao (1973) *Bayesian Inference in Statistical Analysis* (Adison Wesley, Reading, Mass.).

Finkelstein, Michael (1966) "The application of statistical decision theory to the jury selection cases", *Harvard Law Review*, 80, 338–375.

Finkelstein, Michael (1973) "Regression models in administrative hearings", *Harvard Law Review*, 86, 1442–1475.

Finkelstein, Michael and William Fairley (1970) "A Bayesian approach to identification evidence", *Harvard Law Review*, 83, 489–517.

Klein, Roger and Stephen Brown (1984) "Model selection when there is 'minimal' prior information", forthcoming in *Econometrica*.

Leamer, E.E. (1978) *Specification Searches: Ad Hoc Inference with Nonexperimental Data* (John Wiley & Sons, Inc., New York).

Lindley, D. (1957) "A statistical paradox", *Biometrika*, 44, 187–192.

Shafer, G. (1982). "Lindley's paradox" (with invited comments and rejoindor), *Journal of the American Statistical Association*, 325–351.

Shannon, C.E. (1948) "A mathematical theory of communication", *The Bell System Technical Journal*, 378–423.

Tribe, Laurence (1971a) "Trial by mathematics: Precision and ritual in the legal process", *Harvard Law Review*, 84, 1329–1393.

Tribe, Laurence (1971b) "A further critique of mathematical proof", *Harvard Law Review*, 84, 1810–1820.

Zellner, A. (1971) *An Introduction to Bayesian Inference in Econometrics* (John Wiley & Sons, Inc., New York).

Zellner, A. (1977) "Maximal data information prior distribution", in: A. Aykac and C. Brumat, eds., *New Developments in the Application of Bayesian Methods* (North-Holland Publishing Co., Amsterdam) pp. 211–232.

Chapter 11

BAYESIAN ANALYSIS OF CENSORED OR GROUPED DATA FROM PARETO POPULATIONS

BARRY C. ARNOLD and S. JAMES PRESS

University of California at Riverside

1. Introduction

The Pareto distribution has an excellent pedigree as a component in stochastic models for various socio-economic phenomena. Classical estimation techniques have been studied extensively for some time (see Arnold, 1983, for a guide to the relevant literature). Bayesian procedures were developed more recently (see Sinha and Howlader, 1980, and Arnold and Press, 1983). Unfortunately, much data which is fitted by Pareto distributions comes to us in grouped, and/or truncated, or censored form. It is therefore appropriate to develop estimation techniques which specifically take this frequently occurring grouping into account. A combination of censored and ungrouped observations can be expected in certain income data sets. The precise observations would correspond to documented incomes. The censored observations would correspond to reported undocumented incomes. For taxation purposes such reported incomes are frequently underreported, i.e. we know only that the income was at least the reported level.

In large samples the effect of grouping may be expected to be negligible, and may therefore be ignored; in small or moderate size samples, however, we must accept the grouped nature of the data and develop estimates which are precisely appropriate. Aigner and Goldberger (1970) have provided analyses of this nature in the classical, i.e. non-Bayesian framework. Zellner (1971) considers Bayesian analysis of grouped data from a classical Pareto distribution with known scale parameter. In this chapter we develop other Bayesian procedures that are suitable for grouped data. As in Arnold and Press (1983) we approach the prior distribution assessment problem by judiciously selecting families of prior distributions

Bayesian Inference and Decision Techniques, edited by P. Goel and A. Zellner

that are analytically tractable and whose hyperparameters have intuitive meaning and are consequently relatively easy to assess. In addition, we require that they not lead to anomalous posterior densities. Such an attitude clearly represents a somewhat eclectic Bayesian spirit in the interest of statistical pragmatism. Examples of priors which, in the present context, lead to anomalous posterior densities are certain diffuse priors and a large class which is an expansion of the class introduced by Lwin (1972) (herein called generalized Lwin priors). We will catalog the Bayes estimators that result from these "ill-chosen" families of priors. Tinkering even further with the rules required by the Bayes Theorem, we consider briefly certain data-dependent priors which are appropriate for certain specific forms of truncation and/or grouping, which lead to empirical Bayes estimators.

Throughout the chapter we adopt the convention that lower case letters represent realizations of the random variables defined by the corresponding upper case letters (i.e. u is a realization of U, $\boldsymbol{x}$ of $\boldsymbol{X}$, etc.) The letter f is used as a generic form for densities. It does not necessarily represent the same function whenever it appears. Specifically, an expression such as $f(\boldsymbol{x}|\sigma, \alpha)$ is to be read as: the density of the random variable $\boldsymbol{X}$, given σ and α, evaluated at the particular value $\boldsymbol{x}$. We use the symbol $x_{(i)}$ to denote the ith smallest quantity in the set $x_1, x_2, \ldots, x_n$.

2. Classical Pareto data

Suppose that $X_1, X_2, \ldots, X_n$ are independent identically distributed (i.i.d.) random variables (r.v.s) with common two-parameter classical Pareto density:

$$f(x|\sigma, \alpha) = \alpha\sigma^{\alpha}x^{-(\alpha+1)}I(x > \sigma), \tag{2.1}$$

where $I(a > b) = 1$ if $a > b$, and is 0, otherwise. Both σ (the scale parameter) and α (the shape parameter) are positive. Suppose that we cannot observe the actual data point X_i, but rather we are informed that it has a value in some interval $B_i = (a_i, b_i)$, where $a_i < b_i$. The B_i's might not all be disjoint, although in most grouped data situations there are only a finite number of disjoint B_i's, and often they partition the positive real line. Since, from (2.1), $P(X < t) = 1 - (\sigma/t)^{\alpha}$, $t > \sigma$, the likelihood function corresponding to such a grouped (or imprecise) data set is of the form:

$$L(\sigma, \alpha) = \prod_{i=1}^{n}\left\{\min\left[\left(\frac{\sigma}{a_i}\right)^{\alpha}, 1\right] - \min\left[\left(\frac{\sigma}{b_i}\right)^{\alpha}, 1\right]\right\}. \tag{2.2}$$

Note that this likelihood is non-zero for every $\alpha > 0$ and for every $\sigma < b_{(1)}$. Arnold and Press (1983) studied several possible prior distribution families for (σ, α) for the case of ungrouped data. It is reasonable to investigate how well these families perform with grouped data. We divide our discussion into three cases likely to be of distinct interest: (i) α unknown, σ known; (ii) α known, σ unknown; and (iii) both α and σ unknown.

2.1. σ known

2.1.1. Grouped data

Suppose we have grouped data and the scale parameter is known. For the unknown shape parameter we adopt the gamma distribution family of prior densities:

$$f(\alpha) = \alpha^{c-1}e^{-d\alpha}d^c/\Gamma(c), \quad \alpha > 0, \tag{2.3}$$

where $c > 0$ and $d > 0$. Multiplying (2.2) and (2.3) gives the posterior density:

$$f(\alpha|\boldsymbol{a},\boldsymbol{b},\sigma) \propto \alpha^{c-1}e^{-d\alpha}\prod_{i=1}^{n}\left\{\min\left[\left(\frac{\sigma}{a_i}\right)^{\alpha},1\right]-\min\left[\left(\frac{\sigma}{b_i}\right)^{\alpha},1\right]\right\},$$

where $\propto$ denotes proportionality, $\boldsymbol{a} \equiv (a_1, a_2, \ldots, a_n)$, and $\boldsymbol{b} \equiv (b_1, b_2, \ldots, b_n)$. Some simplification is possible since σ is known. We may, without loss of generality, assume that $a_{(1)} \geq \sigma$, and rewrite the posterior density of α as

$$\begin{aligned} f(\alpha|\boldsymbol{a},\boldsymbol{b},\sigma) &\propto \alpha^{c-1}e^{-d\alpha}\prod_{i=1}^{n}\left[\left(\frac{\sigma}{a_i}\right)^{\alpha}-\left(\frac{\sigma}{b_i}\right)^{\alpha}\right] \\ &\propto \alpha^{c-1}e^{-(d+u)\alpha}\prod_{i=1}^{n}\left[1-\left(\frac{a_i}{b_i}\right)^{\alpha}\right], \end{aligned} \tag{2.4}$$

where $u = \sum_{i=1}^{n}\log(a_i/\sigma)$. By expanding the product in (2.4) and writing $(a_i/b_i)^{\alpha} = \exp\{-\alpha\log(b_i/a_i)\}$ we may verify that this posterior density is of the form:

$$f(\alpha|\boldsymbol{a},\boldsymbol{b},\sigma) \propto \alpha^{c-1}\sum_{j=0}^{k}(-1)^j e^{-d_j\alpha},$$

where k is a positive integer, $c > 0$, and the d_j's are (not necessarily distinct) positive real numbers. If $f_j(\alpha|c,d_j) \equiv d_j^c\alpha^{c-1}e^{-d_j\alpha}/\Gamma(\alpha)$ denotes a gamma density with parameters c and d_j, it is clear that the posterior density (2.4) can be written as a linear combination of gamma densities, i.e.

$$f(\alpha|\boldsymbol{a},\boldsymbol{b},\sigma) = \sum_{j=0}^{k}q_j f_j(\alpha|c,d_j), \tag{2.5}$$

where

$$\sum_{j=0}^{k}q_j = 1$$

and

$$q_j = (-1)^j d_j^{-c}\Big/\left[\sum_{j=0}^{k}(-1)^j d_j^{-c}\right].$$

Thus, the class of densities which corresponds to linear combinations of gamma densities with common shape parameters forms a conjugate prior family for the likelihood (2.2). This broader class could be used if the available prior information did not conform well to a gamma distribution.

A special case in which considerable simplification is possible occurs when the grouping intervals are of constant logarithmic width. That is, whenever there exists a $w>1$ such that for every i, $b_i = wa_i$. In such a case the posterior density (2.4) can be rewritten as

$$f(\alpha|\boldsymbol{a}, w, \sigma) \propto \alpha^{c-1} e^{-(d+u)\alpha}(1-w^{-\alpha})^n,$$

or

$$f(\alpha|\boldsymbol{a}, w, \sigma) \propto \sum_{j=0}^{n} \binom{n}{j}(-1)^j \alpha^{c-1} \exp\{-[d+u+j(\log w)]\alpha\}.$$

This is in the form (2.5) with $d_j = d+u+j(\log w)$, $k=n$, and

$$q_j = d_j^{-c}(-1)^j\binom{n}{j}\Bigg/\left[\sum_{j=0}^{n} d_j^{-c}(-1)^j\binom{n}{j}\right].$$

For certain forms of grouping, the gamma family of prior densities is not the most tractable. Suppose that the grouping is of such a nature that there exists a number $w>1$ such that every $b_i/a_i = w$ and, in addition, every $a_i = \sigma w^{k_i}$ for some integer k_i. If, when we observe $X_i \in (\sigma w^{k_i}, \sigma w^{k_i+1})$, we define $Y_i = k_i$, then our data can be conveniently thought of as being n independent geometric random variables. For geometric random variables, a conjugate prior family is provided by the beta distribution family.

The likelihood of this grouped Pareto data is thus (using (2.2) with $a_i = \sigma w^{k_i}$, $b_i = wa_i$):

$$\begin{aligned} L(\sigma, \alpha) &= \prod_{i=1}^{n} (w^{-\alpha})^{k_i}(1-w^{-\alpha}) \\ &= w^{-\alpha K}(1-w^{-\alpha})^n, \end{aligned} \tag{2.6}$$

where $K = \sum_{i=1}^{n} k_i$. As our family of prior densities we consider those for which $\mathscr{L}(w^{-\alpha}) = B(r,s)$. Transforming gives the family of prior densities for α:

$$f(\alpha) \propto w^{-\alpha r}(1-w^{-\alpha})^{s-1}, \quad r>0,\ s>0. \tag{2.7}$$

The resulting posterior density is of the same form:

$$f(\alpha|w, K, \sigma) \propto w^{-\alpha(r+K)}[1-w^{-\alpha}]^{n+s-1}. \tag{2.8}$$

Thus,

$$\mathscr{L}(w^{-\alpha}|w, K, \sigma) = B(r+K, n+s).$$

The posterior mean of α is given by

$$E(\alpha|w, K, \sigma) = \int_0^1 \left(\frac{-\log u}{\log w} \right) \frac{u^{r+K-1}(1-u)^{n+s-1}}{B(r+K, n+s)} \, du.$$

If we change variables from u to $(1-u)$, then expand $\log(1-u)$ in a series and integrate term by term, we obtain:

$$E(\alpha|w, K, \sigma) = \frac{1}{\log w} \sum_{j=1}^{\infty} \frac{1}{j} \frac{B(r+K, n+s+j)}{B(r+K, n+s)},$$

a slowly converging series.

If instead we wish to estimate the survival function, we seek $E\{\min[(t/\sigma)^{-\alpha}, 1]|w, K, \sigma\}$. Noting that

$$\left(\frac{t}{\sigma}\right)^{-\alpha} = w^{-[\log_w(t/\sigma)]\alpha}$$

and using (2.8), we obtain as our estimated survival function:

$$\min\left(\frac{B(r+K+\log_w(t/\sigma), s+n)}{B(r+K, s+n)}, 1 \right).$$

Diffuse prior information in this situation corresponds to the choice $r = s = 0$.

Another, possibly more common, form of grouping would involve reporting of quantities $N_0, N_1, N_2, \ldots, N_{M+1}$, where N_i is the number of observations in the interval $(\sigma w^i, \sigma w^{i+1})$, $i = 1, 2, \ldots, M$, while N_{M+1} is the number of observations greater than σw^M. The corresponding likelihood is of the form:

$$L(\sigma, \alpha) = [1 - w^{-\alpha}]^{n - n_{M+1}} (w^{-\alpha})^{\sum_{i=1}^{M+1} i n_i},$$

where $n = \sum_{i=1}^{M+1} n_i$. The form of this likelihood suggests adoption of a prior in which $\mathscr{L}(w^{-\alpha}) = B(r, s)$. This yields a posterior density in which for $\tilde{\boldsymbol{n}} \equiv (n_1, \ldots, n_{M+1})'$:

$$\mathscr{L}(w^{-\alpha}|w, \tilde{\boldsymbol{n}}, \sigma) = B\left(r + \sum_{i=1}^{M+1} i n_i, s + n - n_{M+1} \right).$$

2.1.2. Censored data

The case of censoring, which corresponds to grouping with semi-infinite intervals (i.e. $b_i = \infty$ for each i), is much more tractable. If we set $b_i = \infty$ in (2.4) we find the posterior gamma density for α:

$$f(\alpha|\boldsymbol{a}, \sigma) \propto \alpha^{c-1} e^{-(d+u)\alpha}. \tag{2.9}$$

It may be observed that the shape of the prior density survives intact and that the posterior density is merely a rescaling of the prior density.

2.1.3. Mixed censored and uncensored data

Data which include some precise observations (i.e. ungrouped), and some censored observations, present no special problems. For example, if we have available m i.i.d. observations from the density (2.1), say $X_1,\ldots, X_m$, and, in addition, n censored observations, $X_{m+1} > a_1,\ldots, X_{m+n} > a_n$, the corresponding likelihood is

$$L(\alpha) = \prod_{i=1}^{m} \alpha\sigma^{\alpha} x_i^{-(\alpha+1)} \prod_{j=1}^{n} \left(\frac{\sigma}{a_j}\right)^{\alpha} I(x_{(1)} > \sigma) I(a_{(1)} \geq \sigma)$$

$$= \alpha^m \sigma^{(n+m)\alpha} \left(\prod_{i=1}^{m} x_i\right)^{-(\alpha+1)} \left(\prod_{j=1}^{n} a_j\right)^{-\alpha} I(x_{(1)} > \sigma) I(a_{(1)} \geq \sigma).$$

Adopting a gamma prior density for α in (2.3) we obtain the gamma posterior density:

$$f(\alpha \mid \boldsymbol{x}, \boldsymbol{a}, \sigma) \propto \alpha^{c+m-1} e^{-(u+v+d)\alpha}, \tag{2.10}$$

where $\boldsymbol{x} = (x_1, x_2,\ldots, x_m)$, $u = \sum_{i=1}^{m} \log(x_i/\sigma)$, and $v = \sum_{j=1}^{n} \log(a_j/\sigma)$. This result includes as special cases the ungrouped data result (found perhaps earliest in Muniruzzaman, 1968) and the completely censored result, (2.9). The posterior mean (the Bayes estimator corresponding to squared error loss) is thus:

$$\alpha^* = \frac{m+c}{u+v+d}. \tag{2.11}$$

Bayesian credibility intervals are readily found by using standard chi-square tables. Diffuse (vague) prior information is associated with the choice $c = d = 0$ in (2.3).

If, in the mixed data case, we wish to estimate the survival probability $\min[(t/\sigma)^{-\alpha}, 1]$ under squared error loss, we need only to compute the conditional expectation of $\min[(t/\sigma)^{-\alpha}, 1]$ given $\boldsymbol{x}$, $\boldsymbol{a}$, and σ. It is of remarkably simple form, namely

$$\min\left[(u+v+d)^{m+c}(u+v+d+\log(t/\sigma))^{-(m+c)}, 1\right].$$

2.2. α known

Although it is perhaps unlikely that α will be known and σ unknown, the case will be considered for completeness. The corresponding analysis for ungrouped uncensored data was first presented by Lwin (1972). If we view the likelihood (2.2) as a function of σ, we may observe that it will be non-zero if and only if $\sigma < b_{(1)}$. When $\sigma \in (0, b_{(1)})$ the likelihood unfortunately is still quite complicated. It can be noted from (2.2) that its functional form changes as σ increases past each a_i. Adoption of a power function prior density for σ, i.e.

$$f(\sigma \mid \delta, \sigma_0) = \delta\sigma^{\delta-1}\sigma_0^{-\delta}, \quad \delta > 0,\ \sigma_0 > \sigma, \tag{2.12}$$

yields a posterior density of similar character of the likelihood viewed as a function of σ. If (2.12) obtains, we write $\mathscr{L}(\sigma) = PF(\delta, \sigma_0)$. (The priors (2.12) are the conjugate priors for σ in the case of ungrouped data.) In the case of censored data we have $b_i = \infty$ for every i. Our likelihood, from (2.2), now looks like

$$L(\sigma) = \prod_{i=1}^{n} \min\left[\left(\frac{\sigma}{a_i}\right)^{\alpha}, 1\right]. \tag{2.13}$$

For convenience we will assume that all the a_i's are distinct (the more general case can then be resolved by limiting arguments). Denoting, as usual, the ordered a_i's by $a_{(1)} < a_{(2)} < \cdots < a_{(n)}$ we may rewrite (2.13) in the form:

$$L(\sigma) = \begin{cases} \dfrac{\sigma^{(n-j)\alpha}}{\left(\prod\limits_{i=j+1}^{n} a_{(i)}\right)^{\alpha}}, & \text{if } a_{(j)} < \sigma < a_{(j+1)},\ j = 0, 1, 2, \ldots, n-1, \\ 1, & \text{if } \sigma > a_{(n)}, \end{cases} \tag{2.14}$$

where, by convention, $a_{(0)} = 0$.

If we adopt the power function prior (2.12), then the posterior density is given by

$$f(\sigma | \boldsymbol{a}, \alpha) \propto \sigma^{\delta - 1} L(\sigma) I(0 < \sigma < \sigma_0), \tag{2.15}$$

where $L(\sigma)$ is given in (2.14). Although this is not a simple density it is possible, though tedious, to evaluate the posterior mean. It is:

$$\sigma^* = \begin{cases} \sigma_0(\delta + n\alpha)/(\delta + n\alpha + 1), & \text{if } \sigma_0 < a_{(1)}, \\ \dfrac{\sum\limits_{k=1}^{j} b(\delta+1, k) + c(\delta+1, j)}{\sum\limits_{k=1}^{j} b(\delta, k) + c(\delta, j)}, & \text{if } a_{(j)} < \sigma_0 < a_{(j+1)},\ j = 1, 2, \ldots, n, \end{cases} \tag{2.16}$$

where

$$b(\delta, k) = \frac{a_{(k)}^{\delta + (n-k+1)\alpha} - a_{(k-1)}^{\delta + (n-k+1)\alpha}}{[\delta + (n-k+1)\alpha]\left(\prod\limits_{i=k}^{n} a_{(i)}\right)^{\alpha}} \tag{2.16a}$$

and

$$c(\delta, j) = \frac{\sigma_0^{\delta + (n-j)\alpha} - a_{(j)}^{\delta + (n-j)\alpha}}{[\delta + (n-j)\alpha]\left(\prod\limits_{i=j+1}^{n} a_{(i)}\right)^{\alpha}},$$

where by convention $a_{(n+1)} = \infty$. For $\alpha + \delta - 1 > 0$, which is typically the case, the posterior mode is at σ_0 since $\sigma^{\delta-1}L(\sigma)$ is a monotone increasing function of σ.

If, in addition to the censored data, we have some precise observations, no significant complications are encountered. Thus, if we observe $X_1, X_2, \ldots, X_m$ exactly and $\{X_{m+1} > a_1\}, \ldots, \{X_{m+n} > a_n\}$ in censored form, the corresponding likelihood is

$$L(\sigma) = \alpha^n \sigma^{m\alpha}\left(\prod_{i=1}^m x_i\right)^{-(\alpha+1)} I(x_{(1)} > \sigma) \prod_{i=1}^n \left\{\min\left[\left(\frac{\sigma}{a_i}\right)^\alpha, 1\right]\right\}.$$

Adoption of the power function prior (2.12) leads to a posterior density of character similar to (2.15). Specifically, we have

$$f(\sigma|\boldsymbol{x}, \boldsymbol{a}, \alpha) \propto \sigma^{\delta+m\alpha-1}L(\sigma)I(0 < \sigma < \min(\sigma_0, x_{(1)})),$$

where, again, $L(\sigma)$ is of the form (2.14). The corresponding Bayes estimator is of the form (2.16) with δ replaced by $\delta + \alpha m$ and σ_0 replaced by $\min\{\sigma_0, x_{(1)}\}$. The mode of the posterior distribution is at $\min\{\sigma_0, x_{(1)}\}$, provided $(m+1)\alpha + \delta - 1 > 0$.

Vague prior information can be accommodated by setting $\delta = 0$ and letting $\sigma_0 \to \infty$.

2.3. *Both scale and shape parameters unknown*

In most practical situations both α and σ will be unknown. The choice of a suitable joint prior density in such a setting is a little more complicated than in the one-parameter settings described above. For ungrouped data, Lwin (1972) proposed a family of natural conjugate prior densities which was modified by Arnold and Press (1983). Such generalized Lwin priors assume the form

$$f(\sigma, \alpha) \propto \alpha^\delta \sigma^{\lambda\alpha-1}\mu^{-\alpha}, \quad \alpha > 0, 0 < \sigma < \sigma_0, \tag{2.17}$$

where μ, δ, λ, and σ_0 are positive hyperparameters and where, to guarantee integrability, $0 < \sigma_0^\lambda < \mu$. A more instructive description of such priors is provided by the joint statements:

$$\mathscr{L}(\alpha) = \gamma(\delta, \log\mu - \lambda\log\sigma_0)$$

and

$$\mathscr{L}(\sigma|\alpha) = PF(\lambda\alpha, \sigma_0).$$

Arnold and Press (1983) suggest the term "power-gamma" for such generalized Lwin priors, and we write

$$\mathscr{L}(\sigma, \alpha) = P\gamma(\delta, \lambda, \mu, \sigma_0).$$

With ungrouped data, $X_1, \ldots, X_n$, the posterior density of (σ, α) is of the same power-gamma form; specifically:

$$\mathscr{L}((\sigma, \alpha)|\boldsymbol{x}) = P\gamma\left(n+\delta, n+\lambda, \mu\left(\prod_{i=1}^n x_i\right), \min\{\sigma_0, x_{(1)}\}\right). \tag{2.18}$$

The corresponding marginal posterior distribution of α is of the gamma form, but the marginal posterior distribution of σ turns out to be of a more complicated bimodal form. There is an infinite posterior mode for σ at 0 no matter how much or what kind of ungrouped data are collected. For this reason, Arnold and Press (1983) advocate the use of independent gamma and power function priors for α and σ.

2.3.1. *Grouped data*

What if we have grouped data? For general grouping the likelihood (2.2), thought of as a function of σ and α, does not produce a convenient joint prior density family for (α, σ). Even if we were to adopt independent priors for α and σ we find, in the light of the discussion at the beginning of section 2.2, a rather complicated posterior distribution. We can proceed a little further in the case of censoring, i.e. $b_i = \infty$ for every i in (2.2). The likelihood assumes the form:

$$L(\sigma, \alpha) = \prod_{i=1}^{n} \min\left[\left(\frac{\sigma}{a_i}\right)^{\alpha}, 1\right],$$

or, in terms of the ordered a_i's:

$$L(\sigma, \alpha) = \begin{cases} \dfrac{\sigma^{(n-j)\alpha}}{\left(\prod\limits_{i=j+1}^{n} a_{(i)}\right)^{\alpha}}, & \text{if } a_{(j)} < \sigma < a_{(j+1)},\ j = 0, 1, 2, \ldots, n-1, \\ 1, & \text{if } \sigma > a_{(n)}, \end{cases}$$

where again $a_{(0)} = 0$.

Because of their mathematical tractability in the uncensored ungrouped setting, we first consider power-gamma or generalized Lwin priors of the form (2.17). The simplicity encountered with ungrouped data now disappears, however. The posterior density has the following non-trivial form, for $\sigma_0 \in (a_{(k)}, a_{(k+1)})$:

$$f(\sigma, \alpha | \boldsymbol{a}) \propto \begin{cases} \dfrac{\alpha^{\delta}\sigma^{(\lambda+n-j)\alpha-1}\mu^{\alpha}}{\left(\prod\limits_{i=j+1}^{n} a_{(i)}\right)^{\alpha}} I(\alpha > 0), \quad a_{(j)} < \sigma < a_{(j+1)}, \\ \qquad\qquad\qquad\qquad\qquad\qquad\qquad j = 0, 1, \ldots, n-1, \\ \dfrac{\alpha^{\delta}\sigma^{(\lambda+n-k)\alpha-1}\mu^{\alpha}}{\left(\prod\limits_{i=k+1}^{n} a_{(i)}\right)^{\alpha}} I(\alpha > 0), \quad a_{(k)} < \sigma < \sigma_0. \end{cases}$$

What if we adopt independent gamma and power function priors for α and σ, as proposed by Arnold and Press (1983) for ungrouped data? Instead of (2.17) our prior density will have the form:

$$f(\sigma,\alpha)\propto\alpha^{c-1}e^{-d\alpha}\sigma^{\delta-1},\quad \alpha>0,\ 0<\sigma<\sigma_0. \tag{2.19}$$

Combining this with the likelihood yields a posterior density which is no easier to deal with than (2.18). The precise form of the posterior density is, if $\sigma\in(a_{(k)},a_{(k+1)})$:

$$f(\sigma,\alpha|\boldsymbol{a})\propto\begin{cases}\dfrac{\alpha^{c-1}e^{-d\alpha}\sigma^{\delta+(n-j)\alpha-1}}{\left(\prod_{i=j+1}^{n}a_{(i)}\right)^{\alpha}}I(\alpha>0), & a_{(j)}<\sigma<a_{(j+1)},\\ & j=0,1,2,\dots,n-1,\\ \dfrac{\alpha^{c-1}e^{-d\alpha}\sigma^{\delta+(n-k)\alpha-1}}{\left(\prod_{i=k+1}^{n}a_{(i)}\right)^{\alpha}}I(\alpha>0), & a_{(k)}<\sigma<\sigma_0.\end{cases} \tag{2.20}$$

The corresponding marginal densities are

$$f(\alpha|\boldsymbol{a})\propto\alpha^{c-1}e^{-d\alpha}\left[\sum_{j=1}^{k}b(\delta,j)+c(\delta,k)\right]I(\alpha>0),$$

where $b(\delta,j)$ and $c(\delta,k)$ are given by (2.16a) and

$$f(\sigma|\boldsymbol{a})\propto\begin{cases}\sigma^{\delta-1}\left[d-(n-j)\log\sigma+\sum_{i=1}^{n}\log a_{(i)}\right]^{-c}, & a_{(j)}<\sigma<a_{(j+1)},\\ & j=0,1,2,\dots,k-1,\\ \sigma^{\delta-1}\left[d-(n-k)\log\sigma+\sum_{i=k+1}^{n}\log a_{(i)}\right]^{-c}, & a_{(k)}<\sigma<\sigma_0.\end{cases}$$

As in the corresponding case with ungrouped data, $f(\sigma|\boldsymbol{a})$ will be unimodal if $\delta>1$. However, now the mode is at σ_0. This mode is unchanged no matter how much censored data is observed! This is not surprising since a similar situation was observed even when α was assumed known, for $\alpha+\delta>1$.

2.3.2. Grouped and ungrouped data

If some ungrouped data are available in addition to the censored data, the mode anomoly described in the previous subsection disappears. Thus, assume m observations $X_1, X_2,\dots, X_m$ are available from the density (2.1) together with n censored observations $X_{m+1}>a_1$, $X_{m+2}>a_2,\dots$, $X_{m+n}>a_n$. Again assume independent prior densities for α and σ, i.e. a prior given by (2.19). Now the posterior density is a slight modification of (2.20). Specifically, if $\min(x_{(1)}, \sigma_0) \in (a_{(k)}, a_{(k+1)})$,

$$f(\sigma, \alpha | \boldsymbol{a}, \boldsymbol{x}) \propto \begin{cases} \dfrac{\alpha^{m+c-1} e^{-d\alpha} \sigma^{\delta+(m+n-j)\alpha-1}}{\left(\prod_{i=1}^{m} x_i\right)^{\alpha} \left(\prod_{i=j+1}^{n} a_{(i)}\right)^{\alpha}} I(\alpha > 0), & a_{(j)} < \sigma < a_{(j+1)}, \\ & j = 0, 1, 2, \ldots, k-1, \\ \dfrac{\alpha^{m+c-1} e^{-d\alpha} \sigma^{\delta+(m+n-k)\alpha-1}}{\left(\prod_{i=1}^{m} x_i\right)^{\alpha} \left(\prod_{i=k+1}^{n} a_{(i)}\right)^{\alpha}} I(\alpha > 0), & \\ a_{(k)} < \sigma < \min\{x_{(1)}, \sigma_0\}, & \end{cases}$$

with marginals

$$f(\alpha | \boldsymbol{a}, \boldsymbol{x}) \propto \frac{\alpha^{m+c-1} e^{-d\alpha}}{\left(\prod_{i=1}^{m} x_i\right)^{\alpha}} \left\{ \sum_{j=0}^{k-1} \frac{a_{(j+1)}^{\delta+(m+n-j)\alpha} - a_{(j)}^{\delta+(m+n-j)\alpha}}{[\delta + (m+n-j)\alpha]\left(\prod_{i=j+1}^{n} a_{(i)}\right)^{\alpha}} \right.$$

$$\left. + \frac{\left[\min\{x_{(1)}, \sigma_0\}\right]^{\delta+(m+n-k)\alpha} - a_{(k)}^{\delta+(m+n-k)\alpha}}{[\delta + (m+n-k)\alpha]\left(\prod_{i=k+1}^{n} a_{(i)}\right)^{\alpha}} \right\}$$

and

$$f(\sigma | \boldsymbol{a}, \boldsymbol{x}) \propto \begin{cases} \sigma^{\delta-1}\left[d - (m+n-j)\log\sigma + \sum_{i=1}^{m} \log x_i + \sum_{i=j+1}^{n} \log a_{(i)}\right]^{-(c+m)}, \\ \qquad a_{(j)} < \sigma < a_{(j+1)}, \; j = 0, 1, 2, \ldots, k-1, \\ \sigma^{\delta-1}\left[d - (m+n-k)\log\sigma + \sum_{i=1}^{m} \log x_i + \sum_{i=k+1}^{n} \log a_{(i)}\right]^{-(c+m)}, \\ \qquad a_{(k)} < \sigma < \min\{x_{(1)}, \sigma_0\}. \end{cases}$$

Here the posterior marginal density for σ is more reasonable. If $\delta > 1$, it is unimodal at $\min\{x_{(1)}, \sigma_0\}$. The complicated form of the posterior marginal density for α precludes interpretation.

An alternative, pragmatic procedure would probably involve acting as if σ were known to be equal to $X_{(1)}$ and then using the posterior analysis for known σ, and not using $X_{(1)}$ in the subsequent analysis.

3. More general Pareto distributions

If we introduce a location parameter μ in (2.1) we obtain what is called by Arnold (1983), a Pareto (II) distribution. The corresponding density is

$$f(x|\mu,\sigma,\alpha)=\frac{\alpha}{\sigma}\left(1+\frac{x-\mu}{\sigma}\right)^{-(\alpha+1)}I(x>\mu).$$

It is difficult to imagine a much better estimator for μ than $x_{(1)}$. We propose to use that estimator and then act, without loss of generality, as if μ were known to be 0, and the sample size had been reduced by one. Thus, we are concerned with the two-parameter family:

$$f(x|\sigma,\alpha)=\frac{\alpha}{\sigma}\left(1+\frac{x}{\sigma}\right)^{-(\alpha+1)}I(x>0). \tag{3.1}$$

A sample $X_1,\dots, X_n$ unfortunately admits of no reduction via sufficiency, beyond the order statistics. We can thus envision difficulties with Bayesian analysis of either grouped or ungrouped data; unless of course σ is known. In that happy situation progress can be made.

3.1. σ known (Pareto (II))

If we have m ungrouped observations from (3.1) say $X_1,\dots, X_m$, together with n censored observations $X_{m+1}>a_1,\dots, X_{m+n}>a_n$, the corresponding likelihood is

$$L(\alpha)=\left(\frac{\alpha}{\sigma}\right)^m\left[\prod_{i=1}^{m}\left(1+\frac{x_i}{\sigma}\right)\right]^{-(\alpha+1)}\left[\prod_{i=1}^{n}\left(1+\frac{a_i}{\sigma}\right)\right]^{-\alpha}.$$

A gamma prior density for α immediately suggests itself, and the resulting posterior density is again of the gamma form, i.e.

$$\mathscr{L}(\alpha|\boldsymbol{x},\boldsymbol{a},\sigma)=\gamma(m+c,u+v+d), \tag{3.2}$$

where

$$u=\sum_{i=1}^{n}\log\left(1+\frac{a_i}{\sigma}\right)$$

and

$$v=\sum_{i=1}^{m}\log\left(1+\frac{x_i}{\sigma}\right).$$

This result is included in Cozzolino (1974). In his paper the distribution in question is labeled the Lomax distribution. If we consider general grouped data, i.e. $a_i<x_i<b_i$, $i=1,2,\dots,n$, a posterior density of the form (2.5) is encountered. Some simplification occurs if $\sigma+b_i=w(\sigma+a_i)$, $i=1,2,\dots,n$.

3.2. σ and α both unknown (Pareto (II))

In this situation little can be done unless there are some ungrouped observations. If there are some such X_i's then, as proposed in Arnold and Press (1983), we may estimate σ using these X_i's based on the method of moments. Thus, we compute

$$\sigma_{\mathrm{M}} = \bar{x}(s^2 + \bar{x}^2)/(s^2 - \bar{x}^2)$$

where $\bar{x} = (1/m)\sum_{i=1}^{m} x_i$ and $s^2 = (1/m)\sum_{i=1}^{m}(x_i - \bar{x})^2$. We would then estimate α based on censored and ungrouped data using the density (3.2) with $\sigma = \sigma_{\mathrm{M}}$. While these estimators might not be efficient, they would at least be consistent.

3.3. Other generalized Pareto distributions

For samples from densities of the forms

$$f(x|\mu,\sigma,\gamma) = \frac{1}{\gamma\sigma}\left(\frac{x-\mu}{\sigma}\right)^{(1/\gamma)-1}\left[1+\left(\frac{x-\mu}{\sigma}\right)^{1/\gamma}\right]^{-2} I(x>\mu) \tag{3.3}$$

and

$$f(x|\mu,\sigma,\gamma,\alpha) = \frac{\alpha}{\gamma\sigma}\left(\frac{x-\mu}{\sigma}\right)^{(1/\gamma)-1}\left[1+\left(\frac{x-\mu}{\sigma}\right)^{1/\gamma}\right]^{-(\alpha+1)} I(x>\mu), \tag{3.4}$$

a partial Bayes approach is feasible (cf. Arnold and Press, 1983, who treat ungrouped data). With samples from (3.3) one estimates μ by $x_{(1)}$, the minimum of the available ungrouped data. The parameter γ is then estimable using the method of moments (using ungrouped data), and finally σ is estimated using a power function prior and utilizing both censored and ungrouped observations. For samples from (3.4) we estimate μ by $x_{(1)}$, use the method of moments for σ and γ, and then estimate α using a gamma prior.

4. Multivariate Pareto data

Multivariate Pareto distributions can be expected to be applicable to studies involving, for example, (1) incomes of employed husband–wife pairs, (2) incomes of individuals at several time points, or (3) incomes from several sources.

Suppose that observations are available from Mardia's type I k-dimensional Pareto distribution which has joint survival function of the form (Mardia, 1962):

$$P(\boldsymbol{X} > \boldsymbol{x}) = \left(1 + \sum_{i=1}^{k}\left(\frac{x_i - \mu_i}{\sigma_i}\right)\right)^{-\alpha} I(\boldsymbol{x} > \boldsymbol{\mu}). \tag{4.1}$$

Again, we focus attention on ungrouped and censored data. Thus, suppose that we have available m observations $\boldsymbol{X}^{(1)},\ldots,\boldsymbol{X}^{(m)}$ from the distribution (4.1), and in addition, n censored observations of the form $(\boldsymbol{X}^{(m+1)}>\boldsymbol{a}^{(1)}),(\boldsymbol{X}^{(m+2)}>\boldsymbol{a}^{(2)}),\ldots,(\boldsymbol{X}^{(m+n)}>\boldsymbol{a}^{(n)})$. Since marginal minima and moment estimators will undoubtedly be used for estimating $\boldsymbol{\mu}$ and $\boldsymbol{\sigma}$, we will assume they are known and, without loss of generality, assume that $\boldsymbol{\mu}=\boldsymbol{0}$ and $\boldsymbol{\sigma}=\boldsymbol{1}$. Thus, our data have common joint survival function of the form:

$$P(\boldsymbol{X}>\boldsymbol{x})=\left(1+\sum_{i=1}^{k}x_i\right)^{-\alpha}I(\boldsymbol{x}>\boldsymbol{0}).$$

It is convenient to define

$$T_j=\left(1+\sum_{i=1}^{k}X_i^{(j)}\right),\quad j=1,2,\ldots,m,$$

and

$$u_j=\left(1+\sum_{i=1}^{k}a_i^{(j)}\right),\quad 1,2,\ldots,n.$$

With this notation, the likelihood of the combined ungrouped and censored sample is

$$L(\alpha)=[\alpha(\alpha+1)\ldots(\alpha+k-1)]^m\left(\prod_{j=1}^{m}t_j\right)^{-(\alpha+k)}\left(\prod_{j=1}^{n}u_j\right)^{-\alpha}I(\alpha>0).$$

If we adopt the gamma prior (2.3), we may write the resulting posterior density for α in the form:

$$f(\alpha|\boldsymbol{t},\boldsymbol{u})\propto\alpha^{m+c-1}\left[\prod_{i=1}^{k-1}(\alpha+i)\right]^m e^{-(z+y+d)\alpha},$$

where

$$z=\sum_{j=1}^{m}\log t_j\quad\text{and}\quad y=\sum_{j=1}^{n}\log u_j.$$

If we observe that

$$\frac{\partial^2}{\partial\alpha^2}\log f(\alpha|\boldsymbol{t},\boldsymbol{u})=-\frac{m+c-1}{\alpha^2}-m\sum_{i=1}^{k-1}\frac{1}{(\alpha+i)^2}<0,$$

for every $\alpha>0$, it follows that there is a unique value of α, say α^*, for which

$$\frac{\partial}{\partial\alpha}\log f(\alpha|\boldsymbol{t},\boldsymbol{u})=0. \tag{4.2}$$

The value of α^* is the location of the unique mode of $f(\alpha|\boldsymbol{t},\boldsymbol{u})$. To determine α^*

we must solve (4.2), i.e.

$$\frac{m+c-1}{a^*} + m\sum_{i=1}^{k-1}\left(\frac{1}{\alpha^*+i}\right) = z+y+d.$$

If we define p_l by

$$\left[\prod_{i=1}^{k-1}(\alpha+i)\right]^m \equiv \sum_{l=0}^{m(k-1)} p_l\alpha^l,$$

then

$$f(\alpha|\boldsymbol{t},\boldsymbol{u}) \propto \sum_{l=0}^{m(k-1)} p_l\alpha^{m+l+c-1}\mathrm{e}^{-(z+y+d)\alpha}.$$

The corresponding posterior mean is given by

$$E(\alpha|\boldsymbol{t},\boldsymbol{u}) = \left(\frac{m+c}{z+y+d}\right) + \frac{1}{z+y+d}\left[\sum_{l=1}^{m(k-1)} lg_l\right]\Bigg/\left[\sum_{l=1}^{m(k-1)} g_l\right], \tag{4.3}$$

where

$$g_l \equiv p_l\Gamma(m+l+c)(z+y+d)^{-l}.$$

For reasonable sample sizes, i.e. m large, the posterior mean is well approximated as

$$\alpha^* = \left(\frac{m+c}{z+y+d}\right). \tag{4.4}$$

In this case one also might consider a rich conjugate family of priors of the form

$$f(\alpha) \propto P(\alpha)\mathrm{e}^{-d\alpha}I(\alpha>0),$$

where $d>0$ and $P(\alpha)$ is a polynomial in α. Provided that the degree of $P(\alpha)$ is not too high, the posterior mean is reasonably computable.

For $k=2$, considerable simplification occurs in (4.3). Moreover, in this case, an analytic expression for the posterior mode is available, namely

$$\alpha^* = \frac{2m+c-1-z-y-d+\sqrt{\begin{array}{c}(2m+c-1-z-y-d)^2\\ +4(z+y+d)(m+c-1)\end{array}}}{2(z+y+d)}$$

5. Remarks

(1) Dyer (1981) discusses the structural distribution of the survival probability $\min((t/\sigma)^{-\alpha},1)$ based on an ungrouped sample of size n from a classical Pareto

distribution with both σ and α unknown. Although philosophically distinct, this structural distribution is mathematically equivalent to the posterior distribution assuming a diffuse prior. His analysis can be extended readily to cover the case of an ungrouped sample assuming a generalized Lwin prior (2.17).

(2) Rather than use (4.3) to estimate α, one might consider using the much-easier-to-compute average of the k estimates obtained from the k marginal data sets using (3.2); cf. Arnold and Press (1983). Thus, one might use

$$\alpha^* = \frac{1}{k} \sum_{i=1}^{k} \left(\frac{m+c}{u_i + v_i + d} \right), \tag{5.1}$$

where

$$u_i = \sum_{j=1}^{n} \log(1 + a_i^{(j)})$$

and

$$v_i = \sum_{j=1}^{m} \log(1 + x_i^{(j)}).$$

Instead, if we used the harmonic mean of the marginal estimates, we would obtain the much simpler expression:

$$\tilde{\tilde{\alpha}} = \frac{m+c}{d + \bar{u} + \bar{v}}. \tag{5.2}$$

For large values of m and n, these expressions will usually not differ significantly from (4.4).

(3) Multivariate income data is often presented in tables for which a multivariate log-geometric distribution would be appropriate (the corresponding univariate case was discussed in section 2.1). In this setting a prior distribution for α which specifies $\mathscr{L}(w^{-\alpha}) = B(a, b)$ may be felt appropriate. The corresponding posterior distribution is unfortunately quite complicated.

(4) Zellner (1971) includes an example involving grouped classical Pareto data with known σ, and a diffuse prior distribution for α. The corresponding posterior density (of the form (2.4) with $c = 0$) is integrated numerically.

References

Aigner, D.J. and A.S. Goldberger (1970) "Estimation of Pareto's law from grouped observations", *Journal of the American Statistical Association*, 65, 712–723.

Arnold, B.C. (1983) *Pareto Distributions* (International Co-operative Publishing House, Fairland, Maryland).

Arnold, B.C. and S.J. Press (1983) "Bayesian inference for Pareto populations", *Journal of Econometrics*, 21, 287–306.

Cozzolino, J.M. (1974) "Conjugate distributions for incomplete observations", *Journal of the American Statistical Association*, 69, 264–266.

Dyer, D. (1981) "Structural probability bounds for the strong Pareto law", *Canadian Journal of Statistics*, 9, 71–77.

Lwin, T. (1972) "Estimation of the tail of the Paretian law", *Skand. Aktuar.*, 55, 170–178.

Mardia, K.V. (1962) "Multivariate Pareto distributions", *Annals of Mathematical Statistics*, 33, 1008–1015.

Muniruzzaman, A.N.M. (1968) "Estimation of the parameter of Pareto distribution–A Bayesian approach", Statistics for Development, Proceedings, Institute of Statistical Research and Training, University of Dacca.

Sinha, S.K. and H.A. Howlader (1980) "On the sampling distributions of the Bayesian estimators of the Pareto parameter with proper and improper priors and associated goodness of fit", Technical Report 103, Department of Statistics, University of Manitoba, Winnipeg, Canada.

Zellner, A. (1971) *An Introduction to Bayesian Inference in Econometrics* (John Wiley and Sons, Inc., New York).

PART V

PROBABILITY ASSESSMENT

Chapter 12

SUBJECTIVE-PROBABILITY ASSESSMENT METHODS FOR MULTIVARIATE-*t* AND MATRIX-*t* MODELS

J.M. DICKEY

State University of New York at Albany

A. PHILIP DAWID

University College London

JOSEPH B. KADANE*

Carnegie-Mellon University

1. Introduction

The matrix-*t* or matrix-Student density is proportional to a negative power of the determinant of a quadratic matrix polynomial:

$$p(\boldsymbol{Y}) = k_{\delta}(p,q)^{-1}|\boldsymbol{H}|^{-(1/2)q}|\boldsymbol{K}|^{-(1/2)p} \times |\boldsymbol{I}+(\boldsymbol{Y}-\boldsymbol{M})'\boldsymbol{H}^{-1}(\boldsymbol{Y}-\boldsymbol{M})\boldsymbol{K}^{-1}|^{-(1/2)(\delta+p+q-1)}, \tag{1.1}$$

for all real matrices $\boldsymbol{Y}(p \times q)$, where

$$k_{\delta}(p,q) = \pi^{(1/2)pq}\Gamma_p[(1/2)(\delta+p-1)]/\Gamma_p[(1/2)(\delta+p+q-1)], \tag{1.2}$$

with

$$\Gamma_p(\lambda) = \pi^{p(p-1)/4}\prod_{i=1}^{p}\Gamma(\lambda-(1/2)i+1/2).$$

*The authors wish to thank the editor, Prem Goel, and referees for helpful suggestions. Portions of this work were reported to the Eighteenth Meeting of the NBER–NSF Seminar on Bayesian Inference in Econometrics, The Rand Corporation, Santa Monica, California, March 1979. Research supported by U.S. National Science Foundation, Grant MCS-8301335.

Bayesian Inference and Decision Techniques, edited by P. Goel and A. Zellner

The multivariate-t distribution is the special case of a single row or column ($p=1$ or $q=1$). These distributions appear in the frequentist approach to normal regression as the distribution of the studentized error, both the error in the least squares estimate of the coefficients matrix and the error in the corresponding predictor of a future data array (Cornish, 1954; Kshirsagar, 1960; Kiefer and Schwartz, 1965). In the Bayesian conjugate-prior and diffuse-prior analyses for the same sampling models, it arises as the marginal prior or posterior distribution of the unknown coefficients matrix, and also as the predictive distribution of a future data array (Geisser and Cornfield, 1963; Ando and Kaufman, 1965; Geisser, 1965; Dickey 1967, section 4; Zellner, 1971, ch. 8; Press, 1972, section 8.6).

Under interpretations of Bayesian analysis in which probability enters as a model of subjective expert uncertainty, there is a need to specify a prior distribution as an expression of an inferer's opinion prior to his knowledge of the data. In addition, subjective-probability models can be useful in real decision problems where properly modeled statistical data are not available, and an expert's opinion must be assessed directly (posterior to any available indirect data) for immediate report, prediction, or decision-analytic use. The subjective-probability models used in such cases may or may not have reference to an underlying traditional sampling model (Dickey, 1980).

It is our purpose to provide methods to elicit and fit subjective-probability models of multivariate-t and matrix-t form. We propose interactive algorithms involving conversational computing and, as in an earlier paper on assessing multivariate-t models (Kadane et al., 1980), our methods will be heuristically reasonable rather than formally optimal. Properties of the matrix-t distribution which relate it to the multivariate-t distributions will be relied on heavily to exploit assessment methods devised for multivariate-t models.

We first review the matrix-t family with its synthetic representations, as a generalization of the multivariate-t distributions (section 2). We review the assessment methods for multivariate-t distributions (section 3), and then give methods for a general matrix-t distribution (section 4).

We give a new method for assessing a multivariate-t model structured through its location and scale parameters to account for concomitant and factor variables, predictors, and control variables. We then exploit these multivariate-t methods in the assessment of matrix-t models that include similar linear location constraints and covariance constraints (section 5). These models are derived as conditional distributions from a joint matrix-t distribution and also can be interpreted as Bayesian predictive distributions in a conjugate-prior analysis of a multivariate normal regression sampling model.

We conclude with selected applications (section 6), treating as special cases the previously outstanding problems: "How to assess a hierarchical prior distribution" and "How to assess a prior covariance matrix in multivariate normal sampling".

2. The matrix-*t* family

We shall use the notational scheme of Dawid (1981). The reader is referred there for a more detailed development of properties of the matrix-*t* family. See also Dickey (1967).

We say that $\boldsymbol{T}(p\times q)$ has the *matrix-t distribution* $T_\delta(\boldsymbol{H},\boldsymbol{K})$, with δ *degrees of freedom* and *spread matrices* $\boldsymbol{H}$ and $\boldsymbol{K}$ (with $\boldsymbol{H}(p\times p)$, $\boldsymbol{K}(q\times q)$ nonnegative definite symmetric) if $\boldsymbol{T}$ can be represented as

$$\underset{(p\times q)}{\boldsymbol{T}} = \underset{(p\times r)}{\boldsymbol{A}}\ \underset{(r\times r)}{\boldsymbol{U}}\ \underset{(r\times s)}{\boldsymbol{X}}\ \underset{(s\times q)}{\boldsymbol{B}}, \tag{2.1}$$

where $\boldsymbol{A}$ and $\boldsymbol{B}$ are fixed, with $\boldsymbol{AA'}=\boldsymbol{H}$ and $\boldsymbol{B'B}=\boldsymbol{K}$; $\boldsymbol{X}$ has independent standard normal entries; and, independently of $\boldsymbol{X}$, $(\boldsymbol{UU'})^{-1}$ has the standard Wishart distribution on $\delta+r-1$ degrees of freedom. Thus, in obvious notation, submatrices $\boldsymbol{T}_{ij}$ of partitioned $T_\delta(H,K)$ are marginally $T_\delta(\boldsymbol{H}_{ii},\boldsymbol{K}_{jj})$ for the same value δ.

It may be shown that for the transpose matrix $\boldsymbol{T}'\sim T_\delta(\boldsymbol{K},\boldsymbol{H})$, and that $\boldsymbol{Y}=\boldsymbol{M}+\boldsymbol{T}$ has density (1.1) when $\boldsymbol{H}$ and $\boldsymbol{K}$ are both nonsingular.

The familiar multivariate-*t* family is obtained as the special case of $\boldsymbol{T}=\boldsymbol{t}$ having a single p-element column:

$$\begin{aligned} \boldsymbol{t} &\sim T_\delta(\boldsymbol{H},\delta) \\ &\sim N^{(p)}(\boldsymbol{0};\boldsymbol{H})/(\chi^2_\delta/\delta)^{1/2}, \end{aligned} \tag{2.2}$$

and analogously for the case of a single row. The density for the translated p-vector $\boldsymbol{y}\sim\boldsymbol{m}+\boldsymbol{t}$ is then:

$$\begin{aligned} p(\boldsymbol{y}) &= \Gamma[(1/2)(\delta+p)]/[\Gamma((1/2)\delta)\pi^{(1/2)p}] \\ &\quad\times|\boldsymbol{H}|^{-(1/2)}\delta^{-(1/2)p}[1+\delta^{-1}(\boldsymbol{y}-\boldsymbol{m})'\boldsymbol{H}^{-1}(\boldsymbol{y}-\boldsymbol{m})]^{-(1/2)(\delta+p)}. \end{aligned} \tag{2.3}$$

The familiar univariate standard Student-*t* on δ degrees of freedom is $T_\delta(1,\delta)$.

The following lemma derives the marginal and conditional distributions.

Lemma 2.1 (Dickey, 1967). If $\boldsymbol{T}\sim T_\delta(\boldsymbol{H},\boldsymbol{K})$ and if $\boldsymbol{T}=(\boldsymbol{T}_1,\boldsymbol{T}_2)$ and $\boldsymbol{K}=(\boldsymbol{K}_{ij})(i,j=1,2)$ are conformably partitioned, with each $\boldsymbol{T}_i\, p\times q_i$ $(i=1,2)$, then $\boldsymbol{T}_1\sim T_\delta(\boldsymbol{H},\boldsymbol{K}_{11})$ and

$$\boldsymbol{T}_2|\boldsymbol{T}_1\sim \boldsymbol{T}_1\boldsymbol{K}_{11}^-\boldsymbol{K}_{12}+T_{\delta+\mathrm{rank}(\boldsymbol{K}_{11})}(\boldsymbol{H}+\boldsymbol{T}_1\boldsymbol{K}_{11}^-\boldsymbol{T}_1',\ \boldsymbol{K}_{22.1}), \tag{2.4}$$

where $\boldsymbol{K}_{22.1}=\boldsymbol{K}_{22}-\boldsymbol{K}_{21}\boldsymbol{K}_{11}^-\boldsymbol{K}_{12}$. (Note that $\boldsymbol{T}_1$ can have a singular distribution.) Since the transpose is also matrix-*t*, analogous statements hold for the partition $\boldsymbol{T}=(\boldsymbol{S}_1',\boldsymbol{S}_2')'$.

The matrix-normal distributions, of which the matrix-t is a mixture, are themselves a limiting case of the matrix-t. Namely, the matrix $\boldsymbol{Z} \sim \boldsymbol{AXC} \sim N^{(p \times q)}(\boldsymbol{H}, \boldsymbol{L})$, $\boldsymbol{H} = \boldsymbol{AA}'$, $\boldsymbol{L} = \boldsymbol{C}'\boldsymbol{C}$, with density

$$p(\boldsymbol{Z}) = (2\pi)^{-(1/2)pq} |\boldsymbol{H}|^{-(1/2)q} |\boldsymbol{L}|^{-(1/2)p} \operatorname{etr}(-(1/2)\boldsymbol{Z}'\boldsymbol{H}^{-1}\boldsymbol{Z}\boldsymbol{L}^{-1}), \tag{2.5}$$

has the distribution $T_\delta(\boldsymbol{H}, \delta\boldsymbol{L})$ for $\delta \to \infty$. Thus, many of our results are directly applicable to the matrix normal. The reader can easily verify that the matrix normal is a mere arrangement of the coordinates of a *multivariate*-normal random vector, as follows. Denoting the vertical array formed by stacking columns of $\boldsymbol{Z}$ by $\operatorname{vec}(\boldsymbol{Z}) = (z_{11}, z_{21}, \dots, z_{p1}, z_{12}, \dots, z_{pq})'$, we have

$$\operatorname{vec}(\boldsymbol{Z}) \sim N^{(pq)}(\boldsymbol{0};\ \boldsymbol{L} \otimes \boldsymbol{H}). \tag{2.6}$$

Hence, a matrix-normal model can be assessed directly in stacked form as a multivariate-normal model. Thus, one might suspect that the matrix-t may be merely an analogous rearrangement of the multivariate-t random vector. Indeed, as shown by Dawid (1981, section 6), if the second-moments matrix of $\operatorname{vec}(\boldsymbol{T})$ exists where $\boldsymbol{T} \sim T_\delta(\boldsymbol{H}, \boldsymbol{K})$, then it is proportional to the Kronecker product matrix $\boldsymbol{K} \otimes \boldsymbol{H}$. However, the matrix-$t$ is not just a rearrangement of the multivariate-t, as the following argument shows.

Consider a matrix-t distributed array consisting of two column vectors, $\boldsymbol{T} = (\boldsymbol{t}_1, \boldsymbol{t}_2) \sim T_\delta(\boldsymbol{H}, \boldsymbol{K})(p \times 2)$. Then $\operatorname{vec}(\boldsymbol{T}) = (\boldsymbol{t}_1', \boldsymbol{t}_2')'$ is a single column vector of dimension $2p$. If $\operatorname{vec}(\boldsymbol{T})$ were jointly multivariate-t, its distribution would have the same degrees-of-freedom parameter as the marginal distribution of one of its subvectors, which is δ. Then, two ways of using lemma 2.1, on the joint multivariate-t and on the joint matrix-t, would yield two different degrees-of-freedom values for the conditional distribution of $\boldsymbol{t}_2$ given $\boldsymbol{t}_1$: namely, $\delta + p \neq \delta + 1$ for $p > 1$. Hence, $\operatorname{vec}(\boldsymbol{T})$ cannot be multivariate-t for $p > 1$, and thus new elicitation strategies must be devised for matrix-t models that are not merely relabeled multivariate-t strategies. Our matrix-t strategies will, however, exploit multivariate-t strategies as building blocks.

3. Multivariate-t assessment

Methods which can be used to specify a general subjective multivariate-t distribution are given elsewhere (Kadane et al., 1980; Dickey, 1980), and space is not permitted here for other than a mere outline. The problem is how to elicit and fit interactively a model:

$$\boldsymbol{y} \sim \boldsymbol{m} + T_\delta(\boldsymbol{H}, \delta). \tag{3.1}$$

The methods transform the problem into questions of how to assess univariate marginal and conditional distributions which are Student-t with parameters that are known scaler functions of the multivariate parameters of (3.1), by lemma 2.1.

How then can we elicit a location and scale transformed univariate-t distribution, say denoted by

$$y \sim m + h^{1/2}t, \tag{3.2}$$

where $t \sim T_\delta(1,\delta)$? The idea is to elicit quantiles of univariate y and to relate functions of them to the univariate parameters. The shape parameter δ is determined by ratios of such quantile differences; the scale $h^{1/2}$ by ratios of differences of y quantiles to differences of corresponding tabled t quantiles; and m is the median of y.

The methods for univariate-t models (3.2) applied to each coordinate distribution,

$$y_i \sim m_i + T_\delta(h_{ii},\delta), \quad i=1,\ldots,p, \tag{3.3}$$

yield all the parameters of multivariate $\boldsymbol{y}$ (3.1), except the off-diagonal elements of the joint scale matrix, h_{ij} $(i \neq j)$. Thus, a separate approach is made to assess the full matrix $\boldsymbol{H}$. Positive-definiteness or nonnegative-definiteness is guaranteed for $\boldsymbol{H}$ by the assessment strategy, although this aspect was omitted by Winkler (1980). (Note that the variance–covariance matrix, when it exists ($\delta > 2$), is proportional to $\boldsymbol{H}$, namely $\mathrm{Cov}(\boldsymbol{y}) = [\delta/(\delta-2)]\boldsymbol{H}$.)

The two methods are available, and in each method the matrix $\boldsymbol{H}$ is constructed by successive bordering of principal submatrices. The positive-definiteness (or nonnegative-definiteness) of $\boldsymbol{H}$ is then automatically guaranteed by the definiteness of the submatrix resulting from each step. This bordering strategy of assuring definiteness for H is, in fact, available for both the following methods, provided the spread value as assessed at each bordering step is compelled to satisfy a simple inequality. (Kadane et al., 1980, emphasize this for only one of the methods.) Any such method has the potential disadvantage in practice that the final assessed matrix $\boldsymbol{H}$ can depend on the order in which the bordering is done.

The first available method for assessing off-diagonal elements of $\boldsymbol{H}$ is the *marginal method*. This is based on the marginal Student-t distribution of a linear form in $\boldsymbol{y}$ (3.1):

$$\boldsymbol{a}'\boldsymbol{y} \sim \boldsymbol{a}'\boldsymbol{m} + T_\delta(\boldsymbol{a}'\boldsymbol{H}\boldsymbol{a},\delta). \tag{3.4}$$

For example, $y_i - y_j$ has scale parameter $h_{ii} + h_{jj} - 2h_{ij}$, which can be solved for h_{ij} if h_{ii} and h_{jj} have already been assessed. This has the advantage of simplicity and the disadvantage that a linear combination of $\boldsymbol{y}$ coordinates may be difficult for the expert to consider if it does not arise naturally in the application context. Such linear combinations are natural, however, in an important class of problems, as we shall see.

The second method is the *conditioning method*. In this method one elicits conditional medians given subvectors of $\boldsymbol{y}$, one assumes a linear form of conditional location parameter, as in lemma 2.1, and one solves for the desired off-diagonal vector subarray. Kadane et al. (1980) would report to the expert an accumulating series of "data" values, $y_1^0, y_2^0, \ldots, y_i^0, \ldots, y_p^0$, and at each step i elicit his conditional medians for all remaining quantities, $\text{med}(y_j | y_1^0, \ldots, y_i^0)$ all $j > i$. This conditioning requires consideration by the expert of hypothetical (or even real) data. It would seem wise to do this in a carefully planned succession of data values, never requiring the expert to forget any values once they are introduced. The difficulties raised in this method by the distinction between pretending and believing are comparable to the difficulties in the general marginal method associated with asserting opinions concerning differences or sums.

4. Matrix-t assessment

Similarly to the multivariate-t, we first give methods for assessing an unconstrained matrix-t distribution. Eventually we shall turn to models having concomitant variables. The model $\boldsymbol{Y} \sim \boldsymbol{M} + T_\delta(\boldsymbol{H}, \boldsymbol{K})$ has parameters $\boldsymbol{M}$, δ, $\boldsymbol{H}$, and $\boldsymbol{K}$ to be assessed. By a subtraction, after eliciting the element-wise medians $\boldsymbol{M}$ (or after fitting a structured $\boldsymbol{M}$ to elicited element-wise medians), assume

$$\boldsymbol{M} = \boldsymbol{0}. \tag{4.1}$$

Of the remaining parameters, the degrees of freedom δ can again be assessed by considering elements of $\boldsymbol{Y}$. We assess the two spread matrices in what follows.

4.1. The two-vector method

Lemma 4.1. The matrix-t family is redundantly parameterized, the pair $\boldsymbol{H}$ and $\boldsymbol{K}$ being identified only up to a swapping scalar multiple,

$$T_\delta(\boldsymbol{H}, \boldsymbol{K}) \sim T_\delta(c\boldsymbol{H}, c^{-1}\boldsymbol{K}), \tag{4.2}$$

for every constant $c > 0$.

Thus, one cannot hope to assess both $\boldsymbol{H}$ and $\boldsymbol{K}$ merely by questioning the expert concerning $\boldsymbol{Y}$. Additional exogenous information will be required if, in addition to the distribution of $\boldsymbol{Y}$, one must also have $\boldsymbol{H}$ and $\boldsymbol{K}$ explicitly. For example, if the quantity $\boldsymbol{b'Kb}$ is provided for some nonzero vector $\boldsymbol{b}$, say $\boldsymbol{b'Kb} = k_{11}$ for $\boldsymbol{b} = (1,0,\ldots,0)'$, then the following result, which enables assessment of the distribution of $\boldsymbol{Y}$, will also thereby yield $\boldsymbol{H}$ and $\boldsymbol{K}$. We shall see below that, in the presence of concomitant variables, the assumed functional form of $\boldsymbol{H}$ may carry the exogenous information.

Theorem 4.2. Within the matrix-t family, a distribution is identified by the marginal distributions of two vectors. If $\boldsymbol{Y} \sim T_\delta(\boldsymbol{H}, \boldsymbol{K})$, then

$$\begin{aligned} \boldsymbol{a'Y} &\sim T_\delta(\boldsymbol{a'Ha}, \boldsymbol{K}) \\ &\sim T_\delta(1, (\boldsymbol{a'Ha})\boldsymbol{K}) \end{aligned} \tag{4.3}$$

and

$$\begin{aligned} \boldsymbol{Yb} &\sim T_\delta(\boldsymbol{H}, \boldsymbol{b}'\boldsymbol{Kb}) \\ &\sim T_\delta((\boldsymbol{b}'\boldsymbol{Kb})\boldsymbol{H}, 1). \end{aligned} \tag{4.4}$$

These are the respective multivariate-t distributions, $T_\delta(\delta, \delta^{-1}(\boldsymbol{a}'\boldsymbol{Ha})\boldsymbol{K})$ [with transpose $T_\delta(\delta^{-1}(\boldsymbol{a}'\boldsymbol{Ha})\boldsymbol{K}, \delta)$] and $T_\delta(\delta^{-1}(\boldsymbol{b}'\boldsymbol{Kb}), \delta)$.

Corollary 4.3. If $\boldsymbol{Y} \sim T_\delta(\boldsymbol{H}, \boldsymbol{K})$, then

$$\begin{aligned} \boldsymbol{a}'\boldsymbol{Yb} &\sim T_\delta((\boldsymbol{a}'\boldsymbol{Ha})(\boldsymbol{b}'\boldsymbol{Kb}), 1) \\ &\sim (\boldsymbol{a}'\boldsymbol{Ha})(\boldsymbol{b}'\boldsymbol{Kb})t/\delta, \end{aligned} \tag{4.5}$$

where t is the usual standard scalar Student-t on δ degrees of freedom.

To assess $\boldsymbol{H}$ and $\boldsymbol{K}$ when $\boldsymbol{b}'\boldsymbol{Kb}$ is given, first assess $(\boldsymbol{b}'\boldsymbol{Kb})\boldsymbol{H}$ by (4.4), and hence $\boldsymbol{H}$, and then assess $(\boldsymbol{a}'\boldsymbol{Ha})\boldsymbol{K}$ by (4.3), and hence $\boldsymbol{K}$. If one is only interested in the distribution itself, then one can merely adopt the convention $\boldsymbol{b}'\boldsymbol{Kb} = 1$, or δ. Assessment by consideration of one of the rows and one of the columns is of particular interest.

Corollary 4.4. If $\boldsymbol{Y} \sim T_\delta(\boldsymbol{H}, \boldsymbol{K})$ and $\boldsymbol{Y} = (\boldsymbol{u}_1, \ldots, \boldsymbol{u}_p)' = (\boldsymbol{v}_1, \ldots, \boldsymbol{v}_q)$, then each row $\boldsymbol{u}_i'$ has the marginal distribution

$$\boldsymbol{u}_i \sim T_\delta(\boldsymbol{K}, h_{ii}) \sim T_\delta(\delta^{-1}h_{ii}\boldsymbol{K}, \delta), \tag{4.6}$$

and each column $\boldsymbol{v}_j$ has the marginal distribution

$$\boldsymbol{v}_j \sim T_\delta(\boldsymbol{H}, k_{jj}) \sim T_\delta(\delta^{-1}k_{jj}\boldsymbol{H}, \delta). \tag{4.7}$$

All one need do is assess the multivariate-t distribution (4.6) of a single row, say $\boldsymbol{u}_1'$, and the multivariate-t distribution (4.7) of a single column, say $\boldsymbol{v}_1$. This can be done by our "marginal" method directly. However, the "conditioning" method might appear to have a special difficulty associated with its use in this context. Two marginal distributions are to be assessed, not the joint distribution of $\boldsymbol{u}_1$ and $\boldsymbol{v}_1$. The philosophy of the conditioning method requires that the expert consider a single succession of hypothetical data values, and that he not be required artificially to forget values part way through the task, say after assessing the distribution of $\boldsymbol{u}_1$ and before assessing the distribution of $\boldsymbol{v}_1$. Fortunately, the matrix-t model has conditional independence properties, as follows, under which such forgetting is natural and hence need not be imposed artificially in the method. (Of course, the model may fail to describe well the expert's uncertainties.)

Theorem 4.5. Consider doubly partitioned $\boldsymbol{Y}(p \times q) = \boldsymbol{Y}_{ij}: i = 1,2;\ j = 1,2$, with each $\boldsymbol{Y}_{ij}(p_i \times q_j)$, $p_1 + p_2 = p$, and $q_1 + q_2 = q$. Partition $\boldsymbol{H}$ and $\boldsymbol{K}$ conformably with $\boldsymbol{Y}$. If $\boldsymbol{Y} \sim T_\delta(\boldsymbol{H}, \boldsymbol{K})$, then the conditional distribution of $\boldsymbol{Y}_{12}$ given $\boldsymbol{Y}_{11}$ and

$\boldsymbol{Y}_{21}$ is translated matrix-t:

$$\boldsymbol{Y}_{12}|\boldsymbol{Y}_{11}, \boldsymbol{Y}_{21} \sim \boldsymbol{Y}_{11}\boldsymbol{K}_{11}^{-1}\boldsymbol{K}_{12} + T_{\delta+q_1}\left(\boldsymbol{H}_{11} + \boldsymbol{Y}_{11}\boldsymbol{K}_{11}^{-1}\boldsymbol{Y}_{11}', \boldsymbol{K}_{22.1}\right). \tag{4.8}$$

Since this depends only on $\boldsymbol{Y}_{11}$, we have conditional independence of $\boldsymbol{Y}_{12}$ and $\boldsymbol{Y}_{21}$ given $\boldsymbol{Y}_{11}$.

Proof. Apply lemma 2.1 twice, obtaining first $\boldsymbol{Y}_{12}, \boldsymbol{Y}_{22}|\boldsymbol{Y}_{11}, \boldsymbol{Y}_{21}$, and then marginalizing this for the conditional distribution of $\boldsymbol{Y}_{12}$ given $\boldsymbol{Y}_{11}$ and $\boldsymbol{Y}_{21}$.

Theorem 4.5 shows that, in particular, when assessing $\boldsymbol{v}_1$ after $\boldsymbol{u}_1$, the assessment values used for $\boldsymbol{u}_1$, other than that for y_{11} which is common to both vectors, are irrelevant to the assessed distributions for components of $\boldsymbol{v}_1$.

4.2. Other methods

It is clear by lemma 2.1 that a matrix-t distribution can be assessed as the succession of conditional multivariate-t distributions of its columns, since for $\boldsymbol{Y} = (\boldsymbol{v}_1, \ldots, \boldsymbol{v}_q)$,

$$p(\boldsymbol{Y}) = p(\boldsymbol{v}_1)p(\boldsymbol{v}_2|\boldsymbol{v}_1) \cdots p(\boldsymbol{v}_q|\boldsymbol{v}_1, \ldots, \boldsymbol{v}_{q-1}). \tag{4.9}$$

(Similarly, for rows.) These component multivariate-t distributions can be assesed by the marginal method or the conditioning method. The philosophy of not-forgetting in the conditioning method can easily be followed in this application. This "conditioning" method for the matrix-t rivals the preceding "two-vector" method for convenience in the case where one of the dimensions of the random matrix is low. The case of two rows will play a prominent role in the presence of an underlying sampling model in a later section.

Further powerful methods are easily derived which combine methods. For example, one can alternate between different methods for different submatrices of the matrix-t. Or assessments by various methods can be combined by averaging to obtain a possibly more accurate model of an expert's opinion, and similarly, for different orders of eliciting.

5. Concomitant and factor structures

5.1. Concomitant structure in the multivariate-t

A powerful class of subjective-probability models can take account of concomitant and factor variables for prediction and control by including a function in the location parameter. In particular, for multivariate-t deviates and a location function linear in matrix $X(n \times r)$, consider

$$Y \sim Xb + T_{\delta}(H, \delta). \tag{5.1}$$

The row vectors of X can be a known function, $x_i' = x'(z_i)$, of concomitant variable values z_i, $i = 1, \ldots, n$.

More generally, the scale parameter, which here is the spread matrix H, can also depend on the concomitant information. For example, we might have the matrix H take the form of a quadratic function of X,

$$H = H(X) = s^2(J + XUX'). \tag{5.2}$$

This model, (5.1) and (5.2), has a justification as a conditional distribution from an arbitrary joint matrix-t distribution, as follows. Note in lemma 2.1 above the linear form of the conditional center and the quadratic form of the conditional spread. Thus, consider a joint distribution,

$$(X, y) \sim T_{\nu}(P, Q), \tag{5.3}$$

in which a common center for the rows is assumed, by a translation, to lie at the origin (without loss of generality). Then marginally,

$$X \sim T_{\nu}(P, Q_{11}), \tag{5.4}$$

and (5.1) and (5.2) hold as the conditional distribution for $y|X$, with

$$b = Q_{11}^{-}Q_{12}, \tag{5.5}$$

$$\delta = \nu + \text{rank}(Q_{11}), \tag{5.6}$$

and

$$H(X) = \delta^{-1}Q_{22.1}(P + XQ_{11}^{-}X'). \tag{5.7}$$

Note the possibility for singular Q_{11}.

Of course, there are many contexts in which it would not be natural to think of X as random. The form (5.1) and (5.2) for y based on X arises also as the Bayesian predictive distribution in a conjugate-prior analysis for the familiar multiple-regression sampling model, where X may be controlled, or otherwise considered nonrandom. In the usual notation, we have the sampling model for y based on $X, \boldsymbol{\beta}, \sigma$:

$$y|\boldsymbol{\beta}, \sigma \sim N^{(n)}(X\boldsymbol{\beta};\ \sigma^2 J), \tag{5.8}$$

where $J = J_n$ is known, and the corresponding conjugate prior family:

$$\boldsymbol{\beta}|\sigma \sim N^{(r)}(b;\ \sigma^2 U) \tag{5.9}$$

$$\sigma^2 \sim s^2 \cdot (\delta/\chi_{\delta}^2)^{1/2}, \tag{5.10}$$

for which marginally,

$$\boldsymbol{\beta} \sim b + T_{\delta}(s^2 U, \delta). \tag{5.11}$$

This yields (5.1) as the prior predictive distribution for future data $\boldsymbol{y}$ based on $\boldsymbol{X}$ alone, with $\boldsymbol{H}$ as given by (5.2).

Purely as a subjective-probability model for $\boldsymbol{y}$ given $\boldsymbol{X}$, the model (5.1) with (5.2) has great advantages. It can depend on concomitant variables through the arbitrarily structured matrix $\boldsymbol{X}$. It is parameterized by the bounded-dimensional quantities $\boldsymbol{b}(r\times 1)$, $\boldsymbol{U}(r\times r)$, s^2, and δ, and it can refer to a sequence $\boldsymbol{y}=(y_1,\ldots,y_n)'$ of arbitrary and indefinitely extendible length n. In effect, once the model is fully asssessed, the uncertainty would be modeled concerning an infinite process. The model can be updated to account for observed data by probability conditioning, either by conditioning in the multivariate-t distribution for $\boldsymbol{y}$ based on $\boldsymbol{X}$, or equivalently by the usual Bayesian prior-posterior analysis for $\boldsymbol{\beta}$ and σ. Also, the model can be assessed without the expert needing to think about the metaphysical quantities $\boldsymbol{\beta}$ and σ, which are merely dummy mixing variables in the marginal model for $\boldsymbol{y}$ given $\boldsymbol{X}$.

5.2. Assessment

To assess such a model, it seems natural to assess first the location coefficients $\boldsymbol{b}$ by specifying the row vectors of $\boldsymbol{X}$ to the expert, eliciting his medians for the y-coordinates, and fitting by least squares or a similar criterion. Then δ and $\boldsymbol{H}$ can be assessed by the methods of the preceding section, taking $p=n$ and taking either $\boldsymbol{m}=\boldsymbol{Xb}$ or the raw elicited medians. Dickey (1980) argued that the least squares fit for $\boldsymbol{b}$ should be done by generalized least squares, with the elicitation-error variances of the elicited medians taken proportional to the spreads h_{ii} in the model itself. This will require an obvious iteration if a multivariate-t or matrix-t assessment method is used involving a standardization $\boldsymbol{m}=\boldsymbol{0}$ based on fitted median values.

As noted above, the spread matrix too can depend on the concomitant information, $\boldsymbol{H}=\boldsymbol{H}(\boldsymbol{X})$. Kadane et al. (1980) concentrated on the special case of an imbedded normal linear sampling model. Their method of assessment for this model assesses an n-dimensional multivariate-t distribution with spread matrix $\boldsymbol{H}$ in unconstrained form, and then fits the matrix quadratic form (5.2) by a projection:

$$\boldsymbol{U}=(\boldsymbol{X}'\boldsymbol{X})^{-1}\boldsymbol{X}'(s^{-2}\boldsymbol{H}-\boldsymbol{J})\boldsymbol{X}(\boldsymbol{X}'\boldsymbol{X})^{-1}. \tag{5.12}$$

The parameter s^2 is assessed separately by consideration of a difference, y_1-y_2, for repeated sampling under fixed concomitant-variable values.

A new marginal method takes advantage of the model's concomitant structure, as follows. After first assessing $\boldsymbol{b}$, s^2, and δ similarly to the conditioning method, one then assesses the spread matrix $\boldsymbol{U}$ of random ("dummy") $\boldsymbol{\beta}$ by building it up in the proposed way as a positive definite matrix. The elements of $\boldsymbol{U}$ can be obtained by assessing the scalar spread $h(\boldsymbol{x})$ of $y|\boldsymbol{x}$ for strategically chosen

Table 12.1. Strategy for assessing $\boldsymbol{U}$ by $h(\boldsymbol{x})$ (eq. (5.13)).

x_1	x_2	x_3	$\boldsymbol{x}'\boldsymbol{U}\boldsymbol{x} - u_{11}$
1	1	0	u_{22}
1	0	1	u_{33}
1	1	1	$u_{22} + u_{33} + 2u_{23}$

points $\boldsymbol{x}$:

$$h(\boldsymbol{x}) = s^2(j_x + \boldsymbol{x}'\boldsymbol{U}\boldsymbol{x}), \tag{5.13}$$

where j_x is the appropriate diagonal element of known $\boldsymbol{J}$. Table 12.1 exemplifies such a strategy for $\boldsymbol{x} = (x_1, x_2, x_3)'$, assuming, as usual, that a general-mean term is accommodated by having a constant first coordinate $x_1 \equiv 1$, and that the corresponding coefficient β_1 is uncorrelated with any other β_i, so $u_{1i} = 0$ for $i \neq 1$. This can be accomplished by the following lemma, after an initial direct elicitation of the spread-minimizing point $\boldsymbol{x}^0$ and a redefinition of $\boldsymbol{x}$ as the difference $\boldsymbol{x} - \boldsymbol{x}^0$.

Lemma 5.1. Each diagonal element h_{jj} of $\boldsymbol{H}(\boldsymbol{X})$ (5.2) depends on $\boldsymbol{X}$ only through the corresponding jth row of $\boldsymbol{X}$. Define the minimum-spread array of $\boldsymbol{H}(\boldsymbol{X})$ as that value $\boldsymbol{X}^0$ for $\boldsymbol{X} = (\boldsymbol{1}_n, \boldsymbol{X}^*)$ which minimizes each diagonal element of $\boldsymbol{H}(\boldsymbol{X})$ among all $\boldsymbol{X}$ of this form. Then the minimum-spread array has a common row vector, and furthermore it will take the form

$$\boldsymbol{X}^0 = \boldsymbol{1}_n(1,0,\ldots,0) \tag{5.14}$$

if and only if

$$\boldsymbol{U} = \begin{pmatrix} u_{11} & \boldsymbol{0}' \\ \boldsymbol{0} & \boldsymbol{U}^* \end{pmatrix} \tag{5.15}$$

and hence $\boldsymbol{H}(\boldsymbol{X})$ decomposes:

$$\boldsymbol{X}\boldsymbol{U}\boldsymbol{X}' = u_{11}\boldsymbol{1}_n\boldsymbol{1}_n' + \boldsymbol{X}^*\boldsymbol{U}^*\boldsymbol{X}^{*\prime}. \tag{5.16}$$

Proof. If for some $i \neq 1$, $u_{i1} \neq 0$, then the partial derivatives of h_{jj} with respect to x_i would not vanish at $(1,0,\ldots,0)$.

Garthwaite (1982) developed a method based on eliciting further $\boldsymbol{x}$-points of constrained-minimal y-spread. Both his method and the marginal method proposed here are complicated by any functional constraints, for example a polynomial dependence between the coordinates of $\boldsymbol{x}$.

5.3. Subjective response surfaces

One possible application of the work here and that reported in Kadane et al. (1980) is to the design of industrial products and similar areas where a linear form

$x'\beta$ is used to summarize the long-run influence of concomitant conditioning variables in an efficient way. Often, this can be achieved by defining variables x and y as transformations of primary variables z and w, say $x(z)$ and $y(w)$, where $y(w)$ is invertible. The function $x(z)'\beta$ of the primary conditioning variable z is then what is known as a "response surface" and can have a polynomial form or other form linear in given functions x of z.

The surfaces which are assessed here, $x(z)'b$, are "subjective response surfaces". They model the expert's opinion concerning real measurements and refer only indirectly, if at all, to an ideal long-run frequency relation $x(z)'\beta$. Indeed, a subjective probability model need not be constrained to the special form of a Bayesian "predictive" distribution based on a regression sampling model. However, a difficulty can arise if the subjectively assessed surfaces are based on linear regressor variables $x(z)$ which are essentially different in shape from those functions of z discovered later by an exploratory analysis of a new objective (experimental or observational) data set. For example, z and z^2 versus $1/z$ and $\sin(z)$. Then there would seem to be no way under the principle of coherence that the data can be used to update the assessed prior distribution. In practice, if the expert's own functions $x(z)$ can only be used to fit the data poorly, it may be advisable to confront him with the data-determined functions $x(z)$ to see whether his prior opinion can be reassessed to involve these new functions. If so, the actual data y-values could next be used to update his "revised prior", which would itself already depend on the data through the discovered functional forms $x(\cdot)$.

5.4. *Concomitant structure in the matrix-t*

We contribute a method for working with multiple-valued subjective response surfaces. In our experience these tend to be more common in practice than single surfaces. Only rarely is the performance of an industrial product, for example, summarized in a single quantity without loss of important information. Quality acceptance criteria in the petrochemical industry involve batteries of tests with thresholds on several measured properties. Our joint model allows for correlation between several response variables. This is achieved by the use of two scale matrices. The location function, however, is elicited and fit much as if there were separate single-surface models.

Analagously to our previous generalization of the multivariate-t, a generalization of the matrix-t model of considerable potential practical importance has the center parameter taking a matrix linear form. Consider

$$Y \sim XB + T_\delta(H, K), \tag{5.17}$$

where again $X(n \times r)$ is an observed or controlled "design" matrix function of concomitant variables, and the coefficients matrix $B(r \times q)$ is a new parameter of the subjective distribution which replaces the parameter M of (4.1). This formulation allows cases in which the matrix $\boldsymbol{M}$ is subject to linear constraints of the form $\boldsymbol{CM} = \boldsymbol{0}$, where $\boldsymbol{CX} = \boldsymbol{0}$. Again, the concomitant variables may or may not be conceived as contributing to the spread-matrix parameter $\boldsymbol{H}$. We take particular interest in a matrix quadratic function:

$$\boldsymbol{H} = H(\boldsymbol{X}) = \boldsymbol{J} + \boldsymbol{XUX}'. \tag{5.18}$$

This is just a higher dimensional generalization of the concomitant-structured multivariate-t model, (5.1) and (5.2). Matrices $\boldsymbol{Y}(n \times q)$, $\boldsymbol{B}(r \times q)$, and $\boldsymbol{K}(q \times q)$ replace the vectors and scalar, $\boldsymbol{y}$, $\boldsymbol{b}$, and δs^2, respectively. Our motivation of (5.17), (5.18) from a joint distribution $(\boldsymbol{X}, \boldsymbol{Y}) \sim T_\nu(\boldsymbol{P}, \boldsymbol{Q})$ has $\delta = \nu + \text{rank}(\boldsymbol{Q}_{11})$:

$$\boldsymbol{B} = \boldsymbol{Q}_{11}^{-}\boldsymbol{Q}_{12}, \tag{5.19}$$

$$\boldsymbol{H}(\boldsymbol{X}) = \boldsymbol{P} + \boldsymbol{XQ}_{11}^{-}\boldsymbol{X}', \tag{5.20}$$

and

$$\boldsymbol{K} = \boldsymbol{Q}_{22.1}. \tag{5.21}$$

Note that now $\boldsymbol{K}$ is not necessarily a scalar.

To derive the model as a predictive distribution in a Bayesian analysis, define the normal multivariate-regression sampling model for $\boldsymbol{Y}(n \times q)$, based on $\boldsymbol{X}(n \times r)$, $\boldsymbol{\Gamma}(r \times q)$, $\boldsymbol{\Sigma}(q \times q)$, and known $\boldsymbol{J}(n \times n)$:

$$\boldsymbol{Y} | \boldsymbol{\Gamma}, \boldsymbol{\Sigma} \sim \boldsymbol{X\Gamma} + N^{(n \times q)}(\boldsymbol{J}, \boldsymbol{\Sigma}). \tag{5.22}$$

The conjugate prior family has

$$\boldsymbol{\Gamma} | \boldsymbol{\Sigma} \sim \boldsymbol{B} + N^{(r \times q)}(\boldsymbol{U}, \boldsymbol{\Sigma}) \tag{5.23}$$

and

$$\boldsymbol{\Sigma} \sim \text{IW}_\delta(\delta \boldsymbol{S}), \tag{5.24}$$

where we use Dawid's (1981) notation for the inverse-Wishart distribution, $\text{IW}_\delta(\boldsymbol{C}'\boldsymbol{C}) \sim \boldsymbol{C}'\boldsymbol{W}_\nu^{-1}\boldsymbol{C}$, when $\boldsymbol{W}_\nu$ has the standard $p \times p$ Wishart distribution on $\nu = \delta + p - 1$ degrees of freedom. This convention for the degrees of freedom parameter permits invariance of the value δ to marginalization under partitions of the inverse-Wishart matrix.

The marginal prior distribution of the "dummy" $\boldsymbol{\Gamma}$ is matrix-t,

$$\boldsymbol{\Gamma} \sim \boldsymbol{B} + T_\delta(\boldsymbol{U}, \delta \boldsymbol{S}); \tag{5.25}$$

and conditionally,

$$\boldsymbol{\Sigma} | \boldsymbol{\Gamma} \sim \text{IW}_{\delta + \text{rank}(\boldsymbol{Q}_{11})}\left[\delta \boldsymbol{S} + (\boldsymbol{\Gamma} - \boldsymbol{B})'\boldsymbol{U}(\boldsymbol{\Gamma} - \boldsymbol{B})\right]. \tag{5.26}$$

To obtain the predictive distribution of data $\boldsymbol{Y}$ based on $\boldsymbol{X}$, first condition on only the sampling scale $\boldsymbol{\Sigma}$:

$$\boldsymbol{Y} | \boldsymbol{\Sigma} \sim \boldsymbol{XB} + N^{(n \times q)}(\boldsymbol{H}(\boldsymbol{X}), \boldsymbol{\Sigma}), \tag{5.27}$$

where $\boldsymbol{H}(\boldsymbol{X})$ is given by (5.18). Then by averaging out $\boldsymbol{\Sigma}$ under (5.24), we obtain the marginal distribution (5.17) with (5.18) and

$$K = \sigma S.$$

Statements made in sections 5.1 and 5.2 concerning the model for $q=1$ also apply here for the more general case. Updating to account for observed data can proceed by conditioning in the matrix-t (4.17) or, equivalently, by Bayes' Theorem applied to the prior distribution (5.23) and (5.24). This yields the posterior predictive distribution of further future data.

Lemma 5.2. The conjugate posterior distribution given data X and Y has parameters

$$B^* = B + W(Y - XB), \tag{5.28}$$

$$\delta^* = \delta + \text{rank}(H), \tag{5.29}$$

$$U^* = U - WXU, \tag{5.30}$$

$$S^* = \delta^{*-1}\left[\delta S + (Y - XB)'H^-(Y - XB)\right], \tag{5.31}$$

where H^- is any generalized inverse of H (5.18) and W is any matrix satisfying $WH = UX'$. Since one can show that the column space of XU is contained in that of H, then

$$W = UX'H^-. \tag{5.32}$$

Proof. The joint distribution of $Y - X\Gamma$ and Γ, given Σ, easily yields the distribution of Γ given Y, Σ (by reference to Dempster, 1969, Theorem 12.4.4), $\Gamma | Y, \Sigma \sim B^* + N^{(r\times q)}(U^*, \Sigma)$. Also, from (5.24) and (5.27), $\Sigma | Y \sim \text{IW}_{\delta^*}(\delta^* S^*)$.

Note that if U is nonsingular, then

$$U^* = (U^{-1} + X'J^-X)^{-1}. \tag{5.33}$$

Also, if the column space of X is contained in that of J (for example if J is nonsingular), we can take

$$W = U^*X'J^-; \tag{5.34}$$

and so if (5.33) holds, we have $B^* = B + (U^{-1} + X'J^-X)^{-1}X'J^-(Y - XB)$.

5.5. Assessment

We recommend the following steps to assess a matrix-t model with concomitant structure (5.17) and (5.18).

Step 1. Assess the coefficients matrix B by first eliciting element-wise medians $Y_{(0.50)}$ of Y, and then fitting for B by least squares,

$$B = (X'X)^{-1}X'Y_{(0.50)}, \tag{5.35}$$

or by weighted least squares, or otherwise.

Step 2. Assess the degrees of freedom δ by assessing δ for a few elements and then "averaging".

Step 3. The matrix random quantity $\boldsymbol{Y}$, homogeneous after subtracting the center matrix $\boldsymbol{M}$, has structured spread matrix $\boldsymbol{H}(\boldsymbol{X})$ (5.18), parameterized by $\boldsymbol{U}$, and unstructured spread matrix $\boldsymbol{K} = \delta \boldsymbol{S}$.

(a) To determine $\boldsymbol{U}$ and $\boldsymbol{S}$ by the two-vector method (section 4.1), proceed as follows:

(i) First assess $\boldsymbol{S}$ by assessing the associated multivariate-t model,

$$\boldsymbol{Y}'\boldsymbol{a} \sim T_{\delta}((\boldsymbol{a}'\boldsymbol{J}\boldsymbol{a})\boldsymbol{S}, \delta), \tag{5.36}$$

where $\boldsymbol{a} = (1, -1, 0, \ldots, 0)'$; that is, the difference of two rows of $\boldsymbol{Y}$ has a distribution depending on $\boldsymbol{H}$ only through the known matrix $\boldsymbol{J}$, and hence the assessment yields $\boldsymbol{S}$ directly. The distribution (5.36) requires that the first two rows of $\boldsymbol{X}$ be set equal, as in repeated performance tests of motor oil of a single composition.

(ii) Then, using $\boldsymbol{S}$, assess the multivariate-t model,

$$\boldsymbol{Y}\boldsymbol{b} \sim T_{\delta}((\boldsymbol{b}'\boldsymbol{S}\boldsymbol{b})\boldsymbol{H}(\boldsymbol{X}), \delta). \tag{5.37}$$

This can be done by either the "marginal" or the "conditioning" method for structured $\boldsymbol{H}(\boldsymbol{X})$.

(b) If one of the matrices, $\boldsymbol{H}$ or $\boldsymbol{K}$, is low dimensional, the expert may prefer to assess (5.17) by partitioning $\boldsymbol{Y}$ and using lemma 2.1 to factorize its distribution into conditional multivariate-t distributions (4.9).

6. Applications

6.1. Simple multivariate normal sampling

Various Bayesian multivariate analyses in the literature can now be carried out in practice by use of the methods of this chapter to assess the requisite prior distributions. In particular, consider the analysis of a sample of n vectors from a multivariate normal distribution. With a conjugate prior distribution (Ando and Kaufman, 1965), the joint predictive distribution of any number of future data vectors is matrix t, as detailed below.

This sampling model is the special case of our general normal linear sampling model (5.22) with $r = 1$, $\boldsymbol{X} = \boldsymbol{I}_n$, $\boldsymbol{\Gamma} = \boldsymbol{\mu}'(1 \times q)$, and $\boldsymbol{J} = I_n$. The matrix $Y = (y_1, y_2, \ldots, y_n)'$consists of independent row vectors identically sampled according to the multivariate normal distribution:

$$\boldsymbol{y} | \boldsymbol{\mu}, \boldsymbol{\Sigma} \sim N^{(q)}(\boldsymbol{\mu}; \boldsymbol{\Sigma}). \tag{6.1}$$

The corresponding conjugate prior family, from $\boldsymbol{B} = \boldsymbol{m}'$, $U = l^{-1}$, has the form:

$$\boldsymbol{\mu}|\boldsymbol{\Sigma} \sim N^{(q)}(\boldsymbol{m};\, l^{-1}\boldsymbol{\Sigma}), \tag{6.2a}$$

$$\boldsymbol{\Sigma} \sim \mathrm{IW}_\delta(\delta \boldsymbol{S}), \tag{6.2b}$$

by which marginally,

$$\boldsymbol{\mu} \sim \boldsymbol{m} + T(l^{-1}\boldsymbol{S}, \delta). \tag{6.3}$$

The predictive distribution is

$$\boldsymbol{Y} \sim \boldsymbol{1}_n \boldsymbol{m}' + T_\delta(\boldsymbol{H}, \delta \boldsymbol{S}), \tag{6.4}$$

where $\boldsymbol{H} = \boldsymbol{I}_n + l^{-1}\boldsymbol{1}_n\boldsymbol{1}_n'$.

Both the prior and the predictive families are thus parameterized by the quadruple $\boldsymbol{m}, \delta, l, \boldsymbol{S}$. To assess these parameters, one can work with the matrix-t predictive distribution, using any of our assessment methods. Perhaps the distribution of a single future vector has the most immediate interest and concrete interpretability,

$$y_1 \sim \boldsymbol{m} + T_\delta((1 + l^{-1})\boldsymbol{S}, \delta). \tag{6.5}$$

Note, however, that l and $\boldsymbol{S}$ are not identified, i.e. not distinguished in this distribution. According to the matrix-t "conditioning" method, one should proceed also to assess the conditional distribution of a second future vector given the first:

$$y_2|y_1 \sim \boldsymbol{m} + (l+1)^{-1}(y_1 - \boldsymbol{m}) + T_{\delta+1}(\boldsymbol{W}(y_1),\, \delta + 1), \tag{6.6a}$$

where

$$\boldsymbol{W}(y_1) = \frac{(l+2)}{(\delta+1)(l+1)}\left\{\delta \boldsymbol{S} + l(l+1)^{-1}(y_1 - \boldsymbol{m})(y_1 - \boldsymbol{m})'\right\}. \tag{6.6b}$$

In order to establish l, as distinct from $\boldsymbol{S}$, one need merely assess the center of this distribution. But the conditional dispersion matrix $\boldsymbol{W}$ can be assessed, as well, to provide a diagnostic check on the fit of the model.

6.2. Estimating the multivariate normal variance matrix

An outstanding problem in Bayesian statistics is the one of estimating a variance–covariance matrix. This problem is also vexing to traditional statisticians in cases when the data have missing values or a small enough sample size that the sample moments matrix is unstable, e.g. likely not to be positive definite. Despairing of being able to choose a whole positive-definite conjugate prior variance matrix $\boldsymbol{S}$, as in (6.2b), Bayesian theorists have opted to impose constraints on the prior parameters (see, for example, Dickey, Lindley and Press,

1985). The consequent loss of flexibility and threatened lack of realism is a positive disadvantage of these alternative approaches.

Under the simple multivariate normal sampling model of the preceding section, the mean of $\boldsymbol{\Sigma}$ could provide one form of estimate. The conjugate inverse Wishart distribution (6.2b) yields the posterior mean $E(\boldsymbol{\Sigma}|\boldsymbol{Y}) = (\delta^* - 2)^{-1}\delta^*\boldsymbol{S}^*$ (Kaufman, 1967). The relevant posterior values (5.31) are $\delta^* = \delta + n$, and $\boldsymbol{S}^* = \delta^{*-1}\{\delta\boldsymbol{S} + (\boldsymbol{Y} - \boldsymbol{1}_n\boldsymbol{m}')'\boldsymbol{H}^{-1}(\boldsymbol{Y} - \boldsymbol{1}_n\boldsymbol{m}')\}$, where $\boldsymbol{H}^{-1} = \boldsymbol{I}_n - (l+n)^{-1}\boldsymbol{1}_n\boldsymbol{1}_n'$. Hence,

$$\begin{aligned} E(\sigma_{jj'}|\boldsymbol{Y}) &= \{\delta^*/(\delta^* - 2)\}s^*_{jj'} \\ &= (\delta + n - 2)^{-1}\{(\delta - 2)E(\sigma_{jj'}) \\ &\quad + n\sum_{i=1}^{n}(y_{ij} - m_j)(y_{ij'} - m_{j'})/n \\ &\quad - n[n/(l+n)](\bar{y}_{.j} - m_j)(\bar{y}_{.j'} - m_{j'})\}, \end{aligned}$$

where $E(\sigma_{jj'}) = \{\delta/(\delta - 2)\}s_{jj'}$ and $\bar{y}_{.j} = \sum_{i=1}^{n} y_{ij}/n$.

Most of the theory in the literature on estimating a normal variance matrix focuses on the simpler problem in which the mean vector is known, $\boldsymbol{\mu} \equiv \boldsymbol{m}$. The methods here adapt immediately to this case by the device of taking $l \to \infty$, for which $\boldsymbol{H} \to \boldsymbol{I}_n$. Then only one future vector $\boldsymbol{y}_1$ need be considered to assess δ and $\boldsymbol{S}$.

6.3. *Constrained forms*

In general, methods developed for assessing a distribution in a particular parameterized family can easily be adapted to apply to a new family obtained by restricting the range of the parameters or by a functional constraint reducing their dimensionality. Assessment questions which had been intended to determine parameters which no longer exist can be used to provide consistency checks on the new model. Such diagnostics may or may not be reported to the expert whose degrees of belief are being assessed in an interactive computing environment. If so desired, estimates of unconstrained parameters can then be used to "overdetermine" the new parameters in a least squares or other fitting procedure. This approach is promising for the assessment of hierarchical prior distributions, as introduced by Lindley and Smith (1972).

Included in constrained forms are various types of subjective dispersion matrix: Toeplitz forms; Markov property; exchangeability. Exchangeability, for example, applies when the assessor does not incorporate any information distinguishing the quantities whose joint distribution is being assessed. In the simple multivariate sampling context considered here, exchangeability would imply a common mean coordinate $\boldsymbol{m} = m\boldsymbol{1}_q$ and the intraclass form for the dispersion parameter $S =$

$k\{I_q + \rho(1-\rho)^{-1}1_q1'_q\}$. The parameters to be assessed are then reduced to five scalars, m, δ, l, k, and ρ.

A weaker assumption than exchangeability, which often seems realistic in applications, is to assume that appropriate coordinate-scale transformations exist under which the coordinates of the vector y (a single column y_1) would be exchangeable. If the original variables are z, then $z = z_0 + \mathrm{Diag}(s)y$. Elicitations can then be carried out in terms of z, while the internal assessment procedures are fit in terms of y to determine m, δ, l, k, and ρ.

One way of assessing the transformation z_0 and s will also establish $m = 0$ and $k = 1$, and that is to first elicit the z_0 coordinates as medians of the z coordinates, and the s coordinates as their individual dispersions. This is a promising method in applications which involve many coordinate variables which the expert wishes to distinguish in his prior distribution only as far as their medians and dispersions. It offers a great savings in the joint assessment effort, perhaps without oversimplifying the prior structure for z.

References

Ando, A. and G.W. Kaufman "Bayesian analysis of the independent multi-normal process – neither mean nor precision known", *J. Amer. Statistical Assoc.*, 60, 347–358.

Cornish, E.A. (1954) "The multivariate t distribution", *Australian J. Physics*, 7, 531–542.

Dawid, A.P. (1981) "Some matrix-variate distribution theory: Notational considerations and a Bayesian application", *Biometrika*, 68, 265–274.

de Finetti, Bruno (1937) "La prevision: Ses lois logiques, ses sources subjectives", *Annales de l'Institut Henri Poincaré*, 1937, 1–68. English translation in: H.E. Kyburg, Jr. and H.E. Smokler, eds., *Studies in Subjective Probability* (Wiley, New York, 1964).

Dempster, A.P. (1969) *Elements of Continuous Multivariate Analysis* (Addison-Wesley, Reading, Mass.).

Dickey, J.M. (1976) "Matricvariate generalizations of the multivariate t distribution and the inverted multivariate t distribution", *Annals of Mathematical Statistics*, 38, 511–518.

Dickey, J.M. (1980) "Beliefs about beliefs, a theory for stochastic assessments of subjective probabilities", in: J.M. Bernardo, M.H. DeGroot, D.V. Lindley and A.F.M. Smith, eds., *Bayesian Statistics*, Proceedings of the First Valencia Meeting, Valencia, Spain (Valencia University Press).

Dickey, J.M. and C-H. (1985) "Direct subjective-probability modeling using ellipsoidal distributions", in: J.M. Bernardo, M.H. DeGroot, D.V. Lindley and A.F.M. Smith, eds., *Bayesian Statistics*, Proceedings of the Second Valencia Meeting, Valencia, Spain (Elsevier Science Publishers and Valencia University Press).

Dickey, J.M., D.V. Lindley and S.J. Press (1985) "Bayesian estimation of the dispersion matrix of a multivariate normal distribution", *Commun. Statist.–Theor. Meth.*, 14 (5) 1019–1034.

Garthwaite, P.H. (1982) "Prior-distribution elicitation for a normal linear model", Ph.D. dissertation, Dept. of Statistics, University College of Wales, Aberystwyth, U.K.

Geisser, Seymour (1965) "Bayesian estimation in multivariate analysis", *Annals of Mathematical Statistics*, 36, 150–159.

Geisser, Seymour and Jerome Cornfield (1963) "Posterior distributions for multivariate normal parameters", *J. Roy. Statist. Soc.*, Ser. B 25, 368–376.

Kadane, J.B., J.M. Dickey, R.L. Winkler, W.S. Smith and S.C. Peters (1980) "Interactive elicitation of opinion for a normal linear model", *J. Amer. Statist. Assoc.*, 75 (Dec. 1980) 845–854.

Kaufman, G.M. (1967) "Some Bayesian moment formulae", Report No. 6710, Center for Operations Research and Econometrics, Catholic University of Louvain. Heverlee, Belgium.

Kiefer, J. and R. Schwartz (1965) "Admissible Bayes Character of T^2, R^2, and other fully invariant tests for classical multivariate normal problems", *Annals of Mathematical Statistics*, 36, 747–770.

Kshirsagar, A.M. (1960) "Some extensions of the multivariate t distribution and the multivariate generalization of the distribution of the regression coefficient", *Proc. Cambridge Phil. Soc.*, 57, 80–85.

Lindley, D.V. and A.F.M. Smith (1972) "Bayes estimates for the linear model", *J. Roy. Statist. Soc.*, Ser. B 34, 1–41 (with discussion).

Lindley, D.V., A. Tversky and R.V. Brown (1979) "On the reconciliation of probability assessments", *J. Roy. Statist. Soc.*, Ser. A 142, 146–180 (with discussion).

Press, S.J. (1972) *Applied Multivariate Analysis* (Holt, Rinehart and Winston, Inc., New York).

Tversky, Amos and Daniel Kahneman (1974) "Judgement under uncertainty: Heuristics and biases", *Science*, 185, 1124–1131.

Winkler, R.L. (1980) "Prior information, predictive distribution, and Bayesian model-building", in: A. Zellner, ed., *Studies in Bayesian Econometrics and Statistics, in Honor of Harold Jeffreys* (North-Holland Publishing Co., Amsterdam), pp. 95–109.

Zellner, A. (1971) *An Introduction to Bayesian Inference in Econometrics* (John Wiley & Sons, New York).

Chapter 13

SEPARATING BELIEFS

MICHAEL GOLDSTEIN*

University of Hull

1. Introduction

de Finetti (1972) quotes the following statements in Fisher (1956):

> Although in the deduction of statements of certainty it is legitimate to draw inferences from some of the axioms available while ignoring others, or, in other words, to base a valid argument on a chosen subset only of the available axioms, no such liberty can be taken with statements of uncertainty, where it is essential to take the whole of the data into account, though some part of it may be shown on examination to be irrelevant, and not to affect the result.

In subjectivist theory, the data consist of everything that has an effect on your beliefs, namely all observations and thoughts up to the point where your belief statements are announced. The most popular brand of subjectivism is Bayesian theory. Here, it is claimed that in principle you may anticipate all the forms of evidence that you might receive. Furthermore, you may assess all these outcomes in terms of prior and conditional probabilities. Further yet, on observing an outcome, Bayes' Theorem will express the way that your beliefs should change.

There are various logical oddities about this procedure, and we will briefly return to them in section 4. For now, let us identify one point on which most people would agree. In any realistic information-gathering situation, you cannot anticipate all of the possible relevant combinations of information which you might receive. In terms of Fisher's statement above, you are attempting to base your statements of uncertainty on a subset of the data, namely that which you

*I am grateful to the referee for suggesting many clarifying comments.

Bayesian Inference and Decision Techniques, edited by P. Goel and A. Zellner

happened to consider a priori. Thus, the relationship between such statements and your actual assessment of the evidence is undefined.

Many Bayesians would reply that Bayesian theory should be viewed as a model for rational beliefs. This implies that we have not yet reached the bedrock of subjectivistic theory, as a model can only have meaning if it is a model for something which is well defined prior to the model. In this chapter we analyse the bedrock and demonstrate that Bayesian procedures, when carefully defined, do play a valid part in the general subjectivist analysis. We also point to a much wider class of procedures which play exactly the same limited but valuable role in the analysis as the Bayesian procedures. We will also define the wider context within which the Bayesian formalism should be interpreted.

In order to do this we develop a fundamental aspect of the subjectivist language, namely the ability to separate the beliefs of an individual into subsystems which may be analysed as separate systems. As a preliminary, section 2 outlines the ingredients of the subjectivist theory deriving from de Finetti's concept of prevision and the application of this concept in Goldstein (1981).

Then in section 3, the notion of separation of beliefs is introduced. Essentially this notion provides the conditions under which the effect of new information upon your beliefs may be determined purely by considering the effect of the information on certain limited aspects of your beliefs. The general conditions for such a separation are derived. We then demonstrate that these conditions may always be realised, purely by considering any limited but well-defined quantitative aspects of the proposed information.

Finally, in section 4 we demonstrate that the standard Bayesian theory may be considered to be a special case of the separation of beliefs outlined in section 3. This section provides an exact account of the extent to which Bayesian theory can be justified within the more fundamental framework.

2. Ingredients of the subjectivist theory

Subjective theory describes the properties of an objective language. We make two demands on this language, namely that we can say what we want and we can "not say" what we do not want. (Imagine a language in which there is no word for "table" and no possibility of adding such a word to the language. Conversely, imagine a language in which any mention of a "table" had to be accompanied by a detailed description of the room in which it was located.) In our context, the first condition is that our statements should be applicable to all uncertain quantities. The second condition is that our statements should be applied directly to the quantities of interest, rather than indirectly through intermediary quantities. This is achieved by basing our language on the notion of prevision rather than probability.

Thus, your basic statement of belief concerns a random quantity X (i.e. a quantity whose value you do not know). Your beliefs concerning the value of X

are translated into a single number, the fair price, expectation or *prevision* of X, $P(X)$, which is defined, in de Finetti (1974), to be the value $\bar{x}$ which you would choose if, having made this choice, you were to suffer a penalty, L, given by $L = (X - \bar{x})^2/K$, where K is a suitable constant defining the units of loss. de Finetti demonstrates that the coherence condition of avoiding sure loss implies that $P(X+Y) = P(X) + P(Y)$. If H is an event, then identifying the event with the indicator function of the event (i.e. so that $H=1$ if H occurs, $H=0$ otherwise), we may call $P(H)$ the probability of H. The notation makes no distinction between "probabilities" and "expectations". Assessing a probability distribution for X is equivalent, within our framework, to assessing the prevision of every bounded function of X. You may attempt to make all these assessments if you wish. However, as with the "table" in the "room", you retain the option of just discussing the table, i.e. X itself, if you wish.

2.1. *Your current beliefs*

de Finetti (1974) suggests the following representation for your beliefs concerning a general collection of random quantities $\{X_i\}$. Begin with the affine vector space $\mathscr{L}$ of random quantities X. (Each X is represented by a vector and linear combinations by linear combinations.) Thus, a possible element of $\mathscr{L}$ might be one-half the number of doctors in Asia minus 7000 times the amount of money Richard Nixon will make by 1990. $\mathscr{L}$ may be thought of as the set of all random quantities for which the previsions are determined by specifying the previsions of the quantities $\{X_i\}$. There is no logical restriction on the quantities X_i which are allowed. However, it is clear that in practice there will be some quantities which it is not very helpful to consider in combination, and in section 3 we will examine the conditions under which, for example, the above beliefs might be treated quite separately as "doctor" beliefs and "Richard Nixon" beliefs.

The first step is to represent your beliefs about the relationships between the quantities, i.e. to consider $\mathscr{L}$ as an inner product space $\mathscr{A}$ in which the inner product and norm are defined by

$$(X,Y) = P(X - P(X))(Y - P(Y)), \quad \|X\| = (X,X)^{1/2}.$$

We restrict attention to random quantities X with $P(X^2) < \infty$. de Finetti considers equivalence classes of random quantities differing by a constant value, but we will find it more helpful to adjust each random quantity X in $\mathscr{L}$ by subtracting the prior prevision of X, so that each element of $\mathscr{A}$ has prior prevision zero. The space $\mathscr{A}$ need not contain all the quantities for which you are prepared to state a prevision. Rather, it will contain some subset of your choice for which you are interested in explicitly revising your beliefs as you receive new information.

Representation of prior beliefs in terms of probability distributions imposes the requirement that the previsions of all bounded functions of all considered quantities (and products) must be specified. In our representation, you specify the previsions only of those functions which you explicitly introduce into $\mathscr{A}$ as random quantities (and products).

A note on alternative inner products

The above inner product is appropriate for measuring uncertainty (and thus changes in uncertainty) as large values of $\|X\|$ correspond to random quantities about whose values there is great uncertainty. However, this inner product, because it has already corrected for the value of the prevision, is not very useful for discussing the actual values taken by the previsions.

de Finetti suggests an alternative inner product which we also find useful. Suppose $\mathscr{L}$ is a general vector space of random quantities (i.e. we have not corrected each $X \in \mathscr{L}$ so that $P(X)=0$). Now define $\mathscr{A}^*$ to be the inner product space corresponding to the inner product

$$(X, Y)^* = P(XY).$$

If we denote by X_0 the random quantity whose value is identically one, then for $X \in \mathscr{A}^*$, $P(X) = (X, X_0)^* X_0$, i.e. the prevision of each $X \in \mathscr{A}^*$ is the projection of X into $[X_0]$, the subspace of $\mathscr{A}^*$ spanned by X_0.

Of course the two inner products we have discussed are not intrinsically different. The obvious relationship is to simply consider the space $\mathscr{A}$ to be equivalent to the subspace $[X_0]^{\perp}$ of $\mathscr{A}^*$, i.e. the orthogonal complement of $[X_0]$ in $\mathscr{A}^*$.

2.2. Posterior prevision

You are about to receive new information. This will cause you to revise the values you assign for the previsions of each element of $\mathscr{A}$. How you will do this may be a very complicated process. However, assuming that you have decided that you will make such future previsions, then to each $X \in \mathscr{A}$, there corresponds a quantity $P_D(X)$, namely the value that you will assert for $P(X)$ having observed and analysed the evidence. The subscript D is used in an analogous manner to that of a conditioning event. However, there is no implication that D may be decomposed into an exclusive partition of possibilities. What D corresponds to is a description of what is to be done before the revised prevision is announced. (Thus, for an amateur meteorologist trying to make his own forecast of tomorrow's weather, D might correspond to a walk in the hills, a study of recent satellite photographs, listening to the latest radio forecast, or all of the above, or simply all that he has learned by some specified time.)

Thus, $P_D(X)$ is the random quantity which denotes the prevision for X that you will actually make in some well-specified situation, and the defining characteristic is not the reasons underlying the specification, but rather the circumstances under which the specification will be made. We introduce the posterior prevision for essentially the same reason as we introduce the prior prevision, namely to concentrate attention on what is important in an economical fashion. Again, just as with prior previsions, we begin by imposing no constraints on the random quantity $P_D(X)$, and then investigate the coherence implications for this quantity.

If we accept that current previsions are subjectively assessed, and that previsions made in the future have precisely the same logical status, but are simply made on the basis of different information, then we should also accept that posterior previsions will themselves be subjectively assessed, at the appropriate future time. It is logically hopeless (and uninteresting) to construct a "normative" subjective theory which claims to force an individual to make various judgements in the future. The objective relationships between beliefs now and beliefs in the future lie not in the realm of logic, but in the realm of belief, i.e. prevision.

Just as coherence imposes one basic rule for prevision, namely $P(X+Y) = P(X)+P(Y)$, from which the general theory follows, coherence imposes one basic condition on $P_D(X)$, namely for any bounded X, that

$$P(P_D(X)) = P(X). \tag{1}$$

A "sure loser" argument for this relation is given in Goldstein (1983), for any proposed information. Thus, for example, a doctor cannot say both that "my 'expectation' for how long you will live is 40 years" and also that "my 'expectation' for the value I will state as my 'expectation' for your lifetime after a forthcoming medical examination is 5 years". Relation (1) is analogous to the relation $EX = E(E(X|Y))$. However, the latter relation does not cover the statements made by the doctor, as that would presuppose first that the doctor can specify beliefs over every possible medical examination he might choose to give (a practical and logical impossibility), secondly that he will assign countably additive prior probability distributions (an arbitrary convention), and finally that conditional probabilities as given by Bayes' Theorem can be interpreted as posterior probabilities (a logical misunderstanding).

Some comments on posterior previsions

(1) Is the quantity $P_D(X)$ well defined? Yes, if you take the trouble to "well-define" it. Thus, you might decide that at 3 p.m. tomorrow you will write down your value for $P(X)$, or that after a certain well-defined experiment is concluded you will spend at most a further week considering your results and then write down your value for $P(X)$, or at some time point chosen by a well-defined random process within a finite time interval, whose value you may or may not know in advance, you will write down your value for $P(X)$. (The suggestion that

you write these values down is simply to ensure unambiguous numerical values for these quantities.)

Indeed, we might turn this question around. The fact that conditional probability statements in Bayesian theory do not appear to require a careful establishing of circumstances suggests that they are not directly related to your beliefs under any well-defined circumstances.

(2) Can you meaningfully assign uncertainties to $P_D(X)$? Yes, $P_D(X)$ is a well-defined random quantity, so that previsions for this quantity have precisely the same logical status as previsions for any other random quantity. It might be objected that because you are considering beliefs about your future beliefs, whose values you yourself will determine, that this introduces some element of circularity into the procedure. However, in most cases for which previsions are assigned, the protagonists have some measure of control over the outcome. For example, a doctor might give a prevision for your remaining lifetime as 5 years. It is in no way paradoxical that the doctor will attempt to help you live beyond this. Rather, his prevision is an implicit assessment of his belief that such help will prove successful.

(3) Is it useful to consider $P_D(X)$? At a practical level, we will have to wait and see. However, it is intuitively clear that complicated information is often better characterised by the "diagnosis" rather than the "symptoms". (A doctor could far more easily explain what he expects to find out about your health than he could outline the precise pathways which would lead him to the respective diagnosis.) On a more fundamental level, it is the quantities $P_D(X)$ which will be observed and whose values are actually of interest to you. We might instead find it puzzling if you find it is useful to consider conditional probabilities. Finally, within the subjectivist theory itself, such quantities are surely necessary. How can we consider the question as to what form your future beliefs can take unless these beliefs are explicitly introduced into the language? We might (though I believe we will not) find that this is a redundant concept, which is logically determined by other quantities (such as conditional beliefs). This is an issue for the theory to decide, not for us to make a value judgement over.

2.3. *The effect of new information on your beliefs*

We have suggested that the effect of new information may be assessed for each X by considering your actual future prevision $P_D(X)$. However, the content of the theory is not the individual previsions, which you yourself must specify, but the relationship between the collection of previsions that you specify, implied by coherence. In our context, theory therefore concerns the change in your stated beliefs, i.e. the change in the inner product on $\mathscr{A}$. We must therefore consider, for each $X, Y \in \mathscr{A}$, the value $(X, Y)_D$ which is the random quantity which denotes the actual value you will assign for (X, Y) when you have assessed the data. The relationship between the random quantities $(X, Y)_D$, and the posterior previsions is as follows.

Theorem. For $X, Y \in \mathscr{A}$

$$P(X, Y)_D = (X, Y) - P(XP_D(Y)). \tag{2}$$

Proof. First, note that if Z is any random quantity whose value will certainly be revealed when D is considered, then

$$P_D(ZY) = ZP_D(Y). \tag{3}$$

Thus, for any two random quantities, X and Y,

$$P_D(X)P_D(Y) = P_D(XP_D(Y)), \tag{4}$$

so that, from (1):

$$P(P_D(X)P_D(Y)) = P(XP_D(Y)) = P(YP_D(X)) \tag{5}$$

Thus, using (1), (4), and (5) we have:

$$\begin{aligned} P(X, Y)_D &= P(P_D((X - P_D(X)))(Y - P_D(Y))) \\ &= P(X - P_D(X))(Y - P_D(Y)) \\ &= P(XY) - P(X \cdot P_D(Y)). \end{aligned}$$

Thus, to assess your opinions about the change in the inner product, you need to assess all the previsions of form $P(X \cdot P_D(Y))$. This is precisely equivalent to extending the inner product space $\mathscr{A}$ by introducing into $\mathscr{A}$ all of the random quantities $P_D(X)$, for $X \in \mathscr{A}$.

Suppose that you specify all previsions of form $P(XP_D(Y))$. Recall that each $Y \in \mathscr{A}$ has $P(Y) = 0$, so by (1) $P(P_D(Y)) = 0$ for each Y, so that each element $P_D(Y)$ has zero prevision. Notice from eq. (5) that specifying $P(X \cdot P_D(Y))$ also specifies the value $P(P_D(X)P_D(Y))$. Thus, the elements $P_D(Y)$ may be considered to be elements of the completely specified inner product space $\hat{\mathscr{A}}$ spanned by all of the elements $\{X\}$, $\{P_D(X)\}$. We may write $\hat{\mathscr{A}} = \mathscr{A} + \mathscr{A}_D$, where $\mathscr{A}_D$ is the "belief space", i.e. the inner product space spanned by the elements $P_D(X)$, for each $X \in \mathscr{A}$.

Note: Posterior prevision as projection

Much of the structure of the theory of belief concerns the properties of projection. Thus, to make a first connection notice that what we have termed posterior prevision may be considered as the orthogonal projection operator from $\hat{\mathscr{A}}$ into $\mathscr{A}_D$.

First, it is clear that P_D can be considered as a linear operator, as if $Y, Z \in \hat{\mathscr{A}}$, $P_D(Y + Z) = P_D(Y) + P_D(Z)$. (We do exactly the same with prevision itself. Prevision is a linear functional on a set of random quantities and we pass interchangeably from the role as a linear functional to the values for specified random quantities. For posterior prevision, we transform X into a random quantity $P_D(X)$ and P_D is now a linear operator.) Now, suppose $W \in \hat{\mathscr{A}}$. $W = X + V$, where $X \in \mathscr{A}$, $V = P_D(Y)$ for some $Y \in \mathscr{A}$. Thus, $P_D(W) = P_D(X) + P_D(V)$. Now $P_D(V) = V$. For $X \in \mathscr{A}$, note that for any $Z \in \mathscr{A}_D$:

$$\begin{aligned}(X - P_D(X), Z) &= (X - P_D(X), P_D(U)), \quad \text{for some } U \in \mathscr{A},\\ &= 0, \qquad \text{by (5).}\end{aligned}$$

Thus, $X - P_D(X)$ is orthogonal to each element of $\mathscr{A}_D$, i.e. $P_D(X)$ is the orthogonal projection of X into the subspace $\mathscr{A}_D$.

2.4. *Summarising the effect of new information*

The quantities we wish to assess are of the form $P(X, Y)_D$. Working in the space $\hat{\mathscr{A}}$, relation (2) implies $P(X, Y)_D = (X, Y) - (X, P_D(Y))$. However, the space $\hat{\mathscr{A}}$ was introduced to express changes in belief in $\mathscr{A}$. Now, we can express the above relationship purely within the space $\mathscr{A}$ by defining the projection operator $P_{\mathscr{A}}: \hat{\mathscr{A}} \to \mathscr{A}$, which is the orthogonal projection into the subspace $\mathscr{A}$. Thus, we introduce the operator $T: \mathscr{A} \to \mathscr{A}$, defined for $X \in \mathscr{A}$, by $TX = P_{\mathscr{A}} P_D(X)$, and from (2) we have:

$$P(X, Y)_D = (X, Y) - (X, TY).$$

Thus, the operator T summarises the expected change in inner product over $\mathscr{A}$ induced by the data. In Goldstein (1981) it is shown that T is a bounded self-adjoint operator and an argument is put forward to suggest that T will typically be compact. Also explained is the precise way in which the eigenstructure of the operator T summarises the effect on $\mathscr{A}$ of information.

Essentially if T is compact, there is a set of orthonormal eigenvectors $\{X_i\}$ with eigenvalues $\lambda_i \downarrow 0$. Each $X \in \mathscr{A}$ can be written $X = \sum (X, X_i) X_i$, and the amount you expect to learn about each X, i.e. the ratio

$$\frac{P(\|X\|_D^2)}{\|X\|^2} = D(X),$$

can be expressed purely in terms of the values $D(X_i) = 1 - \lambda_i$, and the inner product (X, X_i). Thus, the data D impose an orthogonal coordinate frame on $\mathscr{A}$ which summarises all that you expect to learn.

3. Separating beliefs

In section 2 we discussed some of the organizational principles of the subjectivist theory. What we must now add is the global coherence principle for the theory.

Coherence is essentially the property that if you consider a problem in different ways, then these ways will agree. This is displayed, at the level of individual previsions, as the linearity property, i.e. if $X = Y + Z$, then you may consider your prevision for X directly or by separately considering your previsions for Y and Z. Now the focus of the theory is not on the individual previsions, but on the relationships between the whole collection of prevision statements that you make, and more specifically on the ways in which your collection of prevision statements are modified by fresh evidence. Thus, the fundamental application of coherence within the subjective theory lies in decomposing the entire collection of beliefs, as represented by $\mathscr{A}$, into subcollections of beliefs which have the following coherence property. You may consider your changes in belief over $\mathscr{A}$ directly, or by separately considering your changes in belief over the subcollections of beliefs.

In section 3.1 we discuss the general conditions under which such a separation of beliefs can be made. In section 3.2 we show how these conditions relate to the general problem of incorporating observational evidence explicitly within the subjectivist analysis.

3.1. Separation of beliefs

We define a general separation of beliefs as follows. The inner product spaces, U and V, separate $\mathscr{A}$ with respect to a specified procedure D if $\mathscr{A} \subset U + V$, and for every $X, Y \in \mathscr{A}$,

$$P(X, Y)_D = P(P_u X, P_u Y)_D + P(P_v X, P_v Y)_D.$$

where P_u and P_v are the projection operators into the subspaces U and V, respectively. (For any random quantities W, Z, $(W, Z)_D$ is the value you will assign for the inner product, i.e. $P_D(W - P_D(W))(Z - P_D(Z))$, after having assessed the evidence D. At present $(W, Z)_D$ is a random quantity.) In other words, U and V separate $\mathscr{A}$ with respect to D if considering how your beliefs will change concerning the elements of U and the elements of V, considered separately, uniquely specifies the expected change in your beliefs over $\mathscr{A}$. Intuitively, this corresponds to the belief that the data will provide essentially different kinds of information about the elements of U and of V. For example, if $\mathscr{A}$ contains quantities related to total fuel expenditure over the year, and U and V are summer and winter expenditure, then it is possible that detailed reports on summer and winter weather conditions might be used to separately assess summer and winter fuel expenditure. At the opposite extreme, the data might include the fuel expenditure totals for the year, so that all elements of $\mathscr{A}$ are precisely known and each $X, Y \in \mathscr{A}$ will have $(X, Y)_D \equiv 0$. However, there may still be considerable uncertainty concerning the division of the expenditure into summer and winter costs, so the inner products on U and V will be non-zero. Notice, in particular, that $\mathscr{A}$ will typically be a proper subspace of $U + V$. Also, as for $\mathscr{A}$, we will assume that the elements of U and V each have zero prevision.

The general sufficient condition to ensure that U and V separate $\mathscr{A}$ is given in the following theorem. (The condition is not strictly necessary, unless the stronger condition is imposed that U and V separate every subspace of $U+V$ with respect to D. However, except for artificially constructed examples, the condition will apply in the general case of interest.)

Theorem. A sufficient condition for U and V to separate $\mathscr{A}$ with respect to D is that: (i) U is orthogonal to V; (ii) U_D is orthogonal to V_D; and (iii) $\mathscr{A} \subset U+V$. (As above, U_D and V_D are the inner product spaces spanned by elements $\{P_D(u)\}, \{P_D(v)\}, u \in U, v \in V$, respectively.)

Proof. For any $X, Y \in \mathscr{A}$, from (2) and (5) we have the usual conditional covariance formula:

$$P((X,Y)_D) = P(XY) - P(P_D(X) \cdot P_D(Y)).$$

As $\mathscr{A} \subset U+V$ and U and V are orthogonal, $X = P_u X + P_v X$ and $Y = P_u Y + P_v Y$. Thus,

$$\begin{aligned} P(X,Y)_D &= P((P_u(X)+P_v(X))(P_u(Y)+P_v(Y)) \\ &\quad - P((P_D P_u(X)+P_D P_v(X))(P_D P_u(Y)+P_D P_v(Y)) \\ &= P(P_u(X)P_u(Y)) - P(P_D P_u(X) P_D P_u(Y)) \\ &\quad + P(P_v(X)P_v(Y)) - P(P_D P_v(X))(P_D P_v(Y)) \\ &\qquad\qquad\qquad\qquad (\text{as } U \perp V,\ U_D \perp V_D) \\ &= P((P_u(X), P_u(Y))_D) + P((P_v(X), P_v(Y))_D). \end{aligned}$$

The above theorem provides the conditions under which there is nothing to be gained by considering the two belief systems, U and V, together. This provides a rationale for separating out obviously unrelated problems (such as where U is statements about doctors, V is statements about Richard Nixon). More interestingly, we will see in the next subsection that this lays the foundation for a subjectivist description of statistical practice.

Notice that we can express the separation of beliefs in terms of the relationship between the operator T on $\mathscr{A}$, as defined in section 2.4, and the corresponding operators T_u and T_v on U and V, defined as $T_u = P_u P_D P_u$ and $T_v = P_v P_D P_v$, as follows:

If U and V separate $\mathscr{A}$ with respect to D, then

$$T = P_{\mathscr{A}} P_D = P_{\mathscr{A}}(T_u + T_v). \tag{6}$$

Notice also the effect of the separation of beliefs on the uncertainties of the individual elements of $\mathscr{A}$. Orthogonality of U and V implies:

$$\|X\|^2 = \|P_u X\|^2 + \|P_v X\|^2. \tag{7}$$

Orthogonality of U_D and V_D implies:

$$P(\|X\|_D^2) = P(\|P_u X\|_D^2) + P(\|P_v X\|_D^2). \tag{8}$$

Recall that the effect of data is to "shrink the elements" of $\mathscr{A}$, that $P(\|Y\|_D^2) \leq \|Y\|^2$ for any Y, and that we may judge the value of the information for Y by how much the value of $P(\|Y\|_D^2)$ is reduced below that of $\|Y\|^2$. (This is discussed in Goldstein, 1981.)

Thus, we have separated the uncertainty about X into uncertainties concerning $P_u X$ and $P_v X$, and also separated the expected gain in information about X into the gains in information about $P_u X$ and $P_v X$. For example, if you can assess completely the operator T_u (i.e. the effect of information on U), then you have simplified your assessment problem, as all you must now consider is the effect of new information on V, and for each element $X \in \mathscr{A}$ there is less uncertainty about $P_v X$ than about X (as $\|P_v X\| \leq \|X\|$). Furthermore, if $P(\|P_u X\|_D^2)$ is much smaller than $\|P_u X\|^2$ you can already assert that the data will be informative for X, as all that a consideration of V can produce is additional information, i.e. a further reduction in $\|X\|$.

3.2. Separation of beliefs by observable random quantities

In the introduction we quoted Fisher's dictum that statements of uncertainty must take into account all of the evidence. However, we will now argue that you can validly discuss the information from limited aspects of the evidence (namely that which you are prepared to anticipate and enumerate), without pretending that this is the totality of the evidence on which you intend to base your conclusions. Informally, we might say that although a statement of fact may only be fully understood in the context in which it arises, it is meaningful to consider the fact and the context separately. This meaning derives from a systematic application of the separation of beliefs.

To consider a specific example, suppose that a doctor is about to examine a patient. The space $\mathscr{A}$ contains various quantities related to the health of the patient. The proposed examination is complicated, perhaps open ended, but the doctor can presumably identify certain quantities, such as temperature, blood pressure, blood group, which he will measure during the examination. Thus, suppose that the doctor identifies a set B of random quantities whose values will definitely be revealed before the revised previsions over $\mathscr{A}$ are announced. Thus, B might by Y_1 (temperature), Y_2 (blood pressure) and a series of indicator functions $Y_3, \ldots$ (for the various blood groups). Call these random quantities logical components of D, as it is a logical, not a subjective, truth that these values will be revealed. (See the note at the end of this subsection on how logical components can be formed in general.)

The role of B in your revision of belief is as follows. If you complete the inner product on the elements of B and $\mathscr{A}$, then this will generate a separation of beliefs over $\mathscr{A}$. Thus, suppose that you adjust each $Y \in B$ so that $P(Y)=0$. Suppose for each Y_i and $Y_j \in B$, $X \in \mathscr{A}$ that you specify $P(Y_iY_j)$ and $P(XY_i)$. You have now created two new inner product spaces; first $\mathscr{B}$, the space spanned by the elements of B, with inner product as for $\mathscr{A}$, and secondly the space $\mathscr{A} + \mathscr{B}$. The separation of beliefs is as follows.

Theorem. If $\mathscr{B}$ is spanned by logical components of D, $\mathscr{A}^* = \mathscr{A} + \mathscr{B}$, and $\mathscr{B}^c$ is the orthogonal complement of $\mathscr{B}$ in the space $\mathscr{A}^*$, then $\mathscr{B}$ and $\mathscr{B}^c$ separate $\mathscr{A}$ with respect to D.

Proof. Clearly $\mathscr{A} \subset \mathscr{B} + \mathscr{B}^c$. All that we must show is that $\mathscr{B}_D$ is orthogonal to $(\mathscr{B}^c)_D$. For any element $b \in \mathscr{B}$, the value of b will be known with certainty when your revised previsions are announced.

Thus, for $b \in \mathscr{B}$, $P_D(b) = b$, so that for any $c \in \mathscr{B}^c$:

$$\begin{aligned} P(P_D(b)P_D(c)) &= P(bP_D(c)) \\ &= P(P_D(bc)) \\ &= P(bc) = 0, \end{aligned}$$

as b and c are orthogonal. Thus, $\mathscr{B}_D, (\mathscr{B}^c)_D$ are orthogonal and the result follows.

We may thus give an unambiguous interpretation of the reduction in uncertainty over $\mathscr{A}$ from observing $\mathscr{B}$. Initially your uncertainty concerning each $X \in \mathscr{A}$ consists of two parts: the uncertainty of that component of X in $\mathscr{B}$, and the uncertainty of X orthogonal to $\mathscr{B}$. Observing the values of all elements of $\mathscr{B}$ will reduce the uncertainty of the first component to zero (i.e. $\|P_{\mathscr{B}}X\|_D \equiv 0$ in eq. (8) above). What you must still assess is the reduction in uncertainty orthogonal to $\mathscr{B}$(i.e. $\|P_{\mathscr{B}^c}X\|_D$). You must assess this subjectively, when you have examined all the evidence D, in precisely the same way that you would assess prior uncertainty. However, you are now considering quantities $\{P_{\mathscr{B}^c}X\}$, for each of which there is less prior uncertainty (i.e. $\|P_{\mathscr{B}^c}X\| < \|X\|$).

Essentially you are writing, for each $X \in \mathscr{A}$,

$$P_D(X) = P_{\mathscr{B}}(X) + Z, \tag{9}$$

where Z is a random quantity satisfying $P(Z)=0$ (as $PP_D(X) = P(P_{\mathscr{B}}(X)) = 0$), and $P(YZ)=0$ for all $Y \in \mathscr{B}$. Eq. (9) expresses a relationship similar to that of the general linear model, in that the various quantities of interest (i.e. the quantities $P_D(X)$) are linear combinations of various observable quantities Y_i and a further quantity Z, uncorrelated with each Y_i. An individual, such as a doctor, who makes many similar judgements on the basis of comparable information may make explicit use of the linear model structure to relate past "diagnoses" and the corresponding Y_i values, via exchangeability, to his current judgements. (I hope to consider this problem in more detail in a further paper.) Of course, this does not mean that there is a "true" linear model. This is simply an efficient arrangement of your belief statements.

It is clear that the more quantities you introduce into $\mathscr{B}$, the larger will be the value $\|P_{\mathscr{B}}X\|$, so the more information you will receive. However, each quantity that you enter into $\mathscr{B}$ involves a completion of the inner product to include that quantity, which may involve considerable effort on your part. All that the subjectivist theory states is that you may enter as many quantities into $\mathscr{B}$ as you find helpful, and that the completion of the inner product automatically will define a separation of belief. Thus, for example, if Y was temperature of the patient, you might introduce Y, Y^2, and $I_{(Y>c)}, I_{(Y<d)}$, for some constants c and d. You might feel able to introduce far more functions of Y. What you should not do is to specify some convenient but ad hoc joint probability distribution for Y and all the elements of $\mathscr{A}$ as an artificial device simply in order to claim that every function of Y is in $\mathscr{B}$. Your assertion of a prevision statement is a fact which should not be falsified in the same way that the data itself should not be falsified.

Notice, finally, that in the decomposition of T given in eq. (6) we may now give the explicit form for that part of T accounted for by $\mathscr{B}$. The operator $T_{\mathscr{B}} = P_{\mathscr{B}}P_D P_{\mathscr{B}} = P_{\mathscr{B}}$ in this case, so that the operator which represents the information from "observing $\mathscr{B}$" is simply $T^* = P_{\mathscr{A}}T_{\mathscr{B}} = P_{\mathscr{A}}P_{\mathscr{B}}$.

Recall that the eigenstructure of T^* conveys the information about the elements of $\mathscr{A}$ which you expect to obtain by observing $\mathscr{B}$. (See Goldstein, 1981, section 5.) Note that for computational purposes it is often more straightforward to assess the eigenstructure of $T^{**} = P_{\mathscr{B}}P_{\mathscr{A}}\mathscr{B} \to \mathscr{B}$.

T^* and T^{**} have the same eigenstructure, as if

$$T^{**}a = \lambda a, \quad \lambda \neq 0,$$

then

$$P_{\mathscr{A}}T^{**}a = (P_{\mathscr{A}}P_{\mathscr{B}})P_{\mathscr{A}}a = T^*P_{\mathscr{A}}a = \lambda P_{\mathscr{A}}a$$

so that the non-zero eigenvalues of T^* and T^{**} are the same. Furthermore, the eigenvectors of T^{**} are simply the projections of the eigenvectors of T^* and vice versa. In particular, this simplifies the assessment if $\mathscr{B}$ is of considerably smaller dimension than $\mathscr{A}$ (i.e. you are learning about many quantities, from limited observation). A special case of this interchange of operators was given in Goldstein (1981, section 7.2) for the case where $\mathscr{A}$ contained all functions of an "unknown" probability distribution and $\mathscr{B}$ was spanned by polynomials, of order n, in a sample observation.

Notice, in particular, how we naturally pass between posterior beliefs and projection into subspaces. We will complete the identification in section 4, when we incorporate conditional beliefs into the structure.

A note on observable data

One of the most important, though most often ignored, aspects of the concept of a random quantity is the requirement that in principle the value of the quantity may be determined by some process. This in itself leads to subtle problems of definition as to what is meant by determinable. These problems are compounded in situations where certain random quantities are mutually exclusive in the sense that if the value of X is determined, then the value of Y cannot be determined. For example, both X and Y might be determined by testing a certain object to destruction, though in different ways to determine each of X and Y. Does it now make sense to consider the prevision of the product XY?

These issues are central to our development as we have introduced a set of random quantities whose values will be determined before a further set of revised previsions are revealed. In order to develop our results concerning the separation of beliefs induced by the evidence, it was a necessary requirement that for each such quantity Y, $P_D(Y) = Y$. This is not a statement of belief but a statement asserting a logical identity between two specified random quantities. Thus, it must be a logical truth that the value of Y will be revealed. The problem may be illustrated as follows. Suppose that a doctor, in his prospective examination of the patient, believes that he will measure Y, the blood pressure, defined in some operational way, of the patient. Now it is clear that he cannot be logically sure that he will measure the patient's blood pressure. He might forget, or any number of other factors might intervene. All he can logically assert is that at the end of the examination one of two events will have occurred. First, he may have made such a measurement so that the value of Y is known. Call this event H_y. Alternatively, he may not have made the measurement (event $\bar{H}_y$). Thus, a random quantity whose value will definitely be known at the time that previsions are revised is the quantity Y_d which is defined to be Y if H_y occurs and zero if $\bar{H}_y$ occurs. We may use the notation $Y_d = YH_y$ (H_y is here the indicator function for the event H_y) as long as it is understood that this is simply a helpful way to think about Y_d. The quantity Y, the blood pressure measurement that might or might not be made, is a concept. The act of "deadlining" Y, by setting it to zero if it is not observed before some specified circumstance, turns the concept of blood pressure into the random quantity, Y_d, satisfying our logical requirements.

It is now clear that the previous requirement that the value of the random quantity will be certainly revealed may be applied quite generally to any quantity that the doctor thinks he might measure during the examination. If Z is any random quantity, the deadlined version $Z_d = ZH_Z$, where H_Z is as defined above, will be certainly revealed. Furthermore, our problems about the compatability of observations have disappeared. For example, if U and V are reactions to two different drugs, which cannot both be administered in the same day, and the doctor is not yet sure which drug he will administer, then the deadline random quantities, U_d and V_d, will both be certainly revealed. Furthermore, the product of U_d and V_d is a well-defined quantity, namely zero.

4. Conditional prevision and posterior prevision

The considerations of section 3 seem to be different from Bayesian theory as usually conceived. In Bayesian theory you usually make a variety of prior probability statements and conditional probability, or likelihood statements, and calculate various posterior probability statements using Bayes' Theorem. This is claimed to express the way that your beliefs will change if you observe the corresponding evidence. This claim is so central to the methodology that it is important to consider carefully the relationship between the conditional probabilities resulting from Bayes' Theorem and the actual beliefs you will hold after you have considered the data.

The logical relationship is clear. de Finetti (1972, pp. 193–194) distinguishes two concepts which could be used in interpreting $P(E \perp H)$. First, it might be "assumed" knowledge, i.e. $P(E|H)$ is the probability that we would regard as fair for a bet on E to be made immediately but to become operative only if H occurs. Secondly, H might be acquired knowledge, so that $P(E|H)$ represents a bet made on E, after H has been learned, on terms that then appear fair. de Finetti discusses these two concepts. He concludes as follows: "What emerges is this: only the predictive interpretation (according to which H is a proposition assumed, not acquired) is free of inextricable difficulties."

The logical difficulty is that the requirement that you observe the event H, and nothing else, cannot be given a substantive meaning. (Does "and nothing else" refer to an absence of further thought or of further evidence, and is this absence purely fortuitous or by design?) Even this consideration is secondary, however, for in practice it is always clear that we observe more than any partition.

Now, in most applications of Bayesian theory the predictive interpretation is taken by verbal juggling to represent actual posterior belief based on acquired information. We are being led astray, because we are trying to ascribe a logical status to the relationship between beliefs stated before and after the observation of information, whereas this relationship exists in the realm of belief (i.e. prevision) itself.

We will now derive the precise relationship between conditional previsions and actual posterior beliefs. This will turn out to be simply a special case of the relationships derived in section 3. Thus, Bayesian theory, as carefully interpreted, is seen not as an alternative to our suggested procedures, but simply as a special case, of no more or less intrinsic interest than any other case.

4.1. Reformulating the definition of conditional prevision

For any random quantity X, and event H, your prevision of X conditional on H, $P(X|H)$, is the value $\bar{x}$ you would choose if, having made this choice, you were to suffer a penalty or loss HL, where $L=(X-\bar{x})^2/k^2$, and $H=1$ if H is revealed to be true, $H=0$ if false (de Finetti, 1974). It is puzzling that we should need such a definition. The motivating force behind our choice of prevision rather than probability as the starting point for the theory is that we do not want to consider random quantities as reducible to some ultimate partition of possible values, and we certainly do not want to enforce a requirement that beliefs about X must be expressed in terms of beliefs about the partition. In short, events, as represented by indicator functions, are no more or less special than any other random quantities.

Why should we now require a definition which makes events special again? The arguments against requiring evidence to be reducible to a partition are considerably stronger than the arguments against requiring the primary random quantities to be so reducible. The random quantities for which you express previsions can, within limits, be selected by you as quantities which, if you wish, you may represent in terms of partitions. However, you have no similar control over the nature of the evidence which will cause you to change your beliefs, as this is not a selection from the evidence you observe, but the totality of this evidence.

Is the new distinction between events and random quantities fundamental? We may ask this more precisely as follows. Does the definition of conditional prevision embody a new concept, quite distinct from prevision (and meaningful only for events)? Or, is the new definition simply an implicit manipulation of the properties already inherent in prevision (and thus extendible to any random quantities)? It may seem that the first of these possibilities is indeed the case, as we appear to have introduced an essentially new penalty function by considering HL instead of L. However, this cannot be the case. The fundamental property of conditional prevision that de Finetti demonstrates is that $P(HX)=P(H)P(X|H)$, i.e. coherence implies an explicit relationship between a conditional prevision and certain related previsions. But, in de Finetti (1974, ch. 3), it is shown that a coherent set of previsions only determines uniquely the previsions of random quantities within the linear span of the set of random quantities over which prevision has already been declared.

Thus, we need only to determine the (linear) operation within the concept of prevision itself upon which the definition of conditional prevision is implicitly based. We may then base our definition explicitly upon this construction and extend it immediately to all random quantities, eliminating the special status of events.

The key to reformulating the definition of conditional prevision is simply to notice that while you appear to be considering a single penalty $L=H(X-\bar{x})^2$, there is also, implicitly, a second penalty for you to consider, namely $L_c=H_c(X-\bar{x}_c)^2$, where H_c is the complement of H.

Thus, an equivalent requirement would be that you must specify two values, $\bar{x}$ and $\bar{x}_c$, and you will incur a penalty

$$L^* = H(X - \bar{x})^2 + H_c(X - \bar{x}_c)^2.$$

Using the relations $H + H_c = 1$, $HH_c = 0$, $H^2 = H$, and $H_c^2 = H_c$, the above penalty is identical to the penalty

$$L^* = (X - \bar{x}H - \bar{x}_c H_c)^2.$$

The implications of the above penalty function are immediate from the geometric interpretation of prevision as projection as given in the comments in section 2.1.

Within this framework, the prevision of X is the element $P(X)X_0$ which is the projection of X into the subspace of $\mathscr{A}^*$ spanned by the element $X_0 \equiv 1$. From the form of L^*, it is clear that the preferred choice of the values $\bar{x}$ and $\bar{x}_c$ is equivalent to the choice of the element $\bar{x}H + \bar{x}_c H_c$ which is the projection of X into $[H, H_c]$ the subspace of $(\mathscr{A}^* + [H, H_c])$ spanned by H and H_c. Thus, we can write

$$X = \bar{x}H + \bar{x}_c H_c + Y,$$

where Y is orthogonal to H and to H_c. Thus, $(X, H) = \bar{x}(H, H)$ (as $(H_c, H) = (Y, H) = 0$) which gives immediately that $P(XH) = \bar{x}P(H)$ and $P(XH_c) = \bar{x}_c P(H_c)$, which is the rule of conditional probability.

Thus, directly from the definition, conditioning on a partition is simply projection into the subspace spanned by the indicator functions. (This is why conglomerability works over a finite partition $\{H_1, \ldots, H_k\}$. The constant function X_0 is a member of $[H_1, \ldots, H_k]$ the subspace spanned by the indicator functions $H_1, \ldots, H_k$. Thus, if you determine $P(X)$ by projecting X onto $[X_0]$ directly, or by first projecting X into $[H_1, \ldots, H_k]$ and then projecting into $[X_0]$, you will clearly obtain the same result in each case.)

Thus, the definition of conditional prevision for X which appears to involve a partition, may be extended to any set of random quantities B, simply as projecting X into the subspace spanned by the elements of B. This is precisely the procedure we discussed in section 3.2, and the purpose is the same, whether for random quantities or for a partition, namely to induce a separation of beliefs. We now examine the separation of beliefs induced by a partition.

4.2. Bayesian procedures and the separation of belief

We now demonstrate the way in which the Bayesian procedure induces a separation of belief. Thus, suppose that $[H_1, \ldots, H_k] = H$ is a logical partition of D. (That is to say, each H_i is one or zero, $\sum H_i = 1$, and the value of each H_i will be known when $P_D(X)$ is announced. For example, H_i might be the event that the patient is observed to have blood group i, $i = 1, \ldots, k-1$, and H_k might be the event that the blood group is not observed. Again we are identifying the indicator function with the corresponding event.) You now assess the values $P(H_i)$ and $P(H_iX_j)$ for each H_i and $X_j \in \mathscr{A}$. This specification is sufficient to complete the inner product on H and $\mathscr{A}$ in the sense of section 3.2. We may thus construct the composite space $\hat{\mathscr{A}} = \mathscr{A} + \hat{H}$, where $\hat{H}$ is the inner product space spanned by the quantities H_i^* (where $H_i^* = H_i - P(H_i)$). Thus, the spaces $\hat{H}$ and $\hat{H}^c$ (the orthogonal complement of $\hat{H}$ in $\hat{\mathscr{A}}$) separate $\mathscr{A}$ with respect to D.

What does this mean? First we can write each $X \in \mathscr{A}$ as

$$X = V + W,$$

where $V = P_{\hat{H}}(X)$ and $W = P_{\hat{H}^c}(X)$, the orthogonal projections of X into $\hat{H}$ and $\hat{H}^c$, respectively. From our reformulation of conditional prevision in section 4.1 we see that

$$V = \sum_{i=1}^{k} P(X|H_i)H_i,$$

where $P(X|H_i) = P(XH_i)/P(H_i)$. (Strictly, $V = \sum_{i=1}^{k} P(X|H_i)H_i^*$, but as $H_i^* = H_i - P(H_i)$, and $\sum P(X|H_i)P(H_i) = P(X) = 0$, as $X \in \mathscr{A}$, the expressions are the same.) Furthermore, W is orthogonal to $\hat{H}$, i.e. $P(H_iW) = 0$ for each i, so that $P(W|H_i) = 0$ for each event H_i in the partition H.

Secondly, for each $X \in A$,

$$P_D(X) = V + Z,$$

where $P_D(W) = Z$, as $P_D(V) = V$. Again Z is orthogonal to $\hat{H}$, i.e. is independent of the partition H. Thus, your actual posterior belief $P_D(X)$ is the sum of two quantities. The first is numerically equal to the value of the conditional prevision corresponding to the event H_i which is actually observed. The second quantity has zero prevision and is unrelated to the partition.

Finally, the total information gained about $\mathscr{A}$ (i.e. the complete change in the inner product) is the sum of two aspects, namely the information gained by the Bayesian analysis (i.e. assessing each element V) and the information gained by considering quantities unrelated to the partition, each aspect being considered separately.

We may thus conclude that a Bayesian analysis, properly formulated, will always induce a separation of beliefs so that the information gained may be directly related to your actual changes in belief (i.e. if $\|V\|^2 = P(\sum P(X|H_i)H_i)^2 = \sum P^2(X|H_i)P(H_i)$ is large, then $P(\|X\|_D^2)$ will be small as $P(\|X\|_D^2) = \|X\|^2 - \|P_D X\|^2 \leq \|X\|^2 - \|V\|^2$).

However, there is no implication that your conditional beliefs should actually equal your future beliefs. That would correspond to assessing each element Z as zero for no better reason than that $P(Z) = 0$. In other words, we are not claiming a strictly normative role for the subjective theory but simply attributing to it a helpful role which is how, in practice, it is viewed anyway. It is clearly helpful to assess as much of your uncertainty in advance as possible, for reasons of intellectual clarity, future planning and communication of your opinions to others. This is precisely what the theory allows.

It is very important to remember, however, that the present section is simply a particular case of the development of section 3.2. There is nothing "special" about conditioning on events, and all of the properties we have discussed apply equally to the separation of beliefs induced by any set of observable random quantities. Just as the notion of prevision makes no distinction between probability and expectation, the separation of beliefs is equally meaningful when applied to indicator functions of events or to general random quantities.

References

de Finetti, B. (1972) *Probability Induction and Statistics* (Wiley, London).

de Finetti, B. (1974) *Theory of Probability*, Vol. 1 (Wiley, London).

Fisher, R.A. (1956) *Statistical Methods and Scientific Inference* (Hofner, New York).

Goldstein, M. (1981) "Revising previsions: A geometric interpretation", *J.R. Statist. Soc.*, B. 43, 105–130.

Goldstein, M. (1983) "The prevision of a prevision", *J. Amer. Statist. Assoc.*, 78, 817–819.

Chapter 14

BID–ASK SPREADS FOR SUBJECTIVE PROBABILITIES

EDWARD E. LEAMER*

University of California

1. Introduction

de Finetti's most popular contribution may be his demonstration (1937) that prices of lottery tickets must be probabilities in the sense of being non-negative, normalized, and additive set functions. He observes that a bookie who offers to buy and sell lottery tickets for carelessly selected prices can be made a sure loser by a skilled customer who selects the right combination of bets. A system of lottery prices which allows certain losses is said to be "incoherent", and de Finetti demonstrates that prices of lottery tickets with payment equal to one are coherent if and only if they are probabilities.

The demonstration that coherent prices are probabilities can be a remarkably seductive introduction to subjective probability, especially to economists who are not trained to deal with abstract concepts such as "degrees of belief", but who comfortably imagine the monetary value of "life" or "health" or "marriage" to be revealed by decisions they suppose influence same. It is to them unsurprising that individuals value Toyota automobiles differently, and it is correspondingly unsurprising that individuals value lottery tickets differently. Thus, de Finetti's notion of probability as the price of a lottery ticket has the appeal of the concrete, and the sweet aroma of the subjective.

This initial enthusiasm of economists dims with further contemplation, however, since prices and valuations are ordinarily thought to differ. In a competitive market setting in which one can buy or sell an unlimited quantity of a commodity at the going price, individuals will engage in exchange until their marginal

*Supported by NSF Grant SES82-07532. Helpful comments from Daniel Friedman, Marc Robinson, and Peter Walley and many useful discussions with John Riley are gratefully acknowledged.

Bayesian Inference and Decision Techniques, edited by P. Goel and A. Zellner

valuations are equal to the price. There is accordingly little information in the market price about differences in the individuals' pre-exchange valuations. In a non-competitive setting with few buyers, few sellers or both, the rate of exchange can be expected to lie between the buyer's valuation and the seller's valuation, but rarely will coincide with either.

It is therefore to be suspected that the price of a lottery ticket will not equal anyone's true valuation of the ticket. Though de Finetti demonstrates that prices of lottery tickets must be probabilities in a mathematical sense, he chooses to ignore the possibility that observed prices may not reveal true valuations. He writes (1937, p. 102):

> Let us suppose than an individual is *obliged* to evaluate the rate p at which he would be ready to exchange the possession of an arbitrary sum S (positive or negative) dependent on the occurrence of a given event E, for the possession of the sum pS; we will say *by definition* that this number *p is the measure of the degree of probability* attributed by the individual considered to the event E, or, *more simply, that p is the probability* of E... (emphasis added).

The italics have been added to emphasize first that de Finetti assumes the individual is "obliged" to play this game, and second, that the price p is transformed from being a "measure of the degree of probability" to being "the probability". A footnote mentions the problems of non-linear utility and unequal information, and in the latter case observes: "This would bring us to game-theoretic situations". Section 2 amplifies this footnote, and concludes that, viewed as a method of elicitation, de Finetti's betting game is deficient since the price setter is unlikely to reveal his true valuation and since the game may be very costly to the elicitor. Subsequent sections consider other elicitation games, including several in which the subject is asked to announce a bid–ask spread of prices.

In order to comment on the relationship between values and prices it is necessary to select a model which describes how values are translated into prices. The model used in this chapter is the traditional one in which probabilities, utilities, and expected utility maximization are taken as primitives. It is assumed that there are two individuals, "you" and "I", who are playing various elicitation games. Information about θ, the probability of some uncertain event, comes in the form of binomial random variables r_i with parameters θ and n_i. For most of the games, it is assumed that r_0, n_0, n_1, and n_2 are common knowledge, but r_1 is known only to you and r_2 is known only to me. Otherwise you and I use the diffuse prior $f(\theta) \propto (1-\theta)^{-1}\theta^{-1}$. Another interesting possibility which is left to the reader to explore is that you and I have different prior distributions.

In the case of de Finetti's elicitation game, you name a price p_1 at which you would be willing to buy or sell lottery tickets. For most of the other games, you select a price p_1^* at which you would sell lottery tickets, with no commitment also to buy. These games could be altered to allow a buying price as well, but without additional content. The basic issues which are addressed are whether an elicitation game is accurate, in the sense that your optimal choice of p_1 or p_1^* conforms to your expected value $E_1(\theta) = E(\theta | r_0, r_1)$, and whether an elicitation game is economical, in the sense that my expected cost of playing the game is small.

In section 2 it is shown that the de Finnetti elicitation game may treat you unfairly in the sense that you may expect to lose. In that case, the elicitation is not economical since I will have to make a prepayment to induce you to play. The game is also inaccurate since, if you do decide to play, you will select a price which generally will not conform to your true valuation. The literature on the elicitation of probabilities not surprisingly uses other revelation games. Instead of forcing you to announce a price at which you are obligated to buy or sell unlimited amounts of the commodity, I announce a "demand curve" $S(p_2)$ indicating the number of tickets that I am committed to purchase at each price. You are allowed to act like a discriminating monopolist, and your net gain as a function of your selling price p_1^* is the area under the demand curve $\int_{p_2=p_1^*}^{\infty}(p_2 - \theta)S(p_2)\,\mathrm{d}p_2$. Provided your valuation is independent of the quantity sold (i.e. linear utility), you have a clear incentive to announce your true valuation $p_1^* = E_1(\theta)$. Your announced valuation determines the transactions which occur but not the prices, which as Savage (1971) notes, is the essence of Marschak's (1964) value elicitation procedure, further explained in section 3. Vickrey (1961) provides the same result in the context of a discussion of sealed-bid second-price auctions.

Although this game can be expected to elicit your true value of a lottery ticket, it can be quite uneconomical since the conditions under which it is accurate are also the conditions under which I can expect to make payments to you. This elicitation game is therefore limited to settings in which my value of your information exceeds my cost of elicitation, and your expected gain of revealing your information to me exceeds your value of keeping the information private or revealing it to someone else.

If the personal valuations of lottery tickets could be elicited economically and accurately, and if such valuations were stable over time, Bayesian inference would be the only proper method of private and public inference. But the elicitation schemes just mentioned can be both uneconomical and inaccurate; moreover, as even Savage (1971, p. 20) remarks, subjects from whom valuations are being elicited "experience wavering and indecision that cannot be identified with mere indifference". There is accordingly an uncomfortable gap between theorists who assume sharp, stable probabilities which can be elicited economically and accurately, and practitioners who cannot seem to find them. Bayesian critics who use the epithet "personal" need not be taken seriously. But the charge that Bayesian inference is "whimsical" because probabilities are fuzzy is very serious indeed, and is the core reason why Bayesian inference is not often used in practice.

Bayesian inference may be rescued from the serious charge of being whimsical if it makes use of intervals of probabilities, since intervals of probabilities can be elicited economically, and can be more stable than sharp probabilities. As a matter of fact, most exchanges of lottery tickets (e.g. insurance contracts) involve bid–ask spreads. Several different models for intervals of probabilities are presented in section 4. These models are used to explore the accuracy and economy of various elicitation games and also to explain why subjects experience wavering and indecision. First, it is observed that elicitation games need not have unique compelling strategic equilibria, and the indeterminacy of strategic equilibrium implies indeterminacy of prices. Hence, an individual who conforms in all respects to the assumptions of statistical decision theory can nonetheless experience "indecision". Secondly, it is noted that my purchase decision partially reveals my information, and you properly anticipate this information by naming a price which is your expected value contingent on making a sale. This price would generally exceed your current expected value by an amount which increases with the amount of my private information. A third model for bid–ask spreads allows you to gather more information if you are willing to bear the sampling costs. The lowest price at which you would sell without gathering more information (e.g. thinking harder) is the asking price or upper probability. The next two models describe wavering caused by variable memory. When naming a selling price at this instant, you will want to avoid undercutting prices which you previously announced or which you will announce in the future when your memory is different. You accordingly select a selling price which is higher than your current valuation. In yet another model with variable memory, you are asked to name a (stable) price at which you will sell now or in the future, though in later periods you may reduce the price if desired. This price, too, will overstate your current valuation. Last, some remarks are made about middlemen who seem to select bid–ask intervals to balance sales against purchases.

The final section offers the suggestion that authors of scientific reports be expected to select bid and asking prices for lotteries contingent on interesting and decidable events, and be required to back up those claims with currency issued by the government or by the journal. The intent of this scheme is to provide a clear incentive for authors to report what they believe, but only what they believe.

Intervals of probabilities have an intellectual history dating at least to Keynes (1921) who rejects measurement error as the source of indeterminacy but instead takes partially ordered probabilities as primitives. In the editorial forward, Braithwaite claims that this notion of indeterminate probabilities "leads to intolerable difficulties without any compensating advantages". In this chapter, intervals of probabilities arise from limitations of elicitation schemes, but sharp probabilities are always assumed to exist. This assumption is made without philosophical commitment and is used primarily because there exists no compelling theory of decision based on indeterminate probabilities which could be used to derive strategies for the elicitation games considered here.

Axioms for upper and lower probabilities are reported in Koopman (1940a and 1940b) and in Good (1962). de Finetti's notion of coherence is used to derive axioms by Smith (1961), Williams (1976), and Walley (1981). Your avoidance of unprofitable exchanges implies few restrictions on your choice of upper and lower probabilities. For example, a selling price in excess of one dollar for a lottery ticket that pays one dollar is "coherent" in the sense of de Finetti, since it does not open up the possibility of sure losses. Sharper results are produced if marketability is imposed as well, i.e. if certainly profitable tickets are excluded. Profitability and marketability are shown by Smith (1961), Williams (1976), and Walley (1981) to imply that the lower (upper) probability is a lower (upper) envelope of probability measures.

A rather different approach is taken by Shafer (1976,1982) who actively markets Dempster's (1967) rule of combination as an alternative to Bayes Rule for lower probabilities (belief functions). This rule is criticized for incoherence by Williams (1978). For a reply see Shafer (1981).

At a more practical level, Chamberlain and Leamer (1976) and Leamer (1978,1982) select families of prior distribution for the normal linear regression problem, and identify the corresponding sets of posterior locations.

2. The elicitation game of de Finetti

When playing de Finetti's elicitation game you announce a price p_1 at which you will buy or sell lottery tickets that pay one dollar if an uncertain event E occurs. I then select S, the number of lottery tickets to be purchased, where negative S means that I sell tickets to you. If event E occurs, you pay S dollars to me, yielding a net reward $(p_1 - 1)S$. If E does not occur, no further payments are made, and you merely keep your receipts $p_1 S$.

Information about $\theta = \Pr(E)$ is assumed to come from the observation of random variables $r_i (i = 0,1,2)$ drawn independently from binomial distributions with parameters θ and n_i. The random variable r_0 is common knowledge, but r_1 is known only to you and r_2 is known only to me. In addition, we are both aware of the sample sizes n_0, n_1, and n_2, and both use the diffuse prior for θ, $f(\theta) \propto \theta^{-1}(1-\theta)^{-1}$. I receive one other piece of information – namely your price p_1. Because of difficulties in modeling the game-theoretic aspects of partial revelation of r_1 by announcement of p_1, the discussion here generally assumes that p_1 is informationless, i.e. the conditional densities satisfy $f(\theta|r_2, r_0) = f(\theta|r_2, r_0, p_1)$. Some discussion is devoted also to the case of complete revelation of r_1, i.e. with $f(\theta|r_2, r_0, p_1) = f(\theta|r_2, r_0, r_1)$.

My expected return as a function of p_1 is $\pi_2(1-p_1)S - (1-\pi_2)p_1 S = (\pi_2 - p_1)S$, where $\pi_2 = E(\theta|r_0, r_2, p_1)$ which is equal to $(r_0 + r_2)/(n_0 + n_2)$ if p_1 is informationless. Since this is linear in S, I will try to buy or sell unlimited amounts of lottery tickets, and expect an unlimited return. This extreme result can be avoided by mechanically limiting the size of S or by assuming risk aversion, with utility a concave function of rewards. If, in particular, my utility is $U = \ln(Y + W)$, where Y is initial wealth and W is the lottery's net return, then my expected utility is

$$E(\ln(Y+W)|r_0, r_2, p_1) = \pi_2 \ln(Y + (1-p_1)S) + (1-\pi_2)\ln(Y - p_1 S),$$

which is maximized if

$$S = Y(\pi_2 - p_1)/(1 - p_1)p_1. \tag{1}$$

By selecting $S = 0$, I can ensure that I am as well off as I would have been had I not played the game. This value of S is selected when the lottery is "fair", $\pi_2 = p_1$, a result which indicates that I have no demand for insurance. A value of $S \neq 0$ reveals that I expect to be made better off, because my evaluation π_2 differs from the price p_1.

You are not necessarily in such an advantageous position. If it is assumed (optimistically) that p_1 is informationless, then your expected gain, conditional on the true probability, θ, is

$$\begin{aligned} E[(-\theta(1-p_1)S + (1-\theta)p_1S)|\theta, p_1] &= E[(p_1 - \theta)S|\theta, p_1] \\ &= YE[(p_1-\theta)(\pi_2 - p_1)/(1-p_1)p_1|\theta, p_1] \\ &= Y(p_1 - \theta)(((r_0 + n_2\theta)/(n_0 + n_2)) - p_1)/(1-p_1)p_1, \end{aligned} \tag{2}$$

which is derived using $\pi_2 = (r_0 + r_2)/(n_0 + n_2)$ and $E(r_2|\theta) = n_2\theta$. Your choice of p_1 maximizes the expected value of (2) with θ coming from a beta distribution with parameters $r_0 + r_1$ and $n_0 + n_1$. The numerator is maximized if p_1 is set to

$$p_1 = \frac{n_0(r_0/n_0) + (n_0 + 2n_2)E_1(\theta)}{n_0 + (n_0 + 2n_2)},$$

which is a compromise between the "common knowledge" estimate r_0/n_0 and your true opinion $E_1(\theta) \equiv E(\theta)|r_0, r_1)$. The effect of the denominator $p_1(1-p_1)$ in (2) is to shrink this value toward one-half to avoid the large losses generated by extreme prices, if the numerator is negative, but to push the value of p_1 away from one-half if the numerator is positive. Your price p_1 is optimally set to your probability, $(r_0 + r_1)/(n_0 + n_1)$, only if your private estimate r_1/n_1 and the common estimate r_0/n_0 are both one-half.

Your optimal p_1 need not imply positive expected gain for you. A bit of tedious algebra produces the following condition for positive expected gain:

$$\left(\frac{r_0}{n_0} - \frac{r_1}{n_1}\right)^2 - 4E_1(\theta)(1 - E_1(\theta))\frac{(n_0 + n_1)^2(n_0 + n_2)n_2}{(n_0 + n_1 + 1)n_0^2 n_1^2} > 0.$$

It is apparent from this expression that you are more likely to profit from this game when:

(1) There is a substantial discrepancy between your private estimate r_1/n_1 and the public estimate r_0/n_0.

(2) There is an adequate amount of public information n_0.

(3) Your private information n_1 is substantial.

(4) My private information n_2 is limited.

(5) Your estimate of θ, $E_1(\theta)$, is extreme.

In summary, if you believe that your price p_1 reveals nothing about your private information r_1, then you will not generally set p_1 equal to your true probability and, even for an optimally chosen price, you cannot generally expect to make money. The same logic applies when p_1 fully reveals r_1 since this is equivalent to your having no private information, $n_1 = 0$. Then you can expect to lose unless I also have no private information, $n_2 = 0$, in which case $p_1 = r_0/n_0$ and no transaction occurs.

3. Marschak's elicitation game

The two defects of de Finetti's elicitation game are that you may refuse to play, and when you do play I cannot expect to elicit your true probability. An alternative which may elicit your true probability is a variant of a sealed-bid second-price auction. Suppose, as in Marschak (1964), you are given the opportunity to submit an offer price to sell a lottery ticket. If your offer price p_1^* is lower than a transaction price p_2, selected by me and currently unknown to you, then you sell the ticket at the price p_2 yielding a profit of $p_2 - \theta$. Otherwise no transaction occurs. This is the same from your standpoint as a sealed-bid second-price auction in which you and I offer to sell a lottery ticket at prices p_1^* and p_2, respectively, and the party offering the lower price makes the sale at the second lowest price.

If you have linear utility, your problem is to select the asking price p_1^* that maximizes your expected profit:

$$\begin{aligned} &E_1(p_2 - \theta | p_2 > p_1^*)\Pr(p_2 > p_1^*) \\ &\quad = \int \left[\int_{p_2 = p_1^*}^{\infty} (p_2 - \theta) f_1(p_2 | \theta) \, \mathrm{d}p_2 \right] f_1(\theta) \, \mathrm{d}\theta \\ &\quad = \int_{p_2 = p_1^*}^{\infty} \left[\int (p_0 - \theta) f_1(\theta | p_2) \, \mathrm{d}\theta \right] f_1(p_2) \, \mathrm{d}p_2 \\ &\quad = \int_{p_2 = p_1^*}^{\infty} [p_2 - E_1(\theta | p_2)] \, f_1(p_2) \, \mathrm{d}p_2, \end{aligned}$$

where $E_1(\theta | p_2)$ is your expected value of θ conditional on the transaction price p_2. If $E_1(\theta | p_2)$ has derivative less than one with respect to p_2, then $p_2 - E_1(\theta | p_2)$ is monotone increasing in p_2, and equal to zero for some value of p_2. The integral over p_2 is maximized by selecting p_1 equal to this value. Thus, your bid price should be a solution to

$$p_1^* = E_1(\theta | p_2 = p_1^*),$$

namely your bid price is equal to your expected value of θ conditional on the transaction price being equal to your bid price. If you believe my transaction price p_2 to be informationless, then $E_1(\theta | p_2) = E_1(\theta)$, and your asking price p_1^* is optimally set equal to your true probability $E_1(\theta)$. In order to use this scheme for the elicitation of your probability, I must therefore convince you that the transaction price is informationless. If it is, I can expect to lose money and my desire to elicit your probability can turn out to be a costly passion.

4. Bid–ask spreads for subjective probabilities

The elicitation schemes just discussed are either inaccurate or uneconomical, or both. The game of de Finetti is inaccurate in the sense that it cannot be expected to elicit your probability, and it may be uneconomical since I may have to make a prepayment to induce you to play. Marschak's elicitation game is accurate only if I can convince you that the transaction price has a distribution independent of θ, but then it is certainly uneconomical since I can expect on the average to make payments to you.

In this section we consider elicitation games in which you are allowed to name separate buying and selling prices. These games will certainly not require me to make a prepayment to you, since you can avoid any possibility of loss by selecting a buying price of zero and a selling price of one. These games turn out to be economical also in the sense that I do not expect to make payments to you when playing the game. The accuracy with which these games elicit your probability is open to question, however. In circumstances identified subsequently they can be accurate.

The point of this section is not only to comment on the usefulness of various elicitation games, but also to offer explanations of the existence of bid–ask spreads (intervals of probabilities). For that reason, in several sections the basic structure of the model is changed. In one case, you are allowed to gather more information, and your interval of prices is the complement of the set of prices at which you would agree to transact without further information. In another case, your memory is variable, and you select a bid–ask spread to control against inconsistent behavior over time. Finally, you are allowed to transact with a sequence of customers, and operate merely as a profit-making middleman.

4.1. Indecision from strategic indeterminacy

One reason why you may have an interval of prices for a lottery ticket is that the elicitation game does not admit a single compelling price-selection strategy. The set of acceptable price-selection strategies will generally imply an interval of acceptable prices. A conservative attitude to this price indeterminacy is to select the lowest price as the buying price and the highest price as the selling price.

For example, suppose, as in section 3, that your offer price p_1^* is a solution to the sealed-bid second-price auction, $p_1^* = E_1(\theta | p_2 = p_1^*)$, where p_2 is my transaction price. Suppose r_i is a binomial random variable with parameters θ and n_i, and suppose r_0, n_0, n_1, and n_2 are common knowledge, but r_1 is known only to you and r_2 is known only to me. If I select the transaction price p_2 in such a way that it reveals my information r_2, $r_2 = \phi(p_2)$, then your expected value is $E(\theta | p_2, r_0, r_1) = (r_0 + r_1 + \phi(p_2))/(n_0 + n_1 + n_2)$, and your optimal bid price is the solution to

$$p_1^* = \left[r_0 + r_1 + \phi(p_1^*)\right] / [n_0 + n_1 + n_2]. \tag{3}$$

However, the solution to this equation depends on the function ϕ. If you thought that I would select the transaction price equal to my expected value, $p_2 = (r_0 + r_2)/(n_0 + n_2)$, then you could solve for r_2 given p_2 as $\phi(p_2) = (n_0 + n_2)p_2 - r_0$. Then inserting this into (3) implies $p_1^* = r_1/n_1$, namely you select p_1^* equal to your private estimate. If, on the other hand, you thought that I would set the transaction price equal to my private estimate, r_2/n_2, then a similar calculation reveals that your optimal bid is your true expected value $(r_0 + r_1)/(n_0 + n_1)$.

In order for you to decide on your offer price, it is therefore clear that you will have to do some thinking about my choice of transaction price. Suppose, for purposes of argument, that the elicitation game is a sealed-bid second-price auction, so that you and I are playing the same game. Then it seems sensible to suppose that each of us uses the same bidding function $p_1^* = f(r_1, n_1, n_2, r_0, n_0)$ and $p_2^* = f(r_2, n_2, n_1, r_0, n_0)$ with inverses $r_1 = \phi(p_1^*, n_1, n_2, r_0, n_0)$ and $r_2 = \phi(p_2^*, n_2, n_1, r_0, n_0)$. If these functions are inserted into (3), we can solve for ϕ:

$$(n_0 + n_1 + n_2)p_1^* = r_0 + \phi(p_1^*, n_1, n_2, r_0, n_0) + \phi(p_1^*, n_2, n_1, r_0, n_0). \quad (4)$$

In the symmetric case, with n_1 and n_2 equal, this implies a unique ϕ: $2\phi = (n_0 + n_1 + n_2)p_1 - r_0$, which implies the offer prices $p_1^* = (r_0 + 2r_i)/(n_0 + n_1 + n_2)$, a special case of a result in Milgrom and Weber (1982). In words, you bid as if you and I had exactly the same private information. In the asymmetric case, with $n_1 \neq n_2$, this logic does not imply a unique choice of ϕ. For example, any ϕ of the following form satisfies (4):

$$\phi = [(n_0 + n_1 + n_2)p - r_0]/2 + g(n_1, n_2)h(n_0, r_0),$$

where h is any function and g satisfies, $g(x, y) = -g(y, x)$ and $g(x, x) = 0$. This implies a very high degree of indeterminacy for the actual bids, since any offer price can be said to be consistent with this notion of equilibrium, provided ϕ is suitably selected. You can confidently surmise that $0 \leq r_2 \leq n_2$ (ϕ must be so restricted), but otherwise your imputed value of r_2 can be anything. A conservative choice for p_1^* is then the minimum price consistent with some form of strategic equilibrium, $p_1^* = (r_0 + r_1 + n_2)/(n_0 + n_1 + n_2)$, where you act as if I observed nothing but favorable experiments. Incidentally, it is not an equilibrium for us both to act in this conservative fashion.

4.2. *Asking prices that anticipate purchase decisions*

Your asking price p_1^* would more efficiently serve your profit-making ambitions if it could be based on my information as well as yours. Although you cannot completely observe my information, my decision to purchase a ticket does reveal my relative optimism, and you may want to select an asking price which anticipates this information. If you do not, you will suffer what is known in the auction literature (e.g. Milgrom and Weber, 1982) as the "winner's curse": contingent on offering the winning bid, you expect to make losses.

A clear example of the correction for winner's curse occurs in a sealed-bid first-price auction, in which you name an offer price p_1^* and I name a reservation price p_2. If your offer price is less than my reservation price, then a sale is made at your price p_1^*, yielding you the profit $p_1^* - \theta$. Your problem is to select p_1^* to maximize

$$E_1(p_1^* - \theta | p_1^* < p_2)\mathrm{Pr}_1(p_1^* < p_2) = [p_1^* - E_1(\theta | p_1^* < p_2)]\mathrm{Pr}_1(p_1^* < p_2).$$

Provided the derivative of $E_1(\theta | p_1^* < p_2)$ with respect to p_1^* is less than one, the value of p_1^* such that $p_1^* = E_1(\theta | p_1^* < p_2)$ divides the set of possible values of p_1^* into low ones, which earn you expected losses, and high ones, which earn you expected gains. The price you select will surely be higher than the breakeven price satisfying $p_1^* = E_1(\theta | p_1^* < p_2)$, and this breakeven price is likely to exceed your true probability $E_1(\theta)$ since the information that my selling price exceeds yours will generally make you revise your opinion upward. The amount by which you will shade your breakeven point over your true probability $E_1(\theta)$ depends on how much information my price p_2 conveys to you. If I have a relatively large amount of private information, and if my reservation price is equal to my probability $E_2(\theta)$, then your breakeven point will be considerably above $E_1(\theta)$.

4.3. *Asking prices to avoid elicitation costs*

The third model of bid–ask spreads for subjective probabilities is based on the assumption that you can, at a cost, decide to gather more information before you announce your buying and selling prices. For example, if I offer you \$100 for your shirt, you will certainly be willing to sell it to me; but if I offer only \$10, you may need to do a bit of thinking before you decide to sell, where thinking may include calling the department store for a price quotation. The lowest price at which you would be willing to sell a lottery ticket, without doing any more thinking, is defined in this section to be the upper probability.

As before, suppose that you have information about $\theta = \mathrm{Pr}(E)$ in the form of the random variables r_0 and r_1. For transaction fee f, you are given the opportunity to sell a lottery ticket with payoff S for the price $p_2 S$. If you desire, before making a decision you may select a sample of size n_1' and observe another binomial random variable r_1' with parameters θ and n_1'. The observation of this random variable is called "thinking" to refer to the possibility that r_1' is stored in your memory. If the price p_2 is high enough, and if the transaction fee f is low enough, your decision will be sufficiently clear that you will sell without further "thinking". The minimum price at which you would sell without "thinking" is your upper probability p_1^*.

If you do not "think", and if you sell the ticket, your expected gain is

$$G_0 = \left(p_2 - \frac{r_0 + r_1}{n_0 + n_1} \right) S - f.$$

If you select sample size n_1', your expected gain is

$$G_1 = \left(Sp_2 - SE_1\left((r_0 + r_1 + r_1')/(n_0 + n_1 + n_1')|H\right) - f\right)\Pr(H) - cn_1',$$

where H is the condition $S(p - (r_0 + r_1 + r_1')/(n_0 + n_1 + n_1')) - f > 0$. If both G_0 and G_1 are negative, you refuse to sell. If G_1 is positive, and if G_0 exceeds G_1, you sell without "thinking". If G_2 is positive, and if G_2 exceeds G_1, then you think before acting. The upper probability is then

$$p_1^*(r_0 + r_1, n_0 + n_1, f/S, c/S) = \min(p_2 | G_0 > 0, G_0 > G_1).$$

The exact dependence of p_1^* on $r_0 + r_1, n_0 + n_1$, f/S, and c/S is complicated, but several qualitative observations can be made.

(a) If the thinking costs c/S are high enough, there will be no integer value of n_1 at which G_1 is positive. Then p_1^* is the minimum value for which G_0 is positive, that is

$$p_1^* = (r_0 + r_1)/(n_0 + n_1) + (f/S).$$

This value of p_1^* can be elicited by a sealed-bid second-price auction of a lottery ticket with stakes sufficiently low that "thinking" is discouraged, but such stakes will also increase f/S and drive p_1^* to one regardless of the true probability $(r_0 + r_1)/(n_0 + n_1)$. I know of no device for making f/S small and c/S large which does not make use of implausible assumptions about the eliciter's knowledge of f and c. There is accordingly no way to elicit $(r_0 + r_1)/(n_0 + n_1)$.

(b) On the other hand, as the stakes S become large, the usefulness of some "thinking" increases, and the value of p_1^* converges to one.

(c) The more informed you are (n_0 large) the less likely are you to need additional information, and consequently the lower is p_1^*, though p_1^* in any case is not less than $((r_0 + r_1)/(n_0 + n_1)) + (f/S)$.

4.4. Wavering from variable memory

Wavering would be a characteristic of an individual whose memory is brief and variable. Suppose at time t you have in your consciousness the information r_0, n_0, r_{1t}, and n_1, where r_{1t} is a binomial random variable with parameters θ and n_1, distributed independently of $r_{1t'}$ for $t \neq t'$. If, as in the de Finetti elicitation game, you were to announce a willingness to buy or sell lottery tickets for the price $(r_0 + r_{1t})/(n_0 + n_1)$ at time t, you would be assured a loss by an opponent who purchased a ticket at one price and later sold it back to you for a higher price. This risk can be minimized if you name an asking price in excess of your current valuation and a bid price lower than your current valuation.

If there are N periods in which prices are announced, the correction that is required to adjust for this form of "wavering" is exactly the same correction for "winner's curse" that a single individual must make if he is bidding against $N - 1$ opponents. Namely the asking price in period t is the expected value of θ given

r_{1t} and given the information that $r_{1t} < r_{1t'}$, for all $t \neq t'$. A minimal amount of coordination is consequently required, since at each instance in time you may act as if you were an independent agent competing with your other "selves" who have appeared before or will appear later.

Suppose that a sealed-bid second-price auction is used to elicit the selling price p_{1t}^* so that in the absence of "wavering and indecision" the current expected value of θ, $(r_0 + r_{1t})/(n_0 + n_1)$, would be the selling price. If p_2 is the informationless hidden offer price with distribution $f(p_2)$, then contingent on θ and selling price p_1^* your expected gain is

$$G(p_1^*, \theta) = \int_{p_2 = p_1^*}^{\infty} (p_2 - \theta) f(p_2) \, \mathrm{d}p_2.$$

If you were to make a single offer price, then the expected value of this function is maximized when $p_1^* = E_1(\theta)$. But your problem is to select a function $p_{1t}^*(r_{1t})$ to maximize $E(G)$, with $p_1^* = \min_t(p_{1t}^*)$. Whatever function p_{1t}^* you select, p_1^* is a random variable with distribution dependent on θ. The joint distribution of p_1^* and θ can equally well be written as $f(p_1^* | \theta) f(\theta)$ or $f(\theta | p_1^*) f(p_1^*)$. If the latter form is selected, the expected value of G can be written as $E(E(G(p_1^*, \theta) | p_1^*))$. The interior expected value is maximized when $p_1^* = E(\theta | p_1^*)$, which is satisfied when $p_{1t}^* = E(\theta | r_t;\ r_t < r_t',$ all $t \neq t')$. This follows from the fact that $p_{1t}^*(r_{1t})$ is monotone increasing in r_{1t}, and $p_1^* = \min(p_{1t}^*)$ is informationally equivalent to $r = \min(r_{1t})$. In words, the correction for variable memory is to report a price at each moment of time equal to your expected value conditional on the fact that your current information is less favorable to the value of the ticket than your information in every other period.

4.5. *Precommitted prices with variable memory*

The model of variable memory in the previous section corrects the current price for the fact that competing offers will be issued in the future at a time when your state of mind may be rather different. The announced prices are more stable than your opinions, but the prices nonetheless waver. Wavering is a most unappealing feature of subjective probabilities, and it is appropriate to define an elicitation game that penalizes price variability. An analogy is a store which publishes list prices in catalogs, but which reserves the right to lower the price if conditions change enough to justify the advertising, labelling, and other costs associated with a price change.

To model this, a sequence of sealed-bid second-price auctions may be assumed with your gain in period t equal to

$$G(p_{1t}^*, \theta) = \int_{p_{2t} = p_{1t}^*}^{\infty} (p_{2t} - \theta) f(p_{2t}) \, \mathrm{d}p_{2t},$$

where p_{2t} is an informationless and hidden transaction price with distribution

$f(p_{2t})$. Currently, you have available the information r_0 and r_1, but at future time t you will have r_0 and r_{1t}, where r_0 is constant memory and r_1 is variable. Your problem is to select a "list price" p_1^* at which you will sell at any time unless you bear a cost c to lower the price (have a sale).

Two alternative models of memory could characterize your behavior in time period t. You could act like an entirely different person and treat p_1^* as informationless, or you could recognize the need for intertemporal coordination and treat p_1^* as if it revealed r_1. The former assumption at first seems questionable since it supposes two different levels of consciousness. Initially, you recognize the need for intertemporal coordination when you select the stable price p_1^*, but thenceforward you lose all awareness of time. An efficiency argument can be made for this kind of memory if intertemporal awareness can be achieved only at some positive cost. When you do suffer the cost of attaining intertemporal awareness, you will want to take steps to assure that decisions in other periods will conform to the dictates of the intertemporal optimization problem. If the adjustment cost c is under your control, you can partially commit yourself in other periods by selecting a non-zero value for c. Among the many examples of this kind of behavior are mandatory payroll deductions for retirement, and zero-interest Christmas clubs in savings banks, and even vengence genes which commit you to apparently irrational acts of violence. Examples of a somewhat different nature are team decisions in which a central coordinator establishes rewards for team members to elicit the best degree of teamwork – if you kill the attacking mastadon, then you can have the best girl in the tribe.

For the problem considered here, your choice of p_1^* and c when you are intertemporally aware must anticipate your later myopic decisions. At the myopic time period t you select the variable price $p_{1t} = E(\theta | r_0, r_{1t})$, instead of the "recommended" price p_1^* if $G(p_{1t}^*, p_{1t}^*) - c$ exceeds $G(p_1^*, p_{1t}^*)$. This occurs when r_{1t} is low enough, say $r_{1t} < r'(c, p_1^*)$. Thus, when you are intertemporally aware you need to select p_1^* and c to maximize

$$E\big(G(p_1^*, \theta) | r_0, r_1, r_{1t} \geq r'\big)\Pr\big(r_{1t} \geq r'\big)$$
$$+ E\big(G(p_{1t}^*, \theta) - c | r_0, r_1, r_{1t} < r'\big)\Pr\big(r_{1t} < r'\big),$$

where the expected value operator refers both to θ and r_{1t}. The choice of p_1^* and c that maximizes this expression is complicated, but it can be expected that c will be large when the current information n_1 is large compared with the future information n_{1t}, since it would then be better to have a committed price than a wavering one. Also, the recommended price p_1^* will be higher than the current expected value $E(\theta | r_0, r_1)$ partially to allow flexibility in the future.

4.6. Bid – ask spreads for intermediaries

Organized markets for lotteries usually are operated by middlemen who select bid and ask prices such that sales and purchases are virtually identical. The middleman then bears no risk and makes profits out of the price spread. The motivation for this balancing of purchases and sales is not so clear-cut as it might seem. Why should the choice of the bid and asked prices be different depending on the number of customers? How is the middleman's decision problem different for the second customer than the first? One possibility is that the middleman is risk averse, and a contract with the first customer opens up risk that might be insured against with the second customer. Another possibility is that the decision of the first customer yields information to the middleman which is used to determine prices for the second customer. For example, a bookie who finds himself always selling but never buying lottery tickets, properly infers that he has undervalued the ticket, and accordingly raises his prices. This will make it appear that he is balancing purchases with sales, though that is neither the goal nor the outcome. In particular, if the middleman is risk neutral, and if he must quote the same prices to all customers, then his bid and ask prices are independent of the number of customers.

In actual market settings, there is a competition to play the role of middleman. Various kinds of institutional arrangements could be imagined to decide the outcome of the competition, but most seem to involve complicated informational transfers which are difficult to model, especially when there are many players in the game.

5. A proposal for scientific reporting

The interpretation of inferences reported in scientific journals is made difficult by the muddled incentives of the authors. An economist, after reviewing the data, forms a judgment about, say, the probability that the Treasury bill rate will rise over a one-year period in which the rate of money growth increases. Readers of research results would like to know what that probability is, but no writer of research results has an incentive currently to reveal it. The probability elicitation devices we have just discussed might be helpful. The writer of a report could be required to select a sharp price to buy or sell lottery tickets à la de Finetti, or name an offer price à la Marschak, or announce a bid–ask spread. But because of the unfairness of de Finetti's game to the price setter, authors are not likely to be forthcoming if they are compelled to buy or sell lottery tickets at a price of their choosing. And journals are unlikely to be willing to play Marschak's game because of the resources needed to support it, and because there is no disincentive to publication by relatively uninformed writers.

The publication of bid–ask spreads for lotteries contingent on interesting and decidable events is feasible, however, and might have a remarkable effect on the quality of empirical work that gets published. Authors would be expected to propose new lotteries or to offer new bid–ask spreads for existing lotteries. Journals would make editorial decisions concerning the degree of interest in new lotteries and would commit journal space only if they anticipate sufficient interest.

Lower selling prices or higher buying prices would be automatically published. If the analysis which led to these prices were sufficiently amusing and if the author were willing, an article describing the method could be published as well.

Subscribers would be allowed to contract with authors at the stated prices, though authors would be free to retract prices at any time, in which case the price reverts to the most recent outstanding price. The journal holds the stakes until the issue is decided, and reports all contracts, and all payments.

The ideal outcome would be a narrow interval of prices accurately representing the available social information. Whether this is the best institutional arrangement for achieving this goal can be questioned since complicated strategic considerations will affect the announcement of the bid and asking prices.

Another problem with this half-baked scheme is that the wealthy, rather than the informed, may "make the market". This is a short-run problem since in the long run resources transfer from those mistaken to those correct. A journal, concerned about initial disparities in wealth among its subscribers and harassed by government officials who generally frown on lotteries, could invent its own currency, say "smarts", and endow initially each subscriber with a thousand smarts. I suppose the subscriber with the most smarts at the end of each year could be awarded a free subscription, though statisticians might wish to test whether the distribution of smarts reveals underlying ability differences or could as well have resulted from flips of a coin.

The great benefit of a scheme like this is it allows clear determination of professional progress as the probability intervals get narrower and move toward one or zero. Empirical work which led to great revisions in the upper and lower probabilities could be highly rewarded. Today we undervalue empirical enterprises, basically because we have no way of measuring their contributions.

To start the ball rolling, I hereby announce my asking price $P^* = 0.02$ and my bid price $P_* = 0.00001$ for the following lottery: "A journal listed in the 1980 Social Science Citation index will adopt by 1990 the scheme just described". These prices are in effect until further notice.

References

Chamberlain, G. and E.E. Leamer (1976) "Matrix weighted averages and posterior bounds", *Journal of the Royal Statistical Society B*, 38, 73–84.

de Finetti, Bruno (1937) "La prevision: Ses lois loqiques, ses sources subjectives", *Annals de l'Institut Henri Poincare*, 7, 1–68; English translation in: Henry E. Kyburg, Jr. and Howard E. Smokler, eds., *Studies in Subjective Probability* (John Wiley and Sons, New York, 1964).

Dempster, A. (1967) "Upper and lower probabilities induced by multivalued mapping", *Annals of Mathematical Statistics*, 28, 325–339.

Good, I.J. (1962) "Subjective probability as the measure of a non-measurable set", in: E. Nagel et al., eds., *Logic, Methodology and Philosophy of Science: Proceedings of the* 1960 *Congress* (Stanford University Press), pp. 319–329.

Keynes, J.M. (1921) *A Treatise on Probability* (Macmillan, London).

Koopman, B.O. (1940a) "The axioms and algebra of intuitive probability", *Annals of Mathematics*, 41, 269–292.

Koopman, B.O. (1940b) "The bases of probability", *Bulletin of the American Mathematical Society*, 46, 763–774.

Leamer, E.E. (1978) *Specification Searches* (Wiley, New York).

Leamer, E.E. (1982) "Sets of posterior means with bounded variance priors", *Econometrica*, 5, 725–736.

Marschak, Jacob (1964) "Actual versus consistant decision behavior", *Behavioral Science*, 9, 103–110.

Milgrom, Paul R. and R.J. Weber (1982) "A theory of auctions and competitive bidding", *Econometrica*, 50, 1089–1122.

Raiffa, H. and R. Schlaifer (1961) *Applied Statistical Decision Theory* (Harvard University Press).

Riley, J. and W. Samuelson (1981) "Optimal auctions", *American Economic Review*, 71, 381–392.

Savage, L.J. (1971) "Elicitation of personal probabilities and expectations", *Journal of the American Statistical Association*, 66, 565–614.

Shafer, G. (1976) *A Mathematical Theory of Evidence* (Princeton University Press).

Shafer, G. (1981) "Constructive probability", *Synthese*, 48, 1–60.

Shafer, G. (1982) "Lindley's paradox", *Journal of the American Statistical Association*, 77, 325–351, with comments by D.V. Lindley, Morris H. DeGroot, A.D. Dempster, I.J. Good, Bruce M. Hill, and Robert E. Kass.

Smith, Cedric A.B. (1961) "Consistency in statistical inference and decision", *Journal of the Royal Statistical Society B*, No. 1, 1–25.

Vickrey, W. (1961) "Counterspeculation, auctions, and competitive sealed tenders", *Journal of Finance*, 16, 8–37.

Walley, P. (1981) "Coherent lower (and upper) probabilities", University of Warwick Discussion Paper.

Williams, P.A. (1976) "Indeterminate probabilities", in: M. Przelecki, K. Szaniawski and R. Wojcicki, eds., *Formal Methods in the Methodology of Empirical Science* (Ossolineum and D. Reidel Publishing Co.), pp. 229–246.

Williams, P.A. (1978) "On a new theory of epistemic probability", *British Journal of Philosophy of Science*, 29, 375–387.

Winkler, Robert J.E. (1969) "Scoring roles and the evaluation of probability assessors", *Journal of the American Statistical Association*, 64, 1073–1078.

Wolfenson, M. and T.L. Fine (1982) "Bayes-like decision making with upper and lower probabilities", *Journal of the American Statistical Association*, 77, 80–83.

Chapter 15

ON ASSESSING PRIOR DISTRIBUTIONS AND BAYESIAN REGRESSION ANALYSIS WITH *g*-PRIOR DISTRIBUTIONS

ARNOLD ZELLNER*

University of Chicago

1. Introduction

In this chapter we consider the normal linear multiple regression model (MRM) and prior information about values of its parameters. As is well known, Bayesian analysis of the MRM when there is little prior information about parameter values has appeared in Jeffreys (1967), Box and Tiao (1973), Zellner (1971), and elsewhere. In cases in which there is prior information about parameter values, prior distributions to represent such information can be and have been employed in analyses of the MRM – see, for example, Raiffa and Schlaifer (1961), Shiller (1973), Leamer (1978), Zellner (1971), Zellner and Geisel (1970), Zellner and Richard (1973), and Zellner and Williams (1973).

Procedures for assessing informative prior distributions for the MRM's parameters have been put forward by Kadane (1980), Kadane et al. (1980), Winkler (1967, 1977), and Zellner (1985). In many instances these procedures will be extremely valuable in enhancing understanding of the particular MRM under consideration and in formulating an appropriate prior distribution for its parameters. In the present chapter another approach for assessing a prior distribution for the parameters of the MRM is put forward. It involves the use of Muth's (1961) rational expectations hypothesis and leads to relatively simple prior distributions that are called "*g*-priors." It will be seen that *g*-priors are a special form of a

*This is a revised version of an earlier paper presented to the Econometrics and Statistics Colloquium, University of Chicago on 27 February 1980 and at the Econometric Society Meeting, Denver, Colorado, September 1980. Research financed by NSF Grant SES 7913414 and by income from the H.G.B. Alexander Endowment Fund, Graduate School of Business, University of Chicago.

Bayesian Inference and Decision Techniques, edited by P. Goel and A. Zellner

natural conjugate prior distribution for the MRM's parameters and can be viewed as reference informative prior distributions.

The plan of the paper is as follows. In section 2 a g-prior distribution is assessed and compared with other prior distributions for the MRM's parameters. The properties of posterior distributions based on a particular g-prior are discussed in section 3. The risk function and average risk of the posterior mean of the regression coefficient vector relative to a quadratic loss function and a g-prior are derived and compared with those of the usual least squares estimator in section 4. In section 5 selected posterior odds ratios involving g-prior distributions are presented and analyzed. Finally, some concluding remarks are presented in section 6.

2. Prior distributions for the multiple regression model (MRM)

The standard MRM can be written as

$$y = X\boldsymbol{\beta} + \boldsymbol{u}, \tag{2.1}$$

where $\boldsymbol{y}$ is an $n\times 1$ vector of observations on the dependent variable, X is an $n\times k$ given matrix, assumed of rank k, $\boldsymbol{\beta}$ is a $k\times 1$ vector of regression parameters with unknown values, and $\boldsymbol{u}$ is an $n\times 1$ vector of disturbance or error terms. It is assumed that the elements of $\boldsymbol{u}$ have been independently drawn from a normal distribution with zero mean and finite unknown variance σ^2. The case of normal error terms will be treated herein, but it should be appreciated that other distributional assumptions can also be utilized – see, for example, Zellner (1976). The likelihood function for the MRM is given by

$$\begin{aligned} l(\boldsymbol{\beta},\sigma|y,X) &\propto \sigma^{-n}\exp\{-(y-X\boldsymbol{\beta})'(y-X\boldsymbol{\beta})/2\sigma^2\} \\ &\propto \sigma^{-n}\exp\{-[\nu s^2+(\boldsymbol{\beta}-\hat{\boldsymbol{\beta}})'X'X(\boldsymbol{\beta}-\hat{\boldsymbol{\beta}})]/2\sigma^2\}, \end{aligned} \tag{2.2}$$

where $\hat{\boldsymbol{\beta}} = (X'X)^{-1}X'y$, $\nu s^2 = (y-X\hat{\boldsymbol{\beta}})'(y-X\hat{\boldsymbol{\beta}})$, and $\nu = n-k$.

We now turn to discuss several prior distributions for $\boldsymbol{\beta}$ and σ. First, in cases in which little is known about the values of $\boldsymbol{\beta}$ and σ, Jeffreys (1967, pp. 147ff.) recommends the following "diffuse" or "vague" prior distribution, which is widely employed – see, for example, Lindley (1965), DeGroot (1970), Zellner (1971), and Box and Tiao (1973):

$$p(\boldsymbol{\beta},\sigma)\propto 1/\sigma, \tag{2.3}$$

that is, the elements of $\boldsymbol{\beta}$ and $\log\sigma$ are assumed independently and uniformly distributed.[1] The prior in (2.3) can be obtained from Jeffreys' (1967, p. 180) general rule, $|\text{Inf}|^{1/2}$ for generating invariant diffuse prior distributions, where

[1] Jeffreys (1967, p. 122) explains that the ranges of the parameters in (2.2) must be finite in order to maintain his Convention 3. However, the ranges can be very large so that using $-\infty<\beta_i<\infty$, $i=1,2,\ldots,k$, and $0<\sigma<\infty$ in integrations will usually provide accurate results.

Inf is the Fisher information matrix. In the present instance the rule is applied separately to $\boldsymbol{\beta}$ and to σ to yield (2.3). Also, the diffuse prior in (2.3) is produced by the maximal data information procedure described in Zellner (1971, ch. 2, and 1977).

If we consider combining the diffuse prior in (2.3) with the likelihood function in (2.2) *before* we take the data, then we know that the posterior distribution for $\boldsymbol{\beta}$ given σ will be normal with mean $\hat{\boldsymbol{\beta}} = (X'X)^{-1}X'y$ and covariance matrix $(X'X)^{-1}\sigma^2$. Thus, we know beforehand that the elements of $\boldsymbol{\beta}$ will be correlated in the posterior distribution in the usual case in which $X'X$ is not a diagonal matrix and know the form of the covariance matrix.[2]

A second prior distribution for $\boldsymbol{\beta}$ and σ is the natural conjugate prior distribution of Raiffa and Schlaifer (1961). It is given by

$$p(\boldsymbol{\beta},\sigma) = p_{\mathrm{N}}(\boldsymbol{\beta}|\sigma)p_{\mathrm{IG}}(\sigma), \tag{2.4a}$$

with

$$p_{\mathrm{N}}(\boldsymbol{\beta}|\sigma) \propto \sigma^{-k}\exp\{-(\boldsymbol{\beta}-\bar{\boldsymbol{\beta}})'A(\boldsymbol{\beta}-\bar{\boldsymbol{\beta}})/2\sigma^2\} \tag{2.4b}$$

and

$$p_{\mathrm{IG}}(\sigma) \propto \sigma^{-(\nu_0+1)}\exp\{-\nu_0 s_0^2/2\sigma^2\}, \tag{2.4c}$$

where $p_{\mathrm{N}}(\boldsymbol{\beta}|\sigma)$ is a multivariate normal pdf for $\boldsymbol{\beta}$ given σ with prior mean $\bar{\boldsymbol{\beta}}$ and prior covariance matrix $A^{-1}\sigma^2$ (the matrix A is assumed to be positive definite symmetric) and $p_{\mathrm{IG}}(\sigma)$ in (2.4c) is in the inverted gamma form with positive parameters ν_0 and s_0^2.

If (2.4) is combined with the likelihood function in (2.2), the posterior pdf for $\boldsymbol{\beta}$ and σ is of well-known form. In particular, the posterior mean for $\boldsymbol{\beta}$, denoted by $\bar{\bar{\boldsymbol{\beta}}}$ is given by

$$\begin{aligned}\bar{\bar{\boldsymbol{\beta}}} &= (A+X'X)^{-1}(A\bar{\boldsymbol{\beta}}+X'X\hat{\boldsymbol{\beta}})\\ &= \bar{\boldsymbol{\beta}} + \left[I-(A+X'X)^{-1}A\right](\hat{\boldsymbol{\beta}}-\bar{\boldsymbol{\beta}}),\end{aligned} \tag{2.5}$$

where $\hat{\boldsymbol{\beta}} = (X'X)^{-1}X'y$ and $\bar{\boldsymbol{\beta}}$ is the prior mean of $\boldsymbol{\beta}$.[3] Thus, $\bar{\bar{\boldsymbol{\beta}}}$ can be viewed as a "shrinkage" estimate. Also, the relation of $\bar{\bar{\boldsymbol{\beta}}}$ to "ridge-regression" estimates is well known. Furthermore, it is known that $\bar{\bar{\boldsymbol{\beta}}}$, viewed as an estimator, is optimal relative to quadratic loss functions in the sense that it minimizes average risk and thus is admissible – see Giles and Rayner (1979) for an analysis of the sampling properties of $\bar{\bar{\boldsymbol{\beta}}}$ relative to those of $\hat{\boldsymbol{\beta}}$. In addition, in large samples, the posterior pdf for $\boldsymbol{\beta}$ is normal, centered at $\hat{\boldsymbol{\beta}}$, the maximum likelihood estimate, and has a covariance matrix proportional to $(X'X)^{-1}$.

[2] If the elements of X can be chosen by an investigator, he can control the form of the posterior covariance matrix for $\boldsymbol{\beta}$.

[3] Using the natural conjugate prior distribution in (2.4), the posterior covariance matrix for $\boldsymbol{\beta}$ is proportional to $(A+X'X)^{-1}$. Since the elements of A are of order 1, while those of $X'X$ are usually of order n, the form of $X'X$ will have a pronounced influence on the form of the posterior covariance matrix for $\boldsymbol{\beta}$. See Zellner and Williams (1973) for an application illustrating this point.

While procedures for assessing the natural conjugate prior in (2.4) have been developed – see Winkler (1967, 1977), Kadane et al. (1980) and Zellner (1985) – for many regression models it is difficult to evaluate the prior covariances for the elements of $\boldsymbol{\beta}$. For such situations and others, we propose the following procedure for assessing a reference informative prior distribution. Before observing $\boldsymbol{y}$, consider a "conceptual" or "imaginary" sample, $\boldsymbol{y}_0$, an $n\times 1$ vector, assumed generated by

$$\boldsymbol{y}_0 = X\boldsymbol{\beta} + \boldsymbol{u}_0, \tag{2.6}$$

where $\boldsymbol{u}_0$ is assumed drawn from $N(\boldsymbol{0}, \sigma_0^2 I_n)$. In (2.6), X is the same design matrix appearing in (2.2) since it is assumed that the MRM is appropriate for the given values of the independent variables in X. However, since $\boldsymbol{y}_0$ is a conceptual sample, we allow σ_0^2 to be different from σ^2 and assume $\sigma^2 = g\sigma_0^2$, $0 < g < \infty$, where g will initially be assumed given. Then $\sqrt{g}\,\boldsymbol{u}_0 = \boldsymbol{\varepsilon}$ is $N(0, \sigma^2 I_n)$ and with a diffuse prior for $\boldsymbol{\beta}$ and σ, $p(\boldsymbol{\beta}, \sigma) \propto 1/\sigma$, the "posterior" pdf based on the conceptual sample in (2.6) is:

$$\begin{aligned} p(\boldsymbol{\beta}, \sigma | D_0) &\propto \sigma^{-(n+1)} \exp\{-g(\boldsymbol{y}_0 - X\boldsymbol{\beta})'(\boldsymbol{y}_0 - X\boldsymbol{\beta})/2\sigma^2\} \\ &\propto \sigma^{-(n+1)} \exp\{-g[\nu s_0^2 + (\boldsymbol{\beta} - \hat{\boldsymbol{\beta}}_0)'X'X(\boldsymbol{\beta} - \hat{\boldsymbol{\beta}}_0)]/2\sigma^2\}, \end{aligned} \tag{2.7}$$

where D_0 denotes the sample information in (2.6) and the diffuse prior assumptions, $\nu = n - k$, $\hat{\boldsymbol{\beta}}_0 = (X'X)^{-1}X'\boldsymbol{y}_0$ and $\nu s_0^2 = (\boldsymbol{y}_0 - X\hat{\boldsymbol{\beta}}_0)'(\boldsymbol{y}_0 - X\hat{\boldsymbol{\beta}}_0)$.

Now suppose that we have given, anticipated values for $\boldsymbol{\beta}$ and σ^2, denoted by $\boldsymbol{\beta}_a$ and σ_a^2, respectively, in addition to (2.6). Application of Muth's (1961) rational expectations hypothesis leads us to take

$$\boldsymbol{\beta}_a = E(\boldsymbol{\beta} | D_0) = \hat{\boldsymbol{\beta}}_0 \tag{2.8}$$

and

$$\sigma_a^2 = E(\sigma^2 | D_0) = \nu g s_0^2/\nu - 2, \tag{2.9}$$

where $E(\boldsymbol{\beta}|D_0)$ and $E(\sigma^2|D_0)$ are posterior means derived from (2.7). From (2.9), $g s_0^2 = (\nu - 2)\sigma_a^2/\nu \equiv \bar{\sigma}_a^2$ and the joint reference informative g-prior distribution is:

$$\begin{aligned} p(\boldsymbol{\beta}, \sigma | \boldsymbol{\theta}_0) &\propto \sigma^{-(\nu+1)} \exp\{-\nu\bar{\sigma}_a^2/2\sigma^2\} \\ &\quad \times \sigma^{-k} \exp\{-g(\boldsymbol{\beta} - \boldsymbol{\beta}_a)'X'X(\boldsymbol{\beta} - \boldsymbol{\beta}_a)/2\sigma^2\}, \end{aligned} \tag{2.10}$$

where $\boldsymbol{\theta}_0' = (\boldsymbol{\beta}_a', \bar{\sigma}_a, g, \nu)$. The marginal prior pdfs for σ and $\boldsymbol{\beta}$ are:

$$p(\sigma | \bar{\sigma}_a, \nu) \propto \sigma^{-(\nu+1)} \exp\{-\nu\bar{\sigma}_a^2/2\sigma^2\} \tag{2.11}$$

and

$$p(\boldsymbol{\beta} | \boldsymbol{\beta}_a, g, \nu) \propto \{\nu\bar{\sigma}_a^2 + g(\boldsymbol{\beta} - \boldsymbol{\beta}_a)'X'X(\boldsymbol{\beta} - \boldsymbol{\beta}_a)\}^{-(\nu+k)/2}. \tag{2.12}$$

It is seen that the joint prior in (2.10) is in the inverted-gamma-normal form with the marginal prior for σ in (2.11) inverted gamma and that for $\boldsymbol{\beta}$ in (2.12) multivariate Student-t. From (2.12), $E\beta = \beta_a$ and $E(\beta - \beta_a)(\beta - \beta_a)' = (X'X)^{-1}\nu\bar{\sigma}_a^2/g(\nu - 2) = (X'X)^{-1}\sigma_a^2/g$ and from (2.11), $E\sigma^2 = \nu\bar{\sigma}_a^2/(\nu - 2) = \sigma_a^2$.

Thus, given β_a, σ_a^2, (2.6) and use of rational expectations, the complete prior for β and σ in (2.10) has been assessed and is in the natural conjugate form (2.4) with $\bar{\beta} = \beta_a$, $A = gX'X$, $\nu_0 s_0^2 = \nu\bar{\sigma}_a^2$ and $\nu_0 = \nu$. Furthermore, note that:

(1) When g's value is unknown, a prior pdf for g, uninformative or informative, say, in the gamma form, can be introduced and g can be integrated out.

(2) One can rewrite (2.6) as $y_0 = X_0\beta + u_0$, with y_0 and u_0 each of dimension $n_0 \times 1$ and X_0 of dimension $n_0 \times k$ and proceed to derive a prior pdf for the regression parameters. This involves the assumption that the form of the regression model is the same for both design matrices X_0 and X.

(3) It is possible to have $g = g(n)$, e.g. $g \propto 1/n$ or $g \propto \log n/n$ assumptions that control the dependence of the prior precision on n.

Having indicated how to assess g-prior distributions, we now turn to discuss properties of posterior distributions based on them.

3. Posterior distribution for β and σ based on a g-prior distribution

In this section we employ a particular g-prior pdf, namely $p(\beta, \sigma) = h(\sigma)f(\beta|\sigma, g)$, with $h(\sigma) \propto 1/\sigma$ and $f(\beta|\sigma, g) \propto \sigma^{-k}\exp\{-g(\beta - \bar{\beta})'X'X(\beta - \bar{\beta})$, to obtain the following joint posterior pdf for β and σ:

$$\begin{aligned} p(\beta, \sigma|D) &\propto p_g(\beta, \sigma)l(\beta, \sigma|y) \\ &\propto \sigma^{-(n+k+1)}\exp\{-[y - X\beta)'(y - X\beta) \\ &\qquad + g(\beta - \bar{\beta})'X'X(\beta - \bar{\beta})]/2\sigma^2\}, \end{aligned} \tag{3.1}$$

where D denotes the data and prior information. If we let $w' = (y' \vdots g^{1/2}\bar{\beta}'X')$ and $W' = (X' \vdots g^{1/2}X')$, then the terms in square brackets in the exponential can be expressed as

$$(w - W\beta)'(w - W\beta) = (w - W\bar{\bar{\beta}})'(w - W\bar{\bar{\beta}}) + (\beta - \bar{\bar{\beta}})'W'W(\beta - \bar{\bar{\beta}}),$$

where $\bar{\bar{\beta}} = (W'W)^{-1}W'w$. Thus, (3.1) can be expressed conveniently as

$$\begin{aligned} p(\beta, \sigma|D) \propto \sigma^{-(n+k+1)}\exp\{-[(w - W\bar{\bar{\beta}})'(w - W\bar{\bar{\beta}}) \\ + (\beta - \bar{\bar{\beta}})'W'W(\beta - \bar{\bar{\beta}})]/2\sigma^2\}, \end{aligned} \tag{3.2}$$

where

$$\bar{\bar{\beta}} = (W'W)^{-1}W'w = (\hat{\beta} + g\bar{\beta})/(1 + g), \tag{3.3}$$

with $\hat{\beta} = (X'X)^{-1}X'y$.

The quantity $\bar{\bar{\beta}}$ in (3.3) is the mean of the posterior pdf. It is seen to be a simple average of $\hat{\beta}$, the least squares quantity, and $\bar{\beta}$, the prior mean vector. Furthermore, the covariance matrix of the conditional normal posterior pdf for β given σ, denoted by $V(\boldsymbol{\beta}|\sigma, D)$, is

$$V(\boldsymbol{\beta}|\sigma, D) = (W'W)^{-1}\sigma^2 = (X'X)^{-1}\sigma^2/(1+g), \tag{3.4}$$

a very simple result which shows how the size of g affects the spread of the conditional posterior pdf for $\boldsymbol{\beta}$ given σ.

The marginal posterior pdf for $\boldsymbol{\beta}$, obtained from (3.2) by integrating with respect to σ, is

$$p(\boldsymbol{\beta}|D) \propto \left\{(\boldsymbol{w} - W\bar{\bar{\boldsymbol{\beta}}})'(\boldsymbol{w} - W\bar{\bar{\boldsymbol{\beta}}}) + (\boldsymbol{\beta} - \bar{\bar{\boldsymbol{\beta}}})'W'W(\boldsymbol{\beta} - \bar{\bar{\boldsymbol{\beta}}})\right\}^{-(n+k)/2}, \tag{3.5}$$

which is in the multivariate Student-t form with mean vector $\bar{\bar{\boldsymbol{\beta}}}$ shown in (3.3) and covariance matrix $V(\boldsymbol{\beta}|D) = (W'W)^{-1}a^2 = (X'X)^{-1}a^2/(1+g)$, where $(n-2)a^2 \equiv (\boldsymbol{w} - W\bar{\bar{\boldsymbol{\beta}}})'(\boldsymbol{w} - W\bar{\bar{\boldsymbol{\beta}}}) = (\boldsymbol{y} - X\bar{\bar{\boldsymbol{\beta}}})'(\boldsymbol{y} - X\bar{\bar{\boldsymbol{\beta}}}) + g(\bar{\bar{\boldsymbol{\beta}}} - \bar{\boldsymbol{\beta}})'X'X(\bar{\bar{\boldsymbol{\beta}}} - \bar{\boldsymbol{\beta}})$. Furthermore, the marginal posterior pdf for σ, obtained by integrating (3.2) with respect to the elements of $\boldsymbol{\beta}$, is

$$p(\sigma|D) \propto \sigma^{-(n+1)}\exp\left\{-(\boldsymbol{w} - W\bar{\bar{\boldsymbol{\beta}}})'(\boldsymbol{w} - W\bar{\bar{\boldsymbol{\beta}}})/2\sigma^2\right\}, \tag{3.6}$$

a posterior pdf in the inverted gamma form – see, for example, Raiffa and Schlaifer (1961) and Zellner (1971) for a review of its properties. For (3.6), the posterior mean of σ^2 is given by $E(\sigma^2|D) = a^2$, where a^2 has been defined above. Note that the value of a^2 depends on g.

Thus, from what has been presented, it is direct to use a g-prior distribution to obtain simple posterior distributions, much the same as in the case of the use of a natural conjugate prior distribution. The posterior mean of the coefficient vector shown in (3.3) is a simple average of the prior mean vector $\bar{\boldsymbol{\beta}}$ and of the least squares estimate $\hat{\boldsymbol{\beta}}$ with the parameter g involved in the weights. If g is large, the posterior mean is close to the prior mean $\bar{\boldsymbol{\beta}}$, while if g is small, it will be close to the least squares estimate. Also, g enters simply in expressions for the dispersion of the posterior distribution of $\boldsymbol{\beta}$ and in the posterior pdf for σ.

We now turn to consider some sampling properties of the posterior mean in (3.3).

4. Sampling properties of the posterior mean

It is well known that the mean of the posterior distribution, $\bar{\bar{\boldsymbol{\beta}}}$ in (3.3), is an optimal point estimate relative to quadratic loss in the senses of minimizing posterior expected loss and minimizing average (or Bayes) risk for any given value of σ. Below we derive some sampling properties of $\bar{\bar{\boldsymbol{\beta}}}$ including the risk function for $\bar{\bar{\boldsymbol{\beta}}}$ relative to a quadratic loss function and compare them with those of the least squares estimator, results similar to those of Giles and Rayner (1979).

First, the sampling mean of $\bar{\bar{\boldsymbol{\beta}}}$ is given by

$$E\bar{\bar{\boldsymbol{\beta}}} = (\boldsymbol{\beta}_0 + g\bar{\boldsymbol{\beta}})/(1+g), \tag{4.1}$$

Table 15.1. Values of $(1+g^2\lambda)/(1+g)^2$ for various values of λ and g.

λ	$g=0.10$	$g=0.25$	$g=0.50$	$g=0.75$	$g=1.00$
0	0.826	0.640	0.444	0.327	0.250
1	0.835	0.680	0.556	0.510	0.500
2	0.843	0.720	0.667	0.694	0.750
3	0.851	0.760	0.778	0.878	1.000
4	0.860	0.800	0.889	1.061	1.250
5	0.868	0.840	1.000	1.245	1.500
6	0.876	0.880	1.111	1.429	1.750
7	0.884	0.920	1.222	1.612	2.000
8	0.893	0.960	1.333	1.796	2.250
9	0.900	1.000	1.444	1.980	2.500

where $\boldsymbol{\beta}_0$ is the true unknown value of the regression coefficient vector. Thus, the bias of $\bar{\bar{\boldsymbol{\beta}}}$ is

$$\boldsymbol{B}(\bar{\bar{\boldsymbol{\beta}}}) = E\bar{\bar{\boldsymbol{\beta}}} - \boldsymbol{\beta}_0 = g(\bar{\boldsymbol{\beta}} - \boldsymbol{\beta}_0)/(1+g). \tag{4.2}$$

Clearly, the bias approaches zero as $g \to 0$ and/or $\bar{\boldsymbol{\beta}} \to \boldsymbol{\beta}_0$.

Second, the second moment matrix of $\bar{\bar{\boldsymbol{\beta}}} - \boldsymbol{\beta}_0$ is

$$E(\bar{\bar{\boldsymbol{\beta}}} - \boldsymbol{\beta}_0)(\bar{\bar{\boldsymbol{\beta}}} - \boldsymbol{\beta}_0)' = V(\hat{\boldsymbol{\beta}})/(1+g)^2 + \boldsymbol{B}(\bar{\bar{\boldsymbol{\beta}}})\boldsymbol{B}(\bar{\bar{\boldsymbol{\beta}}})', \tag{4.3}$$

with $V(\hat{\boldsymbol{\beta}}) = (X'X)^{-1}\sigma^2$, the covariance matrix of the least squares estimator.

Third, the variance–covariance matrix of $\bar{\bar{\boldsymbol{\beta}}}$ is

$$E(\bar{\bar{\boldsymbol{\beta}}} - E\bar{\bar{\boldsymbol{\beta}}})(\bar{\bar{\boldsymbol{\beta}}} - E\bar{\bar{\boldsymbol{\beta}}})' = V(\hat{\boldsymbol{\beta}})/(1+g)^2. \tag{4.4}$$

Fourth, the risk function for $\bar{\bar{\boldsymbol{\beta}}}$ relative to the loss function $L(\boldsymbol{\beta}, \tilde{\boldsymbol{\beta}}) = (\tilde{\boldsymbol{\beta}} - \boldsymbol{\beta})'(\tilde{\boldsymbol{\beta}} - \boldsymbol{\beta})$ is conveniently evaluated by noting that $E(\bar{\bar{\boldsymbol{\beta}}} - \boldsymbol{\beta})'(\bar{\bar{\boldsymbol{\beta}}} - \boldsymbol{\beta}) = \operatorname{tr}\{E(\bar{\bar{\boldsymbol{\beta}}} - \boldsymbol{\beta})(\bar{\bar{\boldsymbol{\beta}}} - \boldsymbol{\beta})'\}$. It follows from (4.3) that

$$E(\bar{\bar{\boldsymbol{\beta}}} - \boldsymbol{\beta})'(\bar{\bar{\boldsymbol{\beta}}} - \boldsymbol{\beta}) = c\sigma^2(1 + g^2\boldsymbol{\delta}'\boldsymbol{\delta}/c\sigma^2)/(1+g)^2, \tag{4.5}$$

where $c \equiv \operatorname{tr}(X'X)^{-1}$ and $\boldsymbol{\delta} \equiv \bar{\boldsymbol{\beta}} - \boldsymbol{\beta}$.

For comparison with (4.5), the risk function for the least squares estimator is $c\sigma^2$. Thus, it is seen that the risk associated with $\bar{\bar{\boldsymbol{\beta}}}$ will be smaller than that associated with $\hat{\boldsymbol{\beta}}$ if

$$(1 + g^2\boldsymbol{\delta}'\boldsymbol{\delta}/c\sigma^2)/(1+g)^2 < 1. \tag{4.6}$$

Shown in table 15.1 are values of the quantity $(1+g^2\lambda)/(1+g)^2$, where $\lambda \equiv \boldsymbol{\delta}'\boldsymbol{\delta}/c\sigma^2$ for different values of g and λ.[4] From the tabled values, it is seen that $\bar{\bar{\boldsymbol{\beta}}}$ has much lower risk than $\hat{\boldsymbol{\beta}}$ when $\lambda = 0$ and when g is large. However, as λ grows in value, the risk of $\bar{\bar{\boldsymbol{\beta}}}$ grows. It grows more rapidly with increasing λ the

[4]Given a g-prior for $\boldsymbol{\beta}$ and σ, the prior pdf for λ that is in the form of a non-central χ^2 pdf can be employed to make prior inferences about the value of λ.

larger is the value of g. When $\lambda = 1 + 2/g$, the risks associated with $\bar{\bar{\boldsymbol{\beta}}}$ and $\hat{\boldsymbol{\beta}}$ are equal.[5]

Now, using the g-prior distribution for $\boldsymbol{\beta}$ given σ, shown above (3.1), it is possible to compute the average of the risk function, (4.5) for $\bar{\bar{\boldsymbol{\beta}}}$. The resulting expression for average (or Bayes) risk of $\bar{\bar{\boldsymbol{\beta}}}$ given σ, denoted by $\mathrm{AR}(\bar{\bar{\boldsymbol{\beta}}}|\sigma)$ is

$$\begin{aligned}\mathrm{AR}(\bar{\bar{\boldsymbol{\beta}}}|\sigma) &= c\sigma^2(1 + g^2 E\boldsymbol{\delta}'\boldsymbol{\delta}/c\sigma^2)/(1+g)^2 \\ &= c\sigma^2/(1+g),\end{aligned} \tag{4.7}$$

since $E\boldsymbol{\delta}'\boldsymbol{\delta} = E(\boldsymbol{\beta} - \bar{\boldsymbol{\beta}})'(\boldsymbol{\beta} - \bar{\boldsymbol{\beta}}) = c\sigma^2/g$, where $c = \mathrm{tr}(X'X)^{-1}$. Given σ, the average risk of the least squares estimator is

$$\mathrm{AR}(\hat{\boldsymbol{\beta}}|\sigma) = c\sigma^2. \tag{4.8}$$

On comparing (4.7) and (4.8), it is seen that there can be a considerable reduction in average risk associated with the use of $\bar{\bar{\boldsymbol{\beta}}}$ relative to use of $\hat{\boldsymbol{\beta}}$.

Finally, we note that $\bar{\bar{\boldsymbol{\beta}}}$ in (3.3) can be generated as the solution to the following constrained least squares problem. Minimize $(\boldsymbol{y} - X\boldsymbol{\beta})'(\boldsymbol{y} - X\boldsymbol{\beta})$ with respect to $\boldsymbol{\beta}$ subject to $(\boldsymbol{\beta} - \bar{\boldsymbol{\beta}})'X'X(\boldsymbol{\beta} - \bar{\boldsymbol{\beta}}) \le r^2$, with $\bar{\boldsymbol{\beta}}$ and r^2 given. Write $(\boldsymbol{y} - X\boldsymbol{\beta})'(\boldsymbol{y} - X\boldsymbol{\beta}) + g[(\boldsymbol{\beta} - \bar{\boldsymbol{\beta}})'X'X(\boldsymbol{\beta} - \bar{\boldsymbol{\beta}}) - r^2]$, where g is interpreted as a Lagrange multiplier. The value of $\boldsymbol{\beta}$ which sets the derivative of the Lagrangian function with respect to $\boldsymbol{\beta}$ equal to zero is $\bar{\bar{\boldsymbol{\beta}}} = \bar{\boldsymbol{\beta}} + (\hat{\boldsymbol{\beta}} - \bar{\boldsymbol{\beta}})/(1+g) = (\hat{\boldsymbol{\beta}} + g\bar{\boldsymbol{\beta}})/(1+g)$, with $\hat{\boldsymbol{\beta}} = (X'X)^{-1}X'\boldsymbol{y}$.

5. Posterior odds ratios and g-prior distributions

Herein, we present posterior odds ratios for selected pairs of hypotheses about the values of the regression coefficient vector $\boldsymbol{\beta}$.

We first consider the following two hypotheses: H_1: $\boldsymbol{\beta} = \bar{\boldsymbol{\beta}}$ and H_2: $\boldsymbol{\beta} \in R^k$. Under H_1 and H_2 we assume $h(\sigma) \propto 1/\sigma$, and under H_2 we use the following g-prior:

$$p(\boldsymbol{\beta}|\sigma, \bar{\boldsymbol{\beta}}) = (2\pi g/\sigma^2)^{k/2}|X'X|^{1/2}\exp\{-g(\boldsymbol{\beta} - \bar{\boldsymbol{\beta}})'X'X(\boldsymbol{\beta} - \bar{\boldsymbol{\beta}})/2\sigma^2\},$$

where $\bar{\boldsymbol{\beta}}$ is the prior mean vector.[6] With a prior odds ratio of 1:1, the posterior

[5] In terms of prediction of a future value of the dependent variable, $y_\mathrm{f} = \boldsymbol{x}_\mathrm{f}'\boldsymbol{\beta}_0 + u_\mathrm{f}$, with $\boldsymbol{x}_\mathrm{f}'$ given, the least squares forecast error is $e_{\mathrm{LS}} = y_\mathrm{f} - \hat{y}_\mathrm{f} = \boldsymbol{x}_\mathrm{f}'(\hat{\boldsymbol{\beta}} - \boldsymbol{\beta}_0) - u_\mathrm{f}$, while the forecast error associated with the use of $\bar{\bar{\boldsymbol{\beta}}}$ as a point estimator for $\boldsymbol{\beta}$ is $e_\mathrm{g} = \boldsymbol{x}_\mathrm{f}'(\bar{\bar{\boldsymbol{\beta}}} - \boldsymbol{\beta}_0) - u_\mathrm{f}$. Then $Ee_{\mathrm{LS}}^2 = \boldsymbol{x}_\mathrm{f}'V(\hat{\boldsymbol{\beta}})\boldsymbol{x}_\mathrm{f} + \sigma^2$ and $Ee_\mathrm{g}^2 = \boldsymbol{x}_\mathrm{f}'E(\bar{\bar{\boldsymbol{\beta}}} - \boldsymbol{\beta}_0)(\bar{\bar{\boldsymbol{\beta}}} - \boldsymbol{\beta}_0)'\boldsymbol{x}_\mathrm{f} + \sigma^2$, with $E(\bar{\bar{\boldsymbol{\beta}}} - \boldsymbol{\beta}_0)(\bar{\bar{\boldsymbol{\beta}}} - \boldsymbol{\beta}_0)'$ as shown in (4.3). Risk comparisons for the two predictors can be made for any given value of $\boldsymbol{x}_\mathrm{f}$.

[6] Zellner and Siow (1979) and Mayer (1980) have employed prior pdfs depending on $X'X$ in deriving posterior odds ratios for hypotheses about regression coefficients' values but have not employed g-priors that are utilized in the present chapter.

odds ratio, K_{12}, is given by

$$K_{12} = \left(\frac{1+g}{g}\right)^{k/2}\left[\frac{(w-W\bar{\bar{\beta}})'(w-W\bar{\bar{\beta}})}{(y-X\bar{\beta})'(y-X\bar{\beta})}\right]^{n/2}$$

$$= \left(\frac{1+g}{g}\right)^{k/2}\left[\frac{1+\left(\frac{g}{1+g}\right)\frac{k}{\nu}F}{1+\frac{k}{\nu}F}\right]^{n/2}, \tag{5.1}$$

with $\nu = n-k$, $F = (\hat{\beta}-\bar{\beta})'X'X(\hat{\beta}-\bar{\beta})/ks^2$, $\hat{\beta} = (X'X)^{-1}X'y$, $\nu s^2 = (y-X\hat{\beta})'(y-X\hat{\beta})$, $\bar{\bar{\beta}} = (\hat{\beta}+g\bar{\beta})/(1+g)$, $w' = (y' \vdots g^{1/2}\bar{\beta}'X')$ and $W' = (X' \vdots g^{1/2}X')$.

In (5.1) F is the usual F-statistic for testing $\beta = \bar{\beta}$. As F grows in value, $K_{12} \to [g/(1+g)]^{n-k/2}$ which will be small for moderate to large n thus providing support for the less restrictive hypothesis, H_2. Also, as $g \to \infty$, $K_{12} \to 1$, which is reasonable since H_1 and H_2 become equivalent as $g \to \infty$.

As pointed out in Zellner and Vandaele (1975), the following Bayesian "pre-test" estimator is optimal relative to quadratic loss:

$$\tilde{\beta}^* = p_1\bar{\beta} + (1-p_1)\bar{\bar{\beta}}$$

$$= \bar{\beta} + (\bar{\bar{\beta}} - \bar{\beta})/(1+K_{12}), \tag{5.2}$$

where p_1 and $1-p_1$ are posterior probabilities associated with H_1 and H_2, respectively, $K_{12} = p_1/(1-p_1)$ and $\bar{\bar{\beta}} = (\hat{\beta}+g\bar{\beta})/(1+g)$ is the posterior mean under H_2.

We now present a posterior odds ratio for the following two hypotheses: H_A: $\beta = \beta_A$ and H_B: $\beta \in R^k$, using the same priors as above, where β_A is an arbitrary value for β, not necessarily equal to the prior mean as in H_1 above. With prior odds 1 : 1, the posterior odds ratio, K_{AB}, for H_A relative to H_B is

$$K_{AB} = \left(\frac{1+g}{g}\right)^{k/2}\left[\frac{(w-W\bar{\bar{\beta}})'(w-W\bar{\bar{\beta}})}{(y-X\beta_A)'(y-X\beta_A)}\right]^{n/2}$$

$$= \left(\frac{1+g}{g}\right)^{k/2}\left[\frac{1+\left(\frac{g}{1+g}\right)\frac{k}{\nu}F_2}{1+\frac{k}{\nu}F_1}\right]^{n/2}, \tag{5.3}$$

where, in addition to the quantities defined above,

$$F_1 = (\hat{\beta}-\beta_A)'X'X(\hat{\beta}-\beta_A)/ks^2$$

and

$$F_2 = (\hat{\beta}-\bar{\beta})'X'X(\hat{\beta}-\bar{\beta})/ks^2.$$

It is seen that F_1 and F_2 are the usual F-statistics for the hypotheses $\beta = \beta_A$ and $\beta = \bar{\beta}$, respectively. As F_2 grows in value, K_{AB} grows in value lending support to

H_A. Similarly, as F_1 grows in value, K_{AB}'s value declines providing support for H_B. As $g \to \infty$,

$$K_{AB} \to \left[\left(1+\frac{k}{\nu}F_2\right)\Big/\left(1+\frac{k}{\nu}F_1\right)\right]^{n/2},$$

which is the odds ratio for the two simple hypotheses $\boldsymbol{\beta} = \boldsymbol{\beta}_A$ and $\boldsymbol{\beta} = \bar{\boldsymbol{\beta}}$.

As above, it is possible to "average" over the two hypotheses to obtain an estimator that is optimal relative to quadratic loss, namely $\boldsymbol{\beta}^* = \boldsymbol{\beta}_A + (\bar{\bar{\boldsymbol{\beta}}} - \boldsymbol{\beta}_A)/(1 + K_{AB})$.

6. Concluding remarks

In this chapter procedures for analyzing the multiple regression model when prior information about its parameters' values is available have been reviewed. A general procedure for assessing prior pdfs for regression coefficients was presented that produced a prior pdf for $\boldsymbol{\beta}$ with a precision matrix proportional to $X'X$ that is termed a g-prior pdf. Use of the g-prior pdf leads to a posterior mean for the regression coefficient vector that is a simple average of the least squares estimate and the prior mean vector. Other properties of the posterior are relatively simple. Also, sampling properties of the posterior mean, derived from a g-prior pdf have been established and compared with those of the least squares estimator including their risk functions and Bayes risk. In addition, posterior odds ratios involving g-prior pdfs have been derived. From these results it appears that g-prior pdfs will be useful in analyzing the multiple regression model in a number of applied problems.

References

Box, G.E.P. and G.C. Tiao (1973) *Bayesian Inference in Statistical Analysis* (Addison-Wesley Publishing Co., Reading, Mass.).

DeGroot, M.H. (1970) *Optimal Statistical Decisions* (McGraw-Hill Book Co., New York.).

Giles, D.E.A. and A.C. Rayner (1979) "The mean squared errors of the maximum likelihood and natural conjugate Bayes regression estimators", *Journal of Econometrics*, 11, 319–334.

Jeffreys, H. (1967) *Theory of Probability* (Oxford University Press, London).

Kadane, J.B. (1980), "Predictive and structural methods for eliciting prior distributions", in: A. Zellner, ed., *Bayesian Analysis in Econometrics and Statistics in Honor of Harold Jeffreys* (North-Holland Publishing Co., Amsterdam), pp. 89–93.

Kadane, J.B., J.M. Dickey, R.L. Winkler, W.S. Smith and S.C. Peters (1980) "Interactive elicitation of opinion for a normal linear model", *Journal of the American Statistical Association*, 75, 845–854.

Leamer, E.E. (1978) *Specification Searches: Ad Hoc Inference with Nonexperimental Data* (John Wiley & Sons, Inc., New York).

Lindley, D.V. (1965) *Introduction to Probability & Statistics from a Bayesian Viewpoint, Part 2. Inference* (Cambridge University Press).

Mayer, F. (1980) "Specification et choix d'un seuil de signification optimal: Une approche Bayesienne", unpublished doctoral dissertation, Department of Economics, University of Montreal, Montreal, Canada.

Muth, J. (1961) "Rational expectations and the theory of price movements", *Econometrica*, 29, 315–335.

Raiffa, H. and R. Schlaifer (1961) *Applied Statistical Decision Theory* (Graduate School of Business Administration, Harvard University).

Shiller, R.J. (1973) "A distributed lag estimator derived from smoothness priors", *Econometrica*, 41, 775–788.

Winkler, R.L. (1967) "The assessment of prior distributions in Bayesian analysis", *Journal of the American Statistical Association*, 62, 776–800.

Winkler, R.L. (1977) "Prior distributions and model-building in regression analysis", in: A. Aykac and C. Brumat, eds., *New Developments in the Applications of Bayesian Methods* (North-Holland Publishing Co., Amsterdam) pp. 233–242.

Zellner, A. (1971) *An Introduction to Bayesian Inference in Econometrics* (John Wiley & Sons, Inc., New York).

Zellner, A. (1976) "Bayesian and non-Bayesian analysis of the regression model with multivariate Student-*t* errors", *Journal of the American Statistical Association*, 71, 400–405.

Zellner, A. (1977) "Maximal data information prior distributions", in: A. Aykac and C. Brumat, eds., *New Developments in the Applications of Bayesian Methods* (North-Holland Publishing Co., Amsterdam), pp. 211–232.

Zellner, A. (1985) "On assessing informative prior distributions for regression coefficients", in: J. M. Bernardo et al., eds., *Bayesian Statistics*, vol. 2 (North-Holland Publishing Co., Amsterdam), pp. 571–581.

Zellner, A. and M.S. Geisel (1970) "Analysis of distributed lag models with applications to consumption function estimation", *Econometrica*, 38, 865–888.

Zellner, A. and J.F. Richard (1973) "Use of prior information in the analysis and estimation of Cobb–Douglas production function models", *International Economic Review*, 14, 107–119.

Zellner, A. and A. Williams (1973) "Bayesian analysis of the federal reserve–MIT–PENN model's Almon lag consumption function", *Journal of Econometrics*, 1, 267–299.

Zellner, A. and W. Vandaele (1975) "Bayes–Stein estimators for *k*-means, regression and simultaneous equation models", in S.E. Fienberg and A. Zellner, eds., *Studies in Bayesian Econometrics and Statistics in Honor of Leonard J. Savage* (North-Holland Publishing Co., Amsterdam), pp. 627–653.

Zellner, A. and A. Siow (1979) "Posterior odds ratios for selected regression hypotheses", presented at the International Meeting on Bayesian Statistics, Valencia, 28 May–2 June 1979 and published in J.M. Bernardo, M.H. DeGroot, D.V. Lindley and A.F.M. Smith, eds., *Bayesian Statistics* (University Press, Valencia, Spain 1980).

PART VI

UTILITY AND SUBJECTIVE PROBABILITY ASSESSMENT

Chapter 16

COMPARING PROBABILITY FORECASTERS: BASIC BINARY CONCEPTS AND MULTIVARIATE EXTENSIONS

MORRIS H. DEGROOT and STEPHEN E. FIENBERG*

Carnegie-Mellon University

1. Introduction

In the statistical literature several authors have considered the elicitation and evaluation of probability forecasts, especially from a Bayesian perspective. Much of this effort has traditionally been focused on the use of proper scoring rules, a topic on which Professor de Finetti (1962, 1965) was an active contributor. Thus, it is especially pleasing to us to have our review of recent developments in this area appear in a volume honoring de Finetti's contributions to statistics.

In the applied forecasting literature much attention has also been lavished on questions about the evaluation of probability forecasts, and the subjectivist view of probability has been invoked to aggregate probability forecasts over a diverse set of events or statements (e.g. see Fischhoff and McGregor, 1982). One criterion invoked in such evaluations is that of calibration: a set of statements or events is considered and we ask if x percent of those assigned probability x of being correct prove to be correct, for each value of x. From this perspective, weather forecasters generally have been found to perform well (Murphy and Winkler, 1974, 1977). What is especially helpful in the evaluation of such probability forecasters is that they make forecasts about a long sequence of events (e.g. rain on a given day), and thus it makes sense to think about probability functions associated with the forecasts. In this chapter we focus on a criterion for compar-

* This research was supported in part by the National Science Foundation under grant SES-8207295 and by the Office of Naval Research under contract N00014-80-C-0637.

Bayesian Inference and Decision Techniques, edited by P. Goel and A. Zellner

ing forecasters, i.e. refinement, which goes beyond that of calibration (see the related discussion in Dawid, 1986; Winkler, 1977, and Chapter 17 in this volume).

The formal setting we consider is the same as that presented in DeGroot and Fienberg (1982, 1983). Consider two forecasters who at the beginning of each period n in a sequential process $(n=1,2,\ldots)$ must independently (i.e. without collaboration) specify their subjective probabilities that a particular event E_n will occur during the period. Assume that each forecaster in specifying the probability of E_n is aware of the values of various variables that are potentially relevant to the occurrence of E_n, including which of the previous events $E_1, E_2, \ldots, E_{n-1}$ have actually occurred. We wish to compare the two forecasters on the basis of their subjective probabilities of the events $E_1, E_2, \ldots, E_n$ and the subsequent observation of exactly which of those events occurred, for large values of n.

It is possible to think of our forecasters as economists who, at the beginning of each period, must specify their probabilities that the value of a particular commodity will rise during the period, or as medical diagnosticians who specify their probabilities that a patient has a particular disorder (e.g. see Habbema, Hilden and Bjerregaard, 1978, and Hilden, Habbema and Bjerregaard, 1978a, 1978b). As in DeGroot and Fienberg (1982, 1983), however, we present our review of the basic results in the context of two weather forecasters. Day after day, they must specify their subjective probabilities, x and y, that there will be at least a certain amount of rain at some given location during a specified time period in the day, and we refer to the occurrence of this well-specified event as "rain". We make the assumption that x and y are restricted to a given finite set of values $0 = x_0 < x_1 < \cdots < x_k = 1$, and we let $\mathscr{X}$ denote the set $\{x_0, x_1, x_2, \ldots, x_k\}$.

In section 2 we present the basic notation and formal definitions of the concepts of *calibration* and *refinement*. Then we go on to describe the relationship between these concepts and the classical concept of *sufficiency*, and we summarize various results when one well-calibrated forecaster is at least as refined as another, as given in DeGroot and Fienberg (1982, 1983) and DeGroot and Eriksson (1985). In section 3 we present two concepts introduced by Vardeman and Meedan (1983)–*semi-calibration* and *rain* (*or dry*)-*domination*– which impose different restrictions on the probability distributions of the forecasters than do calibration and sufficiency or refinement, and we discuss how these different concepts are related.

In section 4 we introduce the notion of strictly proper scoring rules and describe their use in comparing forecasters. We give a proof of a basic partitioning result for strictly proper scoring rules, described earlier in DeGroot and Fienberg (1983). We also give a result on the relationship between Schur-convex measures of the quality of forecasters and strictly proper scoring rules, as presented in DeGroot and Eriksson (1985).

Sections 2, 3, and 4 primarily review and clarify previously published results. In section 5 we turn to the multivariate or vector-probability forecasting situation in which the events of interest have three or more possible outcomes. For example, in the weather context the event may be tomorrow's maximum temperature and the possible outcomes may be grouped into 5°C intervals. The forecasters are required to announce their subjective probabilities associated with each of the $s \geq 3$ possible outcomes. We present several new results, describing how several of the concepts and results from the binary outcome situation can be extended or generalized to the multivariate setting.

2. Calibration and refinement

We begin by considering two forecasters, A and B, and we define the joint probability function, $g(x, y, \theta)$, for the random variables associated with the ith event E_i in a sequence where x and y are forecaster A's and B's subjective probabilities of rain, respectively, and $\theta = 1$ if the outcome is rain, or $\theta = 0$ otherwise. The overall performance of forecaster A can be characterized by two functions defined on $\mathscr{X}$:

(i) the probability function

$$\nu_A(x) = \sum_{y \in \mathscr{X}} \sum_{\theta} g(x, y, \theta),$$

which gives the probabilities or relative frequencies with which forecaster A makes each possible prediction x;

(ii) the conditional probability function

$$\rho_A(x) = \sum_{y \in \mathscr{X}} g(x, y, 1)/\nu_A(x),$$

which gives the conditional probability or relative frequency of rain given forecaster A's specific prediction x.

The functions $\nu_B(x)$ and $\rho_B(x)$ for forecaster B can be defined similarly. Finally, the long-run frequency of rain is

$$\mu = \sum_{x \in \mathscr{X}} \sum_{y \in \mathscr{X}} g(x, y, 1).$$

A forecaster is said to be *well-calibrated* if $\rho(x) = x$ for all $x \in \mathscr{X}$ such that $\nu(x) \neq 0$. Thus, a forecaster is well-calibrated if the forecaster's predictions can be accepted at face value, i.e. given that the forecaster predicts x, the conditional probability of rain is x. DeGroot and Eriksson (1985) give three different interpretations for the quantities $\nu(x)$ and $\rho(x)$, based on (i) limiting or theoretical values for an infinite sequence of days, (ii) the actual relative frequencies for a finite sequence of n days, and (iii) subjective probabilities of an observer who is comparing the forecasters. Interpretation (iii) is based on the subjective Bayesian approach to probability, as contrasted with the frequentist concept on which interpretations (i) and (ii) are based. Under interpretation (iii), the observer can assign values to $\nu(x)$ and $\rho(x)$ for a forecaster who is going to make his prediction x about some specified event E regardless of the nature of the prior information that the observer has about the forecaster's past behavior. Under interpretations (i) and (ii), it is necessary that the forecaster's behavior be known over a long sequence of events that are regarded as "exchangeable" with the event E. We proceed using interpretation (iii), although we sometimes use language that is associated with (i).

For various reasons, being well-calibrated is usually regarded as a desirable characteristic of a forecaster. For example, Pratt (1962) and Dawid (1982) show that a probability forecaster who is coherent in the sense of de Finetti (1937) must be calibrated almost surely in the mind of the forecaster. However, as Murphy and Winkler (1977), Dawid (1982), and DeGroot and Fienberg (1983) note, even if a forecaster is well-calibrated his predictions are not necessarily accurate in all respects, nor are they necessarily of much use to anyone. For example, the forecaster whose prediction on each day is μ is well-calibrated but clearly useless as a forecaster once we know μ.

Consider two well-calibrated forecasters, A and B, whose predictions are characterized by their probability functions, ν_A and ν_B. In DeGroot and Fienberg (1982, 1983) we introduce a concept of refinement which induces a partial ordering on the class of all well-calibrated forecasters. This concept is defined as follows: A *stochastic transformation* $h(y|x)$ is a non-negative function defined on $\mathscr{X} \times \mathscr{X}$ such that

$$\sum_{y \in \mathscr{X}} h(y|x) = 1, \quad \text{for every } x \in \mathscr{X}. \tag{2.1}$$

Forecaster A is said to be *at least as refined* as forecaster B if there exists a stochastic transformation $h(y|x)$ such that

$$\sum_{x \in \mathscr{X}} h(y|x)\nu_A(x) = \nu_B(y), \quad \text{for } y \in \mathscr{X}, \tag{2.2}$$

$$\sum_{x \in \mathscr{X}} h(y|x)x\nu_A(x) = y\nu_B(y) \quad \text{for } y \in \mathscr{X}. \tag{2.3}$$

Following DeGroot and Eriksson (1985) we denote this relationship by the symbols $A \succsim B$. The stochastic transformation here plays the role of an auxiliary randomization which we could use to generate predictions with distribution $\nu_B(y)$ from A's predictions, as in (2.2). Eq. (2.3) is required in the definition to ensure that the predictions generated by this process are well-calibrated.

If $A \succsim B$ and the probability functions ν_A and ν_B are not identically equal, then A is said to be *more refined* than B. We denote this relationship by the symbols $A \succ B$.

The relationship $A \succsim B$ is both reflexive and transitive, and induces a partial ordering (but not a total ordering) among well-calibrated forecasters. In these terms, the forecaster who makes the same prediction μ each day is *least-refined* in the sense that any other well-calibrated forecaster is at least as refined as he is. It is possible that μ is not one of the allowable predictions $x_0, x_1, \ldots, x_k$. In that case there is a value of i ($i = 0, 1, \ldots, k-1$) such that $x_i < \mu < x_{i+1}$, and DeGroot and Fienberg (1982) show that a forecaster who uses only the values x_i and x_{i+1} as his predictions in such a way that he is well-calibrated will now be least-refined. At the other extreme, the forecaster whose prediction each day is either $x = 0$ or $x = 1$ and who is always correct, is *most-refined* in the sense that he is more refined than any other well-calibrated forecaster.

For any well-calibrated forecaster it must be true that

$$\sum_{x \in \mathscr{X}} x\nu(x) = \mu. \tag{2.4}$$

Since every distribution with mean μ *can be thought of* as the $\nu(x)$ of some well-calibrated forecaster, the comparison of well-calibrated forecasters using the relationship $\succsim$ is equivalent to the problem of the comparison of all distributions on $\mathscr{X}$ with a given mean μ.

The left-hand side of expression (2.3) resembles the form of a conditional expectation. This observation leads to the following result:

Theorem 1 (DeGroot and Eriksson, 1985). The relationship $A \succsim B$ is satisfied if and only if there exist discrete random variables X and Y such that the marginal probability distribution of X is ν_A, the marginal probability distribution of Y is ν_B, and $E(X|Y) = Y$.

We temporarily step back and consider the comparison of two arbitrary forecasters, A and B, who are not necessarily well calibrated. For any given forecaster, let $f(x|\theta)$ denote the conditional probability function of the forecaster's predictions given θ. Thus, $f(x|1)$ can be regarded as the frequency function of the forecaster's predictions on days when rain actually occurs, and $f(x|0)$ as the frequency function on days when rain does not occur. It follows that for $x \in \mathscr{X}$:

$$f(x|1) = \rho(x)\nu(x)/\mu, \tag{2.5}$$

$$f(x|0) = \{1 - \rho(x)\}\nu(x)/(1-\mu). \tag{2.6}$$

Thus, we can also use the probability function $f(x|\theta)$ for $\theta = 1$ and $\theta = 0$ to characterize a forecaster's predictions. Let forecasters A and B have conditional probability functions $f_A(x|\theta)$ and $f_B(y|\theta)$. Following Blackwell's (1951, 1953) work on the comparison of experiments we say that A is *sufficient* for B if there exists a stochastic transformation $h(y|x)$ such that

$$\sum_{x \in \mathscr{X}} h(y|x) f_A(x|\theta) = f_B(y|\theta), \quad \text{for } y \in \mathscr{X} \quad \text{and} \quad \theta = 0, 1. \tag{2.7}$$

Using the relationships (2.5) and (2.6) we can now prove

Theorem 2 (DeGroot and Fienberg, 1982). Consider two forecasters, A and B, whose predictions are characterized by $\nu_A(x)$, $\rho_A(x)$, $\nu_B(x)$, and $\rho_B(x)$. Then A

is sufficient for B if and only if there exists a stochastic transformation h such that

$$\sum_{x \in \mathscr{X}} h(y|x)\nu_A(x) = \nu_B(y) \quad \text{for } y \in \mathscr{X}, \tag{2.8}$$

and

$$\sum_{x \in \mathscr{X}} h(y|x)\rho_A(x)\nu_A(x) = \rho_B(y)\nu_B(y) \quad \text{for } y \in \mathscr{X}. \tag{2.9}$$

Theorem 2 is applicable even when A's and B's forecasts are not probabilities, but are informational predictions that take values in an arbitrary space. Now if we again restrict attention to well-calibrated forecasters, who must use probabilities for their predictions, we get the following corollary:

Corollary 1. Consider two well-calibrated forecasters, A and B. Then $A \succsim B$ if and only if A is sufficient for B.

Thus, we can use results from the comparison of experiments to compare two well-calibrated probability forecasters, A and B. For any well-calibrated forecaster, let F denote the distribution function corresponding to the probability function ν, i.e.

$$F(t) = \sum_{\{x:\, x \in \mathscr{X},\, x \leq t\}} \nu(x), \qquad \text{for } 0 \leq t \leq 1. \tag{2.10}$$

Then another result adapted from the comparison of experiments is as follows:

Theorem 3 (DeGroot and Eriksson, 1985). The relationship $A \succsim B$ is satisfied if and only if

$$\int_0^s F_A(t)\,\mathrm{d}t \geq \int_0^s F_B(t)\,\mathrm{d}t, \quad \text{for all } 0 \leq s \leq 1. \tag{2.11}$$

The relationship (2.11) is one of several equivalent definitions of *second-degree stochastic dominance* (e.g. see Fishburn and Vickson, 1978), and we see that it is equivalent to the relationship $A \succsim B$. This leads to yet another equivalence:

Theorem 4 (Hardy, Littlewood and Polya, 1929, 1934). The relationship $A \succsim B$ is satisfied if and only if, for every continuous, convex function c defined on the unit interval,

$$\sum_{x \in \mathscr{X}} c(x)\nu_A(x) \geq \sum_{x \in \mathscr{X}} c(x)\nu_B(x). \tag{2.12}$$

Finally, there is a simplification of the second-order stochastic dominance relationship that can be expressed only in terms of the probability functions ν_A and ν_B:

Theorem 5 (DeGroot and Fienberg, 1982). The relationship $A \succsim B$ is satisfied if and only if

$$\sum_{i=0}^{j-1} (x_j - x_i)\{\nu_A(x_i) - \nu_B(x_i)\} \geq 0, \quad \text{for } j = 1, \ldots, k-1. \tag{2.13}$$

If at least one inequality in (2.13) is strict, then $A \succ B$.

The proof of theorem 5, given by DeGroot and Eriksson (1985), follows by the application of theorem 4, and is much simpler than the original proof given by DeGroot and Fienberg. In turn, from theorem 5 we can gain insight into the refinement relationship through the following corollary:

Corollary 2. Suppose A and B are well-calibrated forecasters such that

$$\nu_B(x_i) \geq \nu_A(x_i), \quad \text{for } i = 1, \ldots, k-1. \tag{2.14}$$

Then $A \succsim B$. If one of the inequalities in (2.14) is strict, then $A \succ B$.

Thus, in a rough sense, A is more refined than B if A spreads his probabilities into the extremes (in this case to 0 and 1) more than B, i.e. since (2.14) holds, $\nu_B(x_0) \leq \nu_A(x_0)$ and $\nu_B(x_k) \leq \nu_A(x_k)$.

As we have seen from theorem 2, in order to determine whether A is sufficient for B, we need only know the marginal probability functions $f_A(x|\theta)$ of A and B separately. However, in order to determine whether forecaster A is sufficient for both himself and forecaster B together, we must know the joint probability function $\nu(x, y) = g(x, y, 1) + g(x, y, 0)$ of the predictions x and y of A and B, as well as the conditional probability of rain given the two predictions x and y:

$$\rho(x, y) = g(x, y, 1)/\nu(x, y). \tag{2.15}$$

Then we can use the following result:

Theorem 6 (DeGroot and Fienberg, 1983). Forecaster A is sufficient for the pair of forecasters (A, B) if and only if

$$\rho(x, y) = \rho_A(x), \quad \text{for } x \in \mathscr{X} \quad \text{and } y \in \mathscr{X}. \tag{2.16}$$

It follows that if A is sufficient for (A, B), then A is sufficient for B. The converse, however, is not necessarily true.

Suppose now that neither A nor B is sufficient for the other. It becomes natural to ask if we can do better than A or B by using only their predictions. Of course, if we knew the long-run outcome distribution given their joint forecasts, i.e. $\rho(x, y)$, then whenever A predicts x and B predicts y, we could introduce a forecaster C whose forecast is $z = \rho(x, y)$. Forecaster C is then well-calibrated and sufficient for (A, B), and, in fact, for (A, B, C). But, if we were only receiving

A's and B's predictions sequentially, and thus did not know $\rho(x, y)$, we would need to develop a different approach. The applied forecasting literature suggests that combining probability forecasts by "averaging" is effective. There are two ways we might consider such "averaging" in a stochastic setting. One way to try to do this is to choose A's prediction with probability α and B's with probability $1-\alpha$. This results in a "new forecaster" whom we label as $M(\alpha)$. If both A and B are well-calibrated, then it is straightforward to show that $M(\alpha)$ is also well-calibrated. Furthermore, it follows directly from theorem 3 that $M(\alpha) \succsim A$ if and only if $B \succsim A$, and that $M(\alpha) \succsim B$ if and only if $A \succsim B$. Thus, "averaging" by randomly mixing two well-calibrated forecasters does not allow us to improve upon either of them in the refinement sense.

The second way to use the predictions of two forecasters is to average them (or, more generally, take linear combinations). Unfortunately, if A and B are well-calibrated it does not necessarily follow that the average of A's and B's predictions will be well-calibrated. This is most easily seen if we take the simple average of the predictions of the most-refined well-calibrated forecaster, who always correctly forecasts 0 or 1, and the predictions of the least-refined well-calibrated forecaster, who always forecasts μ. Thus, if we wish to improve upon A and B by averaging their predictions, there is the potential loss of calibration. Moreover, even if the result of averaging A's and B's forecasts is well-calibrated, it remains unclear whether the distribution of the average can be more spread than A's and B's distributions.

3. Restricted comparisons

As in many other statistical decision problems, there are essentially two different types of error that the forecaster might make. He might predict a high probability of rain on a day when it does not rain, or he might predict a low probability of rain on a day when it does rain. Some forecasters may control one type of error better than the other. We may thus wish to consider how to compare two forecasters separately for days on which it rains and for days on which it does not (i.e. dry days). This leads to two notions of dominance, introduced by Vardeman and Meeden (1983), which are both forms of *first-degree stochastic dominance* (again see Fishburn and Vickson, 1978).

We say that forecaster A *rain dominates* forecaster B if

$$\sum_{i=0}^{j} f_A(x_i|1) \leq \sum_{i=0}^{j} f_B(x_i|1), \quad \text{for } j=0,1,2,\ldots,k; \tag{3.1}$$

dry dominates forecaster B if

$$\sum_{i=0}^{j} f_A(x_i|0) \geq \sum_{i=0}^{j} f_B(x_i|0), \quad \text{for } j=0,1,2,\ldots,k; \tag{3.2}$$

and *dominates* forecaster B if both (3.1) and (3.2) hold. Condition (3.1) says that on rainy days, A's predictions have a distribution which is stochastically larger than the distribution of B's predictions, and condition (3.2) says that for dry days A's predictions are stochastically smaller than B's predictions.

The relationships of rain domination, dry domination, and domination, like that of refinement (or sufficiency), each induce their own partial ordering amongst forecasters. They also provide alternative ways of demonstrating refinement, as the following result shows:

Theorem 7 (Vardeman and Meeden, 1983). For two well-calibrated forecasters, A and B, if either (i) A rain dominates B or (ii) A dry dominates B, then $A \succsim B$.

The relationship of domination is quite stringent and thus it does not need to be combined with a condition as strong as calibration to imply sufficiency. Following Vardeman and Meeden, we say that a forecaster is *semi-calibrated* if $\rho(x)$ is nondecreasing in x for those values of x with $\nu(x) > 0$.

Theorem 8 (Vardeman and Meeden, 1983). If forecasters A and B are both semi-calibrated and A dominates B, then A is sufficient for B.

Vardeman and Meeden (1983) go on to use the concepts of domination and semi-calibration, along with calibration and refinement, to make comparisons between Bayesian forecasters who use stationary n-step Markov chain representations for the sequence of outcomes θ.

4. Strictly proper scoring rules

It has often been suggested in the statistical literature that a forecaster's predictions over a sequence of days can be evaluated by the use of a *scoring rule* which assigns a numerical value, or score, each day based on the forecaster's prediction x and the observation of whether or not rain occurred, i.e. the observation of θ. One property of the use of such rules, when the forecaster attempts to maximize the expectation of this score, is that if the forecaster's predictions are not restricted to be probabilities, then there is a known transform of the values of x to values which are probabilities (Lindley, 1980). For the class of proper scoring rules described below, the values of x must themselves be probabilities. There is little reason, however, to believe that a forecaster will want to maximize his expected overall score (e.g. see the discussion on this point in DeGroot and Fienberg, 1983, and in Staël von Holtstein, 1970).

Our interest in scoring rules in the context of comparing forecasters is somewhat different. Since we know that the relationship $A \succsim B$ induces only a partial ordering on the class of well-calibrated forecasters, we wish to assign a measure of quality $m(\nu)$ to the probability function ν of every well-calibrated forecaster in order to obtain a total ordering of this class. The values $m(\nu)$ should be assigned in such a way that the better the forecaster, the higher his measure of quality will be. It is natural to interpret this requirement to mean that if $A \gtrsim B$, then $m(\nu_A) \geq m(\nu_B)$, with strict inequality unless the probability functions ν_A and ν_B are identical. Indeed, we need not restrict attention to only well-calibrated forecasters. It is convenient to arrive at the measure $m(\cdot)$ for the comparison of well-calibrated forecasters from the more general approach of the use of scoring rules which are applicable to all forecasters.

We begin by considering an arbitrary scoring rule. Suppose that the forecaster's prediction is x and rain occurs. Then the forecaster receives a score $g_1(x)$. If rain does not occur he receives a score $g_2(x)$. Since we assume that the forecaster desires to maximize his score, it is reasonable to assume that $g_1(x)$ is an increasing function of x and that $g_2(x)$ is a decreasing function of x. If the forecaster's actual subjective probability of rain on a particular day is p, and he makes the prediction x, then his expected score is

$$pg_1(x)+(1-p)g_2(x). \tag{4.1}$$

A *proper scoring rule* is one for which expression (4.1) is maximized when $x = p$. A *strictly proper scoring rule* is one for which $x = p$ is the *only* value of x that maximizes expression (4.1). An interesting discussion of these rules, with historical references, is given by Staël von Holstein (1970, Sec. 3.2).

Examples of strictly proper scoring rules include the quadratic rule (Brier, 1950; de Finetti, 1962, 1965) with $g_1(x) = -(x-1)^2$ and $g_2(x) = -x^2$, and the logarithmic rule (Good, 1952) with $g_1(x) = \log x$ and $g_2(x) = \log(1-x)$. Both of these examples have the symmetry property, $g_1(x) = g_2(1-x)$, but this is not a requirement of strictly proper rules. An example of an *improper* scoring rule is the exponential with $g_1(x) = e^{x/2}$ and $g_2(x) = e^{-x/2}$. Here the values that maximize the expected score are $x = \log\{(1-p)/p\}$, i.e. the log-odds.

If a proper scoring rule is used for all of a forecaster's predictions, then we get an overall score S for the forecaster. Among all days on which the forecaster's prediction is x, the score will be $g_1(x)$ with relative frequency $\rho(x)$, and $g_2(x)$ with relative frequency $1-\rho(x)$. Since the relative frequency of the prediction x for the forecaster is $\nu(x)$, we have that

$$S = \sum_{x \in \mathscr{X}} \nu(x)\{\rho(x)g_1(x)+[1-\rho(x)]\,g_2(x)\}. \tag{4.2}$$

We now come to the major result of this section, which shows that this overall score (4.2) can be partitioned into a component which is a measure of calibration and a component which measures sufficiency or refinement.

Theorem 9 (DeGroot and Fienberg, 1983). If a forecaster's predictions are characterized by the functions $\nu(x)$ and $\rho(x)$, and if a proper scoring rule is specified by the functions $g_1(x)$ and $g_2(x)$, then the forecaster's overall score S

can be expressed in the form $S = S_1 + S_2$, where

$$S_1 = \sum_{x \in \mathscr{X}} \nu(x)\big[\rho(x)[g_1(x) - g_1\{\rho(x)\}] + \{1-\rho(x)\}[g_2(x) - g_2\{\rho(x)\}]\big], \tag{4.3}$$

$$S_2 = \sum_{x \in \mathscr{X}} \nu(x)\phi\{\rho(x)\}, \tag{4.4}$$

and

$$\phi(t) = tg_1(t) + (1-t)g_2(t), \quad \text{for } 0 \le t \le 1. \tag{4.5}$$

If the scoring rule is strictly proper, then $\phi(t)$ is strictly convex and S_1 attains its maximum value only when $\rho(x) = x$ for every value of x such that $\nu(x) > 0$.

Proof. It can be verified directly that if S_1 and S_2 are given by (4.3)–(4.5), then $S_1 + S_2$ is given by expression (4.2) for S. Thus, the first part of the theorem is established. Now suppose that the scoring rule is strictly proper. To see that $\phi(t)$ is convex, note that

$$\phi(t) = \max_{0 \le x \le 1} [tg_1(x) + (1-t)g_2(x)]. \tag{4.6}$$

In (4.6), $\phi(t)$ is represented as the maximum of a family of linear functions of t. Hence, $\phi(t)$ is convex.

If $\phi(t)$ is not strictly convex, then it contains at least one linear segment. This means that there must be one particular linear function that yields the maximum value in (4.6) for all the values of t in some interval. But that is impossible, because we know that at any particular value $t = t_0 (0 \le t_0 \le 1)$ the maximum in (4.6) is attained uniquely by the linear function $tg_1(t_0) + (1-t)g_2(t_0)$. Hence, $\phi(t)$ must be strictly convex.

Finally, since the scoring rule is strictly proper we know that

$$\rho(x)g_1(x) + [1-\rho(x)]g_2(x) \le \rho(x)g_1[\rho(x)] + [1-\rho(x)]g_2[\rho(x)], \tag{4.7}$$

with strict inequality unless $x = \rho(x)$. Hence, it can be seen from (4.3) that S_1 will be negative unless $\rho(x) = x$ for every value of x such that $\nu(x) > 0$, in which case S_1 will be equal to zero. □

As we noted above, S_1 is a measure of the forecaster's calibration, which is zero only for a well-calibrated forecaster and negative otherwise. The component S_2 provides us with our sought-after measure to give a total ordering for well-calibrated forecasters:

$$m(\nu) = \sum_{x \in \mathscr{X}} \phi(x)\nu(x), \tag{4.8}$$

which is S_2 with $\rho(x) = x$. Since $\phi(x)$ is strictly convex, theorem 4 implies that if $A \succsim B$, then $m(\nu_A) \ge m(\nu_B)$, with strict inequality, unless ν_A and ν_B are identical.

DeGroot and Eriksson (1985) note that there is a direct relationship between the total ordering provided by the measure m and the concept of Schur-convexity which plays an important role elsewhere in statistics (Marshall and Olkin, 1979). Consider a function m defined on the class of all probability distributions ν over $\mathscr{X}$ that have a given mean μ. Then m is said to be strictly *Schur-convex* if $m(\nu_A) \geq m(\nu_B)$ whenever the relation (2.11) is satisfied with strict inequality unless $\nu_A = \nu_B$. The following result now follows directly from theorem 4.

Theorem 10 (DeGroot and Eriksson, 1985). Consider a strictly proper scoring rule based on the functions g_1 and g_2, and suppose that a measure of quality m is defined by (4.8) and (4.5). Then m is strictly Schur-convex.

Suppose now that we know the functions $\nu(x)$ and $\rho(x)$ that characterize a particular forecaster's predictions. Is it possible for us to use his predictions, and no other relevant meteorological information, to make our own predictions and to attain a larger value of the score S than the forecaster himself? The following argument generalizes the one given in DeGroot and Fienberg (1983) for the quadratic scoring rule (see also Schervish, 1983).

The forecaster's score is given by expression (4.2). In order for us to make our predictions, we must choose a function that specifies the prediction x that we will use when we learn that the forecaster's prediction is y. In order to give ourselves maximum flexibility, we shall allow ourselves to choose a random prediction x in accordance with some stochastic transformation $h(x|y)$. Based on the results that we have developed thus far, however, we would anticipate that introducing a random element into our prediction can only reduce our effectiveness and that the optimal prediction x will be a deterministic function of y.

Thus, if the forecaster's prediction on a given day is y, then we shall choose our prediction at random from $\mathscr{X}$ in accordance with the conditional distribution $h(x|y)$. With this procedure, our predictions are characterized by the functions:

$$\nu_0(x) = \sum_{y \in \mathscr{X}} h(x|y)\nu(y) \tag{4.9}$$

and

$$\rho_0(x) = \sum_{y \in \mathscr{X}} h(x|y)\rho(y)\nu(y)/\nu_0(x). \tag{4.10}$$

It follows from expressions (4.2), (4.9), and (4.10) after some algebra, that our score is

$$S_0 = \sum_{y \in \mathscr{X}} \nu(y) \sum_{x \in \mathscr{X}} [\rho(y)g_1(x) + (1-\rho(y))g_2(x)]\, h(x|y). \tag{4.11}$$

For each fixed value of y, the summation over x in expression (4.11) yields a weighted average of the quantities

$$\rho(y)g_1(x) + \{1-\rho(y)\}\, g_2(x), \tag{4.12}$$

with weights given by the conditional probabilities $h(x|y)$. Thus, to maximize the weighted average we choose the conditional distribution $h(x|y)$ to put all the probability on the values of x that maximize expression (4.12). If $\rho(y) \in \mathscr{X}$, the maximizing value is $x = \rho(y)$, and we make the forecaster well-calibrated. With this choice, our value of S_2 remains the same as that of the original forecaster, but our value of S_1 is now increased to 0. Savage (1971) shows that (4.12) increases as x moves away from $\rho(y)$ in either direction. Hence, if $\rho(y) \notin \mathscr{X}$, the maximizing value of x must be the nearest neighbor to $\rho(y)$ in $\mathscr{X}$ on one side or the other. This choice makes the forecaster *almost well-calibrated*. Finally, a forecaster is said to be *almost well-calibrated* (relative to the strictly proper scoring rule defined by g_1 and g_2) if for each point $y \in \mathscr{X}$ such that $\nu(y) > 0$, the expression (4.12) is maximized over the points $x \in \mathscr{X}$ when $x = y$. Following Schervish (1983), if we take an arbitrary forecaster B we refer to a second forecaster A who uses this concentrated function $h(x|y)$ to transform the predictions of B into his own as "the almost calibrated version of B". Then our result is:

Theorem 11. Consider a strictly proper scoring rule. Let B be any forecaster and let A be the almost calibrated version of B. If B is not almost well calibrated, then A has a strictly larger score than B.

This theorem can be viewed as providing some motivation for the idea of recalibrating forecasters, as suggested by Lindley, Tversky and Brown (1979).

5. Comparing multivariate forecasters

We now turn to a consideration of forecasting events with $s > 2$ outcomes (e.g. a set of temperature ranges). In such settings the probability forecaster specifies a vector of probabilities $\boldsymbol{x} = (x_1, x_1, \ldots, x_s)$ restricted to a finite set $\mathscr{X}^{(s)}$ of values lying in the $(s-1)$-dimensional simplex, i.e. $x_i \geq 0$ and $\sum_{i=1}^{s} x_i = 1$, and thus is a multivariate or vector probability forecaster. If the conditional probabilities of the s outcomes given the prediction $\boldsymbol{x}$ are represented in vector form by $\rho(\boldsymbol{x}) = [\rho_1(\boldsymbol{x}), \rho_2(\boldsymbol{x}), \ldots, \rho_s(\boldsymbol{x})]$, then the multivariate forecaster is said to be *well-calibrated* whenever $\rho(\boldsymbol{x}) = \boldsymbol{x}$ for all $\boldsymbol{x} \in \mathscr{X}^{(s)}$. Note that this well-calibrated multivariate forecaster is also well-calibrated, in the sense of section 2, for each binary problem formed by combining the s outcomes into two groups; however, a forecaster who is "marginally" well-calibrated for predicting "rain" or "no rain" may no longer be well-calibrated when "rain" is divided into two or more possible outcomes.

More formally, let $\boldsymbol{x} = (x_1, \ldots, x_s)$ and $\rho(\boldsymbol{x}) = [\rho_1(\boldsymbol{x}), \ldots, \rho_s(\boldsymbol{x})]$. Furthermore, let $\mathscr{I} = \{I_1, \ldots, I_k\}$ represent a partition of the set $\{1, \ldots, s\}$ into k nonempty, mutually exclusive, and exhaustive sets $I_1, \ldots, I_k$. Then a forecaster is said to be *marginally well-calibrated with respect to the partition* $\mathcal{I}$ if

$$\sum_{i \in I_j} \rho_i(\boldsymbol{x}) = \sum_{i \in I_j} x_i, \quad \text{for } j=1,\ldots,k \quad \text{and } \boldsymbol{x} \in \mathscr{X}^{(s)}. \tag{5.1}$$

We can also focus on a particular set of the partition $\mathcal{I}$, say I_n, and define

$$\rho_i(\boldsymbol{x}, I_n) = P(\theta = i \mid \theta \in I_n, \quad \text{forecast } \boldsymbol{x}). \tag{5.2}$$

Then we can say that a forecaster is *conditionally well-calibrated given the set* I_n if

$$\rho_i(\boldsymbol{x}, I_n) = \frac{x_i}{\sum_{j \in I_n} x_j}, \quad i \in I_n. \tag{5.3}$$

Moreover, because being well-calibrated in the multivariate sense is a demanding requirement, we might also want to know if a forecaster is well-calibrated for some but not necessarily all values of $\boldsymbol{x}$. Let $\mathscr{X}_0^{(s)}$ denote a proper subset of $\mathscr{X}^{(s)}$. Then we say that a forecaster is *partially well-calibrated on the subset* $\mathscr{X}_0^{(s)}$ if $\rho(\boldsymbol{x}) = \boldsymbol{x}$ for $\boldsymbol{x} \in \mathscr{X}_0^{(s)} \subset \mathscr{X}^{(s)}$. We can now combine these notions of partial, conditional, and marginal calibration in various ways. (In particular, we note that the concept of conditional calibration suggested in DeGroot and Fienberg, 1982, is in fact a combination of conditional and partial calibration as defined here.) We also consider an extension of semi-calibration, introduced in section 3, to the multivariate setting in a special case at the end of this section.

For well-calibrated multivariate forecasters, we can define the concept of refinement by means of a multivariate stochastic transformation $h(\boldsymbol{x}|\boldsymbol{y})$. Consider two well-calibrated forecasters characterized by the probability functions ν_A and ν_B of their predictions. Then we say that A is at least as refined as B if there exists a stochastic transformation h such that:

$$\sum_{\boldsymbol{x} \in \mathscr{X}^{(s)}} h(\boldsymbol{y}|\boldsymbol{x}) \boldsymbol{x} \nu_A(\boldsymbol{x}) = \boldsymbol{y} \nu_B(\boldsymbol{y}), \quad \text{for } \boldsymbol{y} \in \mathscr{X}^{(s)}. \tag{5.4}$$

Note that the analogue of eq. (2.2), i.e.

$$\sum_{\boldsymbol{x} \in \mathscr{X}^{(s)}} h(\boldsymbol{y}|\boldsymbol{x}) \nu_A(\boldsymbol{x}) = \nu_B(\boldsymbol{y}), \quad \text{for } \boldsymbol{y} \in \mathscr{X}^{(s)}, \tag{5.5}$$

is automatically satisfied by summing the s equations in expression (5.4).

In order to demonstrate that one well-calibrated forecaster is more refined than another we may need to find the stochastic transformation. Alternatively we might try to reduce the vector or multivariate forecasting problem to a collection of binary problems through the use of marginal and conditional arguments, a technique that often proves useful in other areas of statistics. Thus, we have attempted to formulate concepts of marginal and conditional refinement that might be of some use in this regard. For example, we can immediately define concepts of marginal refinement with respect to a partition $\mathcal{I}$. The concept of conditional refinement given the set I_n which also appears to be immediate (in a definitional sense) is, however, problematic since it involves conditioning of the vector $\boldsymbol{x}$ on $\theta \in I_n$. These conditional predictions have no operational meaning, because we cannot define them only in terms of the probability distribution ν. Similarly, the concept of partial refinement on the subset $\mathscr{X}_0^{(s)} \subset \mathscr{X}^{(s)}$ also is problematic because two different forecasters typically place different amounts of subjective probability on the set $\mathscr{X}_0^{(s)}$.

At any rate, theorem 2 and corollary 1 from section 2 carry over directly from the binary case, i.e. forecaster A is sufficient for forecaster B if and only if there exists an appropriate stochastic transformation, $h(\boldsymbol{x}|\boldsymbol{y})$. Moreover, suppose we define a multivariate scoring rule, $\boldsymbol{g}(\boldsymbol{x}) = [g_1(\boldsymbol{x}),\ldots, g_s(\boldsymbol{x})]$. If the forecaster's actual subjective probability is $\boldsymbol{p}$ and he makes the prediction $\boldsymbol{x}$, his expected score is

$$\Sigma p_i g_i(\boldsymbol{x}). \tag{5.6}$$

The scoring rule is strictly proper if expression (5.6) is maximized if and only if $\boldsymbol{x} = \boldsymbol{p}$. Then there is a direct multivariate analogue to theorem 9 of section 4, i.e. every strictly proper scoring rule,

$$S = \sum_{\boldsymbol{x} \in \mathscr{X}^{(s)}} \nu(\boldsymbol{x}) \sum_{i=1}^{s} \rho_i(\boldsymbol{x}) g_i(\boldsymbol{x}), \tag{5.7}$$

can be partitioned into two components, i.e. $S = S_1 + S_2$, where

$$S_1 = \sum_{\boldsymbol{x} \in \mathscr{X}^{(s)}} \nu(\boldsymbol{x}) \sum_{i=1}^{s} \rho_i(\boldsymbol{x})\{ g_i(\boldsymbol{x}) - g_i(\rho(\boldsymbol{x}))\}, \tag{5.8}$$

$$S_2 = \sum_{\boldsymbol{x} \in \mathscr{X}^{(s)}} \nu(\boldsymbol{x})\phi\{\rho(\boldsymbol{x})\}, \tag{5.9}$$

and

$$\phi(\boldsymbol{t}) = \sum_{i=1}^{s} t_i g_i(\boldsymbol{t}). \tag{5.10}$$

As in the binary case discussed in section 4, S_1 is zero if the forecaster is well-calibrated and S_2 is a measure of refinement giving a total ordering for well-calibrated forecasters.

The following results are also as expected:

Theorem 12. If a multivariate forecaster A is well-calibrated, (i) A is also marginally well-calibrated with respect to all possible proper partitions $\mathscr{I}$ of $\{1,2,\ldots,s\}$, (ii) A is conditionally well-calibrated given the set $I_n \subset \{1,2,\ldots,s\}$, and (iii) A is partially well-calibrated on all proper subsets $\mathscr{X}_0^{(s)} \subset \mathscr{X}^{(s)}$.

Theorem 13. If A and B are well-calibrated multivariate forecasters, and A is at least as refined as B, then A is also marginally at least as refined as B with respect to all possible proper partitions $\mathscr{I}$.

We have, as yet, been unable to provide a collection of refinement conditions for dichotomies which imply multivariate refinement. Nor have we been able to prove a directly verifiable set of conditions analogous to theorem 5 of section 2. We can, however, give multivariate versions of theorems 1 and 4 by reformulating results of Blackwell (1951, 1953), Sherman (1951), Stein (in unpublished lecture notes), and Strassen (1965).

Theorem 14. Consider two well-calibrated forecasters, A and B. Then A is at least as refined as B if and only if there exist discrete random variables $\boldsymbol{x}$ and $\boldsymbol{y}$, defined on the $(s-1)$-dimensional simplex, such that the marginal probability distribution of $\boldsymbol{X}$ is ν_A, the marginal probability distribution of $\boldsymbol{Y}$ is ν_B, and $E(\boldsymbol{X}|\boldsymbol{Y})=\boldsymbol{Y}$.

Theorem 15. Consider two well-calibrated forecasters, A and B. Then A is at least as refined as B if and only if, for every continuous convex function $c(\boldsymbol{x})$ defined on the $(s-1)$-dimensional simplex,

$$\sum_{\boldsymbol{x}\in\mathscr{X}^{(s)}} c(\boldsymbol{x})\nu_A(\boldsymbol{x}) \geq \sum_{\boldsymbol{x}\in\mathscr{X}^{(s)}} c(\boldsymbol{x})\nu_B(\boldsymbol{x}). \tag{5.11}$$

We note that expression (5.11) in theorem 15 is, in our problem, the same as the condition used by Fishburn and Vickson (1978) for their definition of multivariate second-degree stochastic dominance. They also suggest the application of standard feasibility tests of linear programming to determine the existence of the stochastic transformation which we use to define refinement.

Furthermore, we have the following direct multivariate extension of a result presented in DeGroot and Eriksson (1985) regarding a reverse stochastic transformation.

Theorem 16. Consider two well-calibrated forecasters, A and B. Then A is at least as refined as B if and only if there exists a stochastic transformation $\eta(\boldsymbol{x}|\boldsymbol{y})$ such that

$$\sum_{\boldsymbol{y}\in\mathscr{X}^{(s)}} \eta(\boldsymbol{x}|\boldsymbol{y})\nu_B(\boldsymbol{y}) = \nu_A(\boldsymbol{x}), \quad \text{for } \boldsymbol{x}\in\mathscr{X}^{(s)}, \tag{5.12}$$

and

$$\sum_{\boldsymbol{x}\in\mathscr{X}^{(s)}} \boldsymbol{x}\eta(\boldsymbol{x}|\boldsymbol{y}) = \boldsymbol{y}, \quad \text{for } \boldsymbol{y}\in\mathscr{X}^{(s)}. \tag{5.13}$$

Proof. Suppose that A is at least as refined as B, and let h be a stochastic transformation satisfying (5.4). Define

$$\eta(\boldsymbol{x}|\boldsymbol{y}) = \frac{h(\boldsymbol{y}|\boldsymbol{x})\nu_A(\boldsymbol{x})}{\nu_B(\boldsymbol{y})} \tag{5.14}$$

whenever $\nu_B(\boldsymbol{y})>0$, and define $\eta(\boldsymbol{x}|\boldsymbol{y})$ arbitrarily if $\nu_B(\boldsymbol{y})=0$. Then (5.12) is a direct consequence of this definition and (5.13) follows directly from (5.4). Conversely, suppose that (5.12) and (5.13) are satisfied for some η and define the stochastic transformation h by (5.14). [Note that $h(y|x)$ may be defined arbitrarily if $\nu_A(x)=0$.] Then (5.4) follows directly from (5.12) and (5.13). □

A stochastic transformation η satisfying expression (5.13) is known in the economics literature as a *mean-preserving spread* (see, for example, Rothschild and Stiglitz, 1970, 1973).

An interesting version of the multivariate setting results when the probability of outcome i given that the forecaster predicts x depends only on the forecaster's subjective probability x_i for outcome i. (This is clearly true if the forecaster is well-calibrated.) We say that the forecaster is *local* if

$$\rho_i(x)=\rho_i(x_i), \quad i=1,2,\dots,s. \tag{5.15}$$

If the functions $\rho_i(\cdot)$ are monotonically increasing, then the forecaster is marginally semi-calibrated, i.e. semi-calibrated in the sense of section 3. A special case of locality is linearity, and an interesting question arises: Under what conditions on $\mathscr{X}^{(s)}$ and the ρ_i's is a multivariate forecaster being local equivalent to

$$\rho_i(x)=b_0x_i+b_i, \quad i=1,2,\dots,s, \tag{5.16}$$

where $b_0, b_1,\dots,b_s\geq 0$, and $b_0+b_1+\cdots+b_s=1$? If the functions ρ_i are known to be continuous on the entire simplex, then it can be shown that they must be linear for any local forecaster. We omit the details here.

Suppose we now say that forecaster A *dominates* forecaster B *on the outcome* i if the marginal distribution of the ith prediction component for forecaster A given that outcome i occurs is stochastically larger than the corresponding marginal distribution for B. We know from theorem 7 of section 3 that, if multivariate forecasters A and B are both well-calibrated and A dominates B on all s outcomes, then A is marginally at least as refined as B with respect to each possible outcome. An open question is whether it is possible to use calibration, locality, linearity as in (5.16), or semi-calibration in connection with some version(s) of dominance to imply that one forecaster is sufficient for (or more refined than) another in our full multivariate sense.

References

Blackwell, D. (1951) "Comparison of experiments", in: *Proc. Second Berkeley Symp. Math. Statist. Probability* (University of California Press), pp. 93–102.

Blackwell, D. (1953) "Equivalent comparison of experiments", *Ann. Math. Statist.*, 24, 265–272.

Blackwell, D. and M.A. Girschick (1954) *Theory of Games and Statistical Decisions* (Wiley, New York).

Brier, G.W. (1950) "Verification of forecasts expressed in terms of probability", *Monthly Weather Review*, 78, 1–3.

Dawid, A.P. (1982) "The well-calibrated Bayesian", *J. Amer. Statist. Assoc.*, 77, 605–610.

Dawid, A.P. (1986) "Probability forecasting", in: S. Kotz and N.L. Johnson, eds., *Encyclopedia of Statistical Science*, vol. 7 (Wiley, New York), in press.

de Finetti, B. (1937) "Foresight: Its logical laws, its subjective sources", in: H.E. Kyburg and H.E. Smokler, eds., *Studies in Subjective Probability* (1964). (Wiley, New York), pp. 91–158.

de Finetti, B. (1962) "Does it make sense to speak of 'Good Probability Appraisers?'", in: I.J. Good, ed. *The Scientist Speculates – An Anthology of Partly-Baked Ideas* (Basic Books, New York), pp. 357–363.

de Finetti, B. (1965) "Methods for discriminating levels of partial knowledge concerning a test item", *Brit. J. Math. Statist. Psych.*, 18, 87–123.

DeGroot, M.H. and E.A. Eriksson (1985) "Probability forecasting, stochastic dominance, and the Lorenz curve", in: J. M. Bernardo et al., eds., *Bayesian Statistics*, vol. 2 (North-Holland Publishing Company, Amsterdam). pp. 99–118.

DeGroot, M.H. and S.E. Fienberg (1982) "Assessing probability assessors: Calibration and refinement", in: S.S. Gupta and J.O. Berger, eds., *Statistical Decision Theory and Related Topics III*, Vol. 1 (Academic Press, New York), pp. 291–314.

DeGroot, M.H. and S.E. Fienberg, (1983) "The comparison and evaluation of forecasters", *The Statistician*, 32, 12–22.

Fischhoff, B. and D. MacGregor (1982) "Subjective confidence in forecasts", *J. of Forecasting*, 1, 155–172.

Fishburn, P.C. and R.G. Vickson (1978) "Theoretical foundations of stochastic dominance", in: G.A. Whitmore and M.C. Findley, eds., *Stochastic Dominance* (Lexington Books, Lexington, Mass.), pp. 39–113.

Good, I.J. (1952) "Rational decisions", *J. Roy. Statist. Soc., B* 14, 107–114.

Habbema, J.D., J. Hilden and B. Bjerregaard (1978) "The measurement of performance in probabilistic diagnosis. I. The problem, descriptive tools, and measures based on classification matrices", *Methods of Information in Medicine*, 17, 217–226.

Hardy, G.H., J.E. Littlewood and G. Polya (1929) "Some simple inequalities satisfied by convex functions", *Messenger Math.*, 58, 145–152.

Hardy, G.H., J.E. Littlewood and G. Polya (1934) *Inequalities* (Cambridge University Press).

Hilden, J., D.F. Habbema and B. Bjerregaard (1978a) "The measurement of performance in probabilistic diagnosis. II. Trustworthiness of the exact values of the diagnostic probabilities", *Methods of Information in Medicine*, 17, 226–237.

Hilden, J., D.F. Habbema and B. Bjerregaard (1978b) "The measurement of performance in probabilistic diagnosis. III. Methods based on continuous functions of the diagnostic probabilities", *Methods of Information in Medicine*, 17, 238–246.

Lindley, D.V. (1980) "Scoring rules and the inevitability of probability" (with discussion), *Int. Statist. Rev.*, 50, 1–26.

Lindley, D.V., A. Tversky and R.V. Brown (1979) "On the reconciliation of probability assessments", *J. Roy. Statist. Soc., A* 142, 146–180.

Marshall, A.W. and I. Olkin (1979) *Inequalities: Theory of Majorization and Its Applications* (Academic Press, New York).

Murphy, A.H. and R.L. Winkler (1974) "Subjective probability forecasts in meteorology: Some preliminary results", *Bull. Amer. Meteorol. Soc.*, 55, 1206–1216.

Murphy, A.H. and R.L. Winkler (1977) "Reliability of subjective probability forecasts of precipitation and temperature", *Appl. Statist.*, 26, 41–47.

Pratt, J.W. (1962) "Must subjective probabilities be realized as relative frequencies?", Unpublished seminar paper, Harvard University Graduate School of Business Administration.

Rothschild, M. and J.E. Stiglitz (1970) "Increasing risk: I. A definition", *J. Econ. Theory*, 2, 225–243.

Rothschild, M. and J.E. Stiglitz (1973) "Some further results on the measurement of inequality", *J. Econ. Theory*, 6, 188–204.

Savage, L.J. (1971) "The elicitation of personal probabilities and expectations", *J. Amer. Statist. Assoc.*, 66, 783–801.

Schervish, M.J. (1983) "A general method for comparing probability assessors", Department of Statistics Technical Report No. 275, Carnegie-Mellon University.

Sherman, S. (1951) "On a theorem of Hardy, Littlewood, Polya, and Blackwell", *Proc. Nat. Acad. Sci.*, 37, 826–831.

Staël von Holstein, C.A.S. (1970) *Assessment and Evaluation of Subjective Probability Distributions* (Economic Research Institute, Stockholm School of Economics, Stockholm).

Strassen, V. (1965) "The existence of probability measures with given marginals", *Ann. Math. Statist.*, 36, 423–439.

Vardeman, S. and G. Meeden (1983) "Calibration, sufficiency and domination considerations for Bayesian probability assessors", *J. Amer. Statist. Assoc.*, 78, 808–816.

Winkler, R.L. (1977) "Rewarding expertise in probability assessment", in: H. Jungermann and G. de Zeeuw, eds., *Decision Making and Change in Human Affairs* (Reidel, Dordrecht, Holland), pp. 127–140.

Chapter 17

ON "GOOD PROBABILITY APPRAISERS"

ROBERT L. WINKLER*

Indiana University and Duke University

1. Introduction

Some time ago de Finetti (1962) posed the question: "Does it make sense to speak of 'good probability appraisers'?" The answer, of course, depends on what is meant by "good". If good is interpreted as "right or wrong", then the answer has to be no, since probability appraisers who avoid the extreme probabilities of zero and one can never be said to be right or wrong on the basis of what actually happens. If good is interpreted as "more useful", however, then an answer of yes seems reasonable; otherwise a decision-maker would be indifferent about whom to consult for probability appraising or, in the case of differential costs, would prefer the cheapest appraiser. If I would like to know the chance of rain tomorrow, I would view a weather forecaster as a better probability appraiser than my wife. If I were concerned about the likelihood that some elective surgery would not be successful, I would consult a high-priced physician even though my neighbor would be happy to provide his opinions free of charge. Of course, the more experience the weather forecaster and the physician had at probability appraising, the more comfortable I would feel about their probability forecasts.

In this chapter the issue of "good probability appraisers" is investigated. Various types of goodness are discussed briefly in section 2, and some measures used to evaluate probability forecasts are reviewed in section 3. In section 4 the notion of good probability appraisers is related to measures used to evaluate probability forecasts. Some results from Winkler (1977) concerning properties of

*This research was supported in part by the National Science Foundation under grants ATM8004680, ATM8209713, and IST8018578. The author is grateful to two referees for some helpful suggestions.

Bayesian Inference and Decision Techniques, edited by P. Goel and A. Zellner

evaluation measures are reviewed, extended, and related to the notion of refinement developed by DeGroot and Fienberg (1982). Concluding comments are given in section 5.

2. Types of goodness

Three types of goodness, called coherence, calibration, and expertise, are considered here. The first notion of goodness relates not to the degree to which an appraiser's probabilities correspond to what eventually occurs, but to the theory of subjective probability. The axioms underlying this theory (de Finetti, 1937; Savage, 1954) prescribe that appraisers should be coherent in the sense that their subjective probabilities should obey the usual rules of probability theory. For example, if a weather forecaster states that the probability of rain tomorrow is 0.3, then the probability that it will not rain must be 0.7 for coherence to hold. Of course, violations of coherence can be detected and the appraisers can be asked to modify their probabilities to remove inconsistencies. Thus, for the purposes of this chapter, it is assumed that any inconsistencies are indeed removed and that the probabilities are ultimately coherent. Goodness in the sense of coherence is therefore not an issue here.

Before the other two types of goodness, calibration and expertise, are discussed, it is important to distinguish between ex ante and ex post notions. Decisions about which probability appraiser to hire are necessarily based on how good the appraisers are *expected* to be. After the fact, it is possible to take advantage of hindsight to compare the merits of different appraisers, but this evaluation must be recognized as specific to a given situation. For example, suppose a well-regarded physician judges that the probability is very low that my elective surgery will be unsuccessful, whereas my neighbor thinks that the chance of failure is quite high. If I choose to have the surgery and it turns out to be unsuccessful, then it appears that I would have been better off had I paid more attention to my neighbor's probability. In an ex post sense, I see that my neighbor's forecast would have been more useful than the physician's forecast *for this particular event*. But does this make my neighbor a better probability appraiser? For future medical uncertainties, will I embrace my neighbor's probabilities instead of the physician's probabilities? I doubt it. Of course, if this scenario occurs many times I will gain confidence in my neighbor's judgments while losing faith in the physician. As is the case in statistical inference in general, inferences about probability appraisers based on large samples of ex post evaluations can be more useful than inferences based on small samples.

A second notion of goodness, usually referred to as calibration or reliability, relates to the degree to which an appraiser's probabilities correspond, in a relative frequency sense, to what eventually occurs. Here the ex ante/ex post distinction is important. In an ex ante sense, calibration involves the consideration of expected relative frequencies. If I expect rain to occur on 10 percent of the days for which a weather forecaster says the probability of rain is 0.10, on 20 percent of the days for which the probability is 0.20, and so on, then I view the forecaster as being perfectly calibrated ex ante. Ex post calibration involves actual relative frequencies. After a forecaster provides a series of probabilities for different events and the events are observed to occur or not occur, it is possible to look at the set of events for which the probability was, say, 0.20. For perfect ex post calibration, the relative frequency of occurrence in this set should be 0.20. Dawid (1982) shows that a forecaster who is coherent expects to be well calibrated in an ex post sense. For a general discussion of calibration, see Kadane and Lichtenstein (1982).

A considerable body of evidence indicates that in an ex post sense, weather forecasters are very well calibrated for precipitation probability forecasts (Murphy and Winkler, 1977). In contrast, subjects in psychological experiments often tend to be poorly calibrated (Lichtenstein, Fischhoff and Phillips, 1982). Of course, ex post calibration from past data provides information about ex ante calibration for future probability forecasts. Moreover, if I expect a forecaster not to be calibrated, I can take his or her probabilities and attempt to calibrate them in such a manner that I will view the modified probabilities as perfectly calibrated in an ex ante sense. In so doing I am treating the appraiser's probabilities as information and making inferences on the basis of this information (Morris, 1974, 1977; Lindley, Tversky and Brown, 1979; Lindley, 1982; DeGroot and Fienberg, 1983a).

A third notion of goodness, which I shall call expertise, relates to the degree to which the probability appraiser can approach perfect forecasts. The ideal probability appraiser would always give probabilities of zero or one and would always be correct. This is the extreme degree of expertise because the probability appraiser is able to remove all uncertainty. If such a clairvoyant could be found, his or her forecasts would be extremely valuable. Among less-than-perfect forecasters, some might display more expertise than others. A forecaster who uses only probabilities of 0.90 and 0.10 for an event such as "precipitation tomorrow", uses these two values equally often, and is perfectly calibrated is a better expert than a forecaster who is similar but uses probabilities of 0.70 and 0.30 instead of 0.90 and 0.10.

It is important to note that expertise is more than using extreme probabilities. Otherwise forecasters could fake expertise by using extreme probabilities even when their judgments did not warrant such extreme values. The idea is to be able to use extreme probabilities without sacrificing calibration. For example, weather forecasters could be well-calibrated by taking as their precipitation probability forecasts the past long-run relative frequency of rain at the same location and time of year. The more a weather forecaster is able to move away from this climatological forecast toward the limiting values of zero and one while remaining well-calibrated, the better that forecaster is as a probability appraiser. This is the primary notion of goodness of probability appraisers that is considered in this chapter.

3. Evaluation of probability appraisers

After probabilities have been appraised and the events or uncertain quantities of interest have been observed, the probabilities (and hence the probability appraisers) can be evaluated in light of what actually happened. Scoring rules, which involve the computation of a score based on the probabilities and on the event or value of the uncertain quantity that actually occurs, provide a formal means of evaluating probabilities. In this section, scoring rules for various situations are presented and discussed briefly. For more details regarding the development of scoring rules, see Winkler (1967), Winkler and Murphy (1968), and Savage (1971).

First, consider a single event E. Let p denote the appraiser's probability that E will occur, and let r denote the probability of E actually reported by the expert. (Note that r may not be equal to p.) A scoring rule $S(r)$ gives the probability appraiser a score

$$S(r) = \begin{cases} S_1(r), & \text{if } E \text{ occurs}, \\ S_2(r), & \text{if } E \text{ does not occur}. \end{cases} \tag{1}$$

The probability appraiser's expected score is

$$E_p[S(r)] = pS_1(r) + (1-p)S_2(r), \tag{2}$$

and the scoring rule is called strictly proper if

$$E_p[S(p)] > E_p[S(r)], \quad \text{for } r \neq p. \tag{3}$$

That is, if S is strictly proper, the appraiser can maximize the expected score only by setting $r = p$ (i.e. by reporting honestly).

The primary emphasis in the study of scoring rules has been on strictly proper scoring rules because such rules encourage honesty on the part of the probability appraiser. The class of strictly proper scoring rules includes a wide variety of functions; Savage (1971) indicates the conditions that must be satisfied by a function $S(r)$ for it to be a strictly proper scoring rule, and these conditions are satisfied by many different functions. Three examples encountered frequently in the literature on scoring rules are the quadratic, logarithmic, and spherical scoring rules, which are given in table 17.1.

The concept of a scoring rule can be generalized to a situation involving more than a single event and its complement. Consider a set of mutually exclusive and exhaustive events $\{E_i | i \in I\}$, where I is a finite or countably infinite set. Let p_i denote the appraiser's probability that E_i will occur, let r_i denote the probability of E_i actually reported by the appraiser, and define $p = (p_1, p_2, \ldots)$ and $r = (r_1, r_2, \ldots)$. In this chapter, p and r represent single probabilities when a single

Table 17.1. Quadratic, logarithmic, and spherical scoring rules.

	$S_1(r)$	$S_2(r)$
Quadratic	$-(1-r)^2$	$-r^2$
Logarithmic	$\log r$	$\log(1-r)$
Spherical	$r/[r^2+(1-r)^2]^{1/2}$	$(1-r)/[r^2+(1-r)^2]^{1/2}$

event is of interest, vectors of probabilities when several events are of interest, and mass or density functions when a random variable is of interest. This abuse of notation is used to emphasize the distinction between an appraiser's probability judgments (p) and *reported* probability judgments (r).

If a scoring rule $S(r)$ gives a score

$$S(r) = S_j(r), \quad \text{if } E_j \text{ occurs,} \tag{4}$$

then the probability appraiser's expected score is

$$E_p[S(r)] = \sum_{j \in I} p_j S_j(r). \tag{5}$$

The scoring rule is strictly proper if (3) is satisfied. Note that E_i could represent a value of a discrete random variable and that scoring rules of the form (4) could be used to evaluate probability distributions for discrete random variables as well as probability distributions for unordered sets of qualitative events.

Scoring rules for continuous probability distributions can also be developed. Consider a continuous random variable $\tilde{x}$. Let p denote the appraiser's probability density function for $\tilde{x}$, and let r denote the density function reported by the appraiser. If a scoring rule gives a score

$$S(r) = S_x(r), \quad \text{when } \tilde{x} = x, \tag{6}$$

then the expert's expected score is

$$E_p[S(r)] = \int_{-\infty}^{\infty} S_x(r) p(x) \, \mathrm{d}x. \tag{7}$$

The scoring rule is strictly proper if

$$E_p[S(p)] > E_p[S(r)], \quad \text{for } r \neq p. \tag{8}$$

In Matheson and Winkler (1976) a special class of scoring rules is developed for dealing with probability distributions for real-valued random variables. Let P denote the expert's *cumulative* probability distribution function for x, and let R denote the cumulative probability distribution function reported by the expert to the decision-maker. That is, P and R are the cumulative functions corresponding to p and r, respectively. Now, for any real u, let E be the event that $\tilde{x} \leq u$ and apply a binary scoring rule as given in (1):

$$S_x[R(u)] = \begin{cases} S_1[R(u)], & \text{if } \tilde{x} \leq u, \\ S_2[R(u)], & \text{if } \tilde{x} > u. \end{cases} \tag{9}$$

If $S_x[R(u)]$ is integrated over u, the result is a strictly proper scoring rule of the form

$$S(r)=\int_{-\infty}^{x} S_2[R(u)]\,\mathrm{d}u+\int_{x}^{\infty} S_1[R(u)]\,\mathrm{d}u. \tag{10}$$

To make this result even more general, integrate with respect to a cumulative distribution function G:

$$S^*(r)=\int_{-\infty}^{x} S_2[R(u)]\,\mathrm{d}G(u)+\int_{x}^{\infty} S_1[R(u)]\,\mathrm{d}G(u). \tag{11}$$

Scoring rules of the forms (10) and (11) have properties not shared by all strictly proper scoring rules for probability distributions for real-valued variables, as will be shown in section 4.

4. Evaluation measures and "good probability appraisers"

Three types of goodness were discussed in section 2: coherence, calibration, and expertise. It is assumed that any inconsistencies are detected and removed, implying that the probabilities are coherent. In this section some relationships between scoring rules and the remaining two types of goodness are considered. For example, do proper scoring rules provide higher scores for probability appraisers who are well-calibrated and display considerable expertise than for those who are not so well-calibrated or show less expertise?

4.1. Scoring rules and calibration

As in section 2, it is useful to distinguish between ex ante and ex post considerations. From a motivational viewpoint, the ex ante expected score is the relevant measure, and the idea of proper scoring rules encouraging honest forecasting is related to this ex ante measure. But honest forecasting can be interpreted as ex ante calibration. Another way to interpret (3) and (7) is that probability appraisers must view their reported probabilities as being calibrated ex ante or they could attain higher expected scores. Therefore, proper scoring rules encourage ex ante calibration.

Just as ex ante calibration maximizes a probability appraiser's *expected* score, ex post calibration maximizes the actual long-run *average* score from a strictly proper scoring rule. That is, if a probability appraiser feels that on a large number n of occasions E has the same chance of occurring and consequently reports the same probability value r on those occasions, then the average score over those occasions is maximized if the value that was chosen is equal to x/n, the relative frequency of occurrence of the event in question. If S is strictly proper and x denotes the number of times E occurs over the n occasions, then the average score is

$$\bar{S}_{n,x}(x/n) = (x/n)S_1(r) + [(n-x)/n]S_2(r), \tag{12}$$

which corresponds to (2) with x/n replacing p. The ex post version of (3) is

$$\bar{S}_{n,x}(x/n) > \bar{S}_{n,x}(r), \quad \text{for } r \neq x/n. \tag{13}$$

This result readily generalizes to the case of more than two events (in which case the average score is analogous to (5) with p replaced by a vector of relative frequencies) and the continuous case (in which case the average score is analogous to (7) with p replaced by a histogram of observed values of $\tilde{x}$). By their very nature, as represented by (3) and (13), in both an ex ante sense and an ex post sense proper scoring rules provide better evaluations for well-calibrated appraisers than for poorly-calibrated appraisers, ceteris paribus. As noted in section 2, Dawid (1982) shows that a coherent probability appraiser expects to be well calibrated in an ex post sense.

4.2. Scoring rules and expertise: p known

How are scores affected by expertise? From a motivational viewpoint, the ex ante expected score is once again the relevant measure. The results in the remainder of this section are therefore expressed in terms of expected scores. In this section expected scores after p is known are studied, and expected scores before p is known are considered in section 4.3.

Because proper scoring rules encourage ex ante calibration, it is assumed here and in section 4.3 that the probability appraiser is perfectly calibrated in an ex ante sense. (Alternatively, it could be assumed that the probabilities have been adjusted for any perceived miscalibration and that the expertise associated with the adjusted probabilities is of interest.) Given the aforementioned relationship between ex ante and ex post considerations, the results can just as easily be interpreted in terms of actual average scores and perfect ex post calibration.

First, consider a simple binary situation in which a probability appraiser is asked for the probabilities of an event E and its complement. The scoring rule is then of the form given by (1). If S is strictly proper, then $S_1(r)$ is an increasing function of r and $S_2(r)$ is a decreasing function of r (Winkler, 1977). Thus, one property of a strictly proper S is that if a higher probability is assigned to the event that eventually occurs (either E or its complement), the score obtained increases.

An important property of expected scores from a strictly proper S is that $E_p[S(p)]$ is a convex function of p (Savage, 1971). Thus, as p moves closer to zero or one from the point $p^* \in [0,1]$ at which $E_p[S(p)]$ is minimized, the expected score increases. This movement of p might be thought of as an increase in "sharpness". Thus, assuming that two appraisers are perfectly calibrated, the appraiser with the sharper forecast has a higher expected score.

If S is also symmetric in the sense that

$$S_1(u) = S_2(1-u), \quad \text{for } u \in [0,1], \tag{14}$$

then $p^* = 0.5$ and the "sharpness" can be thought of in an intuitively appealing fashion in terms of increases in $\max\{p, 1-p\}$. The symmetry ensures that the scoring rule is label-invariant (e.g. a weather forecaster's score on a given occasion is the same if E = rain today and $r = 0.3$ as if E = no rain today and $r = 0.7$).

The highest possible expected score, of course, is obtained by the ideal probability appraiser who can completely remove the uncertainty in a situation by setting p equal to 0 or 1. (For a symmetric rule, $E_0[S(0)] = E_1[S(1)]$. For a rule which is not symmetric, the maximum expected score occurs at one but not necessarily both of the endpoints.) It must be emphasized, however, that the appraiser must really feel certain, since a strictly proper scoring rule provides a penalty for reporting 0 or 1 when $0 < p < 1$. The expected score $E_p[S(r)]$ can be expressed as

$$E_p[S(r)] = E_p[S(p)] + C(S, r, p), \tag{15}$$

where $C(S, r, p)$ is maximized (at zero) when $r = p$. This expression for $E_p[S(r)]$ separates the incentives for expertise and calibration. If the appraiser sets $r \neq p$ and is therefore not perfectly calibrated ex ante, then $C(S, r, p)$ is brought into play, and it is not sufficient to consider only $E_p[S(p)]$. [Partitioning of an individual score or an average score and the resulting attributes measured by elements of various partitions have been studied, especially for the quadratic scoring rule (e.g. Murphy, 1973; Yales, 1982).] The exact forms of the convex function $E_p[S(p)]$ that rewards expertise and the function $C(S, r, p)$ that provides incentive for calibration depend on the scoring rule S, of course.

Next, consider a situation in which an appraiser is asked for a probability distribution for a set of n mutually exclusive and exhaustive events $E_1, \ldots, E_n$. The set of probabilities is then of the form $p = (p_1, \ldots, p_n)$, and the set reported by the appraiser is $r = (r_1, \ldots, r_n)$. The scoring rule is of the form given by (4), and the expected score $E_p[S(p)]$ is convex in p if S is strictly proper. A generalization of the symmetry (label-invariance) assumption of (14) to the case of n events provides that for any permutation $i_1, \ldots, i_n$ of the integers from 1 through n, inclusive:

$$S_j(r_1, r_2, \ldots, r_n) = S_{i_k}(r_{i_1}, r_{i_2}, \ldots, r_{i_n}), \tag{16}$$

when $i_k = j$. From convexity and symmetry, $E_p[S(p)]$ is minimized at $p = (n^{-1}, \ldots, n^{-1})$ and maximized at $p = \Delta_i$, for $i = 1, \ldots, n$, where Δ_i denotes an n-dimensional vector with the ith element equal to one and all other elements equal to zero. Thus, the expected score $E_p[S(p)]$ is maximized when sharpness is maximized ($p_i = 1$, for some i) and minimized when sharpness is minimized ($p_i = n^{-1}$, for all i). In general, probability vectors "closer" to some Δ_i will receive higher expected scores than vectors "closer" to the uniform vector. If the label-invariance assumption is relaxed, the maximum expected score may occur at some subset of the n vectors $\Delta_1, \ldots, \Delta_n$, and the minimum expected score may occur at some p other than the uniform vector.

If a real-valued discrete random variable is of interest, a scoring rule of the type just considered can still be used, but if the symmetry assumption in (16) is maintained, the ordering inherent in the possible values of the random variable will be ignored by S. For a given set of probabilities, it makes no difference whether the higher probabilities are associated with extreme values or central values of the variable. Furthermore, this result generalizes to the continuous case. For example, with a strictly proper rule such as $S_x(r) = \log r(x)$ in the continuous case, the triangular density $p(x) = x$, if $0 \leq x \leq 1$, and $p(x) = 2 - x$, if $1 \leq x \leq 2$, has the same $E_p[S(p)]$ as the V-shaped density given by $p(x) = 1 - x$, if $0 \leq x \leq 1$, and $p(x) = x - 1$, if $1 \leq x \leq 2$. Thus, scoring rules that ignore the ordering of values of a random variable do not reward sharpness in a manner consistent with the usual interpretation of sharpness in terms of some measure of dispersion such as the variance.

Suppose, however, that an appraiser is asked for a probability distribution for a real-valued variable $\tilde{x}$ and a scoring rule S of the form (10) based on a strictly proper, symmetric binary rule is used. Since the binary rule is strictly proper, so is S. However, S is not "symmetric" because it depends on whether the probability for values of u other than the observed value x is concentrated close to x or far away from x.

What can be said about the expected score when this type of scoring rule is used? The expected score $E_p[S(p)]$ is location-invariant; it may depend on the shape of the distribution p, but it does not depend on the location of the distribution (Winkler, 1977). For example, a weather forecaster whose distribution for tomorrow's maximum temperature in a particular city is a normal distribution with mean 80°F and standard deviation 3°F has the same expected score as a forecaster in a different city whose distribution for tomorrow's maximum temperature in *that* city is a normal distribution with mean 58°F and standard deviation 3°F.

As a result of the location-invariance of S, in comparing the expected scores for two distributions p_1 and p_2 it can be assumed without loss of generality that the medians of the distributions are equal. (When the medians are not equal, one distribution can be translated so that they are equal.) If the cumulative distribution functions P_1 and P_2 have a common median m and satisfy

$$P_1(u) \leq P_2(u), \quad \text{for all } u \leq m, \tag{17}$$

and

$$P_1(u) \geq P_2(u), \quad \text{for all } u \geq m, \tag{18}$$

then

$$E_{p_1}[S(p_1)] \geq E_{p_2}[S(p_2)] \tag{19}$$

(Winkler, 1977). This gives some indication of how S rewards expertise. From

(17) and (18), p_2 is more "spread out" than p_1. In other words, p_1 exhibits more sharpness than p_2. For example, the variance (and most likely other measures of dispersion) associated with p_1 is less than that associated with p_2. In general, strictly proper scoring rules of the form given by (10) reward expertise in a manner consistent with the notion of desiring less dispersion, although orderings of expected scores cannot always be perfectly consistent with orderings in terms of specific measures of dispersion.

Finally, what if a scoring rule of the form (11) is used? If $\mathrm{d}G(u)$ is not constant in u, then it prevents location-invariance, for a distribution that is sharper than a second distribution may be penalized because it happens to be concentrated where $\mathrm{d}G(u)$ is relatively low whereas the second distribution is concentrated where $\mathrm{d}G(u)$ is relatively high. Thus, S^* rewards expertise in the form of sharpness, but it may reward such expertise in a selective manner by providing extra rewards for distributions concentrated where $\mathrm{d}G(u)$ is relatively high. However, S^* is still strictly proper, and as a result probability appraisers only reduce their expected scores if they try to shift r from p in the direction of higher values of $\mathrm{d}G(u)$. Also, note that whereas appraisers may attempt to arrive at sharper distributions by obtaining additional information, such information will not necessarily result in distributions concentrated where $\mathrm{d}G(u)$ is high. In that sense, the effects of the $\mathrm{d}G(u)$ factor in S^* might be thought of as uncontrollable from the appraiser's standpoint. The role of this factor seems to be simply to encourage the appraiser to use extra care in assigning probabilities where $\mathrm{d}G(u)$ is high.

4.3. Scoring rules and expertise: p not known

The expected scores considered in section 4.2 are expected scores given p. They enable an appraiser (or an observer) to see how the appraiser's expected score varies as p changes and how the expected scores of appraisers differ when the appraisers do not have the same p. Expertise is rewarded in the sense that values of p close to perfect forecasts (e.g. p near zero or one for a single event) provide better expected scores than values of p indicating less sharpness (e.g. p near one-half for a single event).

In many instances, probability appraisers are evaluated before p is known. A decision-maker contemplating the purchase of information from an appraiser must consider the expected value of the information before it is actually obtained. To move back one step and investigate expected scores before p is known, it is necessary to anticipate possible values of p and their likelihoods of occurring. Let $f(p)$ represent the probability distribution of p, which may be discrete (in which case $f(p)$ is a probability) or continuous (in which case $f(p)$ is a density). This distribution could indicate the appraiser's own uncertainty about p before the relevant evidence is seen and p is formulated, or it could indicate an observer's uncertainty about an appraiser's p. Also, f may represent a distribution for a single p on a single occasion or it could represent a distribution of values of p for a number of occasions (e.g. a distribution of precipitation probability forecasts for a weather forecaster over numerous forecasting occasions).

When p is not known but has a distribution f, the expected score from a strictly proper scoring rule S is $E_f\{E_p[S(r)]\}$, where the reported r is viewed as a function of p. From (15) (see also DeGroot and Fienberg, 1983a),

$$E_f\{E_p[S(r)]\} = E_f\{E_p[S(p)]\} + E_f[C(S, r, p)], \tag{20}$$

with $E_f[C(S, r, p)]$ penalizing deviations from perfect calibration. If the appraiser sets $r = p$ for all p and is therefore perfectly calibrated ex ante, then $E_f[C(S, r, p)] = 0$ and the expected score reduces to $E_f\{E_p[S(p)]\}$.

Assuming that the appraiser is perfectly calibrated or that the appraiser's probabilities are adjusted to correct any anticipated miscalibration, the expected score of interest is $E_f\{E_p[S(p)]\}$. Since values of p close to perfect forecasts yield higher values of $E_p[S(p)]$ than do values of p indicating less sharpness, a distribution f that places more probability on values of p close to perfect forecasts and less probability on less sharp forecasts will provide a better expected score $E_f\{E_p[S(p)]\}$. This is the general sense in which expertise is rewarded via $E_f\{E_p[S(p)]\}$.

Of course, there is not a unique measure of sharpness. Just as different measures of dispersion may provide different orderings of a set of distributions, the ordering of distributions f (and therefore appraisers) in terms of $E_f\{E_p[S(p)]\}$ may differ for different scoring rules. Although all strictly proper scoring rules reward sharpness, the specific manner in which this is done varies somewhat.

A partial ordering of perfectly-calibrated probability appraisers can be obtained by using the concept of refinement introduced by DeGroot and Fienberg (1982, 1983a, 1983b). For the case of the probability p of a single event with a finite set $\mathscr{P}$ of values of p, appraiser 1 (with distribution f_1) is said to be at least as refined as appraiser 2 (with distribution f_2) if there exists a stochastic transformation $h(p_2|p_1)$ $[h(p_2|p_1) \geq 0$, for $p_1, p_2 \in \mathscr{P}$, and $\sum_{p_2 \in \mathscr{P}} h(p_2|p_1) = 1$, for $p_1 \in \mathscr{P}]$, satisfying

$$\sum_{p_1 \in \mathscr{P}} h(p_2|p_1) f_1(p_1) = f_2(p_2) \tag{21}$$

and

$$\sum_{p_1 \in \mathscr{P}} p_1 h(p_2|p_1) f_1(p_1) = p_2 f_2(p_2), \tag{22}$$

for all $p_2 \in \mathscr{P}$. In words, a perfectly-calibrated appraiser with the same distribution f_2 as appraiser 2 can be generated artificially by passing appraiser 1's probability through a noisy channel h. If appraiser 1 is at least as refined as appraiser 2 and f_1 and f_2 are not identical, then appraiser 1 is more refined than appraiser 2. Partial orderings of appraisers in terms of refinement are similar in spirit to partial orderings of actions by stochastic dominance in decision theory, and DeGroot and Eriksson (1985) show that refinement is equivalent to second-degree stochastic dominance. Other characterizations of refinement are given in DeGroot and Eriksson (1985) and DeGroot and Fienberg (1983b).

A more refined appraiser is one who demonstrates more sharpness, or expertise. This greater expertise leads to higher expected scores. As shown by DeGroot and Fienberg (1983a), if appraiser 1 is more refined than appraiser 2, then appraiser 1 has a higher expected score than appraiser 2 for any strictly proper scoring rule. When two appraisers cannot be ranked by refinement, then rankings in terms of expected scores may differ for different scoring rules.

The notion of refinement can be generalized beyond the case of a single probability p and a finite set of possible values of p. For example, partial orderings of appraisers can be obtained by comparing their distributions of p when p is a single probability that can take on any value in $[0,1]$ or when p is a vector of probabilities. Total orderings of appraisers in terms of expected scores depend on the specific scoring rule that is used. Each rule can be thought of as defining sharpness in a particular manner. In a broad sense, however, all strictly proper scoring rules reward expertise.

5. Summary and discussion

The notion of "good probability appraisers" is a somewhat elusive notion with several different aspects. At one extreme, with no knowledge about an event E or about two probability appraisers other than the fact that one claims that the probability of E is 0.4 and the other claims that the probability of E is 0.6, it seems impossible to say anything about how good these probability appraisers are. At the other extreme, with considerable past data concerning probabilities provided by the two appraisers and information about their backgrounds and expertise, it may not seem unreasonable to judge one to be a better probability appraiser than the other, at least in the sense of how useful their probability forecasts are judged to be to someone who wants to make inferences about E or decisions related to E.

Three types of goodness have been considered in this chapter: coherence, calibration, and expertise. With respect to coherence, a basic result in the theory of subjective probability is that an appraiser who is not coherent can be used as a "money pump". That is, a series of bets can be set up in such a manner that the appraiser is sure to lose money (de Finetti, 1937). Thus, a probability appraiser should want to be coherent, and it has been assumed in this chapter that any inconsistencies in probability forecasts are detected and removed.

The relationship between the other types of goodness and evaluation measures known as proper scoring rules has been examined in sections 3 and 4. From the definition of proper scoring rules, they encourage ex ante calibration in the sense that a probability appraiser should provide probability forecasts in such a manner as to expect to be calibrated. In an ex post sense, they reward calibration in terms of higher average scores.

The results in section 4 indicate that ex ante, proper scoring rules encourage sharpness of probability forecasts; and ex post, they reward sharpness via higher average scores. Thus, if expertise is thought of in terms of sharpness, then a probability appraiser who wants to receive high scores should try to be calibrated while providing probability forecasts that are as sharp as possible. Some incentive is therefore provided for the acquisition of expertise. Furthermore, from the standpoint of a decision-maker who wants to hire a probability appraiser, linking the appraiser's fee to a proper scoring rule is reasonable to the extent that sharper (yet calibrated) forecasts are more valuable to the decision-maker than less sharp forecasts.

What does all of this imply for the decision-maker who is thinking of hiring two or more probability appraisers and combining their forecasts? Models for combining probability forecasts (e.g. Winkler, 1981) often suggest that adding forecasters leads to sharper forecasts, although dependence among the forecasters can reduce the impact of additional forecasters on sharpness. DeGroot and Fienberg (1983a) discuss the notion of sufficiency of forecasters or combinations of forecasters. To the extent that increasing the number of appraisers results in sharper forecasts, it might be said that two (or more) probability appraisers are "better" than one. Thus, it is possible to speak not just of good probability appraisers, but also of good combinations of probability appraisers.

References

Dawid, A.P. (1982) "The well-calibrated Bayesian", *Journal of the American Statistical Association*, 77, 605–610.

de Finetti, B. (1937) "La prévision: Ses lois logiques, ses sources subjectives", *Annales De L'Institut Henri Poincaré*, 7; translated by H.E. Kyburg, in: H.E. Kyburg and H.E. Smokler, eds., *Studies in Subjective Probability* (Wiley, New York, 1964).

de Finetti, B. (1962) "Does it make sense to speak of 'good probability appraisers'?," in: I.J. Good, ed., *The Scientist Speculates: An Anthology of Partly-Baked Ideas* (Basic Books, New York), pp. 357–363.

DeGroot, M.H. and E.A. Eriksson (1985) "Probability forecasting, stochastic dominance, and the Lorenz curve", in: J.M. Bernardo et al., eds., *Bayesian Statistics*, vol. 2 (North-Holland Publishing Company, Amsterdam), pp. 99–118.

DeGroot, M.H. and S.E. Fienberg (1982) "Assessing probability assessors: Calibration and refinement", in: S.S. Gupta and J.O. Berger, eds., *Statistical Decision Theory and Related Topics III*, Vol. 1 (Academic Press, New York) pp. 291–314.

DeGroot, M.H. and S.E. Fienberg (1983a) "The comparison and evaluation of forecasters", *The Statistician*, 32, 14–22.

DeGroot, M.H. and S.E. Fienberg (1983b) "Comparing probability forecasters: Basic binary concepts and multivariate extensions", unpublished manuscript, Carnegie-Mellon University.

Kadane, J.B. and S. Lichtenstein (1982) "A subjectivist view of calibration", Technical Report No. 233, Department of Statistics, Carnegie-Mellon University, Pittsburgh, PA.

Lichtenstein, S., B. Fischhoff and L.D. Phillips (1982) "Calibration of probabilities: The state of the art to 1980", in: D. Kahneman, P. Slovic and A. Tversky, eds., *Judgment Under Uncertainty: Heuristics and Biases* (Cambridge University Press, New York).

Lindley, D.V. (1982) "The improvement of probability judgments", *Journal of the Royal Statistical Society*, A, 145, 117–126.

Lindley, D.V., A. Tversky and R.V. Brown (1979) "On the reconciliation of probability assessments", *Journal of the Royal Statistical Society*, Series A, 142, 146–180.

Matheson, J.E. and R.L. Winkler (1976) "Scoring rules for continuous probability distributions", *Management Science*, 22, 1087–1096.

Morris, P.A. (1974) "Decision analysis expert use", *Management Science*, 20, 1233–1241.

Morris, P.A. (1977) "Combining expert judgments: A Bayesian approach", *Management Science*, 23, 679–693.

Murphy, A.H. (1973) "A new vector partition of the probability score", *Journal of Applied Meteorology*, 12, 595–600.

Murphy, A.H. and R.L. Winkler (1977) "Reliability of subjective probability forecasts of precipitation and temperature", *Applied Statistics*, 26, 41–47.

Savage, L.J. (1954) *The Foundations of Statistics* (Wiley, New York).

Savage, L.J. (1971) "Elicitation of personal probabilities and expectations", *Journal of the American Statistical Association*, 66, 783–801.

Winkler, R.L. (1967) "The quantification of judgment: Some methodological suggestions", *Journal of the American Statistical Association*, 62, 1105–1120.

Winkler, R.L. (1977) "Rewarding expertise in probability assessment", in: H. Jungermann and G. de Zeeuw, eds., *Decision Making and Change in Human Affairs* (D. Reidel, Dordrecht, Holland), pp. 127–140.

Winkler, R.L. (1981) "Combining probability distributions from dependent information sources", *Management Science*, 27, 479–488.

Winkler, R.L. and A.H. Murphy (1968) "'Good' probability assessors", *Journal of Applied Meteorology*, 7, 751–758.

Yates, J.F. (1982) "External correspondence: Decompositions of the mean probability score", *Organizational Behavior and Human Performance*, 30, 132–156.

Chapter 18

SCORING-RULE PROCEDURES FOR ELICITING SUBJECTIVE PROBABILITY AND UTILITY FUNCTIONS

CHARLES A. HOLT, JR.

University of Virginia

1. Introduction

It is well known that, under fairly general conditions, an individual's beliefs about the possible realizations of a random variable can be represented by a "subjective" probability distribution. In addition, there are sets of axioms (about rational behavior in risky situations) which imply that an individual's optimal decision is a decision which maximizes the expected value of a von Neumann–Morgenstern utility function (see DeGroot, 1970, ch. 6 and 7). In many situations, one individual (the "elicitor") would like to discover the precise form of a subjective probability distribution of another individual (the "subject"). The elicitor might attempt to obtain information about this probability distribution by letting the subject make a forecast of the realization of the random variable of interest with the explicit understanding that the subject will receive a money payoff which depends on the ex post accuracy of the forecast. The subject's forecast is a decision made under uncertainty, and the optimal decision will depend on his subjective probability *and* utility functions. In order to correctly interpret the subject's forecast, the elicitor must determine the subject's utility function and then unscramble the separate effects of his probabilistic beliefs and attitudes towards risk. Thus, accurate elicitation of probability information may require accurate elicitation of utility information. There are many other situations in which one individual may wish to obtain the utility function of another. For example, subjects may be participating in laboratory experiments in which their money payoffs depend on decisions made under uncertainty. If subjects' utility functions could be elicited after the experiment, then it would be possible to analyze the effects of risk aversion on behavior in the experiment.

Bayesian Inference and Decision Techniques, edited by P. Goel and A. Zellner

This chapter presents a procedure which can be used by the elicitor to induce the subject to reveal (1) a probability distribution representing the subject's beliefs about some uncertain event, and (2) a utility function representing a subject's attitudes toward risk.

2. Scoring-rule elicitation procedures: An introduction

This section contains a brief introductory discussion of some approaches to the elicitation of information about probability distributions and utility functions. Probability-elicitation schemes are commonly called "scoring rules". Scoring rules are discussed in Savage (1971). Matheson and Winkler (1974) contains an interesting analysis of scoring rules for continuous probability distributions. (In addition to the literature on eliciting probability distributions, there is a separate literature on the elicitation of a fractile of an unknown distribution; see Thomson (1979) and the references therein.)

The essential nature of a scoring rule can be illustrated with a simple example. Suppose the random variable of interest can have two possible realized values, and the subject's subjective probability of a high value is denoted by p. The elicitor asks the subject to report the probability of a high value, and this reported probability will be denoted by q. Suppose the elicitor can force the subject to accept a "reward" of $\ln(q)$ dollars if the value is subsequently observed to be high, and a reward of $\ln(1-q)$ dollars if the value is observed to be low. The subject is penalized severely with a large *negative* reward if the realized value is an event which was reported to have a relatively low probability. If the subject is risk neutral, then expected utility is maximized by making the reported probability for each event equal to the subject's subjective probability for that event. The fact that a logarithmic reward function induces correct reporting of probabilities is well known; see Savage (1971, section 9.4).

This logarithmic scoring rule can also be used for continuous distributions. Let w^* denote the realization of a continuous random variable. The subject's subjective beliefs are represented by a subjective probability distribution with a density function denoted by $f(w)$ and defined on $[a,b]$. The elicitation procedure is to have the subject report a density function $g(w)$ when it is understood that his monetary reward will be $\ln(g(w^*))$ dollars when w^* is observed. A risk-neutral subject would report a density function which maximizes the following functional:

$$\int_a^b \ln(g(w))f(w)\,\mathrm{d}w. \tag{1}$$

This is a special case of the variational problem discussed in the next section.

Suppose that the subject is not risk neutral and that his attitudes toward risk can be represented by a utility function denoted by $U(\cdot)$. Also, let the subject's monetary reward be a function $\pi(\cdot)$ of the reported density at w^*. Thus, if

$w = w^*$, the composite reward function, denoted by $\psi(\cdot)$, can be expressed as $\psi(g(w^*)) = U(\pi(g(w^*)))$. If the elicitor knows the subject's utility function, then it may be possible to specify $\pi(\cdot)$ so that $\psi(g) = \ln(g)$. As indicated above, the optimal reported density would then coincide with the subject's subjective density. But if practical considerations make it impossible to collect arbitrarily large negative rewards, then there may be no practical $\pi(\cdot)$ function which results in a logarithmic composite reward function. When the composite reward function $\psi(\cdot)$ is not logarithmic, then the optimal reported density may differ from the subject's actual subjective density (although there are other scoring rules that do result in correct reporting of probabilities). If the elicitor knows the subject's utility function up to an affine transformation, then the elicitor will know the $\psi(\cdot)$ function, and the reported density will still convey useful information to the elicitor – even though the reported and actual densities are different. For any $\psi(\cdot)$ function in a fairly general class of concave functions, the formulae in the next section can be used by the elicitor to infer the subject's actual density function from the optimal reported density function.

I am not aware of any scoring-rule procedure which does not require the elicitor to make a specific assumption about the subject's utility function. This is because it is difficult, if not impossible, to take a reported probability distribution and unscramble the separate effects of the subject's risk preferences and probabilistic beliefs. Thus, it would be convenient to have scoring-rule procedures for eliciting utility functions. This is the topic of section 4.

The utility-elicitation procedures presented in this chapter have the desirable property that the elicitor does not have to specify a parametric family of utility functions. These procedures are similar in structure to the scoring-rule probability elicitation procedures discussed previously. First, the subject is asked to announce a function, and the elicitor then uses a random device to compute a monetary reward for the subject according to a probability distribution which depends on the subject's announced function. Of course, the subject must know precisely how the announced function affects the distribution of rewards. Sections 4 and 5 contain formulae for constructing the probability distribution of rewards. These formulae cause the optimal announced function to contain all essential information about the utility function. This enables the elicitor, in a single step, to compute the subject's utility function up to an affine transformation. These elicitation procedures are, in this sense, scoring rules for utility functions.

Applications and limitations of scoring-rule procedures are discussed in section 6, and a method of eliciting probability and utility information from the same subject is discussed. Section 7 contains a summary.

3. Elicitation of subjective probability distributions

Let w^* denote the realization of a random variable $\tilde{w}$. The subjective uncertainty about $\tilde{w}$ is represented by a cumulative probability distribution function $F(w)$

and a density function $f(w)$ such that $F(a)=0$, $F(b)=1$, and $f(\cdot)$ is continuous on (a, b). The elicitation procedure to be analyzed in this section has four stages:

(i) The elicitor announces reward functions denoted by ψ and ϕ.

(ii) The subject reports a probability distribution of a continuous random variable; the reported distribution and density functions are denoted by $G(\cdot)$ and $g(\cdot)$, respectively.

(iii) The realized value of $\tilde{w}$, denoted by w^*, is observed by all.

(iv) The subject receives a monetary reward from the elicitor which may be a function of both $g(w^*)$ and w^*. This reward is computed:

$$\psi(G(w^*), g(w^*), w^*)+\int_a^b \phi(G(w), g(w), w)\,\mathrm{d}w. \tag{2}$$

The notation in (2) could be simplified because $g(w)$ can be determined from $G(w)$. Listing $g(w)$ and $G(w)$ separately will facilitate the use of calculus of variations techniques and will clarify the relationship between the general notation and various special cases. The monetary payoff for the logarithmic scoring rule discussed in the previous section is a special case of (2) in which $\psi=\ln(g(w^*))$ and $\phi=0$. Notice that the reward function in (2) has two parts; the first term depends on the realized value of w, and the second term is a nonstochastic guaranteed payment in the probability elicitation procedure. The specific probability scoring rules discussed in this section have the property that the partial derivatives of ψ and ϕ with respect to G and w are zero, but the added generality implicit in (2) will be useful later.

It follows from the reward structure in (2) that a risk-neutral subject will report a function $G(\cdot)$ which solves the following variational problem:

$$\underset{G(\cdot)\in A}{\text{maximize}} \int_a^b [\psi(G(w), g(w), w)f(w)+\phi(G(w), g(w), w)]\,\mathrm{d}w$$

$$\text{s.t. } G(a)=0, \qquad G(b)=1, \tag{3}$$

where the set A of feasible $G(\cdot)$ functions must be specified.

There is a useful alternative interpretation of (3) when $\phi(G(w), g(w), w)$ is identically zero and the subject is risk averse. The *monetary* reward function is still specified by the elicitor, but the utility of this reward is not specified directly. In this case, $\psi(G(w), g(w), w)$ denotes the subject's *utility* of the monetary reward. Thus, the objective functional in (3) becomes an expected-utility expression. The elicitor cannot infer anything from the reported $G(\cdot)$ function unless the subject's utility function is known (or learned subsequently).

The following theorem can be used to solve the elicitor's problem of inferring the subject's probability distribution $F(\cdot)$ from the reported distribution $G(\cdot)$. The following notation is used. The optimal reported density and distribution functions are denoted by $\hat{g}(\cdot)$ and $\hat{G}(\cdot)$ respectively. In addition, partial derivatives of ψ and ϕ are denoted by appropriate subscripts. Total derivatives with respect to w are denoted by prime superscripts, i.e. $(\psi_g)'\equiv \mathrm{d}/\mathrm{d}w(\psi_g)$ and $(\phi_g)'\equiv \mathrm{d}/\mathrm{d}w(\phi_g)$.

Theorem 1. Let $\hat{G}(w)$ be an interior solution to the variational problem in (3) when the feasible set A is some subset of continuously differentiable functions on $[a, b]$. Suppose that $f(w)$ is differentiable on $[a, b]$ and $\psi(G, g, w)$ and $\phi(G, g, w)$ are continuously differentiable on some open set which contains $\{(G, g, w): G \in A, w \in [a, b]\}$. Then $f(\cdot)$ must satisfy the following first-order, linear differential equation:

$$-\psi_g f'(w) + \left[\psi_G - (\psi_g)'\right] f(w) + \phi_G - (\phi_g)' = 0, \tag{4}$$

where all derivatives are evaluated at the point $(\hat{G}(w), \hat{g}(w), w)$. The relevant particular solution to (4) is determined by the boundary condition that $F(b) = 1$.

Proof. Given the assumed properties of $f(\cdot)$ and the reward functions, the following Euler condition is necessarily satisfied in $[a, b]$:

$$\psi_G f(w) + \phi_G - \frac{\mathrm{d}}{\mathrm{d}w}\left[\psi_g f(w) + \phi_g\right] = 0, \tag{5}$$

where all derivatives are evaluated at $(\hat{G}(w), \hat{g}(w), w)$. The Euler condition (5) is equivalent to the first-order, linear differential equation for $f(\cdot)$ in (4). □

Once the elicitor knows $\hat{G}(\cdot)$, the coefficients of (4) can be calculated. To determine the subject's density function $f(\cdot)$, the elicitor must use the relevant boundary condition, $F(b) = 1$. (Note that $F(a) = 0$ implies no restriction on $f(\cdot)$.)

To use theorem 1, it is necessary to specify the feasible set A and the reward functions. If the integrand in the objective functional (3) is concave in $g(\cdot)$ and $G(\cdot)$ for all $f(\cdot)$, then any local maximizer will be a global maximizer. Besides choosing reward functions with the right concavity properties, it may be desirable to structure the rewards so that the necessary computations are simple.

For example, scoring-rule elicitation procedures often have the property that ψ_G and ϕ_G are identically zero. In this case, the Euler condition in (5) implies that the total derivative of the term in square brackets be zero, or equivalently that the term in square brackets be equal to a constant λ. In this case, the subject's density $f(\cdot)$ can, in theory, be computed as

$$f(w) = \frac{\lambda - \phi_g(\hat{g}(w), w)}{\psi_g(\hat{g}(w), w)}, \quad \text{for } w \in [a, b], \tag{6}$$

where the constant λ which ensures that $F(b) = 1$ is

$$\lambda = \frac{1 + \displaystyle\int_a^b \frac{\phi_g(\hat{g}(w), w)}{\phi_g(\hat{g}(w), w)}\,\mathrm{d}w}{\displaystyle\int_a^b \frac{1}{\psi_g(\hat{g}(w), w)}\,\mathrm{d}w}. \tag{7}$$

In particular, suppose that $\psi(G, g, w) = \ln(g)$, $\phi(G, g, w) = 0$, $f(w) > 0$ on $[a, b]$, and A is the set of continuously differentiable, strictly increasing functions on $[a, b]$. In this case, $\psi_g = 1/g(w)$, $\phi_g = 0$, $\lambda = 1$, and eq. (6) becomes: $f(w) = \hat{g}(w)$. This logarithmic scoring rule is "proper" in the sense that the subject's optimal reported density is the subjective density. Another well-known proper scoring rule, the "quadratic rule", is constructed as $\psi(G, g, w) = 2g$ and $\phi(G, g, w) = -g^2$. In this case it follows from (6) and (7) that $\lambda = 0$ and $f(w) = \hat{g}(w)$. If $f(w) > 0$ on $[a, b]$, then $\hat{g}(w) = f(w)$ is in fact an interior solution for both the logarithmic and quadratic scoring rules, and the integrand of the objective functional (3) is strictly concave in $g(\cdot)$ for all $f(\cdot)$ functions, so $\hat{g}(w) = f(w)$ is a global maximum in each case.

Much of the previous work on probability elicitation is concerned with discovering and comparing scoring rules which induce correct reporting of probability distributions. The implication of theorem 1 is that scoring rules which do not induce correct reporting may also be useful. For example, suppose that $\phi = 0$ and $\psi = [g(w)]^{1-r}$, where $0 < r < 1$. This case will be called the power rule. For this rule, the implication of (6) and (7) is

$$f(w) = \frac{[\hat{g}(w)]^r}{\int_a^b [\hat{g}(x)]^r \mathrm{d}x}, \quad \text{for } w \in [a, b]. \tag{8}$$

Thus, the actual density is easily calculated from the reported density, even though the optimal reported density differs from the subjective density. Notice that the payment is always non-negative for this power rule. Recall that there is no lower bound on the "reward" determined by the logarithmic rule, so there is no way to ensure a non-negative payment to the subject. One advantage of the power rule over the logarithmic rule is that the monetary reward will never be negative.

Both the logarithmic and power rules discussed above have the convenient properties that ϕ is identically zero, $\psi_w = 0$, and ψ exhibits constant relative "risk aversion" with respect to g. In other words,

$$r(g) \equiv -g[\psi_{gg}/\psi_g] \tag{9}$$

is constant. To see why this property is computationally convenient, the reader should differentiate both sides of (5) with respect to w. If $\phi = 0$, $\psi_G = 0$, $\psi_w = 0$, the result of this differentiation can be expressed as

$$-f'(w)/f(w) = r(\hat{g}(w))[-\hat{g}'(w)/\hat{g}(w)]. \tag{10}$$

When $r(g)$ is constant, the ratio $f'(w)/f(w)$ for the actual density function is proportional to the corresponding ratio for the reported density function.

4. Elicitation of utility functions

Suppose that a subject's preferences among money lotteries can be represented by a strictly increasing, differentiable utility function $f(w)$ on $[a, b]$, where w denotes money income. Without loss of generality, let $f(a) = 0$ and $f(b) = 1$. This section contains several alternative mechanisms for inducing the subject to report a function $G(w)$ which reveals all essential information about $f(w)$.

In the previous section the subject's expected utility is determined by two factors: the payoff functions, ψ and ϕ, and the subjective probability distribution for the random variable $\tilde{w}$. The subject's reported function $G(\cdot)$ affects the payoff functions but not the probability distribution. The roles of probability and utility can be reversed by letting the subject's reported $G(\cdot)$ function affect the probability distribution of monetary payoffs, but not the monetary payoff functions. As before, the derivative of $G(\cdot)$ is denoted by $g(\cdot)$. The specific elicitation procedure to be considered has four stages:

(i) The elicitor specifies the functions $\psi(G(w), g(w), w)$ and $\phi(G(w), g(w), w)$ which are revealed to the subject.

(ii) The subject reports a differentiable function $G(\cdot)$ which is strictly increasing on $[a, b]$. In addition, it is required that $G(a)=0$ and $G(b)=1$.

(iii) The elicitor then uses a random device to generate the realization of the random variable with a density function of $\psi(G(w), g(w), w)$ for w on the *open* interval (a, b). The probability that $w=b$ is

$$\int_a^b \phi(G(w), g(w), w)\,\mathrm{d}w, \tag{11}$$

and the probability that $w=a$ is

$$1-\int_a^b \psi(G(w), g(w), w)\,\mathrm{d}w-\int_a^b \phi(G(w), g(w), w)\,\mathrm{d}w. \tag{12}$$

(iv) The subject's *monetary* payoff is the realized value of the random variable obtained in stage (iii).

It follows from the normalization of the utility function $f(\cdot)$ that an expected-utility-maximizing subject will report a function which solves the variational problem in (3). The set A must be selected so that the formulae in (11) and (12) for computing probabilities are non-negative and less than or equal to one for all admissible $G(\cdot)$ functions.

The two scoring rules for the elicitation of utility functions to be discussed in this section have the property that ψ and ϕ depend only on $G(\cdot)$, i.e. $\psi_g=\phi_g=\psi_w=\phi_w=0$. If ψ and ϕ are continuously differentiable in $G(\cdot)$, it follows from eq. (4) in theorem 1 that

$$f(w)=\frac{-\phi_G(\hat{G}(w))}{\psi_G(\hat{G}(w))},\quad \text{for } w\in[a,b]. \tag{13}$$

First consider a power rule which is similar to the power rule for probability elicitation in the previous section. For some r such that $0<r<1$, let

$$\psi(G(w), g(w), w)=\frac{G(w)^{1-r}}{2(b-a)}, \tag{14}$$

$$\phi(G(w), g(w), w)=\frac{1-(1-r)G(w)}{2(b-a)}, \tag{15}$$

and let A be the set of strictly increasing, continuous functions on $[a, b]$. Note that both ψ and ϕ are less than or equal to $1/[2(b-a)]$ because $0 \leq G(w) \leq 1$. Thus, the expressions for probabilities in (11) and (12) are always greater than or equal to zero and less than or equal to one as required. Then it follows from (13) that

$$f(w) = \hat{G}(w)^r, \quad \text{for } w \in [a, b], \tag{16}$$

and hence

$$\hat{G}(w) = f(w)^{1/r}, \quad \text{for } w \in [a, b]. \tag{17}$$

Next consider a quadratic rule which is similar to the quadratic rule for probability elicitation. This rule was suggested to me independently by Morris DeGroot and by an anonymous referee. Let

$$\psi(G(w), g(w), w) = \frac{G(w)}{(b-a)}, \tag{18}$$

$$\phi(G(w), g(w), w) = \frac{1 - G(w)^2}{2(b-a)}, \tag{19}$$

and let A be the set of continuous and strictly increasing functions on $[a, b]$. It follows from (13), (18), and (19) that the optimal reported function $G(\cdot)$ equals the utility function $f(\cdot)$.

This quadratic rule is equivalent to the following procedure. After the subject reports $G(\cdot)$, the elicitor obtains a number p^* which is the realization of a uniform random variable on $[0,1]$, and he independently obtains a number w^* which is the realization of a uniform random variable on the interval $[a, b]$. If $G(w^*) \geq p^*$, the subject receives w^* dollars, and if $G(w^*) < p^*$, the subject receives a money payoff of "b" dollars with probability p^* and a money payoff of "a" dollars with probability $1 - p^*$. The subject's utility is $f(w^*)$ if $G(w^*) \geq p^*$, and the utility is p^* if $G(w^*) \leq p^*$. The subject's expected utility can be calculated as

$$\int_a^b \left[\int_0^{G(w)} f(w)\,\mathrm{d}p + \int_{G(w)}^1 p\,\mathrm{d}p \right] \frac{1}{b-a}\,\mathrm{d}w, \tag{20}$$

or equivalently as

$$\int_a^b \left[\frac{G(w)}{b-a} f(w) + \frac{1 - G(w)^2}{2(b-a)} \right] \mathrm{d}w. \tag{21}$$

Note that the subject's expected utility expression in (21) is equivalent to the objective functional in (3) when ψ and ϕ are defined in (18) and (19), respectively.

It is easy to see why the procedure outlined above yields a "proper" scoring rule for utility functions. The subject is told: "You will receive either w^* or a

$p^*:(1-p^*)$ randomization over b and a, depending on which is better according to your reported utility function." Obviously, the best reply is a utility function which correctly represents the subject's preferences.

Improper scoring rules such as the power rules discussed above can be converted to proper rules if there is a closed-form solution for the optimal reported function $\hat{G}(w)$. For the power rule there is a simple expression for $\hat{G}(w): \hat{G}(w) = f(w)^{1/r}$. If a reward structure $\{\psi(G(w)), \phi(G(w))\}$ results in $\hat{G}(w) = f(w)^{1/r}$, then $\{\psi(G(w)^{1/r}, \phi(G(w)^{1/r})\}$ results in $\hat{G}(w)^{1/r} = f(w)^{1/r}$, or equivalently, $\hat{G}(w) = f(w)$. To verify this, replace $G(w)$ in (14) and (15) with $G(w)^{1/r}$ and use (13) to show that $\hat{G}(w) = f(w)$ for this simple modification of the power rule for utility elicitation. Other improper scoring rules can be converted to proper scoring rules in a similar manner.

5. Scoring rules for discrete cases

It is straightforward to modify the utility and probability scoring rules in the previous sections for discrete situations. Suppose that the realization of a random variable will be one of the n integers $\{1,2,\ldots,n\}$, and the subject's subjective probability that $w^* = i$ is denoted by f_i for $i = 1,\ldots,n$. Thus $\sum_{i=1}^{n} f_i = 1$. The subject is asked to report n numbers denoted by g_i, $i = 1,\ldots,n$, and let $G_i = \sum_{j=1}^{i} g_j$. Then the subject's monetary payoff is determined by the realized value i^* of the random variable. This monetary payoff is

$$\psi(G_{i^*}, g_{i^*}, i^*) + \sum_{i=1}^{n} \phi(G_i, g_i, i),$$

where the ψ and ϕ functions are specified by the elicitor in advance, as was the case for the elicitation mechanism in section 3.

A risk-neutral subject will choose the numbers $\{g_1, g_2,\ldots, g_n\}$ to maximize

$$\sum_{i=1}^{n} [\psi(G_i, g_i, i) f_i + \phi(G_i, g_i, i)] \tag{22}$$

subject to the constraint that $\sum_{i=1}^{n} g_i = 1$. Maximization of (22) results in discrete versions of various scoring rules. For example, if $\psi(G_i, g_i, i) = 2g_i$ and $\phi(G_i, g_i, i) = -g_i^2$, then it can be shown that $\hat{g}_j = f_j$ for $j = 1,\ldots,n$.

Next, suppose that the elicitor desires to obtain a utility function representing the subject's preferences for over the $n+1$ integers: $0,1,\ldots,n$. Let the subject's utility numbers be denoted by $f_0, f_1,\ldots, f_n$, where $f_0 = 0$ and $f_n = 1$. (Note that the f_i numbers will not sum to one as was the case for the probability elicitation procedure.) As before, the subject reports n numbers $(g_1, g_2,\ldots, g_n)$ which sum to one, and $G_i = \sum_{j=1}^{i} g_j$. Then these reported numbers are inserted into the reward functions ψ and ϕ to construct a lottery over the integers $\{0,1,\ldots,n\}$ which determines the subject's monetary payoff w:

$$P_r\{w=0\} = 1 - \sum_{i=1}^{n} [\psi(G_i, g_i, i) + \phi(G_i, g_i, i)],$$

$$P_r\{w=i\} = \psi(G_i, g_i, i), \quad \text{for } 0<i<n,$$

and

$$P_r\{w=n\} = \sum_{i=1}^{n} \phi(G_i, g_i, i).$$

Since $f_0 = 0$, the subject's expected utility is given in (22). For example, if $0<r<1$, $\psi(G_i, g_i, i) = (G_i^{1-r})/2n$, and $\phi(G_i, g_i, i) = [1-(1-r)G_i]/2n$, then it can be shown that the relationship between the $\hat{G}_i$ decisions and the actual utilities is $f_i = \hat{G}_i^r$. This is, of course, the discrete analogue of (16).

6. The uses of scoring-rule procedures

The scoring-rule procedures outlined above should not be used without giving careful consideration to the resulting incentive structure. First, the structure should be easy to explain and understand. The discrete scoring rules in section 5 are surely easier to administer than the continuous versions. Second, unlike the interactive procedures such as those developed by Kadane et al. (1980), single-stage scoring rules do not permit the elicitor to alert the subject to "inconsistent" responses. If single-stage scoring rules used alone are to provide good information, the potential monetary payoffs must be large enough to induce the subject to think carefully about the reported function. Third, there are special problems that could arise with the use of scoring-rule procedures to elicit a utility function because the phrase "utility function" means different things to different people. The utility scoring rules can be administered without mentioning the word "utility" or explaining utility theory to subjects who are unfamiliar with this concept. Because the explanations could bias the subjects' responses, any discussion of utility theory should be avoided if possible. This is consistent with the practice of avoiding "leading" phrases such as "competition" or "maximization" in economics experiments in which subjects are motivated by monetary incentives.

It is important to notice that the elicitation procedures in this chapter have only been analyzed in the context of a single, once-and-for-all interaction between the elicitor and the subject. The analysis of incentives in this chapter must be modified if there is a continuing, multiperiod relationship between the elicitor and the subject. In each case considered in this chapter, the subject is solving a single-stage optimization problem. Such myopic behavior may be appropriate in an experimental situation. For example, suppose that the subjects are participating in an experiment in which they must make decisions under uncertainty.

After the experiment is over, the experimentor could use a utility scoring rule to elicit subjects' utility functions. Then the experimentor could use this utility information (corrected for changes in wealth) to analyze the effects of risk aversion on the subjects' behavior in the experiment. Similarly, a utility elicitation procedure, used after a probability elicitation procedure, may allow the experimentor to adjust the reported probability distribution function for the effects of risk aversion.

It is also possible to elicit probability and utility information simultaneously. To do this, the elicitor specifies both probability and utility scoring rules and informs the subject that a flip of a coin will determine whether the utility scoring rule response or probability scoring rule response will determine the subject's monetary payoff. Then the expected-utility-maximizing pair of responses would be identical to the responses which are optimal for the utility scoring rule and the probability scoring rule *considered separately*, without any adjustment for wealth effects. Thus, theorem 1 could be used to determine a utility function from the utility scoring-rule response. Then this utility function, together with the monetary payoff function for the probability scoring rule, could be used to compute the ψ function which is relevant for using theorem 1 to infer a probability distribution from the response to the probability scoring rule. In other words, the ψ function for a probability scoring rule with $\phi = 0$ is really the composition of a utility function and a monetary payoff function, or in the notation of section 2, $\psi(\cdot) = U(\pi(\cdot))$. If the subject's utility function can be determined, then ψ can be determined ex post and theorem 1 can be used to infer probabilities. This ψ function which is used to determine the probability distribution will not generally be the ψ function for a proper scoring rule, and this is one advantage of having a procedure which makes it possible to use improper scoring rules.

Finally, the elicitation results in this chapter may turn out to have useful theoretical applications. Economic agents are often assumed to face the problem of choosing a function from some set of feasible functions, e.g. insurance contracts, investment strategies, or portfolios. It may be possible, in some contexts, to use theorem 1 to associate the function chosen by an economic agent (i.e. the optimal function) with the agent's utility or probability distribution function. In fact, the problem of inferring a utility function from an investor's asset demand correspondence has recently been solved by Green, Lau and Polemarchakis (1978). In theoretical applications of this nature, the continuous versions of the elicitation results may be just as convenient as the discrete rules in section 5.

7. Conclusion

This chapter considers a situation in which the elicitor wants to discover the precise form of a *function* which characterizes a subject's preferences or probabilistic beliefs. All of the "scoring-rule" elicitation procedures developed have a common structure: the subject's expected utility depends on a function reported by the subject, and the elicitor uses the subject's reported function to calculate the subject's actual (utility or distribution) function. By reversing the roles of probability and utility, the method of eliciting probabilities can be used to elicit utilities. The general scoring-rule procedure outlined in theorem 1 can be used to derive a number of specific scoring rules, some of which are well known and some of which are new.

A method of simultaneously eliciting probability and utility information for the same subject is also discussed. This simultaneous elicitation is important because ex post knowledge of the utility function is necessary to interpret the response in a probability elicitation procedure.

There are two important aspects of designing effective elicitation procedures. The first aspect, which I will call the bounded rationality problem, is to design a procedure which is simple enough so that the assumption of optimal behavior on the part of the subject is reasonable. My concern in this chapter, however, is primarily with the second aspect, which I will call the incentives problem. This problem arises when the information being elicited is not going to be used to determine an optimal decision on the subject's behalf, so the subject must be given a monetary incentive to consider the elicitation procedure carefully. The incentives problem is important in probability elicitation when the subject is an expert whose subjective probabilities will be used by someone else. An example in which the incentives problem is present in utility elicitation is when the elicitor needs to know an expert's utility function in order to correctly interpret the expert's response to a probability elicitation procedure. Another example is the case in which an economist desires to have information about the risk preferences of subjects in a laboratory experiment in order to determine the effects of risk preferences on the subjects' behavior.

Acknowledgment

This research was supported, in part, by the National Science Foundation under grant SES-82-19720.

References

DeGroot, M.H. (1970) *Optimal Statistical Decisions* (McGraw-Hill, New York). 1970.

Green, J.R., L.J. Lau and H.M. Polemarchakis (1978) "A theorem on the identifiability of the von Neumann–Morgenstern utility function from asset demands", *Economic Letters*, 1, 217–220.

Kadane, J.B., J.M. Dickey, R.L. Winkler, W.S. Smith and S.C. Peters (1980) "Interactive elicitation of opinion for a normal linear model", *Journal of American Statistical Association*, 75, 845–854.

Matheson, J.E. and R.L. Winkler (1974) "The elicitation of continuous probability distributions", Technical Report No. 86, Department of Statistics, Stanford University.

Savage, L.J. (1971) "Elicitation of personal probabilities and expectations", *Journal of the American Statistical Association*, 66, 783–801.

Thomson, W. (1979) "Eliciting production possibilities from a well-informed manager", *Journal of Economic Theory*, 20, 360–380.

Chapter 19

FARMERS' UTILITY FUNCTIONS

CLIFFORD HILDRETH and GLEN J. KNOWLES*

University of Minnesota and University of Wisconsin, La Crosse

1. Description of the study

A study of farmers' utility functions and expectation formation was undertaken by the authors in the fall of 1976. Its purposes were to aid in the assessment of the relevance of Bayesian decision theory to production and marketing decisions on farms and to develop information that would be needed to apply the theory.

The study was conducted through the Minnesota Agricultural Experiment Station in cooperation with the Economic Research Service, USDA, and the Southwestern Minnesota Farm Management Association. Interviews of about an hour each were held with 13 farm operators, selected from the Association's membership, on four occasions from November 1976 to December 1977. The respondents were established farmers conducting commercially significant operations. The first three interviews concerned the farm operator's general practices, planning procedures, sources of information, and current price expectations and are reported in Hildreth and Knowles (1982). In the fourth interview, respondents were asked to indicate their decisions in hypothetical circumstances involving financial uncertainty. These responses were used to estimate utility functions reported in section 4 below.

*Research performed under Project 14-40 of the Minnesota Agricultural Experiment Station. The authors are indebted to the Economic Research Service, USDA for partial financial support, and to the Computer Centers of the University of Minnesota and Oklahoma State University for computer time. John W. Pratt and two referees made useful suggestions. Erlin J. Weness, Extension Agent for the Southwestern Minnesota Farm Management Association, was very helpful in organizing the interviewing. The principal debt, however, is to the thirteen busy farm operators who conscientiously and cordially replied to numerous and sometimes naive questions.

Bayesian Inference and Decision Techniques, edited by P. Goel and A. Zellner

Problems encountered in formulating questions and interviewing are described in section 2. Section 3 concerns some properties of the responses. The estimates in section 4 are followed by a brief review of results in section 5. Data used in the analysis of section 4 are compiled in Hildreth and Knowles (1982).

2. Interviewing

The authors have assumed throughout that carefully considered financial choices under uncertainty approximately satisfy Bayesian consistency and therefore have an approximate expected utility representation. This is not firmly established and various contrary views have been advanced.

Such questions are difficult to resolve because actions can be interpreted in many ways. Strict consistency, and sometimes even approximate consistency, requires careful calculations by the decision-maker, so in practice errors must be expected. Also, actions that appear inconsistent by one formulation of a decision problem may be consistent when a different framework is used, e.g. a multivariate instead of a univariate utility function or random rather than fixed initial wealth. It seems to us reasonable to continue to explore the implications of expected utility models both theoretically and empirically and to let an eventual judgment rest on the extent to which researchers believe the models have resulted in improved understanding.

Someone studying behavior under uncertainty can, in principle, obtain observations in three ways. One can observe real-life decision-making; one can conduct experiments which pose contrived decision problems; or one can ask respondents to report decisions they would make in hypothetical situations. All have serious inherent difficulties.

In a Bayesian interpretation of real-life decisions the investigator is hampered by the fact that the decisions are based not just on utility functions. The decision-maker's subjective probabilities and the whole array of contingencies faced by the decision-maker are relevant. An experiment usually involves alternative chance rewards of modest value. This means that any incidental motivation – for example, the subject might feel a desire to impress the investigator, a desire to confuse the investigator, or a desire to vent some immediate frustration – may easily influence or even dominate his behavior. Even if no extraneous influences are present, the contrived problem of modest consequences may not call forth a very careful calculation. Investigators try, of course, to minimize such interferences but how successfully this is done is hard to judge.

In the present study, hypothetical questions seemed the only practical possibility. Using hypothetical questions poses some of the above difficulties and also raises an obvious question as to whether or not reported choices are necessarily those that would actually be invoked if the circumstances were real. We believe our respondents answered conscientiously, but anyone who has changed one's mind once or several times in an important decision will recognize that a reported inclination based on a few moments reflection is not necessarily what a party will do. If the decision is about a hypothetical problem there is still more room for doubt. Some of our conjectures about how this may have affected the interviews reported here are discussed in sections 3 and 4.

Naturally, we wished to formulate a hypothetical situation that would yield as reliable responses as we could obtain. It seemed desirable to recognize the following criteria:

(1) The situation should concern a serious matter to justify and hopefully stimulate a careful calculation.

(2) It should have sufficient plausibility to not invite quick dismissal.

(3) The chance aspect should arise as naturally as possible to avoid the complication of feelings respondents might have against contrived or casino-like gambling.

(4) The probabilities involved should arise in a way that the respondent will accept them without second-guessing.

(5) Unless the investigator is in a position to assess all contingencies faced by the respondent, the hypothetical contingency posed should be statistically independent of the real-life chances that the respondent already faces.

The need for (5) arises from the fact that if prospective gains or losses resulting from an immediate decision are related to gains or losses already in the decision-maker's prospect, then the nature of the relationships must be known to correctly interpret the decision. A simple illustration is given below.

Example. Let $\phi(x)$ be utility of wealth, the utility experienced by a subject if he knows his wealth at some future time will be x. Let X be a random variable whose values are the possible amounts that the subject might realize at the future time if he follows his present plans and fulfills his present commitments. Assuming no changes, the subject's expected utility is $E\phi(X)$ and X is called his initial prospect. If he undertakes new plans, commitments, and/or investments (or modifies old ones), he will have a new prospect Z and a new expected utility $E\phi(Z)$.

$Y = Z - X$ is the modification in his prospect and is called his venture. If he is considering new plans, commitments, and/or investments he has a choice of modifications or ventures and his decision problem is

$$\max_{Y \in \mathscr{Y}} E\phi(X+Y), \tag{2.1}$$

where $\mathscr{Y}$ is the family of possible ventures. For any choice of Y, the resulting expected utility may be written

$$E\phi(X+Y) = \int\int \phi(x+y)\,\mathrm{d}F_{X|Y=y}(x)\,\mathrm{d}F_Y(y). \tag{2.2}$$

It is expected that drawing inferences about properties of ϕ from observed choices of Y will generally be complicated by dependence of choice on the joint distributions of X with the eligible Y's.

If X is independent of Y for $Y \in \mathscr{Y}$, matters are simpler; then $F_{X|Y=y} = F_X$ and one may define the utility of gain as

$$\Psi(y) = E\phi(X+y) = \int \phi(x+y)\,\mathrm{d}F_X(x), \tag{2.3}$$

where gain y is defined as the increment (positive or negative) of future wealth that is due to modification of the initial prospect. Expected utility corresponding to a venture Y is then $E\phi(X+Y) = E\Psi(Y)$ over $Y \in \mathscr{Y}$. Estimating Ψ from observed choices requires less data than are needed for ϕ, but important properties are closely related.[1] Neglect of the initial prospect if it is not independent of the ventures will typically lead to misinterpretations. This is illustrated below for a simple case.

Suppose the subject has constant positive absolute risk aversion.[2] His utility of wealth function may we written (Pratt, 1964),

$$\phi(x) = -\mathrm{e}^{-\lambda x}, \quad \lambda > 0, \tag{2.4}$$

and with initial prospect X, utility of gain is

$$-E\mathrm{e}^{-\lambda(X+y)} = -(E\mathrm{e}^{-\lambda X})\mathrm{e}^{-\lambda y}. \tag{2.5}$$

Since utility functions are determined to positive linear transformations we may take

$$\Psi(y) = -\mathrm{e}^{-\lambda y}. \tag{2.6}$$

An investigator wants to determine λ, the coefficient of (absolute) risk aversion. He asks the subject for the cash equivalent m of a chance to receive y if an unbiased coin shows heads. Since the toss of a coin is independent of the subject's initial prospect, m satisfies

$$0.5\Psi(0) + 0.5\Psi(y) = \Psi(m), \tag{2.7}$$

or, equivalently,

$$\mathrm{e}^{-\lambda m} - 0.5\mathrm{e}^{-\lambda y} = 0.5. \tag{2.8}$$

Knowing m and y but not λ, the investigator solves (2.8) (perhaps by an iteration) and obtains the correct value of λ.

Note that $\lambda = 0$ always solves (2.8). If $m = y/2$, this is the only solution and indicates the subject's risk neutrality for this particular choice. If $m > y/2$, we are interested in the unique nonzero solution.

Suppose the subject is a farmer who agrees with the researcher that the chance of a freeze within a week is 50–50. The researcher asks the cash equivalent m of a chance to receive y if it freezes and proceeds to determine λ in the same manner as with the coin toss. This will yield the correct value if the farmer has no prior

[1] For example, if the nth derivative of ϕ has a particular sign, then $\Psi^{(n)}$ agrees in sign (Hildreth, 1974a, p. 103).

[2] Local absolute risk aversion is defined by $r(x) = -\phi''(x)/\phi'(x)$ for the utility of wealth function $\phi(x)$ or by $\rho(y) = -\Psi''(y)/\Psi'(y)$ for the utility of gain function.

stake in the possibility of a freeze and thus determines m as with a coin toss. Suppose, however, that the farmer has a berry crop worth z that will be lost if there is a freeze.

Let W represent the farmer's initial prospect, except for the berry crop, and assume W is independent of a possible freeze. m will then satisfy

$$0.5E\phi(W+z)+0.5E\phi(W+y)=0.5E\phi(W+z+m)+0.5E\phi(W+m). \tag{2.9}$$

Before considering the venture the farmer's expected utility is $0.5E\phi(W+z)+0.5E\phi(W)$. If he receives m in cash, his expected utility becomes the right-hand side of (2.9); if he receives y contingent on a freeze, his expected utility becomes the left-hand side of (2.9). Rewriting the equality with $\phi(t)=-e^{-\lambda t}$ yields:

$$e^{-\lambda m}(1+e^{-\lambda z})-e^{-\lambda y}-e^{-\lambda z}=0. \tag{2.10}$$

If the investigator knows that the initial prospect consists of a contingency (z if it does not freeze, 0 if it does), and some contingencies, W, independent of freezing, he can obtain the correct λ by solving (2.10). However, if he ignores the initial prospect and solves (2.8) he will clearly be misled. For example, with true $\lambda=0.02$, a researcher who applied (2.8) instead of (2.10) would obtain the incorrect values $\tilde{\lambda}$, given in table 19.1, in the indicated circumstances.

To a careless investigator the farmer appears less risk averse than he really is because the venture involving y has some insurance value for the farmer and this is not recognized by the researcher. In fact, in circumstances 1 and 3 the farmer appears to prefer risk unless his/her initial prospect is taken into account. If the researcher's proposal had been to make y contingent on the absence of a freeze, the $\tilde{\lambda}$ calculated from (2.8) would lie above the true value. Whenever $F_{X|Y=y}$ differs from F_X, interpretations of choices of ventures without accounting for the initial prospect, will, except for rare accidents, be misleading.

In the present study, questioning for utility analysis was based on a hypothetical question believed to be independent of the initial prospect. The following situation was posed to introduce alternative choices. "Suppose a cancer vaccine has been discovered and the principal source is a particular gland of a few cattle. An animal that might furnish the needed antibodies has an enlarged gland at the base of the neck. A veterinary technician has visited your farm and found one steer with the enlarged gland. However, experience has shown that only half of the animals with the enlarged gland have the desired antibodies so there is a

Table 19.1.

Case	z	y	m	$\tilde{\lambda}$
1	10	1	0.547	−0.38
2	1	10	4.000	0.016
3	10	10	5.250	−0.020

50–50 chance that your steer will prove useful. The technician has arranged to return to perform a conclusive test. If positive (the steer is useful) he will pay you the current market price plus \$3000 for the steer. If negative he will leave the steer unharmed to be marketed with the others. As an alternative to the above arrangement, he is willing to pay you the current market price plus \$1000 to buy the steer outright and perform the test at his own facilities."

"Would you prefer to sell the steer at market plus \$1000 or take a 50–50 chance on receiving market plus \$3000?"

Upon receiving the answer, the interviewer adjusted the \$1000 sure return up or down until an amount was reached which the respondent indicated was about equally desirable to a 50–50 chance of \$3000. This was recorded and the \$3000 contingent amount was changed. A new certainty equivalent was located. Certainty equivalents corresponding to seven gambles of this kind were obtained in a personal interview and a postcard requesting five additional comparisons was left with the respondent to be mailed.

As each interview progressed, the hypothetical example faded naturally into the background and most respondents tended to say: "You want to know how much cash I would take in preference to a 50–50 chance of getting y dollars. Well..." The interviewers did form the impression, however, that the example helped set the tone in most cases. It is believed to have helped the panel visualize a naturally arising gamble, important but substantially unrelated to their current production and marketing decisions.

Questions, like the above, in which the respondent indicates a cash amount m that is equivalent in his preferences to a 50–50 chance of receiving y_1 or y_2 are called questions of type V in our discussion. In most cases, y_1 was set at zero. Near the end of type V questioning a variation was introduced which supposed that the farmer could have a conclusive test performed for a \$200 fee and would receive \$8000 if the animal were useful. This determined $y_1 = -200$, $y_2 = \$7800$ and was the only question in that particular interview in which a contingency involving possible loss to the farmer was used. If m is treated as an exact response, the utility of gain function satisfies:

$$0.5\Psi(y_1)+0.5\Psi(y_2) = \Psi(m). \tag{2.11}$$

A few days after the personal interviews were completed, the panelists were contacted by phone and asked a question of essentially the form:[3] "What is the most you would just be willing to pay for a 50–50 chance of winning z dollars?" An answer of m indicated that a 50–50 chance of netting $-m$ or $z-m$ was approximately as desirable as doing nothing, thus leaving the initial prospect unchanged. This type of question was called lottery ticket and designated by L. Of five farmers, we asked additional questions of a form suggested by Ramsey (1964). These R questions were either included after type V questions or involved

[3] The exact phrasing depended on what had seemed to be effective in earlier talks with each particular farmer.

an additional visit. As used in these cases[4] the question involves giving the respondent three numbers, y_1, y_2, and z, and asking him to designate an m such that a 50–50 chance of receiving y_1 or y_2 is just as attractive as a 50–50 chance of receiving z or m (any of the numbers may be either positive or negative).

The answer to an L question may be interpreted as an assertion by the respondent that his utility of gain function satisfies:

$$0.5\Psi(-m)+0.5\Psi(z-m)=\Psi(0). \tag{2.12}$$

For an R question the implicit assertion is

$$0.5\Psi(y_1)+0.5\Psi(y_2)=0.5\Psi(z)+0.5\Psi(m). \tag{2.13}$$

With any particular question and response one may associate a coefficient of absolute risk aversion in a natural way. One obtains a constant absolute risk aversion utility function that exactly fits the one observation. One then associates the resulting value of r with that response. When this is done a response will be said to be risk averse if $r>0$, and to be a risk preference response if $r<0$. As shown by Pratt, $\Psi(t)=-\mathrm{e}^{-\lambda t}$ if $r>0$ and $\Psi(t)=\mathrm{e}^{-\lambda t}$ if $r<0$. $\lambda=r$ in either case. If $\Psi(t)$ is replaced by either $-\mathrm{e}^{-\lambda t}$ or $\mathrm{e}^{-\lambda t}$ in (2.11), one obtains:

$$\mathrm{e}^{-\lambda m}-0.5\,\mathrm{e}^{-\lambda y_1}-0.5\,\mathrm{e}^{\lambda y_2}=0. \tag{2.14}$$

Solving for λ then yields the risk aversion coefficient associated with that particular (x, y, y_1, y_2). For L and R questions the respective equations, obtained from (2.12) and (2.13), are

$$0.5\,\mathrm{e}^{\lambda m}+0.5\,\mathrm{e}^{-\lambda(z-x)}-1=0 \tag{2.15}$$

and

$$\mathrm{e}^{-\lambda m}+\mathrm{e}^{-\lambda z}-\mathrm{e}^{-\lambda y_1}-\mathrm{e}^{-\lambda y_2}=0. \tag{2.16}$$

$\lambda=0$ is always a trivial solution to (2.14), (2.15), or (2.16) and is the only solution when the response is risk neutral. In other cases, one is interested in the unique nonzero solution. Uniqueness is easily shown using the Meyer–Pratt Lemma (Meyer and Pratt, 1968, p. 277). For risk neutral and risk averse responses, calculated values of λ are shown on the charts of section 3 and are used in preliminary inspections of the data. They are tabulated along with the questions and responses in Hildreth and Knowles (1982, table 7).

3. Responses: Some preliminary considerations

Economists have generally assumed that utility of wealth functions exhibit positive and decreasing absolute risk aversion (Arrow, 1971; Stiglitz, 1969). Tesfatsion[5] has shown that decreasing absolute risk aversion for utility of wealth implies decreasing absolute risk aversion for utility of gain. We believe that business decisions will typically reflect these properties. In the present study,

[4] See Knowles (1980, ch. IV).

[5] The original proof is, unfortunately, unpublished. An alternative proof is in Hildreth (1974b, p. 11).

positive risk aversion was assumed a priori as a basis for organizing the analyses reported in section 4. These analyses in turn substantially support decreasing absolute risk aversion. If we confine attention to gambles that are statistically independent of the decision-maker's initial prospect, then positive risk aversion means that businessmen only undertake ventures whose expected net return is positive, i.e. favorable gambles. Negative risk aversion would imply a willingness to undertake unfair gambles, i.e. with negative expected returns. Intuitively, one expects that even with negative risk aversion a decision-maker will not take an unfair gamble if a suitable fair or favorable gamble is available. It is shown in Hildreth (1974a, pp. 901–902) that a businessman who thinks one step ahead and always has access to an assortment of fair or favorable gambles will make business decisions according to a concave utility function even though his basic utility of wealth has a Friedman–Savage concave segment (Friedman and Savage, 1948, p. 297).

In the real world there are many opportunities for gambles that are, at worst, almost fair. Futures markets, options markets, currency markets, and others offer the chance to gamble at nearly fair odds (unfair only by a relatively small commission) even if one regards changes of price in each direction equally likely. If a party has even a small inkling on one of these markets that a particular movement is more likely, then betting on this is a (subjectively) favorable gamble. Since businessmen, including our farm operators, are well aware of at least some such opportunities, any inclination to gamble at unfair odds should be very short-lived. Even if someone would, in principle, be willing to gamble at unfair odds, it is clearly still more advantageous to gamble at fair or favorable odds.

How then does one explain that people are seen to gamble at unfair odds? Of various explanations, two particularly appeal to us. One is that many gamblers derive utility from being in a gambling situation – like some people derive utility from drugs. To fully analyze such behavior requires a multivariate utility function whose arguments include variables reflecting gambling experiences as well as the money outcome (see Smith, 1971; also *Newsweek*, 10 April 1972).

The other instance is one in which the individual recognizes a dominant need for a quick increase in wealth, and this outweighs potential consequences of possible losses of greater actuarial value. The decision-maker is desperate to achieve a particular objective or avoid a particular catastrophe (bankruptcy, loss of home, loss of respectability) and doing this requires immediate increased wealth. If the party has already used available access to favorable and fair bets, unfair are all that remain.

We doubted that any of the chosen panelists were financially desperate or strongly motivated by the desire to gamble. The doubts were confirmed by early interviewing. However, in later interviewing there were hints that fun of gambling or other considerations may have entered a few answers when stakes were low.

Before proceeding to formal analyses of the farmers' responses, the authors classified the data in various ways that might indicate general tendencies, including consistency with the above risk aversion presumptions. For this purpose the risk aversion coefficients described in section 2 were quite useful. Of the 286 total responses elicited, 155 were risk averse, 93 were risk neutral, and 38 were risk preferring.

At first glance, this does not seem strong confirmation of risk aversion. However, connecting the types of response with the circumstances in which each was received conveys a somewhat different impression, as we now discuss.

Theorists have typically assumed that gambles for small stakes are decided on actuarial value, i.e. exhibit risk neutrality, provided the decision-maker is not influenced by nonmonetary aspects of the gamble – where it is held, with whom associated, how it is perceived by others, etc.

This seems reasonable to us. In fact, our interviews combined with introspection and other discussions lead us to believe that many people naturally think first in terms of actuarial value (as nearly as they can ascertain it) and then modify the indicated behavior if an adequate reason appears. An adequate reason might be diminishing marginal utility of wealth, in which case a risk averse response occurs. For this to be an adequate reason a non-negligible sum must be involved. Other adequate reasons might be incidental nonfinancial aspects (such as those mentioned above) of an envisioned venture and these could produce either apparently risk averse or apparently risk preference actions or responses.

Another set of possible reasons for revising actions based on actuarial values of prospective ventures are the relations of the prospective venture to the decision-maker's other contingencies (initial prospect), as illustrated by the berry example in section 2.

Of course, reasons for deviating from actuarial or risk neutral behavior occur in various combinations. In real life the resulting calculations of expected utility may be complicated and involve quite a bit of internal utility searching, information gathering and processing, and expectation formation. These procedures are not psychologically cost free and if one does an actuarial value assessment and believes other considerations are of secondary magnitude (in their effect on utility) then one may as well make the risk neutral choice. Expected utility theory does not ordinarily take account of calculation costs[6] so one may expect observed behavior frequently to deviate from theoretical solutions when economizing on calculation costs is a factor.

In our interviews, risk neutral and risk preference responses tended to occur early in the interviewing. In the early stages type V questions were being asked, possible gains were relatively small, and there was only one question involving a possible loss (of only $200) for the farmer.

As interviewing progressed, possible losses were included in most questions, the sums involved became larger, and the respondents seemed to become more at home with hypothetical questions. All of these factors may well have contributed to the steady elimination of risk preference and risk neutral responses. Hopefully, the later responses involved more careful internal calculations.

[6]For an exception, see Marschak and Radner (1972, ch. 2).

Table 19.2. Risk status and type of question (number of responses).

Question type	Risk status A	N	P	
V	79	70	37	186
L	42	4	1	47
R	34	19	0	53
	155	93	38	286

Some notion of the extent of these effects is given in table 19.2, which shows the number of risk averse (A), risk neutral (N) and risk preference (P) responses to each type of question in the entire set of data.

All type L questions and most type R questions involved possible loss to the respondents. Type L questions generally involved substantially larger sums of money than types V or R. Thirty-six of 47 L questions involved possible gains of \$10 000 or more as compared with 85 of 186 for V and 6 of 53 for R. Type V questions were asked in the first visit in which hypothetical choices were posed, most L questions were asked in later phone contacts, most R questions in a subsequent visit.

In the remainder of this chapter, respondents with 75 percent or more cash returns from beef are designated B_1 to B_7. Those with 25 percent or more returns from corn were originally labeled C_1 to C_7. C_4 declined to participate and C_3 discontinued farming during the study. C_5 and C_6 involved father–son partnerships and the authors questioned whichever partner was available. C_{5f} identifies the father's response on farm C_5 and C_{5s} the son's. Both were available on the final visit and they chose to answer jointly. These responses are designated C_{5fs}; C_{6f} and C_{6s} identify father and son, respectively, on farm C_6.

B_3 was asked eleven R questions and all responses were risk neutral. The interviewer reported that these questions were asked after a number of V questions and that the respondent seemed impatient and not inclined to reflect on the questions. Except for these answers, there were only 8 risk neutral responses to R questions.

It may also be worth observing that five respondents ($C_1, C_2, C_{6f}, C_{6s}, C_7$) account for 30 of the 38 risk preference answers. These respondents tended to give different answers to the same questions when some asked in the initial interview were repeated on postcards and their responses included several violations of the sure thing principle (e.g. C_1 valued a 0.5 chance of 50 less than a 0.5 chance of 8 but later valued both at nothing). This suggests that the interviewers may not have succeeded in making questions clear in some of these cases.

C_{6f} started with risk preference answers, switched to risk averse and then mused: "I guess I just had an impulse to gamble when we started." He did not offer to go back and change answers so the originals were tabulated. C_{6s} explained his high risk premium for a chance to win $3000 by noting that he had a $3000 payment due soon and was not very liquid, so winning that much would avoid the inconvenience of raising it some other way. Side remarks of the farmers were usually interesting to the authors and were not discouraged unless a large part of the intended interview seemed threatened.

The authors have the impression that many of the initial answers were dominated by comparisons of actuarial values and sometimes by associations of the postulated ventures with considerations not intended by the interviewers. B_6 initially gave risk preference answers and then asked to revise them. He said his first impulse was to accept a low price for his animal in order to further the research. When assured that the intended circumstance was that the research would proceed regardless of his decision, he changed to risk neutral answers.

In general, the interviewers tried hard not to guide answers and refrained from pointing out possible inconsistencies for this reason. Discussion of inconsistencies would also have used very scarce time. Panelists were permitted to reconsider and revise answers on their own initiative. The authors believe that discussions of inconsistencies with respondents can be useful, but the arrangements for the present study did not seem to offer a good opportunity.

Another impression was that calculating by the respondents became more careful as questioning progressed. Thinking about the questions seemed to some extent cumulative. The possibility of large gains and the existence of possible losses also seemed to stimulate more considered responses from some panelists. These tendencies are evident in the charts which follow. They are discussed and tabulated more fully in Hildreth and Knowles (1982, pp. 9,10).

Before interviewing, the researchers expected quite a few risk neutral answers to the $200 question and some of the others with low stakes. It was expected that risk averse answers would dominate when there was a stake of several thousand dollars or more. Instead, when a respondent seemed to be primarily motivated by the size of stake the switching point was $5000 to $50 000.

Taking the above evidence and other interviewing impressions into account, the authors believe that important actual business decisions will naturally bring forth more thorough calculations by entrepreneurial decision-makers and that risk aversion will be more persistent. Since a contribution to understanding important business decisions is viewed as the main purpose of the study, the analysis of section 4 is based only on risk averse responses.

To inspect the data for a possible tendency toward decreasing risk aversion some indicator of the general level of gain contemplated in a particular question was needed. Each question posed in this study, and many of those posed in others, involves an assertion by the respondent that

$$\sum_{i}^{m} p_i \psi(y_i) \simeq \sum_{i}^{n} q_j \psi(z_j),$$

where " $\simeq$ " is read "is about as desirable as".

$\sum p_i y_i$ is the actuarial value of the possible outcomes in the venture on the left and $\sum q_j z_j$ is the actuarial value of the possible outcomes of the venture on the right. The number used as an indicator is the simple average of the two actuarial values. This indicator, a, represents the average level of possible gain contemplated by the respondent in answering a particular question. In our questions the p's and q's were either 0.5 or 1 and there were at most two possible outcomes of a venture. Hence, for a type V question:

$$a = 1/2(0.5y_1 + 0.5y_2 + x),$$

and for types L and R, respectively:

$$a = 1/2(0.5x + 0.5(z - x)) = 0.25z - 0.5x,$$

$$a = 1/2(0.5y_1 + 0.5y_2 + 0.5z + 0.5x) = 0.25(y_1 + y_2 + z + y).$$

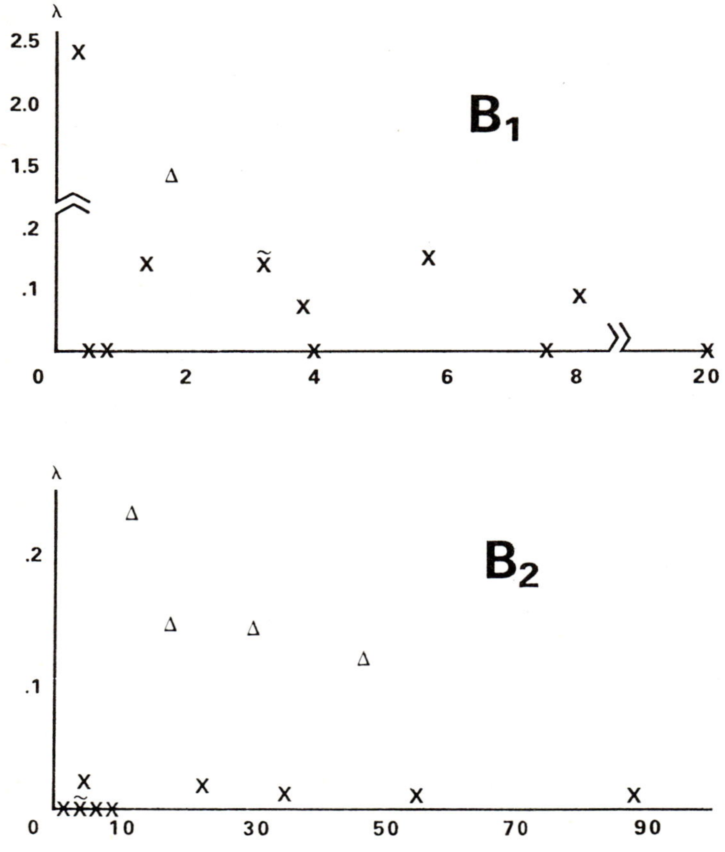

Fig. 19.1. Charts of risk aversion coefficients for individual observations. *Question types*: X designates a type V question; Δ type L question, and 0 a type R question. $\tilde{X}$ is the only type V question involving a possible loss.

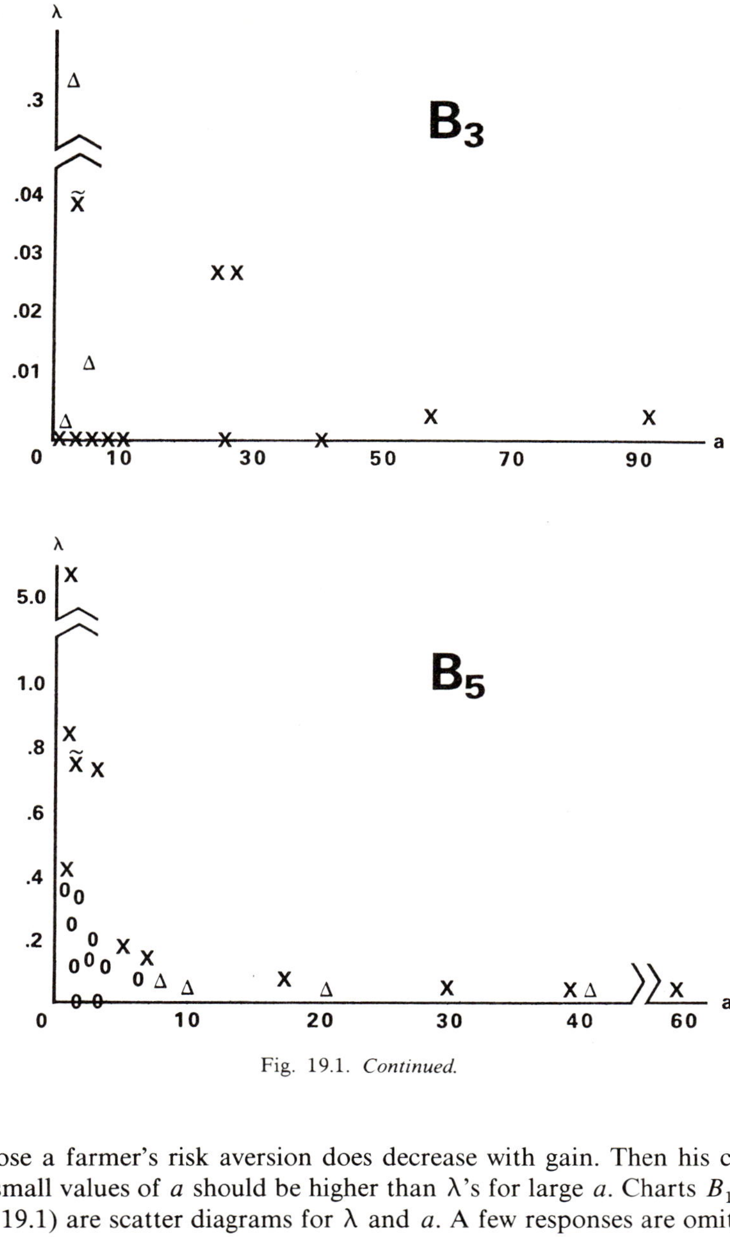

Fig. 19.1. *Continued.*

Suppose a farmer's risk aversion does decrease with gain. Then his calculated λ's for small values of a should be higher than λ's for large a. Charts B_1 through C_7 (fig. 19.1) are scatter diagrams for λ and a. A few responses are omitted from the charts, B_4 was omitted because he had only two nonzero λ's. On the second two type L questions, C_2 said he would not pay anything for the offered opportunities because it was too much like gambling. As noted above, B_3 was distracted during the visit involving type R questions. He replied quickly,

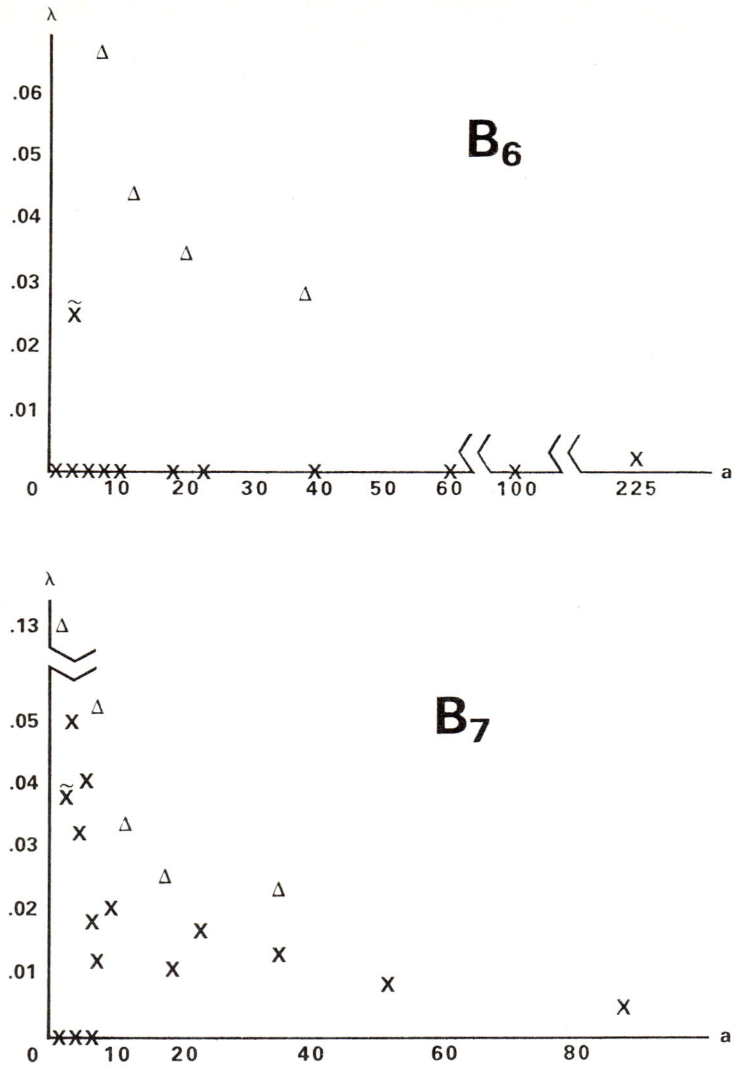

Fig. 19.1. *Continued.*

apparently routinely, and seemed anxious for the interview to end. These eleven observations with $\lambda = 0$ are not plotted.

Responses that indicated risk preference or risk neutrality are plotted along the a-axis in the charts. They are seen to be mostly associated with low a-values and with type V questions. Recall that they were asked early. Neglecting the entries on the a-axis, the downward drift of λ as a increases is striking and persistent – evi-

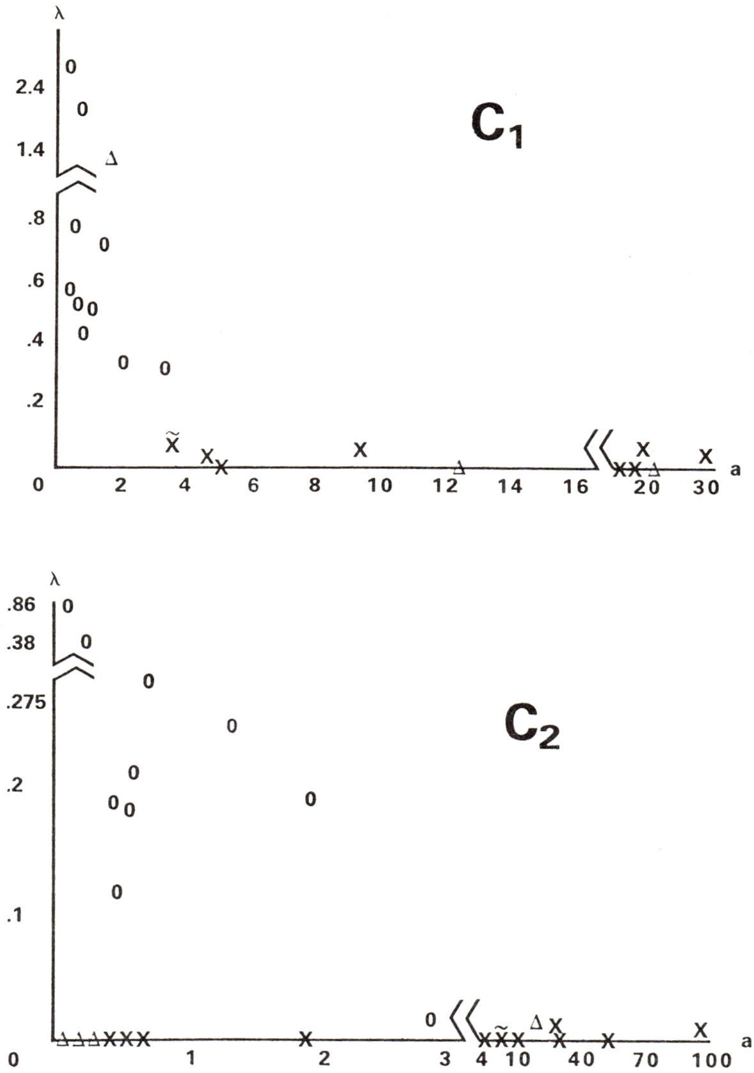

Fig. 19.1. *Continued.*

dent on every chart. The data are certainly consistent with economists' conjectures of decreasing absolute risk aversion.

If one were to approximate points with a free hand curve (or perhaps two curves in cases like B_2, C_{6f}, and C_7 where different questions elicited distinctly different responses) the curves would almost all be convex (note that arrangements like C_7, C_{6f}, and B_6 look more convex when adjustments in scale are taken

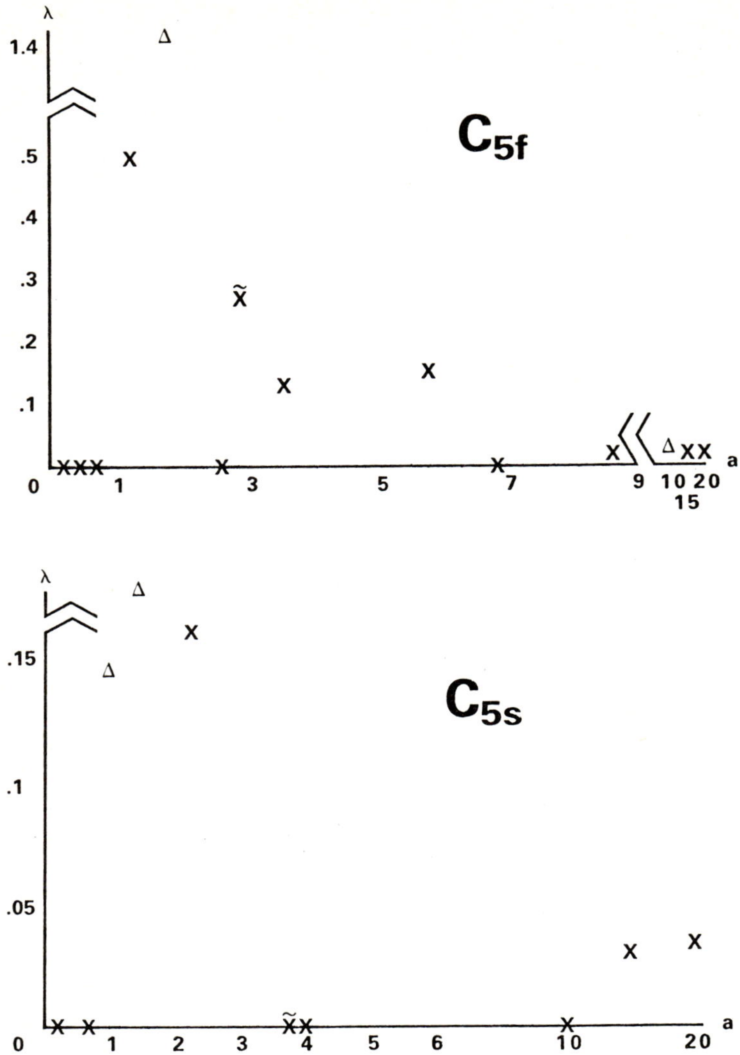

Fig. 19.1. *Continued.*

into account). This reinforces the notion that risk aversion probably remains positive but downward sloping as gain becomes large and suggests that λ (a) might typically be everywhere convex, at least for positive a, the only case for which our study furnishes data.

Some of the charts (B_2, B_7, C_1, and C_7) show a distinct difference between risk aversion corresponding to type V questions and risk aversion corresponding to

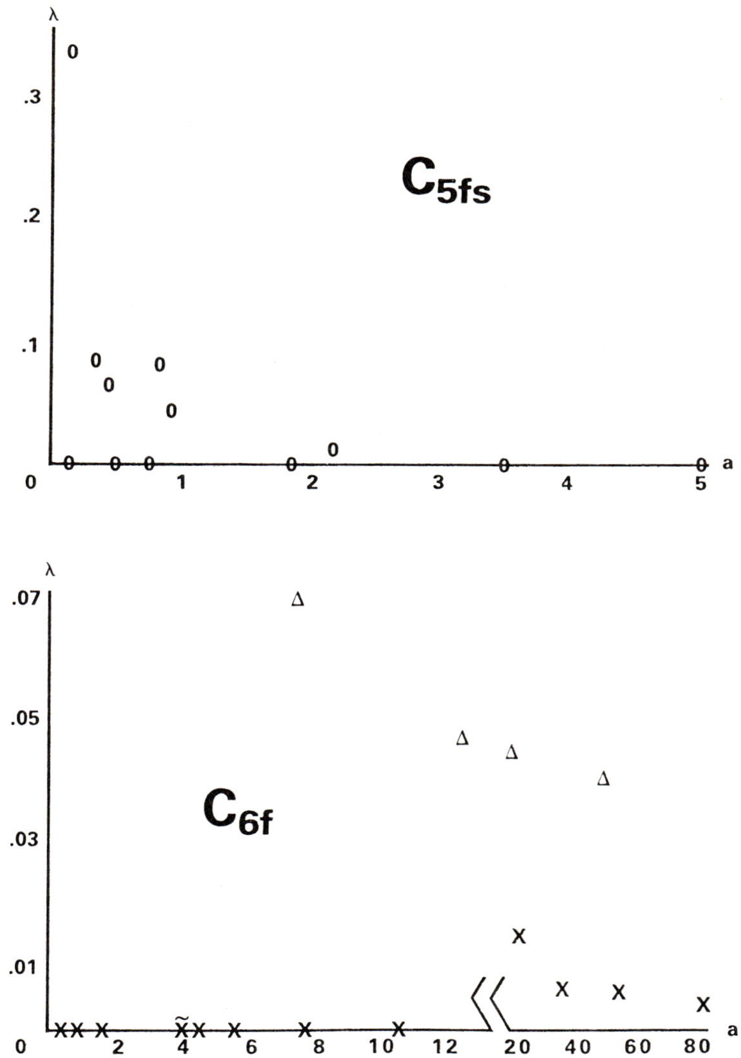

Fig. 19.1. *Continued.*

type L and/or R questions for comparable values of a. On other charts question types look reasonably homogeneous. A possible explanation for differences is that most R and all L questions involve the possibility of loss or cost to the panelist, while a possible loss is involved in only one V question (indicated $\tilde{X}$). As noted before, another possible explanation is that L and R questions were asked after V questions.

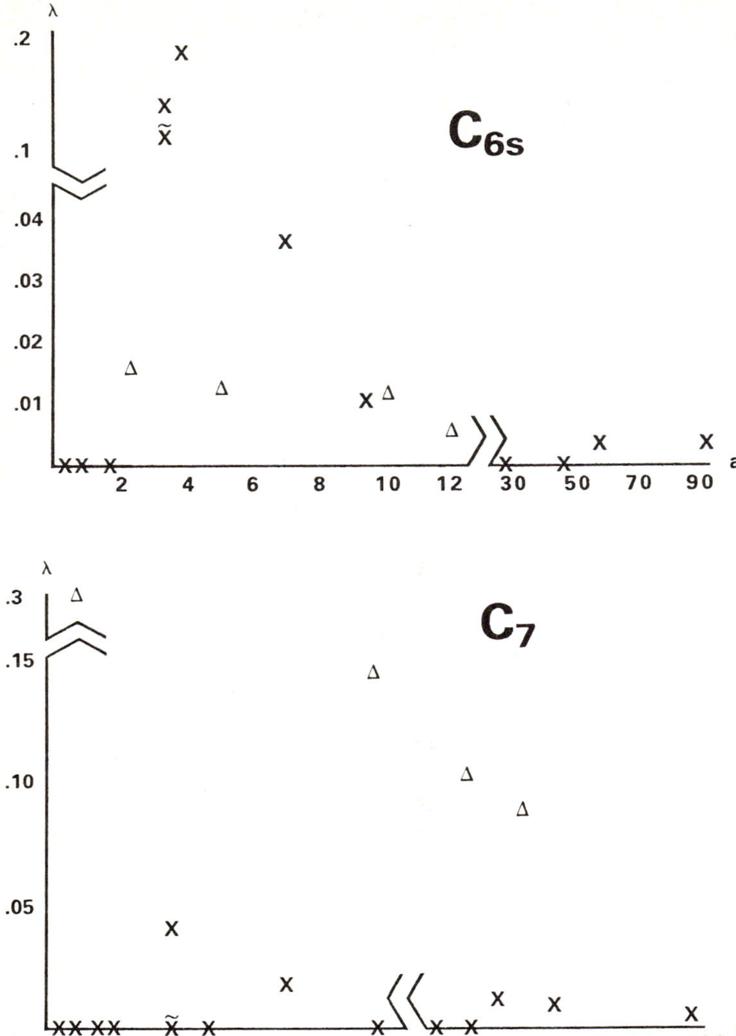

Fig. 19.1. *Continued.*

When V questions were planned, it was not known whether or not other questions would follow. If the initial questioning about hypothetical problems had seemed to disturb the panelists or to produce completely chaotic answers, the later phone and personal visits would not have been undertaken. In planning V questions, it was believed that questions not involving losses would be less troublesome at the outset, but one instance with a possible loss was included to see if this made a marked difference in responses – would even a small possibility of loss stimulate more risk averse responses? Except for B_6 this does not appear to be the case.

4. Estimated utility functions

Other than B_4 and C_{5s} who, respectively, provided only two and five risk averse responses, utility of gain functions were estimated for each respondent. Three parametric forms of utility functions were considered[7] and two approaches to estimation were used. Estimates obtained by least squares were reported in Hildreth and Knowles (1982). Maximum likelihood estimates assuming β-distributed errors are given in table 19.3. Table 19.4 shows that alternative procedures did not drastically change the estimates.

For convenience, constant absolute risk aversion functions,

$$\Psi(y) = -\mathrm{e}^{-\lambda y}, \quad \lambda > 0, \tag{4.1}$$

are called CAR and linear combinations of such functions with positive coefficients

$$\Psi(y) = -\sum_{i=1}^{n} \theta_i \mathrm{e}^{-\lambda_i y}, \quad \lambda_i > 0, \quad \theta_i > 0, \tag{4.2}$$

are called SCAR. The latter show positive but decreasing risk aversion and have other properties that make them promising for empirical work (Hildreth, 1979, pp. 13–14; Testfatsion, 1981). If utility of wealth is SCAR, then utility of gain is also SCAR and $\theta_i = \alpha_i E \mathrm{e}^{-\lambda_i X}$, where X is initial prospect, utility of gain is given by (4.2), and utility of wealth by

$$\psi(x) = -\sum_{1}^{n} \alpha_i \mathrm{e}^{-\lambda_i x}, \quad \alpha_i > 0, \quad \lambda_i > 0. \tag{4.3}$$

CAR and two simple versions of SCAR defined below have been fitted to risk averse responses. CAR is the special case of SCAR_2 obtained by setting $\theta = 0$, SCAR_1 is the special case obtained by setting $\theta = 1$. For SCAR_1 and SCAR_2 we have ordered the terms so that $\lambda_1 > \lambda_2$. Thus, for SCAR_2, θ is the weight of the

[7]In earlier work, Knowles (1980) had also considered the forms:

$$\psi(y) = \log(y - k), \quad k > 0,$$

and

$$\psi(y) = (y - k)^{\delta}, \quad k > 0, \quad 0 < \delta < 1.$$

Neither equation makes sense if $y < -k$. His estimates of $|k|$ were too small to apply to possible losses of a size that need to be considered by our respondents.

Table 19.3. Maximum likelihood estimates based on all risk averse responses.[a]

	Respondent (no. of observations)														
	$B_1(7)$			$B_2(9)$			$B_3(7)$			$B_5(24)$			$B_6(6)$		
Form of equation	CAR	SC_1	SC_2	CAR	SC_1	SC_2	CAR	SC_1	SC_2	CAR	SC_1	SC_2	CAR	SC_1	SC_2
λ_1	0.127	0.196	1.89	0.053	0.064	0.810	0.0016	0.0019	0.031	0.080	0.109	0.109	0.0026	0.047	0.060
λ_2		0.040	0.030		0.010	0.0014		0.00039	0.00030		0.010	0.008		0.0017	0.00041
θ			32.3			233			194			1.25			27.1
ln Likelihood	−8.43	−8.25	−5.20	−38.9	−37.8	−26.4	−17.08	−17.07	−16.9	−46.3	−45.5	−45.4	−25.7	−21.5	−18.4

	Respondent (no. of observations)														
	$B_7(17)$			$C_1(16)$			$C_2(15)$			$C_{5f}(10)$			$C_{5fs}(6)$		
Form of equation	CAR	SC_1	SC_2	CAR	SC_1	SC_2	CAR	SC_1	SC_2	CAR	SC_1	SC_2	CAR	SC_1	SC_2
λ_1	0.013	0.018	0.060	0.029	0.035	3.12	0.0018	0.0029	0.015	0.076	0.107	0.67	0.025	0.023	0.252
λ_2		0.0037	0.0027		0.004	0.024		0.00093	0.00004		0.019	0.012		0.020	0.00057
θ			22.0			155			585			16.2			2916
ln Likelihood	−37.3	−36.9	−35.1	−23.0	−23.0	−11.0	−22.5	−22.5	−22.3	19.8	−19.4	−15.7	−5.96	−5.96	−5.93

	Respondent (no. of observations)								
	$C_{6f}(8)$			$C_{6s}(11)$			$C_7(10)$		
Form of equation	CAR	SC_1	SC_2	CAR	SC_1	SC_2	CAR	SC_1	SC_2
λ_1	0.015	0.021	0.103	0.0031	0.0038	0.019	0.056	0.050	0.55
λ_2		0.004	0.0012		0.00076	0.00008		0.006	0.0029
θ			53			471			202
ln Likelihood	−31.8	−31.2	−25.7	−25.8	−25.8	−25.5	−37.1	−36.1	25.2

[a]Estimates based on data in thousands of dollars.

Table 19.4. Comparisons of maximum likelihood and least squares estimates[a]

	Respondent (no. of observations)													
Estimation method	$B_5(24)$ LS	$B_5(24)$ ML	$B_7(17)$ LS	$B_7(17)$ ML	$C_1(16)$ LS	$C_1(16)$ ML	$C_2(15)$ LS	$C_2(15)$ ML	$C_{5f}(10)$ LS	$C_{5f}(10)$ ML	$C_{6s}(11)$ LS	$C_{6s}(11)$ ML	$C_7(10)$ LS	$C_7(10)$ ML
λ_1	0.048	0.109	0.072	0.060	3.8	3.1	0.019	0.015	0.45	0.67	0.016	0.019	0.57	0.55
λ_2	0.018	0.008	0.0029	0.0027	0.031	0.024	0.00006	0.00004	0.028	0.012	0.00021	0.00008	0.0050	0.0029

[a]Based on $SCAR_2$ functions for all responses furnishing 10 or more observations.

term with the smaller coefficient in the exponent:[8]

$$\text{SCAR}_1\text{: } \psi_1(y) = -\mathrm{e}^{-\lambda_1 y} - \mathrm{e}^{-\lambda_2 y},$$

$$\text{SCAR}_2\text{: } \psi_2(y) = -\mathrm{e}^{-\lambda_1 y} - \theta\mathrm{e}^{-\lambda_2 y}.$$

Each of the 13 clusters of calculated values in table 19.3 pertains to a particular respondent. The numbers in parentheses beside the respondent identifications are the numbers of observations on which the respective calculations are based. In each cluster, columns refer to the three alternative forms for the utility function, rows to parameter estimates.

The sample consists of responses to a series of questions or trials, say $t = 1, \ldots, T$. Suppose t identifies a type V question. Let w_t be the response that would be obtained if the panelist appraised his utilities perfectly, i.e. he responded without error. Then w_t would satisfy:

$$0.5\psi(y_{1t}) + 0.5\psi(y_{2t}) = \psi(w_t). \tag{4.4}$$

For SCAR_2 this is equivalent to

$$\mathrm{e}^{-\lambda_1 w_t} + \theta\mathrm{e}^{-\lambda_2 w_t} = 0.5[\mathrm{e}^{-\lambda_1 y_{1t}} + \theta\mathrm{e}^{-\lambda_2 y_{1t}} + \mathrm{e}^{-\lambda_1 y_{2t}} + \theta\mathrm{e}^{-\lambda_2 y_{2t}}]. \tag{4.5}$$

For an L question the corresponding equation determining w_t is

$$\mathrm{e}^{\lambda_1 w_t} + \theta\mathrm{e}^{\lambda_2 w_t} + \mathrm{e}^{-\lambda_1}(z_t - w_t) + \theta\mathrm{e}^{-\lambda_2}(z_2 - w_t) = 2(1+\theta), \tag{4.6}$$

and for an R question it is

$$\mathrm{e}^{-\lambda_1 w_t} + \theta\mathrm{e}^{-\lambda_2 w_t} = \mathrm{e}^{-\lambda_1 y_{1t}} + \theta\mathrm{e}^{-\lambda_2 y_{1t}} + \mathrm{e}^{-\lambda_1 y_{2t}} + \theta\mathrm{e}^{-\lambda_2 y_{2t}} - \mathrm{e}^{-\lambda_1 z_t} - \theta\mathrm{e}^{-\lambda_2 z_t}.$$

Given y_{1t}, y_{2t}, and/or z_t as specified in the question, the errorless response is a function of the parameters; write it $w_t(\lambda_1, \lambda_2, \theta)$. The observed response m_t is an approximation to $w_t(\lambda_1, \lambda_2, \theta)$. In this study, for an initial parameter point, w_t was obtained from (4.5), (4.6), or (4.7) by an iterative procedure. For a risk neutral respondent answering a V question,

$$w_t = \frac{y_{1t} + y_{2t}}{2},$$

whereas

$$w_t < \frac{y_{1t} + y_{2t}}{2}$$

for a risk averse respondent.

To assume $w_t < y_{1t}$ would violate the sure thing principle. $w_t = y_{1t}$ can be regarded as the limiting case for a succession of increasingly risk averse respondents. It is also the errorless response of a minimaxer.

Thus, for investigators like the present authors who believe business decisions are risk averse and do not violate the sure thing principle, w_t is bounded, namely

$$y_{1t} \le w_t \le \frac{y_{1t} + y_{2t}}{2}. \tag{4.6a}$$

[8]Schlaifer (1971) has considered some properties of SCAR_2 under the label SUMEX.

For an L question,

$$\psi(0)=0.5\psi(-w_t)+0.5(z_t-w_t). \tag{4.7}$$

The errorless response of a risk neutral farmer would be $z_t/2$ and for a minimaxer it would be 0. He would prefer no gain to any gamble that contained a chance of loss. Thus,

$$0\le w_t\le z_t/2 \tag{4.7a}$$

for an L question.

If the question is type R,

$$0.5\psi(y_{1t})+0.5\psi(y_{2t})=0.5\psi(z_t)+0.5\psi(w_t). \tag{4.8}$$

For a risk neutral decision-maker, w_t equates expected gain from the two gambles so $w_t=y_{1t}+y_{2t}-z_t$. For the minimaxer the situation is more complicated. If $z_t>y_{1t}$, then a $(0.5z_t,0.5w_t)$ gamble is preferred to a $(0.5y_{1t},0.5y_{2t})$ gamble as long as $w_t>y_{1t}$ and is regarded as worse if $w_t<y_{1t}$. Considering this as setting one bound along, with risk neutral behavior as setting the other, yields:

$$y_{1t}+y_{2t}-z_t\le w_t\le y_{1t} \tag{4.8a}$$

for an R question with $z_t>y_{2t}$, and yields:

$$y_{1t}\le w_t\le y_{1t}+y_{2t}-z_t \tag{4.8b}$$

for an R question with $y_{2t}>z_t>y_{1t}$. If $z_t=y_{2t}$, the limits coincide and the two gambles become the same.

If $z_t<y_{1t}$, then the minimaxer prefers $(0.5y_{1t},0.5y_{2t})$ to $(0.5z_t,0.5w_t)$ regardless of w_t so the principle applied to other cases does not give an upper limit to w_t. One has

$$y_{1t}+y_{2t}-z_t\le w_t \tag{4.8c}$$

for an R question with $z_t<y_{1t}$. If $z_t=y_{1t}$, then $w_t=y_{2t}$ for any decision-maker, since the gambles become the same.

In our present model, actual response m_t is regarded as a random approximation to errorless response w_t. If one also assumes that carefully considered actual responses are risk averse and do not violate the sure thing principle, then m_t is a random variable confined to an appropriate range depending on question type. m_t is also presumed to have high density in the vicinity of the parameter w_t which solves (4.7) or (4.8). The latter consideration can be effected in a natural way by making w_t the mode of the distribution of m_t.

The β-family of continuous distributions seems a promising choice to permit this choice of range and mode. Accordingly, in the maximum likelihood analysis to be reported, response m_t is assumed to have the β-density

$$f(m_t)=\frac{1}{B(p_t,q_t)}\,\frac{(m_t-a_t)^{p_t-1}(b_t-m_t)^{q_t-1}}{(b_t-a_t)^{p_t+q_t-1}},$$

where $p_t, q_t > 1$, $a_t < m_t < b_t$, $t = 1,\ldots,T$. For type V trials, a_t is taken equal to y_{1t} and $b_t = (y_{1t} + y_{2t})/2$. For type L questions, $a_t = 0$ and $b_t = z_t/2$. With type R questions, when $z_t > y_{2t}$, $a_t = y_{1t} + y_{2t} - z_t$ and $b_t = y_{1t}$. Our samples include no cases where $y_{2t} > z_t > y_{1t}$. When $z_t < y_{1t}$, $a_t = y_{1t} + y_{2t} - z_t$. Since there is no natural b_t in this case, $b_t = y_{2t} + 20(y_{1t} - z_t)$ was chosen after some examination of the data as a sufficiently high upper limit to justify the conjecture that possible biases due to small probabilities above the upper limit would be very small. This choice must be re-examined in future studies and alternatives considered. p_t and q_t vary with trials and also with the assumed utility parameters $(\lambda_1, \lambda_2, \theta)$ in a fashion which requires w_t to be the mode. The computing procedure is sketched in the appendix.

The resulting estimates are in table 19.3. The last row of each cluster gives the maximum value of ln likelihood for the indicated form of utility function. In nine of the 13 cases, the shift from SCAR_1 to SCAR_2 produced a substantially larger increase in ln likelihood than the shift from CAR to SCAR. A comparable result for least squares estimates reported in Hildreth and Knowles (1982) was that R^2 was substantially higher for SCAR_2. This aspect of the results suggests continued exploration of SCAR_2 in future studies. The general pattern of estimates by maximum likelihood and least squares was similar, as shown in table 19.4, where comparisons are made for all respondents furnishing ten or more observations.

5. Future work

When statisticians undertake empirical analyses expected to be of use in the decisions of others, an explicit or implicit statistician–client relation is created. The usefulness of particular parcels of information and/or analysis that might be furnished by the statistician depend on the circumstances of the client and of opportunities for exchanges of information. In cases like the present where the statistician prepares a document that may be seen by various potential users, the relevant circumstances are difficult to know and efficient statistical presentations hard to specify. An initial survey of these problems from a Bayesian viewpoint was made by Hildreth (1963).

In the present context, the intended potential users of the results are farmers, advisers to farmers, and economists who need to forecast farm behavior in order to assess the probable effects of alternative economic policies. After formulation and study of representative decision problems for these groups, it is hoped that a reasonably complete Bayesian approach can be constructed.

Meanwhile the authors regard the results of the present study as mildly encouraging.

The very close fit of SCAR_2, a family of utility functions with several properties specified by underlying theories, clearly justifies continued exploration using this family and possible extensions. The authors' contention that non risk averse responses are mainly due to failures of the questions and interviewing procedures to produce careful calculations dominated by financial considerations has to be carefully studied. Investigation of discrepancies between responses to hypothetical questions and actual decisions is important. The authors believe risk aversion will prove to be more persistent in actual decisions.

Studying actual business decisions will require enough background information on the decision-maker's total prospect to be sure that the investigator can correctly interpret observed decisions. For farm entrepreneurs this suggests a number of intensive studies. For financial reasons and availability of qualified researchers' time this would probably involve just a few farmers at the outset. These would be case studies directed toward the indicated primary objective.

Another type of studies clearly needed are sampling studies with enough entrepreneurs in a given sample to ensure reasonable representativeness of a chosen class of entrepreneurs. Samples should be chosen to ensure reasonably strong statistical inferences about characteristics of the chosen population. Obtaining reliable indications of properties of utility functions for large groups of entrepreneurs clearly poses severe interviewing difficulties. One must obtain highly personal information and be sure that it is validly interpreted.

The present study falls in between these types. More farmers were interviewed than could be studied intensively with available resources. A sample of 13 farmers is not large enough to reliably establish characteristics of an interesting class of entrepreneurs. In retrospect, however, there were offsetting advantages in the size of the panel. A number of interviews could be held with each respondent. The two researchers were able to conduct all of the interviewing. Some farm and personal characteristics of individuals could be recalled from one visit to another. It was not necessary to structure the interviews as rigidly as would have been the case with a large panel.

The study provides substantial additional evidence for diminishing absolute risk aversion by individuals. As utility concepts for corporations and other institutions are sharpened, comparisons will be interesting. The principal discouraging result in the authors' view is the large number of risk preferred and risk neutral responses. This has been primarily explained as failure of the interviewers to elicit a careful calculation. As noted in section 3, some such answers should be expected when stakes are low and in situations in which financial considerations do not sufficiently dominate the respondent's motivation. If, however, risk aversion is characteristic of most entrepreneurs' business decisions, stakes of \$5000 to \$50 000 should not be required to bring it out. This has to be restudied using case studies of actual decisions and different interviewing designs and techniques. Large stakes and possible losses can be introduced earlier in the questioning. Perhaps enough can be learned about other prospects to use business-related questions. Respondents can be asked to rationalize their indicated decisions after which questions designed to check the rationalizations can be employed. This should be an interesting and productive area of future research.

Appendix

The ML computation is sketched below for CAR functions. Extension to more than one utility parameter involves no new principles.

For a particular respondent there is a main iteration on the utility parameter. For each value of the utility parameter that appears in the main iteration, there are two subiterations. The latter determine values of some parameters in the density function corresponding to the utility parameter being considered at that step.

The observations consist of responses m_t and stakes y_{1t}, y_{2t}, and z_t. T is the total number of risk averse responses by the farmer being considered. At the nth stage of the general iteration, x_t is considered to have β-density:

$$f_{nt}(x_t) = \frac{1}{B(p_{nt}, q_{nt})} \frac{(x_t - a_t)^{p_{nt}-1}(b_t - x_t)^{q_{nt}-1}}{(b_t - a_t)^{p_{nt}+q_{nt}-1}}. \tag{A1}$$

Bounds a_t and b_t are calculated from the stakes, as described in section 4. It is assumed that the mode of m_t is the errorless response w_{nt} which is calculated from (4.6), (4.7), or (4.8) with ψ determined by the current value of the utility parameter, $\lambda(n)$.

Making w_{nt} the mode of f_{nt} requires:

$$(w_{nt} - 1)p_{nt} + w_{nt}q_{nt} + (1 - 2w_{nt}) = 0. \tag{A2}$$

If this is the only requirement imposed on p_{nt} and q_{nt}, then $f_{nt}(m_t)$ is unbounded above. The density for this observation, and consequently the likelihood function, has no maximum. Another restriction or source of information must be involved.

It seems to the authors that each respondent might be characterized by an accuracy of response parameter, say α, as well as a risk aversion parameter λ. Two respondents with the same utility functions could answer questions with different degrees of care (this seemed evident at interviews).

It was decided to make such an assumption and to interpret α as a coefficient giving the variance of response as a proportion of the square of the range:

$$\text{Var } m_t = \alpha(b_t - a_t)^2. \tag{A3}$$

For known α, this would require:

$$\alpha(b_t + a_t)^2 = \frac{p_{nt}q_{nt}}{(p_{nt} + q_{nt})^2(p_{nt} + q_{nt} + 1)}. \tag{A4}$$

p_{nt} and q_{nt} could then be determined from (A2) and (A4), thus determining $f_{nt}(m_t)$ for the $\lambda(n)$ under consideration. Proceeding thus for $t = 1, \ldots, T$ determines the value of the likelihood for $1(m, \lambda_{(n)}) = \prod_1^T f_{nt}(m_t)$ at the parameter point $\lambda_{(n)}$.

How should α be determined? For the authors this is a very open question which will receive further study. One could devise questions to make the same w_{nt} relevant for several questions (several values of t) stated in different ways and then estimate α from $(m_t - w_{nt})$ for such cases. Also, one could try to link α to other respondent characteristics.

For likelihood analysis in the present study, α was arbitrarily set at 1/30 for all respondents. For the special case of a uniform distribution, $\alpha = 1/12$. For a unimodal distribution with bounded density, $1/12 > \alpha > 0$. After calculating variances for β-densities of various shapes, the authors chose 1/30 as a guess at a possible typical value.

The calculations of ML estimates is then:

(1) Calculate $a_t b_t$, $t = 1, \ldots, T$.

(2) Choose $\lambda_{(1)}$.

(3) Given $\lambda_{(n)}$, calculate w_{nt}, $t = 1, \ldots, T$.

(4) Determine p_{nt} and q_{nt} from (A2) and (A4).

(5) Determine $L = \ln 1(m, \lambda_{(n)})$, $D_\lambda L(m, \lambda_{(n)})$.

(6) Choose $\lambda_{(n+1)}$ by calculating L on a grid around $\lambda_{(n)}$ or by a Gauss–Newton procedure.

Recall for SCAR_1, $\lambda_{(n)} = (\lambda_{1n}, \lambda_{2n})$, and for SCAR_2, $\lambda_{(n)}$ stands for $(\lambda_{1n}, \lambda_{2n}, \theta_n)$. In calculations to date, convergence has sometimes been facilitated by shifting between Gauss–Newton and grids.

It is hoped that progress can be made in lessening the arbitrariness of $\alpha = 1/30$ and $b_t = y_{2t} + 20(y_{1t} - z_t)$ for R with $z_t < y_{1t}$. However, making variance proportional to squared range does seem reasonable and does allow for heteroscedastic errors. Knowles (1980) notes statistical evidence suggesting heteroscedasticity.

References

Arrow, Kenneth J. (1971) "The theory of risk aversion", in: *Essays in the Theory of Risk-Bearing* (Markham, Chicago), essay 3.

Friedman, Milton and L.J. Savage (1948) "The utility analysis of choices involving risk", *Journal of Political Economy*, 56, 279–304.

Graybill, Franklin A. (1961) *An Introduction to Linear Statistical Models* (McGraw-Hill, New York).

Hildreth, Clifford (1963) "Bayesian statisticians and remote clients", *Econometrica*, 31, 422–438.

Hildreth, Clifford (1974a) "Ventures, bets and initial prospects", in: M.S. Balch, D.L. McFadden and S.Y.Wu, eds., *Essays on Economic Behavior Under Uncertainty*, (North-Holland, Amsterdam).

Hildreth, Clifford (1974b) "Expected utility of uncertain ventures", *Journal of the American Statistical Association*, 69, 9–17.

Hildreth, Clifford (1977) "What do we know about agricultural producer's behavior under price and yield instability?", *American Journal of Agricultural Economics*, 59, 898–902.

Hildreth, Clifford (1979) *An Expected Utility Model of Grain Storage and Hedging by Farmers*, Technical Bulletin 321, University of Minnesota Agricultural Experiment Station.

Hildreth, Clifford and Glenn J. Knowles (1982) *Some Estimates of Farmers' Utility Functions*, Technical Bulletin 335, University of Minnesota Agricultural Experiment Station.

Johnson, N.L. and S. Kotz (1970) *Distributions in Statistics: Continuous Univariate Distributions*, vol. 2 (Houghton-Mifflin, Boston).

Knowles, Glenn J. (1980) "Estimating utility of gain functions for southwest Minnesota farmers", Unpublished Ph.D. Thesis, University of Minnesota.

Marschak, Jacob and Roy Radner (1972) *The Economic Theory of Teams* (Yale University Press, New Haven).

Meyer, Richard F. and John W. Pratt (1968) "The consistent assessment and fairing of preference functions", *IEEE Transactions on Systems Science and Cybernetics*, SSC-4, 270–278.

Pratt, John W. (1964) "Risk aversion in the small and in the large", *Econometrica*, 32, 122–136.

Ramsey, F.P. (1931) "Truth and probability", in: R.M. Braithwaite, ed., *The Foundations of Mathematics and Other Logical Essays*, (Harcourt Brace and Co., New York); reprinted in Kyburg and Smookler (1964, pp. 61–92).

Schlaifer, Robert (1971) *Computer Programs for Elementary Decision Analysis* (Division of Research, Harvard Business School).

Smith, Vernon L. (1971) "Economic theory of wager markets", *Western Economic Journal*, 9, 242–255.

Stiglitz, Joseph E. (1969) "Review of Arrow, *Essays in the Theory of Risk Bearing*", *Econometrica*, 37, 742–743.

Tesfatsion, Leigh (1981) "Dynamic investment, risk aversion, and foresight sensitivity", *Journal of Economic Dynamics and Control*, 3, 65–96.

Chapter 20

SIMPLIFIED ASSESSMENT OF SINGLE AND MULTI-ATTRIBUTE UTILITY FUNCTIONS

GARY KLEIN

Southern Methodist University

HERBERT MOSKOWITZ and SATHIADEV MAHESH

Purdue University

A. RAVINDRAN*

University of Oklahoma

1. Introduction

The measurement of single and multi-attribute utility functions in practice has been a challenge for the decision analyst. The actual measures are subject to modeling errors in functional form, parameter estimation errors, and also to measurement errors due to faulty communication between the analyst and the decision-maker (DM). The process of measurement per se, can be tedious and time consuming (Keeney and Raiffa, 1976). Recently, considerable attention has been focused on simplifying the measurement of both single and multi-attribute utility (MAU) and value (MAV) functions (Farquhar, 1982; Keelin, 1982; Keeney and Raiffa, 1976; Kirkwood and Sarin, 1980). Keeney and Raiffa (1976, ch. 4), for example, discuss the respective importance of determining certain qualitative and quantitative risk characteristics and restrictions of the DM's utility function prior to selecting a specific functional form. Qualitative characteristics include monotonicity, boundedness, continuity, and risk properties, such as risk aversion. Quantitative restrictions are determined by comparing responses to various

*This research was supported in part by a National Science Foundation Grant, no. 8007103A1.

Bayesian Inference and Decision Techniques, edited by P. Goel and A. Zellner

gambles over the attribute in question. Based upon such responses from the DM, a specific utility function is chosen by the analyst from a collection of functions having such characteristics that represent the DM's expressed risk attitudes. When several functions characterize the DM's expressed risk attitudes, the choice of a functional form is then often determined by the function having the best fit or the one which is most mathematically expedient.

A number of researchers have focused on first measuring a value (ordinal utility) function in multi-attribute problems for subsequent conversion into a cardinal utility function, to simplify the assessment process by minimizing the number of responses to hypothetical gambles (Keelin, 1981; Keeney and Raiffa, 1976; Kirkwood and Sarin, 1980; Meyer and Pratt, 1978). Kirkwood and Sarin (1980), for example, introduce a methodology that yields a precise functional form for the measured value function and minimizes the required interaction between the DM and the analyst when certain preference properties are exhibited by the DM. Keelin (1981) develops a general process of value function measurement with implications for utility function properties which are analogous to the risk characteristics and restrictions developed for utility functions.

Combining conditional utility functions on individual attributes into a real-valued multi-attribute utility function can also lead to measurement difficulties. Various approaches exist (e.g. Keeney and Raiffa, 1976; Pekelman and Sen, 1974; Srinivasan and Schocker, 1973) for the determination of appropriate scaling constants to achieve this aggregation. An experimental examination of the more commonly used methods for scaling additive utility functions showed that the methodology used in the measurement process affects the values of the scaling constants (Schoemaker and Waid, 1982).

Irrespective of the methodology selected in assessing a multi-attributed utility function, there exists the problem of satisfying certain conditions specific to a given MAU function. The querying procedures that reveal the DM's preference or risk properties have been well documented (e.g. Hauser and Urban, 1979) but a general process for determining a specific functional form for the conditional utility functions for each attribute have not been devised. Furthermore, a simpler, more meaningful procedure to obtain the scaling constants may be useful. The work described in this chapter is directed toward these goals by substituting some linear and non-linear programs for several of the current procedural steps in assessing utility functions. The proposed methodology is aimed at reducing the time and effort required by the analyst and DM, while providing a more common framework for utility construction. Using a general summed exponential functional form, or any form preferred by the analyst, the construction of single attribute utility functions is performed with the aid of a non-linear programming (NLP) formulation. The NLP estimates the parameters of the general utility form by fitting observed response values to gambles subject to the DM's expressed qualitative and quantitative risk characteristics. Once all single attribute utility functions have been determined, a linear program is used to compute the scaling constants for a general multilinear utility function based on expressed preferences of pairwise consequence vectors.

2. Determining single attribute utility functions

2.1. The summed exponential utility function

When a decision analyst is attempting to map a single dimensional utility function of a DM, he should have a set of readily available acceptable forms for the utility function. In general, a collection of various functions exist that adequately represent all reasonable qualitative restrictions on the risk characteristics of the utility function. For example, the logarithmic $\ln(x+b), b>0$, and other functions, such as $(x+b)^{-c}$, with $c>0$, and $-e^{-ax}+bx$, with $a, b>0$, are some common decreasing risk-averse utility functions on the random variable (attribute) x, for the restrictions on the parameters specified (Hauser and Urban, 1979, p. 173). Loosely speaking, risk is a measure of the DM's reaction to uncertainty. Risk measures, like risk premiums, are used to express a DM's risk characteristics. For example, a risk prone DM prefers choosing a gamble rather than settling for a guaranteed outcome equal to the expected value of the gamble in question. A risk averse DM prefers taking the sure outcome of a gamble's value over the gamble.

If a common set of plausible shapes representing most DM's utility functions may be ascertained or assumed prior to the actual measurement of the DM's utility function, it may be possible to define a single general functional form for the single attribute utility function whose parameters can be adjusted to fit the particular qualitative and quantitative risk characteristics of the DM. This suggested form is a requirement for the use of the NLP defined subsequently, and will also be an aid in testing the consistency of the DM's responses.

For convenience, our subsequent discussion will be arbitrarily limited to monotonically decreasing utility functions, although little change in the methods discussed would be required to accommodate increasing utility functions. The most commonly observed classes of single attribute utility functions are:

(1) Fully increasing risk averse.
(2) Fully decreasing risk averse.
(3) Constantly risk averse.
(4) Fully increasing risk prone.
(5) Fully decreasing risk prone.
(6) Constantly risk prone.
(7) Risk prone for small attribute values and risk averse for large attribute values.
(8) Risk averse for small attribute values and risk prone for large attribute values.

Through appropriate parameterization, a summed exponential function can represent all of the above characteristic types of utility functions. It also allows easy mathematical manipulation relative to other functional forms in terms of model development and optimization.

The summed exponential utility function for an attribute i is represented as

$$u_i(x) = a_i - b_i \exp(c_i x) - d_i \exp(e_i x). \tag{1}$$

The summed exponential satisfies the above eight utility classes for various values of the parameters. Using the risk aversion function

$$r(x) = \frac{-u''(x)}{u'(x)}, \tag{2}$$

it can be verified that for $a = 5$, $b = 1$, $c = 1.8$, $d = 9.7$, and $e = 4.3$, the risk function is negative at $x = 0.91$ and positive at $x = 0.99$. Hence, these parameters yield the seventh utility class listed above. Parameters are obtainable for the other seven classes.

Several recent studies have investigated various functional representations of single and multi-attribute utility functions. Moskowitz, Ravindran, Klein and Eswaren (1982) in a quality control environment addressed the problem of using differing single and bi-attribute functions. They conclude that functional form had little impact on the final quality acceptance plan selected, and less impact than does the selection of scaling constants. A more general study conducted by Keefer and Pollock (1980) outlines a procedure to aid in the construction of a MAU function. Included are the testing of preferential, utility, and probabilistic independence, parameter estimation for a common one-dimensional utility function, selection of the MAU form, and determination of the scaling constants. Sensitivity analysis is performed on a bi-criterion model. Optimal solutions were found to be sensitive to scaling constant selection, parameter estimation of the single attribute utility functions, and the form selected for the multiple attribute utility function. Demonstrations of solution differences through extreme point changes are provided, with the conclusion being, in contrast to Moskowitz et al. (1982), that selecting the proper form of utility function is as important as obtaining accurate probability and utility data. Therefore a general functional form that satisfies the proper qualitative characteristics of the DM's risk attitudes is as important as the fitting of the observed data to the selected function. The ability of the summed exponential to represent a wide spectrum of risk attitudes thus makes it a desirable function for mapping utility functions.

2.2. *Construction of the single attribute utility functions*

2.2.1. *Risk considerations*

The functional form of a one-dimensional utility function should be a construct consistent with the risk characteristics expressed and/or exhibited by the DM.

Risk aversion $r(x)$ and the rate of change of risk aversion are often the measures used as the characteristic constraints when selecting the utility form that will be fit to the observed certainty equivalents to gambles expressed by the DM. von Neumann and Morgenstern (1947) provide the basis for the questions to ask, and Keeney and Raiffa (1976) discuss the appropriate parametric families of utility functions derived from the constraints on the risk characteristics. The measures used to determine the risk properties are the risk premiums at various levels of a given attribute. The risk premium is the difference between the certainty equivalent of a given lottery and the expected (actuarial) value of the same lottery.

In describing our methodology, the lotteries we shall consider will be two outcome gambles, with a 50 percent chance of each occuring. A lottery will be denoted $A_i\langle x, y\rangle$, meaning that attribute i has a 50 percent chance of realizing a value of x, and a 50 percent chance of realizing a value of y. The certainty equivalent of a lottery is a sure amount of an attribute such that the DM is indifferent between choosing the sure amount and the lottery. The risk premium, which is the certainty equivalent less the expected value of the gamble, is denoted as $P(x, h)$ for the lottery $A_i\langle x-h, x+h\rangle$. Two results in Keeney and Raiffa (1976) are as follows:

Theorem 1. For *decreasing* utility functions, the following are equivalent.

(a) A DM is risk averse (prone, neutral).

(b) The risk premium for all non-degenerate lotteries is positive (negative, zero).

(c) The risk aversion function $r(x)$ is positive (negative, zero).

(d) The utility function $U(x)$ is concave (convex, linear).

Theorem 2. The risk aversion function $r(x)$ for a utility function $U(x)$ is increasing (constant, decreasing) iff the risk premium $P(x, h)$ is an increasing (constant, decreasing) function of x for all h.

Given these two results, quantitative restrictions on the DM's utility function are determined by obtaining his/her certainty equivalent for a series of lotteries $\langle x-h, x+h\rangle$ while varying x (the level of the attribute) and maintaining a constant h. If the risk premiums at the various levels are observed to increase, then the observations can be extrapolated to include all possible $\langle x-h, x+h\rangle$ lotteries. With the extrapolation, theorem 2 may be used to decide on which one of the eight types of utility functions is appropriate.

The specification and curve fitting procedure involves three basic steps:

Step 1. Elicit qualitative information on risk properties, including such characteristics as monotonicity, boundedness, and continuity.

Step 2. Select an admissible utility function satisfying these properties.

Step 3. Use the quantitative (certainty equivalent) responses to gambles as observations in fitting the selected utility function.

Step 1 involves considerable questioning of the DM in order to assure proper application of theorem 2. Step 2 requires the availability of appropriate utility functional forms that are consistent with the known or determinable risk properties of the DM. Step 3 requires solving a system of linear or nonlinear equations to obtain the parameters of the utility function that fit the observations.

2.2.2. *A nonlinear programming model*

We propose to reduce the process of determining a proper utility function to two steps, by combining function selection (step 2) with the curve-fitting process. Moreover, our procedure essentially eliminates the analyst's task of selecting an admissible and/or "best" admissible functional form by using a general utility function having a rich variety of risk properties. This may be accomplished with the use of a general summed exponential utility function and a non-linear programming model formulation. The process does not reduce the curve-fitting effort, but aids in reducing the number of responses required of the DM. Consider the following NLP to be solved for each attribute:

$$\min \sum_{j=1}^{r} \left[U(x_j) - Q(x_j)\right]^2 \tag{3a}$$

$$\text{s.t. } r(x_m) \leq \text{or} \geq r(x_{m+1}), \quad \text{for } m = 1,\ldots,q-1, \tag{3b}$$

$$u''(x_m) < \text{or} > \text{zero}, \quad \text{for } m = 1,\ldots,q, \tag{3c}$$

$$\max u'(x) \leq \text{zero}, \tag{3d}$$

where

x_j = the observations (certainty equivalents),
n = the number of observations (x_j) determined by the von Neumann–Morgenstern method,
x_m = the certainty equivalent responses used to determine risk properties,
q = the number of certainty equivalent responses used to determine the qualitative characteristics of the utility function,
$r(x_m)$ = the risk function in (2) evaluated at the mth value of the attribute (i) under consideration,
$u''(x_m)$ = the second derivative of the utility u evaluated at x_m, and
$Q(x_j)$ = the observed utility value at x_j.

It should be noted that the parameters of the utility form selected are the decision variables for this NLP. For the summed exponential, a, b, c, d, and e would be the decision variables. Also, the objective function minimizes the sum of the squared differences between the predicted and observed ability forms. Hence, only the n certainty equivalent (CE) responses for which the utility of the attribute at the CE value is known is used in the objective. The q CE responses are for small lotteries at various attribute levels and the utilities at these levels are not determined.

Equation (3a) represents a least squares curve-fitting criterion. This may be changed as desired. Eq. (3b) represents the decreasing (or increasing) nature of the risk function. This set of equations is generated by the type of questions required in step 1 of the standard approach. Fewer responses may be required, however, because any number of responses may be incorporated into the constraint set. Inconsistent responses to risk premium questions will indicate a form other than the classes provided by the functional form selected. Thus, inconsistent responses will not yield a feasible region for fitting responses and provide automatic consistency checks. However, irregularities or disturbances in the risk function are permitted during the curve-fitting step in the attribute values between responses. This feature may or may not be desirable depending upon the certainty of the analyst about the prior estimates of possible utility forms of the DM. Technically, only three responses are neseccary to define any of the eight utility function classifications that were listed above. If the analyst wishes to validate his prior assumptions regarding the utility function or test the consistency of the DM, more responses are required. As long as a feasible region can be found, the analyst's assumptions are valid and the DM is consistent, but only within the responses provided.

Equation (3c) defines whether the DM is risk prone or risk averse at certain levels of the attribute. Eq. (3d) enforces the monotonically decreasing nature of the utility function. The constraint forms an optimization problem itself, and as such requires special attention. Bracken (1975) discussed the nature of model formulations having optimization problems embedded in the constraint set. When using the summed exponential utility function, the following short-cut procedure can be taken. The maximum of the function in (3c) must occur at either an extreme point or an interior point where the second derivative is equal to zero. Since we are dealing only with a single attribute utility function, one can enumerate all possible maxima, and include a constraint for each. With the summed exponential, this amounts to $u(x)=0$ for x at the lowest attribute value, x at the highest attribute value, and for two interior points, where

$$u''(x)=bc^2\exp(cx)+de^2\exp(ex)=0. \tag{4}$$

The solutions for (4) are

$$x=\frac{\ln(-bc^2/de^2)}{e-c} \quad \text{and} \quad x=\frac{\ln(-de^2/bc^2)}{c-e}.$$

Further constraints in the model may be added to represent upper and lower bounds on the utility, e.g. $u_i(x_{i\,\max})=1$. The proposed method and formulation is simple from the viewpoint of the analyst and the DM, given satisfactory computing support. Even though the formulation is not complex, the NLP may be difficult to solve for a global optimum. Care must therefore be given to the optimization algorithm selected. Our testing was performed using the COMPLEX search method of Box (1965), because of its ability to overcome some local minimum difficulties. This search procedure does not guarantee a global optimum, but a near optimum. A probable optimal fit may be achieved by varying the starting points and comparing the final solutions obtained. Other difficulties involve the starting feasible region. A phase one (preliminary) approach was used to find a starting feasible point, but this was not guaranteed. Another difficulty is the possible non-convexity of the feasible region. This difficulty was mitigated by using different starting points for each optimization problem. This allows an algorithm to search small segments of the entire feasible region, increasing the chance of finding global optimality rather than local optimality.

2.2.3. *Illustration of the NLP approach*

In order to demonstrate the differences between existing approaches and our proposed non-linear programming methodology for determining a DM's utility function, let us consider the multi-attribute problem of selecting a new car. For simplicity, we shall limit the attributes to price, operating costs, size, and styling. For price, let us derive the single attribute utility function by the NLP approach.

The first step will require that several questions regarding the DM's qualitative risk characteristics be asked. A minimum of three is sufficient to employ the procedure. Keeney and Raiffa (1976) recommend a minimum of ten lottery responses for the standard procedure. We will assume the responses shown in table 20.1 for our example. Our lotteries use hypothetical values for the attributes. When reassuring a specific utility function, choices of actual and hypothetical values should all be on the same curve that approximates the actual utility. In some applications, when actual lotteries are used to adjust for "wealth effects", this may be easy when the attribute is money income, but it may be difficult for other applications. This is a problem with some other methods of fitting multi-

Table 20.1. Certainty equivalent responses to obtain qualitative risk characteristics and utility observations.

Lottery no.	Lottery	Certainty equivalent	Risk premium
1	5000, 6000	5675	175
2	8000, 9000	8600	100
3	11000, 12000	11530	30
4	4000, 14000	10000	–
5	4000, 10000	7800	–
6	10000, 14000	12800	–

attribute utility functions. Lotteries 1, 2, and 3 are directed at finding the risk properties, and lotteries 4, 5, and 6 are used to generate three utility value observations using the von Neumann–Morgenstern method. Also included are the extreme values of the attributes, which are assigned utilities of 0 and 1 as utility observations. These observed utility values give us the following least squares objective function, using price in thousands of dollars:

$$\min\left[\{u(4)-1\}^2+\{u(7.8)-0.75\}^2+\{u(10)-0.5\}^2 + \{u(12.8)-0.25\}^2+\{u(14)-0\}^2\right].$$

The selected certainty equivalent responses for lotteries 1, 2, and 3 in table 20.1 yield the following constraints by using expected values from the lotteries (since risk premium is measured as a deviation from expected value).

Set (3b) represents decreasing risk aversion; thus $r(5.5) < r(8.5)$ and $r(8.5) < r(11.5)$. For the summed exponential, the first constraint in set (3a), for example, would thus be

$$\frac{-bc^2\exp(5.5c)-de^2\exp(5.5e)}{-bc\exp(5.5c)-de\exp(5.5e)} < \frac{-bc^2\exp(8.5c)-de^2\exp(8.5e)}{-bc\exp(8.5c)-de\exp(8.5e)}.$$

For set (3c), which is used to represent risk aversion, the constraints are

$$u''(5.5)\le 0,$$

$$u''(8.5)\le 0,$$

$$u''(11.5)\le 0.$$

For the summed exponential, the first constraint in set (3b) would be

$$-bc^2\exp(5.5c)-de^2\exp(5.5e)<0.$$

Set (3d) would be identical regardless of the DM's responses, since it enforces the decreasing nature of the utility function. Using the summed exponential function, the first constraint of set (3c) (see also eq. (4)) would be

$$-bc\exp(4c)-de\exp(4e)\le 0.$$

A locally optimal solution to this NLP is

$$u(x)=2.51-1.002\exp(0.0089x)-0.328\exp(0.105x).$$

2.2.4. Applications

The methodology outlined above employing a general summed exponential utility function and NLP formulation was applied in both a laboratory and field setting.

The experimental setting involved a bi-criterion quality control model developed in Klein et al. (1982), where the utility functions of 73 subjects were measured on each of the two attributes of the model, yielding 146 summed exponential utility functions. With the attributes scaled between zero and one, and fitting five observations in a least squares framework, the average least squares error was 0.0075, with a range from 0.0000 to 0.19735. In this case consistency checks were performed prior to the final curve fitting, and the subjects were requested to revise their responses accordingly.

The second application was in a field study (Crosby and Moskowitz, n.d.) of the utility functions (on profit) of 14 auditors in several "big 8" accounting firms. Again, the proposed method using the summed exponential yielded excellent fits consistent with the risk characteristics of the auditors. Several auditors displayed Friedman–Savage (F–S) type utility functions (Friedman and Savage, 1952). In these cases, although the fits were good using a single set of parameters, a two-piece approximation using different parameters for the convex and concave regions yielded significantly better results, and is therefore recommended for individuals exhibiting F–S type utility functions. This two-piece approximation can still use the NLP presented by approximating the attribute level where the utility shifts risk properties by using several lotteries, then fitting the proper shape on the attribute from its minimum level to shift level, then also from its shift level to maximum level.

A simulation study was also performed to compare the sum of squares fit of the summed exponential solutions to solutions obtained with some other functional forms (i.e. cubic and quadratic) over a variety of conditions. No major differences in fit were noted, although the cubic exhibited a slightly better fit, on the average, under the conditions tested.

3. Obtaining the scaling constants of multi-attribute utility functions

Constructing a multi-attribute utility function from a set of single attribute conditional utility functions involves choosing a proper multi-attribute functional form which should be based on certain independence conditions. Often a common heuristic employed is to determine attribute weights that are then combined additively with the single attribute functions into a multi-attribute utility function.

There are a variety of methods used to obtain the scaling constants of MAU function. Many of these are described and compared in Schoemaker and Waide (1982) for additive MAU functions. For example, Keeney and Raiffa (1976) propose obtaining a set of k independent equations with k unknown scaling constants, which are generated from responses to tradeoffs (certainty considerations) or gambles (uncertainty considerations). This procedure has several potential drawbacks. First, the DM's responses are to tradeoffs or gambles at their best or worst levels. Responding to such extreme conditions is cognitively complex, and ignores any information provided by the DM's previously measured single attribute utility functions. Our procedure, utilizing and extending the LINMAP concepts of Srinivasan and Shocker (1973), simplifies the process of obtaining the scaling constants by requiring only pairwise preferences to consequence vectors in the relevant ranges of interest, and reducing the need for independence testing to determine an appropriate MAU function.

3.1. Forms of the multiple attribute utility function

A reasonable general multi-attribute utility function is the multilinear form, i.e.

$$U(X)=\sum_{i=1}^{m}k_iu_i(x_i)+\sum_{i=1}^{m}\sum_{j>i}^{m}k_{ij}u_i(x_i)u_j(x_j)$$
$$+\sum_{i=1}^{m}\sum_{j>i}^{m}\sum_{l>j}^{m}k_{ij}u_i(x_i)u_j(x_j)u_l(x_l)$$
$$+\cdots+k_{1,2,3,\ldots,m}u_1(x_1)u_2(x_2)\ldots u_m(x_m),$$

where m is the number of attributes, X is the attribute vector, and x_i is the value of the ith attribute, given the set of attributes $X\equiv\{x_1,\ldots,x_m\}$ with $m>2$, the independence conditions for the multilinear utility function are that each attribute x_j be utility independent of its complement $\bar{x}_j$, i.e.

$$\bar{x}_j=X,\quad \text{with } x_j \text{ excluded.}$$

Utility independence implies that the values of all remaining attributes at any given value are independent of the level of the values of all remaining attributes. Operationally, utility independence (UI) of an attribute from all others indicates that the DM can define a utility function on a single attribute independent (without any knowledge) of the values of the remaining attributes.

The multilinear utility function is a generalization of both the multiplicative and additive utility functions. The independence conditions for these special cases are more stringent than for the more general multilinear form. The multiplicative form is similar to the multilinear form, but the conditions that each attribute $x_1,\ldots,x_m$ must be mutually utility independent (MUI) implies the form:

$$U(X)=\sum_{i=1}^{m}k_iu_i(x_i)+k\sum_{i=1}^{m}\sum_{j>i}k_ik_ju_i(x_i)u_j(x_j)$$
$$+k^2\sum_{i=1}^{m}\sum_{j>i}\sum_{l>j}k_ik_jk_lu_i(x_i)u_j(x_j)u_l(x_l)$$
$$+\cdots+k^{m-1}k_1k_2\ldots k_mu_1(x_1)u_2(x_2)\ldots u_m(x_m).$$

If the multiplicative form is appropriate, fewer scaling constants are required. The constraints insuring consistency in the scaling constants, however, are no longer linear. Testing for the utility independence conditions for the multiplicative function involves more questions than those required by the multilinear form.

A further simplification of the multilinear form is the additive form, i.e.

$$U(X) = \sum_{i=1}^{m} k_i u_i(x_i), \quad \text{with all } 0 \le k_i \le 1.$$

The additive form has only one constraint on the scaling constants. The sum of the scaling constants is set equal to one so that $U(X)$ is between zero and one, as is each $u_i(x_i)$. The additive utility function requires additive independence (AI) among all attributes for UI to hold. Additive independence holds if preferences over lotteries on attributes $x_1,\ldots,x_m$ depend only on their marginal probability distributions and not on their joint probability distribution. This property permits the elimination of the interaction terms needed in the other utility forms to accurately represent the utility when taking expectations of the utility function. Scaling constants for an additive utility function are simpler to determine, and also provide much latitude in consistency testing. Unfortunately, additive independence involves more testing than utility independence.

3.2. Proposed linear programming formulation

The proposed method begins with the use of the general multilinear form and stated pairwise preferences by the DM similar to those proposed in Srinivasan and Schocker (1973). From this, scaling constant determination can be formulated into a simple linear program. The framework is to present the DM with a series of pairwise alternate decision vectors and obtain preference responses. Each preference indicated by the DM provides a strong statement about the $U(X)$. That is, $U(X_1)$ is greater than $U(X_2)$ if the consequence vector X_1 is preferred to consequence vector X_2. A series of these statements provides a set of linear inequalities. Let each inequality be expressed as $U(X_1)-U(X_2)+E^- - E^+ = 0$. The E^+ and E^- terms represent differences between the utility of the two consequences. When X_1 is preferred to X_2, the variable E^+ should be ≥ 0, but when the stated preference is violated by the structure of the utility function, E^- will become > 0. Hence, to obtain the most consistent scaling constants for the DM, the sum of the E^- terms for all preference statements is to be minimized.

Construct a formal LP by letting X_{1j} denote the preferred consequence of the pair (X_{1j}, X_{2j}). Thus:

$$\max Z = \sum_{i=1}^{p} E_i^- \tag{5a}$$

$$\text{s.t. } U(X_{1j}) - U(X_{2j}) + E_j^- \geq 0, \quad j = 1, \ldots, p, \tag{5b}$$

$$k_i \leq 1, \quad \text{for } i = 1, \ldots, m,$$

$$k_{ij} = d_{ij} - k_i, \quad \text{for } i = 1, \ldots, m; \quad j = i+1, \ldots, m,$$

$$k_{ijl} = d_{ijl} - d_{ij} - d_{il} - k_{il} - k_j - k_l,$$

$$\vdots \qquad \text{for } i = 1, \ldots, m; \quad j = i+1, \ldots, m, \quad l = j+1, \ldots, m,$$

$$k_{1234\ldots m} = 1 \text{ minus the sum of all other } k \text{ variables}, \tag{5c}$$

$$\text{all } d_i, d_{ij}, d_{ij} \text{ are between 0 and 1},$$

$$d_{ij} \geq k_i,$$

$$d_{ij} \geq k_j,$$

$$\vdots$$

$$d_{1234\ldots m} \geq \text{all } k \text{ variables},$$

where the k variables are the scaling constants for the multilinear MAU function and the d variables represent the utility of all included attributes at their maximum values, and all omitted attributes at their minimum values (Hauser and Urban, 1979), and the E^- variables take the form of slack variables.

Constraint set (5b) is used to establish the direction of the preference reponses as stated by the DM. Set (5c) constrains the scaling of the MAU function between 0 and 1.

3.3. Example

Let us now return to the new car selection example. Using the new general summed exponential and the earlier proposed procedure, assume that all the single attribute utility functions have been determined and are as follows:

$$u_1(\text{price} = P) = 2.51 - 1.002\exp(0.0089P) - 0.328\exp(0.150P),$$

$$u_2(\text{operating cost} = C) = 2.06 - \exp(-0.045C) - 0.087\exp(0.275C),$$

$$u_3(\text{room} = R) = 2.16 - \exp(-0.049R) - 0.18\exp(0.33R),$$

$$u_4(\text{style} = S) = 3.35 - \exp(0.034S) + 1.464\exp(0.3S),$$

where P is price in thousands of dollars, C is operating costs in \$ per mile, R is roominess (expressed as 6 minus the number of passengers), and S is a subjective rating in the range 1 to 10. The utility ($U(X)$) would be represented in multilinear

Table 20.2. Pairwise preference comparisons.

Alternative 1	Alternative 2	Preference response
(7.8, 6.5, 2.5, 3.25)	(10, 2.5, 2.5, 3.25)	1
(10, 2, 3.5, 10)	(10, 2, 4.0, 3.5)	1
(10, 7.5, 4.5, 2)	(12.3, 7.5, 2.5, 2.0)	2
(10, 3.5, 2.5, 6)	(10, 5, 2.5, 3.25)	2
(14, 5, 1.0, 2)	(10, 5, 4.5, 2)	1
(4, 5, 3.5, 6)	(6.5, 5, 3.5, 1)	2
(7.8, 2, 3.5, 10)	(7.8, 6.0, 3.5, 1.5)	2
(7.8, 5, 2.5, 6)	(8, 5, 2.5, 1.0)	2

form as:

$$\begin{aligned} U(\text{price, operating cost, room, style}) &= U(P,C,R,S) = k_1u_1(P)+k_2u_2(G) \\ &+ k_3u_3(R)+k_4u_4(S)+k_{12}u_1(P)u_2(C)+k_{13}u_1(P)u_3(R) \\ &+ k_{14}u_1(P)u_4(S)+k_{23}u_2(C)u_3(R)+k_{24}u_2(C)u_4(S) \\ &+ k_{34}u_3(R)u_4(S)+k_{123}u_1(P)u_2(C)u_3(R)+k_{124}u_1(P)u_2(C)u_4(S) \\ &+ k_{134}u_1(P)u_3(R)u_4(S)+k_{234}u_2(C)u_3(R)u_4(S) \\ &+ k_{1234}u_1(P)u_2(C)u_3(R)u_4(S). \end{aligned}$$

Next the DM is asked to compare the pairs of alternatives and state his preferences, as shown in table 20.2. It probably makes sense to generate the alternatives randomly from among undominate solutions, and insuring a wide spectrum of consequence pairs. The consequences are then used to determine the single attribute utility values leaving only the scaling constants (k values) and error term (E^- values) as decision variables in the linear program. For the example the LP is

$$\min = \sum_{j=1}^{8} E_j^-$$

$$\text{s.t. } U(7,8,6.5,2.5,3.25) - U(10,2.5,2.5,3.25) + E_1^- \geq 0$$

$$\vdots$$

$$U(7.8,5,2.5,6) - U(8,5,2.5,1) + E_8^- \geq 0$$

and the constraints of set (5c) for the multilinear form. The first constraint, for example, would appear as

$$\begin{aligned} &0.265k_1 - 0.2340k_2 + 0.0817k_{12} + 0.1624k_{13} + 0.1609k_{14} - 0.1410k_{23} \\ &\quad - 0.1397k_{24} + 0.05k_{123} + 0.0496k_{124} + 0.098k_{134} - 0.855k_{234} \\ &\quad + 0.030k_{1234} + E_1^- \geq 0 \end{aligned}$$

Since k variables range between -1 and $+1$, variable substitution is made to allow negative values in the LP and then the problem is solved. The result yields

$k_1 = 0.365$, $k_2 = 0.294$, $k_3 = 0.047$, $k_4 = 0.052$, $k_{13} = 0.022$, $k_{24} = 0.219$ and all other scaling constants equal to zero.

Modifications to the procedure presented may include the incorporation of weights into the objective function based upon strength of preferences between alternatives. The strength of preference weights would place large costs on certain preferences and small costs on vague preferences. This would entail obtaining strength-of-preference information from the DM, which would probably reduce the number of pairwise comparisons to be made to achieve a given level of accuracy. For example, if the DM rates his preferences on a scale of 1 to 10 and gives a rating of 10 to preference 1 (table 20.2) and a rating of 1 to all other preferences, then the objective function is

$$\min Z = 10E_1^- + E_2^- + E_3^- + \cdots + E_8^- .$$

The model can be changed to include terms that encourage, but do not enforce, the additive form of the utility function. This can be accomplished by bringing the scaling constants associated with the interactive terms into the objective and multiplying by a penalty cost. The objective is then

$$\min Z = \sum_{i=1}^{8} E_i^- + P(k_{12} + k_{13} + k_{14} + \cdots + k_{1234}),$$

where P is a positive constant.

There exist several advantages to the linear programming method just described. Responses to extreme attribute values are avoided. The information obtained in measuring the single attribute utility functions is incorporated directly into the construction of the MAU function. The cognitive burden on the DM is reduced to pairwise preferences and elimination of the need for testing the DM for UI or MUI. Another advantage of the model formulation is that inconsistencies are accepted with the value of the objective function (Z) providing a measure of inconsistency, i.e. $c = (Z/1 + Z)$ (Srinivasan and Schocker, 1973).

4. Summary

Two mathematical programs to aid in the measurement of single and multi-attributed utility functions have been presented. Each formulation may be used independently or conjointly to help reduce the cognitive burden and interaction between the analyst and the DM. The NLP formulation in conjunction with a general summed exponential function is used to describe and fit a DM's single attribute utility function. The richness of the risk properties of the summed exponential alleviates the analyst's burden of choosing an admissible utility functional form, while providing easy mathematical manipulation in formal decision models. The LP formulation, based on Srinivasan and Schocker (1973), and assuming (but not necessarily restricted to) a multilinear MAU form, permits easy cognitive and mathematical determination of the scaling constants of the MAU function as well as special cases of the multilinear form without stringent testing of the associated independence conditions.

References

Bracken, J. (1975) "Math programs with optimization in the constraint set", *Operations Research*, 23.

Box, M.J. (1965) "A new method of constrained optimization and a comparison with other methods", *Computer Journal*, 8.

Crosby, Michael A. and H. Moskowitz (n.d.) "Determining auditor utility functions via mathematical programming", Unpublished Working Paper, Purdue University.

Farquhar, P.H. (1982) "Utility assessment methods", Graduate School of Administration, University of California, Davis, Working Paper 81-5.

Friedman, M. and L.J. Savage (1952) "The expected utility hypothesis and the measurability of utility", *Journal of Political Economy*, 60.

Hauser, J.R. and G.L. Urban (1979) "Assessment of attribute importances and consumer utility functions", *Journal of Consumer Research*, 4.

Keefer, D.L. and S.M. Pollock (1980) "Approximations and sensitivity in multiobjective resource allocation", *Operations Research*, 28.

Keelin, T. (1981) "A parametric representation of additive value functions", *Management Science*, 27.

Keeney, R.L. and H. Raiffa (1976) *Decisions with Multiple Objectives* (John Wiley and Sons, Inc.).

Kirkwood, C.W. and R. Sarin (1980) "Preference conditions for MAV functions", *Operations Research*, 28.

Klein, G., H. Moskowitz and A. Ravindran (1982) "Prior versus progressive articulation of preference approaches for bicriterion problems", Purdue University, School of Industrial Engineering, Research Memorandum No. 82-1.

Meyer, R.F. and J.W. Pratt (1978) "Consistent assessment and fairing of preference functions", *IEEE Systems Science and Cybernetics*, SSC-4.

Moskowitz, H., A. Ravindran, G. Klein and P.K. Eswaren (1982) "A bicriterion model for acceptance sampling", in: S. Zanakis, ed., *Optimization in Statistics* (North-Holland, Amsterdam).

Pekelman, D. and S.K. Sen (1974) "Mathematical programming models for the determination of attribute weights", *Management Science*, 20.

Schoemaker, P.J.H. and C.C. Waid (1982) "An experimental comparison of different approaches to determining weights in additive utility models", *Management Science*, 28.

Srinivasan, V. and A.D. Shocker (1973) "Estimating the weights for multiple attributes in a composite criterion using pairwise judgements", *Psychometrica*, 38.

von Neumann, J. and D. Morgenstern (1947) *Theory of Games and Economic Behavior*, 2nd edn. (Princeton University Press).

PART VII

ANALYSIS OF TIME SERIES AND GENERAL LINEAR MODELS

Chapter 21

A BAYESIAN ANALYSIS OF TIME SERIES

LYLE D. BROEMELING and SAMIR SHAARAWY*

Oklahoma State University

1. Introduction

This chapter reviews Bayesian contributions to time series analysis in the time domain, presents a way to analyze time series generated by ARMA processes, and compares the Bayesian analysis with the Box–Jenkins analysis of ARMA processes.

In 1970, Box and Jenkins presented a statistical analysis of ARMA processes which has grown in popularity and is today the prevailing methodology of time series analysis. On the other hand, the Bayesian analysis of ARMA processes is still being developed. The basic question is: does the Bayesian approach to time series analysis offer any advantages over the Box–Jenkins methodology? This chapter will show that a Bayesian approach to time series is feasible, practical, and possesses some advantages over the Box–Jenkins methodology. Indeed, it is hopefully demonstrated that Bayes' Theorem offers a unifying way by which ARMA processes may be analyzed.

In what is to follow, Bayesian contributions to time series analysis are reviewed, ARMA processes are explained and defined, a brief review of the Box–Jenkins methodology is presented, a Bayesian analysis is introduced, a numerical study of the Bayesian analysis of MA processes is demonstrated, and finally the concluding section contrasts the Box–Jenkins methodology to the Bayesian approach.

*This research was supported by the Office of Naval Research under contract no. N00014-82-K-0292.

Bayesian Inference and Decision Techniques, edited by P. Goel and A. Zellner

2. Review of Bayesian time series

Of course, the literature on time series is vast and is found in many areas other than statistics. Most of the literature is non-Bayesian and the reader is referred to Box and Jenkins (1970) and Priestley (1981) for the non-Bayesian theory and methodology. There are no books on the Bayesian analysis of time series, although Box and Jenkins (1970) give a brief account, and Zellner (1971) presents the Bayesian analysis of some special parametric models.

Most of the Bayesian developments have occurred within the past two decades, but it seems that Jeffreys (1939) was the first to study spectral theory. It was not until Zellner (1971) that the systematic analysis of time series began. Using Jeffreys' improper prior density, he found the posterior distribution of the parameters of first- and second-order autoregressive processes and the Bayesian predictive distribution of one future observation. He also gave a complete treatment of regression models with autocorrelated errors and of distributed lag models.

It is interesting to note that Box and Jenkins (1970) had earlier introduced a Bayesian analysis of ARMA models; however, their book focused on non-Bayesian methods.

In 1973, Newbold did a Bayesian analysis of a transfer function model of which an ARMA process is a special case. His analysis is based on the t-approximation to the coefficients of the model and is quite similar to the Bayesian analysis proposed earlier by Box and Jenkins (1970). Thus, within two years, the Bayesian analysis of time series had advanced from some special cases of AR (autoregressive) processes to models which include ARMA processes as special cases. The Newbold work is also similar to the 1978 report of Zellner and Reynolds who also used a t-approximation to the coefficients of an ARMA process.

Another important development in the area of AR processes was by Chow (1975), who found the moments of the predictive distribution of future vector observations. Also in connection with vector AR models is Litterman's (1980) way to predict economic variables. By using a clever proper prior distribution for the parameters, he is able to forecast GNP in a way that compares favorably with other forecasting techniques.

One of the most important Bayesian contributions has been recently done by Monahan (1983) who has found a way to work with ARMA processes. Assuming stationarity and invertibility, he has developed a numerical algorithm which produces posterior and predictive inferences for low-order ($p,\ q=0,1,2$) processes. Lahiff (1980) examined the AR(1) process in the same way Monahan worked with ARMA processes. Stationarity of the AR(1) process was assumed by putting a beta prior distribution on the coefficient over the interval $(-1,1)$.

The only published work on identification has been done by Monahan (1983); however, Diaz and Farah (1982) have reported a Bayesian technique which calculates the posterior probability of the possible orders of an AR process. They use a sequence of conditional conjugate prior densities and derive a formula giving the exact posterior probability mass function of the order of the model.

There has been very little work on structural change in time series. For example, Salazar (1982) working with AR(1) and AR(2) processes has found the exact posterior mass function of the change point (assuming one change) and reported similar results for the regression model with autocorrelated errors. Smith (1980) also gave a Bayesian way to detect change in an ARMA process. The above two articles are in the same direction as Tsurumi (1982) who has been working with simultaneous equations in econometrics.

One last reference is Gersch (1982) who uses a smoothness prior approach to model time series with trend and seasonal components. This review contains most of the Bayesian contributions to time series analysis in the time domain, but the time series literature in engineering and economics has not been emphasized. For some additional information of the time series literature in economics see Zellner (1977, 1983).

The engineering literature in communications and control contains a large amount of information some of which has a Bayesian flavor. Indeed, the Kalman (1960) filter can be thought of as the mean of a normal distribution. The reader should consult Harrison and Stevens (1976), who use the linear dynamic model (the state space representation) to forecast future sales, and Smith and Makov (1980), who describe the Bayesian analysis of linear dynamic systems.

After a review of the literature, one may conclude that the Bayesian approach to time series analysis is not a serious competitor to the Box–Jenkins analysis.

Is it possible to develop an efficient Bayesian analysis of ARMA processes which compares favorably with Box–Jenkins? The authors believe that it is and hope the reader will be convinced after reading the remainder of the chapter.

3. ARMA processes

The Bayesian analysis of time series is based on some special parametric models such as regression models with autocorrelated errors, distributed lag model, and autoregressive models. As Box and Jenkins (1970) have shown, a very useful class of models are the ARMA (p,q) processes:

$$\phi(B)y(t)=\theta(B)e(t), \tag{3.1}$$

where $Y(t)$ is the observation at time t, $t=\dots,\ -1,0,1,\dots,e(t)$ is a sequence of independent normally distributed random variables where $e(t)\sim N(0,\tau^{-1})$, $\tau>0$, where τ is unknown. The polynomial operators are

$$\phi(B)=1-\phi_1B-\cdots-\phi_pB^p, \tag{3.2}$$

where the ϕ_i are real unknown coefficients and p is a non-negative integer; and

$$\theta(B)=1-\theta_1B-\cdots-\theta_qB^q, \tag{3.3}$$

where again the θ_j are real unknown coefficients, q is a non-negative integer, and the backshift operator B is such that

$$B^j Y(t) = Y(t-j), \quad j = 0,1,2,\ldots. \tag{3.4}$$

The ARMA (p, q) process is stationary if the roots of $\phi(B) = 0$ lie outside the unit circle in the complex plane and the process is said to be invertible if the roots of $\theta(B) = 0$ are outside of the unit circle. Stationary is interpreted in the strict sense, and Box and Jenkins impose stationarity and invertibility in the analysis of ARMA processes.

Consider a special case, the ARMA (1,1) process with moving average coefficient θ_1, autoregressive coefficient ϕ_1, and residual variance τ^{-1}, then the process is stationary if $|\phi_1| < 1$ and invertible if $|\theta_1| < 1$. Invertibility is imposed so that one may identify a unique model. If invertibility is not imposed, there are two ARMA (1,1) models which gave rise to the same autocorrelation function. The autocorrelation function of a stationary ARMA (p, q) process is

$$\rho(l) = \rho[Y(t), Y(t+l)], \tag{3.5}$$

where $l = 0,1,2,\ldots$. For example, consider an AR(1) or ARMA (1,0) model, then the autocorrelation function is

$$\rho(l) = \phi_1^l, \quad |\phi_1| < 1, \quad l = 0,1,2. \tag{3.6}$$

The autocorrelation function and the partial autocorrelation function play an important role in the Box–Jenkins analysis because they are used to identify a particular ARMA model from the data. If the process is invertible and stationary, the corresponding autocorrelation function is unique, and the sample autocorrelation function is used to identify the particular ARMA process which has generated the data.

After the model has been identified, say as an ARMA (p, q), the moving average parameters $\theta_1, \theta_2, \ldots, \theta_q$, the autoregressive parameters $\phi_1, \phi_2, \ldots, \phi_p$, and the residual variance τ^{-1} are estimated via maximum likelihood or least squares. If $q \geq 1$, the likelihood function is a nonlinear function of the parameters, as is the sum of squares, and one must either use a grid search or a nonlinear least squares algorithm to estimate the parameters. Currently, one may use either a conditional least squares approach (conditional on the initial observations and unobservable residuals) or do a complete (unconditional) least squares estimation. These techniques are described by Priestley (1981, pp. 359–364) for ARMA processes and will be explained in the subsequent Bayesian analysis.

After performing an iterative cycle of diagnostic checking, Box and Jenkins (1970) are ready to forecast future observations. Their method of forecasting is explained by Priestley (1981, pp. 762–773) and is based on the conditional expectation of the future observation given the actual observations. The result is a recursive formula, equation 10.2.8 of Priestley (1981), which allows one to predict one step ahead.

The Box–Jenkins analysis is based on a variety of sampling theory concepts. For example, the identification phase is essentially based on the method of moments, the estimation phase is based on maximum likelihood (using either the exact likelihood or the conditional version) or nonlinear least squares, and the forecasting phase is founded on an estimated conditional expectation.

In the next section we will see how Bayes' Theorem unifies these three phases of a time series analysis.

4. A Bayesian analysis of ARMA processes

A Bayesian analysis of a parametric model consists of three phases: the prior, posterior, and predictive analyses. The Bayesian analysis of ARMA processes will be presented in three parts. First, the exact analysis of an autoregressive process is presented, then an approximate analysis is proposed for pure MA process, and lastly the ARMA class is analyzed.

4.1. Autoregressive processes

The AR(p) process is given by

$$\phi(B)Y(t)=e(t), \tag{4.1}$$

which is the ARMA (p,0) process of (3.1). In particular the AR(1) model is

$$Y(t)=\phi_1 Y(t-1)+e(t) \tag{4.2}$$

and is stationary if $|\phi_1|<1$ and has autocorrelation function (3.6). The usual Bayesian analysis, see Zellner (1971), ignores stationarity and uses a noninformative Jeffreys' prior density,

$$\xi(\phi_1,\tau)\propto\tau^{-1},\quad \phi_1\in R,\quad \tau>0,$$

to express prior information. As we have seen, Zellner did the posterior and predictive analysis for AR(1) and AR(2) models and his work is easily generalized to include AR(p) processes with normal-gamma prior densities.

Suppose n observations are available, say $Y(1)$, $Y(2),\dots,Y(n)$, then the conditional density of $S_{n-p}=[Y(p+1),\dots,Y(n)]$ given the first p observations $S_p=[Y(1),Y(2),\dots,Y(p)]$ is

$$f\left[S_{n-p}|S_p;\ \phi,\tau\right]\propto\tau^{(n-p)/2}\exp\left(-\frac{\tau}{2}\sum_{t=p+1}^{n}[\phi(B)Y(t)]^2\right), \tag{4.3}$$

where $\phi=(\phi_1,\phi_2,\dots,\phi_p)\in R^p$, $\tau>0$.

The posterior analysis is based on combining (4.3) with a normal-gamma prior density where $\phi|\tau\sim N(\mu,\Delta\tau)$ is the conditional prior density of ϕ given τ, where μ is a $p\times1$ vector and Δ a pth order positive definite matrix. The marginal prior density of τ is gamma with parameters α and β, thus the joint posterior distribution of ϕ and τ will also be normal gamma. The marginal posterior distribution of ϕ will be a p-variable multivariate t and the marginal posterior distribution of τ will be gamma.

The above analysis does not assume stationarity nor is it based on the exact likelihood function. Not imposing the assumption of stationarity has its advantages because one cannot be sure that the process is always stationary. Also, even if the process is stationary, the above analysis will not yield misleading inferences about the parameters if the sample size is large.

That the likelihood function (4.3) is conditional on the first p observations should not be a problem if n is large relative to p because it has been shown that under these conditions inferences based on the conditional are almost the same as those based on the exact. See Priestley (1981) for a discussion.

How does one forecast future observations $Y(n+1)$, $Y(n+2), \ldots$ using the Bayesian approach? Zellner gave the one-step-ahead predictive density for first- and second-order processes using a vague prior, but how is it done in general? Broemeling and Land (1984) have determined the exact predictive distribution of k future observations using a normal-gamma prior density (including Jeffreys' vague prior density) and it was shown how one may use the predictive density to forecast future observations. It was shown that the predictive distribution of $Y(n+1)$ is a univariate t, that of $Y(n+2)$ given $Y(n+1)$ is also a univariate t, etc. Also, it was shown how to compute the marginal predictive moments of $Y(n+2)$, $Y(n+3), \ldots$, how to determine point and interval forecasts.

It should be noted that a Bayesian forecast is based on the predictive density of $Y(n+1)$ and not on an estimated conditional expectation of $Y(n+1)$ given the observations, as is done by Box and Jenkins. One must assume the parameters are random (such as in a Bayesian analysis) in order to have a legitimate predictive distribution.

In the above analysis, it is assumed that the order p is known. Thus, what should be done if p is unknown? One could do a backward elimination if one knew the order of the model was no more than say m. That is to say, one would do the above Bayesian analysis, resulting in a t-distribution for the marginal posterior distribution of $\phi_1, \phi_2, \ldots, \phi_m$. Then one could test H_0: $\phi_m = 0$ versus H_1: $\phi_m \neq 0$, using the marginal posterior distribution of ϕ_m (which is a univariate t). If one does not reject H_0, one would test $\phi_{m-1} = 0$ versus $\phi_{m-1} \neq 0$ using the conditional posterior distribution of ϕ_{m-1}, given $\phi_m = 0$. This procedure is continued until one accepts the hypothesis $\phi_p \neq 0$ for some p, $0 \leq p \leq m$. This Bayesian solution to identification is quite different from the Box–Jenkins analysis of using the sample autocorrelation function.

We see that a complete Bayesian analysis of autoregressive processes is easily done and amounts to the same way one would analyze a multiple linear regression problem. Now the problem is to see if one can give a similar solution to the analysis of MA and ARMA processes.

4.2. *Moving average processes*

The qth order moving average process is given by the ARMA $(0, q)$ model of (3.1) namely

$$Y(t) = \theta(B)e(t), \quad t = \dots, -1, 0, 1, \dots, \tag{4.4}$$

and is invertible if the roots of $\theta(B) = 0$ are outside of the unit circle. A moving average process is always stationary and invertibility is imposed so that the model is uniquely identified. Thus, with a MA(1) process, the processes with $\theta_1 \in R$ and $\theta_1^{-1}(\theta_1 \neq 0)$ yield the same autocorrelation function. If $|\theta_1| < 1$, the MA(1) is invertible and equivalent to an infinite-order autoregressive process. Invertibility is a reasonable restriction, but will not be formally imposed with the Bayesian analysis. Instead, an informal imposition of invertibility will be used in a way which is similar to the Box–Jenkins analysis.

The main problem with analyzing a MA process is that the sum of squared residuals and the likelihood are nonlinear functions of the moving average coefficients $\theta_1, \theta_2, \dots, \theta_q$.

Let us consider the conditional likelihood function, given by

$$L(\theta, \tau | S_n) \propto \tau^{n/2} \exp\left(-\frac{\tau}{2} \sum_{t=1}^{n} \left[Y(t) + \sum_{j=1}^{q} \theta_j e(t-j) \right]^2 \right), \tag{4.5}$$

where $\theta = (\theta_1, \theta_2, \dots, \theta_q) \in R^q$, $\tau > 0$, and $e(t-q) = 0$, $t = 1, 2, \dots, q$. Thus, the likelihood is conditional on the first q residuals being zero. The Box–Jenkins analysis is based on either the conditional likelihood or the exact likelihood function where one finds the MLE of the parameters. Using the conditional likelihood is equivalent to minimizing the conditional sum of squared residuals:

$$e(t) = Y(t) + \sum_{j=1}^{q} \theta_j e(t-j), \tag{4.6}$$

which can be computed recursively if one knew θ. Since the sum of squared residuals,

$$\text{ss}(\theta) = \sum_{t=1}^{n} \left[Y(t) + \sum_{j=1}^{q} \theta_j e(t-j) \right]^2, \tag{4.7}$$

is a nonlinear function of θ, one must either use a grid search (usually over the region of invertibility) or employ a nonlinear regression algorithm using starting values produced by the Yule–Walker equations and checked for invertibility. The estimates $\hat{\theta}$ so obtained are, hopefully, conditional maximum likelihood estimates and are used to estimate the moving average coefficients. Priestley (1981, pp. 354–359) gives a good explanation of exact and conditional MLE.

Notice that once one has the conditional MLE $\hat{\theta}$, they may be used to calculate estimated residuals, via eq. (4.6), with $e(t-q)=0,\ t=1,2,\ldots,q$, producing

$$\hat{e}(t)=Y(t)+\sum_{j=1}^{q}\hat{\theta}_j\hat{e}(t-j). \tag{4.8}$$

If these estimated residuals are substituted into (4.5), one has an approximate conditional likelihood:

$$L^*(\theta,\tau|s_n)\propto\tau^{n/2}\exp\left(-\frac{\tau}{2}\sum_{t=1}^{n}\left[Y(t)+\sum_{1}^{q}\theta_j\hat{e}(t-j)\right]^2\right), \tag{4.9}$$

where $\theta\in R^q$, $\tau>0$, and S_n is the vector of n observations.

It is now noted that (4.9) will produce a convenient Bayesian analysis when combined with a normal-gamma prior density. This is so because as a function of θ and τ, the approximate conditional likelihood function (4.9) is a normal-gamma density. If this is a good approximation, the Bayesian analysis of MA processes is the same (normal-gamma analysis) as the analysis of an AR process. That is to say, the approximate marginal posterior distribution of θ will be a q-dimensional multivariate t, the approximate marginal posterior distribution of τ will be gamma, and the approximate predictive distribution of future observations will be characterized by a sequence of conditional t densities.

Using a normal-gamma prior density for θ and τ, Shaarawy and Broemeling (1984) have derived these approximate distributions. For example, the marginal posterior distribution of θ is a t with $n+2\alpha$ degrees of freedom, location vector:

$$E(\theta|S_n)=A^{-1}B, \tag{4.10}$$

precision matrix:

$$Q(\theta|S_n)=(n+2\alpha)A(C-B'A^{-1}B)^{-1}, \tag{4.11}$$

where A is the qth-order matrix,

$$A=P+(a_{ij}), \tag{4.12}$$

B is a $q\times 1$ vector,

$$B=P\mu-(b_j), \tag{4.13}$$

and C is a scalar,

$$C=\mu'P\mu+2\beta+\sum_{1}^{n}Y^2(t). \tag{4.14}$$

Furthermore, μ, P, α, and β are parameters of the prior distribution of θ and τ, where $\theta|\tau\sim N(\mu,P\tau)$ and $\tau\sim G(\alpha,\beta)$. Also,

$$a_{ij}=\sum_{t=1}^{n}\hat{e}(t-i)\hat{e}(t-j),\quad i,j=1,2,\ldots,q, \tag{4.15}$$

and

$$b_j = \sum_{t=1}^{n} Y(t)\hat{e}(t-j), \quad i, j = 1, 2, \ldots, q. \tag{4.16}$$

It can also be shown that the approximate marginal posterior distribution of τ is gamma with parameters $(n+2\alpha)/2$ and β', where

$$2\beta' = C - B'A^{-1}B. \tag{4.17}$$

The parameters of the joint posterior distribution are easily calculated from the data using formulae (4.10)–(4.17) and formula (4.8). To recapitulate, the t-approximation to the marginal posterior distribution of θ is calculated as follows:

(1) Find the conditional least squares estimate $\hat{\theta}$ from (4.7) and check to see if $\hat{\theta}$ is in the region of invertibility.

(2) Calculate the estimated residuals from (4.8).

(3) Calculate the parameters of the posterior distribution of the parameters using formulae (4.10)–(4.17).

After calculating the parameters of the posterior distribution, one may make inferences in the usual way. For example, an HPD region for θ of content $1-\gamma$ is

$$R_\gamma(\theta) = \left\{\theta : \frac{(\theta - A^{-1}B)'A(\theta - A^{-1}B)(n+2\alpha)}{C - B'A^{-1}B} \le qF_{\gamma;q,n+2\alpha}\right\}, \tag{4.18}$$

and one may find HPD regions for any subset of the moving average parameters in the same way. Since θ has a q-dimensional multivariate t distribution, any particular moving average coefficient has a univariate t distribution, approximately.

The above posterior analysis is derived under the assumption that a normal-gamma distribution describes prior information. Thus, if one is quite confident about the values of the parameters, one would specify such prior information by the appropriate choice of μ, P, α, and β; however, if one knows very little, a priori, about the parameters θ and τ, one might employ Jeffreys' prior density:

$$\xi(\theta, \tau) \propto \tau^{-1}, \quad \theta \in R^q, \tau > 0,$$

which will produce a normal-gamma posterior density with parameters given by (4.10)–(4.17), but modified by letting $P \to 0$ $(q \times q)$, $\alpha \to -q/2$, and $\beta \to 0$ (scalar).

As with the AR process, the Bayesian analysis of an MA process allows one to solve the identification problem of choosing the order of the model. If one knows the maximum order of the process is say m, one may use a backward elimination procedure of a sequence of univariate t tests to estimate the order of the process. Refer to section 4.1, and the explanation of the identification of an AR model.

The adequacy of the t-approximation to the marginal posterior distribution of θ will be examined numerically in the next section. As we will see, it provides accurate posterior and predictive inferences and gives results similar to a Box–Jenkins analysis.

This section is concluded with the moving average prediction problem, i.e. if one has n observations, how does one predict $Y(n+1)$, $Y(n+2)$, etc.? One would expect the joint predictive density of future observations to be given approximately by a sequence of conditional t densities. To be more specific we would expect the marginal distribution of $Y(n+1)$ to be a univariate t, the conditional predictive distribution of $Y(n+2)$ given $Y(n+1)$ to be a univariate t, and so on. Broemeling and Land (1984) have determined the parameters of these distributions for autoregressive forecasting, but they apply approximately to the moving average process.

The approximate predictive density of $Y(n+1)$ can be shown to be a univariate t (see Shaarawy and Broemeling, 1984) with $n+2\alpha$ degrees of freedom, location:

$$E[Y(n+1)|S_n] = G^{-1}H, \tag{4.19}$$

and precision:

$$Q[Y(n+1)|S_n] = \frac{(n+2\alpha)G}{M - H'G^{-1}H}, \tag{4.20}$$

where

$$G = 1 - E_2'D^{-1}E_2, \tag{4.21}$$

$$H = -E_2'D^{-1}(P\mu - E_1), \tag{4.22}$$

and

$$M = \mu'P\mu + 2\beta + \sum_1^n Y^2(t) - (P\mu - E_1)'D^{-1}(P\mu - E_1). \tag{4.23}$$

Also,

$$D = P + D_1, \tag{4.24}$$

$$E = P\mu - E_1 - Y(n+1)E_2, \tag{4.25}$$

where D_1 is $q \times q$ with jkth element $\sum_1^n \hat{e}(t-j)\hat{e}(t-k)$, E_1 has jth component $\sum_1^n Y(t)\hat{e}(t-j)$, and E_2 has jth component $\hat{e}(n+1-j)$. The parameters of the normal-gamma prior distribution are μ, P, α, and β.

Thus, if we want to predict the first future observation,

$$G^{-1}H \pm t_{\gamma/2, n+2\gamma} Q^{-1/2}[Y(n+1)|S_n] \tag{4.26}$$

provides a forecast interval with confidence $1-\gamma$.

As with the autoregressive process, the moving average process is analyzed by using the t-distribution to make posterior and predictive inferences. The order of the process is identified with a backward elimination of a series of univariate t-tests, then the parameters of the process are estimated by referring to their marginal posterior distribution (a multivariate t), and finally forecasting is achieved by a series of univariate t predictive densities. All these inferences are approximate in the sense that the normal-gamma posterior distribution and the predictive t-distributions are based on the approximate conditional likelihood (4.9).

4.2. Approximate posterior and predictive inferences with MA processes

How does the t-approximation to the posterior distribution of the parameters and to the predictive distribution of a future observation perform?

In this section several MA processes are considered. Observations are generated from a particular moving average process with known parameters and the posterior mean of ϕ, (4.15), posterior dispersion matrix (4.16), predictive mean (4.29), and predictive dispersion matrix (4.30) are computed for samples of size $n = 30$, 50, 70, and 90. In all cases a vague prior density,

$$\xi(\phi, \tau) \propto \tau^{-1}, \quad \phi \in R^q, \ \tau > 0, \tag{4.27}$$

was employed to express prior information about the parameters. Since the data were generated from a known process it would not be realistic to use a conjugate prior density (unless we were conducting a sensitivity study).

Four examples are presented, two MA(1) processes and two MA(2). Tables 21.1 and 21.2 summarize the posterior and predictive analysis of a MA(1) model with parameters $\phi_1 = -1$ and 0.5, respectively. The error precision is always two. Each of these tables consists of five columns, the sample size, posterior mean of ϕ_1, variance of ϕ_1, coefficient of variation of ϕ_1, the mean of $Y(n+1)$, and the variance of $Y(n+1)$, where $n = 30$, 50, 70, and 90. For example, consider table 21.1, where 91 observations were generated from a MA(1) process with parameters $\phi_1 = -1$ and $\tau = 2$, then the first 30 were used to compute the first row of the

Table 21.1. The t-approximation MA(1) model, $\phi_1 = -1.0$ and $\tau = 2$.

N	$E(\phi_1 \mid Y)$	$V(\phi_1 \mid Y)$	$CV(\phi_1 \mid Y)$	$E(Y(n+1))$	$V(Y(n+1))$
30	−0.8632	0.0374	−0.2240	−0.9345	0.6017
50	−0.9252	0.0206	−0.1550	0.1216	0.5689
70	−1.0327	0.0148	−0.1178	−0.7325	0.5424
90	−1.0486	0.0113	−0.1012	−0.4292	0.5313

Table 21.2. The t-approximation MA(1) model, $\phi_1 = 0.5$ and $\tau = 2$.

N	$E(\phi_1 \mid y)$	$V(\phi_1 \mid y)$	$CV(\phi_1 \mid y)$	$E(Y(n+1))$	$\mathrm{Var}(Y(n+1) \mid y)$
30	0.6894	0.0353	0.2724	0.6851	0.5871
50	0.5982	0.0202	0.2376	0.0403	0.5743
70	0.4880	0.0147	0.2484	0.2714	0.5310
90	0.4782	0.0114	0.2233	0.3425	0.5345

table, the first 50 the second row, and the last row is based on the first 90. Considering the last row, -1.0487 is the mean of the posterior distribution of ϕ_1 using the t-approximation, a vague prior density (4.32) and sample of size 90. Also, the 91st observation has a mean of -0.4292.

The results for the two MA(2) processes are shown in tables 21.3 and 21.4. Table 21.3 shows what happens with the MA(2) process which has parameters $\phi_1 = -0.5$, $\phi_2 = 0.5$, and $\tau = 2$. Again 91 observations were generated from the model, and the first 50 were used to compute the quantities in the second row, namely the posterior mean of ϕ_1, variance of ϕ_1, coefficient of variation of ϕ_1, posterior mean of ϕ_2, variance of ϕ_2, coefficient of variation of ϕ_2, correlation between ϕ_1 and ϕ_2, mean of the predictive distribution of $Y(n+1)$, and the variance of the predictive distribution. Table 21.4 has a similar interpretation.

Note this is not a simulation study of the sampling properties of Bayes estimators, but instead an examination of the behavior of the t-approximation in making inferences about the parameters of the model and forecasting future observations. One must ask: Are the results reasonable? That is, would you use them to make inferences about the model?

Consider table 21.2 in more detail, where $\phi_1 = 0.5$ and $\tau = 2$. We see that as the sample size increases the posterior mean of ϕ_1 becomes closer to 0.5 and the posterior variance decreases from 0.0372 to 0.0116. Using 90 observations one would estimate ϕ_1 by 0.4793 (with a true value of 0.5), and predict $Y(91)$ with 0.5417.

4.4. A comparison with Box and Jenkins methodology

Table 21.5 records the results of a Box–Jenkins analysis performed on an MA(1) with parameters $\tau = 2$ and $\theta_1 = (\phi_1) = 0.5$ using the same data set as that of table 21.2, where the t-approximation was employed to make inferences about the moving average parameter. The first column of table 21.5 indicate the sample size, while the remaining columns give the estimate, the estimated variance of the estimate, the estimated coefficient of variation of the estimate, the forecast of the next observation, and the variance of the forecast. These numbers were calculated with the ETS package of SAS (Statistical Analysis System), but the package uses the MA(1) model $Y(t) = \mu + e(t) + \theta e(t-1)$, where the mean of the observation is μ. By comparing tables 21.2 and 21.5, one sees that the two methods of analysis produce similar results. The Box–Jenkins estimate of θ_1 is 0.4128 when $n = 90$ and the estimated variance is 0.0094, while the forecast of the 91st observation is 0.2718. These Box–Jenkins estimates can be compared to the last row of table 21.2, where one observes that the Bayesian analysis produces a forecast of 0.3425 for the 91st observation.

No attempt is made to conclude that one method is better than the other. From a large scale study of Shaarawy (1984), one may conclude only that the two methods produce similar estimates and forecasts.

Table 21.3. The t-approximation MA(2) model, $\phi_1 = -0.5$, $\phi_2 = 0.5$ and $\tau = 2$.

N	$E(\phi_1 \mid \boldsymbol{Y})$	$V(\phi_1 \mid \boldsymbol{Y})$	$CV(\phi_1 \mid \boldsymbol{Y})$	$E(\phi_2 \mid \boldsymbol{Y})$	$V(\phi_2 \mid \boldsymbol{Y})$	$CV(\phi_2 \mid \boldsymbol{Y})$	$\rho(\phi_1, \phi_2 \mid \boldsymbol{Y})$	$E(Y(n+1) \mid \boldsymbol{Y})$	$V(Y(n+1) \mid \boldsymbol{Y})$
30	−0.4970	0.0360	−0.3818	0.2230	0.0390	0.8856	−0.0170	−0.2300	0.6600
50	−0.4150	0.0220	−0.3574	0.3150	0.0230	0.4815	−0.0020	−0.4900	0.6300
70	−0.5200	0.0149	−0.2347	0.2340	0.0150	0.5234	−0.0200	0.1480	0.5278
90	−0.5360	0.0116	−0.2009	0.3060	0.0117	0.3535	−0.0002	0.0370	0.5310

Table 21.4. The t-approximation MA(2) model, $\phi_1 = 0.3$, $\phi_2 = 0.1$ and $\tau = 2$.

N	$E(\phi_1 \mid \boldsymbol{Y})$	$V(\phi_1 \mid \boldsymbol{Y})$	$CV(\phi_1 \mid \boldsymbol{Y})$	$E(\phi_2 \mid \boldsymbol{Y})$	$V(\phi_2 \mid \boldsymbol{Y})$	$CV(\phi_2 \mid \boldsymbol{Y})$	$\rho(\phi_1, \phi_2 \mid \boldsymbol{Y})$	$E(Y(n+1) \mid \boldsymbol{Y})$	$V(Y(n+1) \mid \boldsymbol{Y})$
30	0.3256	0.0469	0.6651	−0.3100	0.0452	−0.6858	0.1914	0.0532	0.7083
50	0.3618	0.0212	0.4025	−0.1098	0.0212	−1.3260	0.0431	0.2507	0.5885
70	0.2749	0.0147	0.4406	−0.1477	0.0148	−0.8223	0.0145	−0.0885	0.5046
90	0.2753	0.0114	0.3878	−0.0742	0.0115	−1.4453	−0.0005	0.0039	0.5239

Table 21.5. Moderate sample behavior of Box and Jenkins' procedure for $\phi_1 = 0.5$.

N	$\hat{\phi}_1$	$V(\hat{\phi}_1)$	$CV(\hat{\phi}_1)$	$\hat{Y}(n+1)$	$V(\hat{Y}(n+1))$
30	0.6032	0.0230	0.2514	0.7775	0.5232
50	0.5088	0.0156	0.2454	0.1158	0.5842
70	0.4069	0.0126	0.2758	0.1714	0.5271
90	0.4128	0.0094	0.2349	0.2718	0.5291

4.5. ARMA processes

From the numerical study of the Bayesian analysis of moving average processes, it appears that the approximate conditional likelihood (4.9) will produce a Bayesian analysis which gives accurate and convenient posterior and predictive inferences. Can the approximation be extended to cover ARMA processes, (3.1)?

Consider an ARMA (p, q) process (3.1) which is invertible and stationary and suppose there are n observations $S_n = [Y(1), Y(2), \ldots, Y(n)]$, then the residuals are given by

$$e(t) = Y(t) - \sum_1^p \phi_i Y(t-i) + \sum_1^q \phi_j e(t-j), \quad t = 1, 2, \ldots, n. \tag{4.28}$$

By conditioning on the first p observations and letting $\varepsilon(p) = 0 = \varepsilon(p-1) = \cdots = \varepsilon(p-q-1)$, see Tiao and Box (1981, p. 809), one may approximate the exact likelihood by

$$L(\phi, \theta, \tau | S_n) \propto \tau^{(n-p)/2} \exp\left(-\frac{\tau}{2} \sum_{p+1}^{n} e^2(t) \right), \tag{4.29}$$

where $\phi = (\phi_1, \phi_2, \ldots, \phi_p) \in R^p$, $\theta = (\theta_1, \theta_2, \ldots, \theta_q) \in R^q$, and $\tau > 0$.

The conditional likelihood function (4.29) is nonlinear in the parameters θ and ϕ; however, the residuals $e(t)$ of (4.28) and (4.29) can be estimated by conditional least squares estimates:

$$\hat{e}(t) = Y(t) - \sum_1^p \hat{\phi}_i Y(t-i) + \sum_1^q \hat{\theta}_j \hat{e}(t-j), \tag{4.30}$$

and $\hat{\phi}_i$ and $\hat{\theta}_j$ are the nonlinear least squares estimators of $\hat{\phi}_i$ and $\hat{\theta}_j$, respectively, found by minimizing the conditional sum of squares:

$$\mathrm{ss}(\phi, \theta) = \sum_1^n e^2(t),$$

with respect to ϕ and θ over the region of invertibility and stationarity. The conditional least squares estimates are found by a nonlinear regression algorithm and the technique is explained by Harvey (1981, p. 126).

The estimates residuals are then substituted into the conditional likelihood function (4.29) yielding an approximate likelihood function of

$$L^*(\phi;\, \theta, \tau | S_n) \propto \tau^{(n-p)/2} \times \exp\left(-\frac{\tau}{2} \sum_{p+1}^{n} \left[Y(t) - \sum_1^p \phi_i Y(t-i) + \sum_1^q \theta_j \hat{e}(t-j) \right]^2 \right), \tag{4.31}$$

where $\phi \in R^p$, $\theta \in R^q$, and $\tau > 0$.

But, (4.31) as a function of the parameters is a normal-gamma density function which, when combined with a normal-gamma prior density (or a Jeffreys' prior density) for ϕ, θ, and τ, will give a normal-gamma posterior distribution for the parameters and a one-step-ahead predictive t-distribution for future observations. The analysis of an ARMA process is therefore the same as the analysis of an MA process.

In summary, one may say the Bayesian analysis of ARMA processes amounts to about the same thing as doing a multiple linear regression analysis. The one aspect of the analysis which is different is that it is usually necessary to perform a conditional nonlinear regression analysis in order to estimate the residuals (4.28) of the model.

5. Bayesian time series

The results of the preceding section will give one the machinery necessary to do a complete time series analysis as follows. Suppose there are n observations $Y(1)$, $Y(2),\dots,Y(n)$ which have been generated by an ARMA process and that the maximum autoregressive order is M and the maximum moving average order is N, then:

(1) Determine the posterior distribution of $\phi=(\phi_1, \phi_2,\dots,\phi_M)$ and $\theta=(\theta_1, \theta_2,\dots,\theta_N)$. This will be an $M+N$ multivariate t-distribution assuming Jeffreys' prior density:

$$\xi(\phi,\theta,\tau)\propto\tau^{-1},\quad \tau>0,\ \phi\in R^M,\ \theta\in R^N.$$

(2) Determine the order p and q of the process by:

(a) Test $H_0:\theta_N=0$ versus $H_1:\theta_N\neq 0$. Use the marginal posterior distribution of θ_N, which is a univariate t.

(b) If H_0 is not rejected above, test $H_0:\theta_{N-1}=0$ versus $H_1:\theta_{N-1}\neq 0$ with the conditional distribution of ϕ_{N-1} given $\theta_N=0$.

(c) If the above H_0 is not rejected, test $H_0:\theta_{N-2}=0$ versus $H_1:\theta_{N-2}\neq 0$, using the conditional distribution θ_{N-2} given $\theta_N=\theta_{N-1}=0$, etc.

(d) On the other hand, if H_0 of (a) is rejected, test $H_0:\phi_M=0$ versus $H_1:\theta_M\neq 0$ with the marginal posterior distribution of ϕ_M.

(e) And so on.

Continuing in this fashion, one will arrive at a stage where one will reject the hypotheses $\phi_p=0$ and $\theta_q=0$. One concludes the data was generated by an ARMA (p,q) process, where $0\le p\le M$ and $0\le q\le N$.

(3) Assuming the data was generated by an ARMA (p,q) process (perhaps after differencing the series), one determines the parameters of the normal-gamma posterior distribution of $\phi=(\phi_1, \phi_2,\dots,\phi_p)$, $\theta=(\theta_1, \theta_2,\dots,\theta_q)$, and τ. Inferences about the parameters are done in the usual way, using the multivariate t-distribution for ϕ and θ and the gamma distribution for τ. Notice that at this stage one must "fit" the data to an ARMA (p,q) process using the conditional nonlinear least squares algorithm (4.30).

(4) A future observation $Y(n+1)$ is predicted by referring to the Bayesian predictive density of $Y(n+1)$, which will be approximately a univariate t. The parameters of this distribution are easily found in the same way as (4.19) and (4.20) were derived for the MA(q) process. See Shaarawy and Broemeling (1984) and refer to Broemeling and Land (1984) for predicting with AR(p) processes. One may also want to forecast $Y(n+2)$ and observations at higher lags, and one would expect the conditional predictive distribution of $Y(n+2)$, given $Y(n+1)$, to be a univariate t.

The above Bayesian analysis of time series generated by ARMA processes accomplishes the same thing as the Box–Jenkins analysis; however, for the most part, the techniques are quite different. Steps (1) and (2) above identify a particular ARMA process, but the Box–Jenkins analysis uses the sample autocorrelation and partial autocorrelation functions to estimate the orders p and q, while the Bayesian analysis consists of a backward type regression elimination of a series of t-tests. In practice the two methods give about the same results. This has been indicated by some preliminary work of the authors.

The third phase of the above analysis estimates and tests hypotheses about the ϕ, θ, and τ parameters using the normal-gamma posterior distribution, but Box and Jenkins do the same using the asymptotic normal distribution of the estimated parameters. See Harvey (1981, pp. 130–139) for a discussion of the asymptotic properties of the conditional and full maximum likelihood estimators.

The Bayesian uses the Bayesian predictive density of a future observation $Y(n+1)$ in order to forecast. For example, one might use the mean or an interval to forecast one step ahead. This is easily done using the t-tables. See (4.26), the one-step-ahead interval forecast of $Y(n+1)$. With the Box–Jenkins analysis one determines the conditional expectation of $Y(n+1)$ given the observations S_n, say $E[Y(n+1)|S_n]$, which is a function of θ, ϕ, and unobservable residuals. These are estimated by a recursive method explained in Priestley (1981, pp. 762–773). Thus, an estimated conditional expectation forecasts one step ahead (and higher) in the Box–Jenkins analysis, but for the Bayesian the mean of the approximate conditional expectation of $Y(n+1)$ given S_n (which is free of unknown parameters and unobservable residuals) is used to forecast one step ahead. The similarity between these two techniques of forecasting needs to be studied.

The foundation of the Bayesian analysis is the approximate conditional likelihood function which allows one to make inferences about the parameters of the model via a normal-gamma posterior density and to predict future observations via the predictive t-distribution. This type of approximation is similar to the one described in the Bayesian analysis of Box and Jenkins (1970), who expand the residual sum of squares in the likelihood function (see eq. (4.30) and the following) in a Taylor series about the maximum likelihood estimates of θ and ϕ. The same approach was taken by Newbold (1973) and Zellner and Reynolds (1978) in a Bayesian analysis of ARMA processes. The proposed approximation by the authors avoids the need to find the first- and second-order partial derivatives of the residual sum of squares, but we would expect the two types of approximations to produce similar results in large samples.

6. Summary and conclusions

This chapter has reviewed Bayesian contributions to the analysis of time series and has presented a Bayesian analysis of data generated by ARMA processes.

The Bayesian analysis expresses prior information about the moving average autoregressive parameters and the residual precision with a normal-gamma distribution or with a vague Jeffreys' prior density. The conditional likelihood function is approximated by a normal-gamma density; thus, when combined with the prior information will yield posterior inferences (including identification of the appropriate ARMA process) about the coefficients of the process and predictive inferences for future observations.

On the basis of the above results, a Bayesian time series analysis in the time domain was outlined in the previous section and was compared to the Box–Jenkins analysis. The essential ingredients of identification estimation, and prediction are unified with the Bayesian approach.

References

Box, G.E.P. and Gwilym M. Jenkins (1970) *Time Series Analysis, Forecasting and Control* (Holden-Day, San Francisco).

Broemeling, Lyle and Margaret Land (1984) "On forecasting with univariate autoregressive processes", *Communication In Statistics*, 13.

Chow, Gregory C. (1975) "Multiperiod predictions from stochastic difference equations by Bayesian methods", *Studies in Bayesian Econometrics and Statistics* S.E. Fienberg and A. Zellner, eds., (North-Holland, Amsterdam), ch. 8.

Diaz, J. and J. Farah (1982) "Bayesian identification of autoregressive processes", 22-nd NBER-NSF Seminar on Bayesian Inference in Econometrics.

Gersch, Will (1982) "A smoothness priors approach to the modeling of time series with trends and seasonalities", *Proceedings of the Business and Economics Section of the ASA*.

Harrison, P.G. and C.F. Stevens (1976) "Bayesian Forecasting" (with discussion), *Journal of the Royal Statistical Society*, Ser. B 38, 205–245.

Harvey, A.C. (1981) *Time Series Models* (John Wiley and Sons Inc., New York).

Jeffreys, H. (1939) *Theory of Probability* (Oxford University Press, London).

Kalman, R.E. (1960) "A new approach to linear filtering and prediction problems", *Transactions of ASME*, Ser. D. *Journal of Basic Engineering*, 95–108.

Lahiff, Maureen (1980) "Time series forecasting with informative prior distribution", Technical Report No. 111, Department of Statistics, University of Chicago.

Litterman, R.B. (1980) "A Bayesian procedure for forecasting with vector autoregressions", manuscript, Department of Economics, MIT.

Monahan, J.F. (1983) "Fully Bayesian analysis of ARMA time series models", *Journal of Econometrics*, 21, 307–331.

Newbold, Paul (1973) "Bayesian estimation of Box–Jenkins transfer function models", *Journal of Royal Statistical Society*, Ser. B 35, 323–336.

Priestley, M.B. (1981) *Spectral Analysis and Time Series (Academic Press, London)*.

Salazar, Diego (1982) "Structural change in time series", *Journal of Econometrics*, 19, 147–164.
Shaarawy, Samir (1984) "The analysis of moving average processes", unpublished Ph.D. dissertation, Oklahoma State University, Stillwater.
Shaarawy, Samir and Lyle Broemeling (1984) "Bayesian inferences and forecasts with moving average processes", *Communications in Statistics*, 13, 15.
Smith, A.F.M. (1980) "Change point problems: Approaches and applications", in: Bernardo et al., eds., *Bayesian Statistics, proceedings of the First International Meeting* (University Press, Valencia, Spain).
Smith, A.F.M. and Makov, U.E. (1980) "Bayesian detection and estimation of jumps in linear systems", in: O.R.L. Jacobs et al., eds., *Proceedings of the IMA conference on The Analysis and Optimization of Stochastic Systems* (Academic Press, London).
Tiao, G.C. and G.E.P. Box (1981) "Modeling multiple time series with applications", *Journal of the American Statistical Association*, 76, 802–816.
Tsurumi, Hiroki (1982) "A Bayesian and maximum likelihood analysis of a gradual switching regression in a simultaneous equation framework", *Journal of Econometrics*, 19, 165–182.
Zellner, A. (1971) *An Introduction to Bayesian Inference in Econometrics* (John Wiley and Sons, Inc., New York).
Zellner, A. (1977) "Statistical analysis of econometric models", address to the American Statistical Association Meeting, Chicago, 15–18 August; published in *Journal of the American Statistical Association*, 74 (1979) 628–651 (with discussion).
Zellner, A. (1983) "Application of Bayesian analysis in econometrics", *The Statistician*, 32, 23–34.
Zellner, Arnold and Roger A. Reynolds (1978) "Bayesian analysis of ARMA models", presented at the 16th seminar on Bayesian inference in Econometrics, 2–3 June.

Chapter 22

A BAYESIAN APPROACH TO NESTED MISSING-DATA PROBLEMS

CHAN-FU CHEN*

Academia Sinica, Taipei, Taiwan and Lederle Laboratories

1. Introduction

Anderson (1957) has derived the explicit maximum likelihood estimates (MLEs) of parameters of multivariate normal distributions in cases where the missing observations in the data matrix form a nested pattern. Such patterns may arise, for instance, in panel studies because a decision is made not to pursue nonrespondents in subsequent waves for various practical reasons, as discussed recently by Marini, Olsen and Rubin (1980).

Anderson's basic technique is to decompose the original estimation problem into smaller estimation problems by factoring the likelihood function into a product of likelihoods. In this chapter, a Bayesian approach to this nested missing-data problem is exploited by utilizing Anderson's likelihood factorization. Throughout the discussion, normal–Wishart priors are assumed for the population mean vector and the precision matrix of all variates under study. It is shown that, under this type of conjugate priors, the distinct sets of parameters appearing in the factored likelihoods are independent of each other, both a priori and a posteriori – a factorization with "distinct parameters", as Rubin (1974) terms them.

*I wish to thank D.M. Titterington for drawing my attention to the missing-data problems, and the referees for helpful suggestions.

Bayesian Inference and Decision Techniques, edited by P. Goel and A. Zellner

2. Anderson's likelihood factorization

Let $\boldsymbol{w} = (\boldsymbol{x}', \boldsymbol{y}')'$, where $\boldsymbol{x}: r \times 1$ and $\boldsymbol{y}: s \times 1$, be a p-dimensional $(p = r + s)$ random vector with a multivariate normal distribution, $N_p(\boldsymbol{\mu}, \Sigma)$, $\Sigma > 0$, where

$$\boldsymbol{\mu} = \begin{bmatrix} \boldsymbol{\mu}_x \\ \boldsymbol{\mu}_y \end{bmatrix} \quad \text{and} \quad \Sigma = \begin{bmatrix} \Sigma_{xx} & \Sigma_{xy} \\ \Sigma_{yx} & \Sigma_{yy} \end{bmatrix}, \tag{2.1}$$

with $\boldsymbol{\mu}_x = E(\boldsymbol{x})$, $\boldsymbol{\mu}_y = E(\boldsymbol{y})$, $\Sigma_{xx} = \text{Cov}(\boldsymbol{x})$, $\Sigma_{xy} = \Sigma'_{yx} = \text{Cov}(\boldsymbol{x}, \boldsymbol{y})$ and $\Sigma_{yy} = \text{Cov}(\boldsymbol{y})$. Then, the joint density of $\boldsymbol{x}$ and $\boldsymbol{y}$ can be written as the marginal density of $\boldsymbol{x}$ times the conditional density of $\boldsymbol{y}$ given $\boldsymbol{x}$:

$$f(\boldsymbol{x}, \boldsymbol{y} | \boldsymbol{\mu}, \Sigma) = f(\boldsymbol{x} | \boldsymbol{\mu}_x, \Sigma_{xx}) f(\boldsymbol{y} | \boldsymbol{\alpha} + B_{yx}\boldsymbol{x}, \Sigma_{yy\cdot x}), \tag{2.2}$$

where $f(\cdot | \boldsymbol{m}, C)$ denotes the normal density with mean $\boldsymbol{m}$ and covariance matrix C, and

$$\begin{aligned} \boldsymbol{\alpha} &= \boldsymbol{\mu}_y - B_{yx}\boldsymbol{\mu}_x, \\ B_{yx} &= \Sigma_{yx}\Sigma_{xx}^{-1} \\ \Sigma_{yy\cdot x} &= \Sigma_{yy} - \Sigma_{yx}\Sigma_{xy}^{-1}\Sigma_{xy}. \end{aligned} \tag{2.3}$$

Suppose the missing observations in the data matrix form a nested pattern as

$$\begin{aligned} &\boldsymbol{x}_1, \ldots, \boldsymbol{x}_n, \boldsymbol{x}_{n+1}, \ldots, \boldsymbol{x}_N, \\ &\boldsymbol{y}_1, \ldots, \boldsymbol{y}_n. \end{aligned}$$

That is, while n independent observations are available on the whole pair $(\boldsymbol{x}, \boldsymbol{y})$, $N - n$ independent observations are available on $\boldsymbol{x}$ only with the values of the corresponding $\boldsymbol{y}$ missing. Under such a pattern, the likelihood function of $(\boldsymbol{\mu}, \Sigma)$ can be factored, according to (2.2), into a product of likelihoods (Anderson, 1957):

$$\begin{aligned} &\prod_{i=1}^{n} f(\boldsymbol{x}_i, \boldsymbol{y}_i | \boldsymbol{\mu}, \Sigma) \prod_{i=n+1}^{N} f(\boldsymbol{x}_i | \boldsymbol{\mu}_x, \Sigma_{xx}) \\ &\quad = \prod_{i=1}^{N} f(\boldsymbol{x}_i | \boldsymbol{\mu}_x, \Sigma_{xx}) \prod_{i=1}^{n} f(\boldsymbol{y}_i | \boldsymbol{\alpha} + B_{yx}\boldsymbol{x}_i, \Sigma_{yy\cdot x}). \end{aligned} \tag{2.4}$$

Based on (2.3) and the MLEs of $\boldsymbol{\mu}_x$, Σ_{xx}, $\boldsymbol{\alpha}$, B_{yx} and $\Sigma_{yy\cdot x}$ obtained from (2.4), the MLEs of the original parameters $\boldsymbol{\mu}_y$, $\Sigma_{yx} = \Sigma'_{xy}$ and Σ_{yy} can be easily derived (see Anderson, eqs. (6) and (7)).

3. Bayesian inference for $(\boldsymbol{\mu}, \Sigma)$

3.1. Factorization of a normal–Wishart prior

Suppose that the prior distribution of $(\boldsymbol{\mu}, \Sigma)$ defined in (2.1) is of the normal–Wishart type, $\text{NW}_p(\boldsymbol{\theta}, \tau, \Omega, \nu)$: the marginal distribution of $\Lambda \equiv \Sigma^{-1}$ is a

p-dimensional Wishart,

$$\Lambda \sim W_p(\Omega^{-1}, \nu), \tag{3.1}$$

with $E(\Lambda|\Omega, \nu) = \nu\Omega^{-1}$, $\Omega: p \times p > 0$, and $\nu(\geq p)$ degrees of freedom; and, conditional on fixed Σ, the distribution of $\boldsymbol{\mu}$ is a p-dimensional normal,

$$\boldsymbol{\mu}|\Sigma \sim N_p(\boldsymbol{\theta}, \tau^{-1}\Sigma), \tag{3.2}$$

with $\boldsymbol{\theta}: p \times 1$ and $\tau > 0$.

Partition $\boldsymbol{\theta}$ as $\boldsymbol{\mu}$ and both Λ and Ω as Σ in (2.1). The following theorem shows that the distinct sets of parameters appearing in the factored likelihoods (2.4) are independently distributed under the normal–Wishart prior.

Theorem 3.1.[1] Under (3.1) and (3.2), $(\boldsymbol{\mu}_x, \Sigma_{xx})$ and $(\boldsymbol{\alpha}, B_{yx}, \Sigma_{yy\cdot x})$ of (2.3) are independently distributed. Furthermore,

(i) $\Sigma_{xx}^{-1} \sim W_r(\Omega_{xx}^{-1}, \nu - s)$ and $\boldsymbol{\mu}_x$ given Σ_{xx} is $N_r(\boldsymbol{\theta}_x, \tau^{-1}\Sigma_{xx})$;

(ii) $\Sigma_{yy\cdot x}^{-1} \sim W_s(\Omega_{yy\cdot x}^{-1}, \nu)$ and $B \equiv (\boldsymbol{\alpha}, B_{yx})$ given $\Sigma_{yy\cdot x}$ is $N_{s,(r+1)}(\Gamma, \Sigma_{yy\cdot x}, \Psi)$, a matrix-type normal distribution;[2] where $\Omega_{yy\cdot x} = \Omega_{yy} - \Omega_{yx}\Omega_{xx}^{-1}\Omega_{xy}$, $\Gamma = (\boldsymbol{\gamma}, \Gamma_{yx})$, $\Gamma_{yx} = \Omega_{yx}\Omega_{xx}^{-1}$, $\boldsymbol{\gamma} = \boldsymbol{\theta}_y - \Gamma_{yx}\boldsymbol{\theta}_x$, and

$$\Psi = \begin{bmatrix} \tau^{-1} + \boldsymbol{\theta}_x'\Omega_{xx}^{-1}\boldsymbol{\theta}_x & -\boldsymbol{\theta}_x'\Omega_{xx}^{-1} \\ -\Omega_{xx}^{-1}\boldsymbol{\theta}_x & \Omega_{xx}^{-1} \end{bmatrix}. \tag{3.3}$$

Proof. First, from (3.1), we have:

(a) Σ_{xx} and $(B_{yx}, \Sigma_{yy\cdot x})$ are independently distributed;

(b) $\Sigma_{xx}^{-1} \sim W_r(\Omega_{xx}^{-1}, \nu - s)$;

(c) $\Sigma_{yy\cdot x}^{-1} \sim W_s(\Omega_{yy\cdot x}^{-1}, \nu)$; and

(d) conditional on $\Sigma_{yy\cdot x}$, $B_{yx} \sim N_{s,r}(\Gamma_{yx}, \Sigma_{yy\cdot x}, \Omega_{xx}^{-1})$, by noting that $\Sigma_{xx}^{-1} = \Lambda_{xx\cdot y}$, $\Sigma_{yy\cdot x}^{-1} = \Lambda_{yy}$ and $B_{yx} = \Sigma_{yx}\Sigma_{xx}^{-1} = -\Lambda_{yy}^{-1}\Lambda_{yx}$ (cf. Dempster, 1969, theorem 13.4.2 or Srivastava and Khatri, 1979, theorem 3.3.5).

Secondly, from (3.2), $\boldsymbol{\mu}_x$ given Σ_{xx} is $N_r(\boldsymbol{\theta}_x, \tau^{-1}\Sigma_{xx})$, which is independent of $(B_{yx}, \Sigma_{yy\cdot x})$. Also, since $\boldsymbol{\alpha} = \boldsymbol{\mu}_y - B_{yx}\boldsymbol{\mu}_x$, the conditional distribution of $\boldsymbol{\alpha}$, given B_{yx} and $\Sigma_{yy\cdot x}$, is $N_s(\boldsymbol{\theta}_y - B_{yx}\boldsymbol{\theta}_x, \tau^{-1}\Sigma_{yy\cdot x})$, independent of $(\boldsymbol{\mu}_x, \Sigma_{xx})$. This result combined with (d) above yields the desired conditional distribution of B stated in (ii) of theorem 3.1.

[1] This theorem is a generalization of theorem 5.1 given in Chen (1979) for the case $s = 1$. The proofs are essentially the same.

[2] A $p \times n$ random matrix X is said to have a matrix-type normal distribution, $N_{p,n}(\Gamma, \Sigma, \Psi)$, if its pdf is given by

$$p(X) = (2\pi)^{-np/2}|\Sigma|^{-n/2}|\Psi|^{-p/2}\exp\left\{-\tfrac{1}{2}\operatorname{tr}\Sigma^{-1}(X-\Gamma)\Psi^{-1}(X-\Gamma)'\right\}.$$

(See Srivastava and Khatri, 1979, p. 54.)

3.2. Posterior analysis

Combining independent priors (i) and (ii) of theorem 3.1 with the product of likelihoods (2.4), one can easily obtain independent posterior distributions of $(\boldsymbol{\mu}_x, \Sigma_{xx})$ and $(B, \Sigma_{yy\cdot x})$ by the standard Bayesian inference procedure. That is, the joint density

$$\begin{aligned} &P(\boldsymbol{\mu}_x, \Sigma_{xx}, \boldsymbol{\alpha}, B_{yx}, \Sigma_{yy\cdot x}|\text{data}) \\ &\quad = P(\boldsymbol{\mu}_x, \Sigma_{xx}\, \boldsymbol{x}_i,\, i=1,\dots,N)\, P\big(\boldsymbol{\alpha}, B_{yx}, \Sigma_{yy\cdot x}|(\boldsymbol{x}_i, \boldsymbol{y}_i),\, i=1,\dots,n\big) \\ &\quad \propto \left[\prod_{i=1}^{N} f(\boldsymbol{x}_i|\boldsymbol{\mu}_x, \Sigma_{xx})\cdot P(\boldsymbol{\mu}_x, \Sigma_{xx})\right]\left[\prod_{i=1}^{n} f(\boldsymbol{y}_i|\boldsymbol{\alpha} + B_{yx}\boldsymbol{x}_i, \Sigma_{yy\cdot x})\right. \\ &\qquad \left. \times P\big(\boldsymbol{\alpha}, B_{yx}, \Sigma_{yy\cdot x}\big)\right], \end{aligned} \tag{3.4}$$

where $P(\boldsymbol{\mu}_x, \Sigma_{xx})$ and $P(\boldsymbol{\alpha}, B_{yx}, \Sigma_{yy\cdot x})$ are prior densities of $(\boldsymbol{\mu}_x, \Sigma_{xx})$ and $(\boldsymbol{\alpha}, B_{yx}, \Sigma_{yy\cdot x})$, respectively, as indicated in theorem 3.1.

It is evident from (3.4) that, with the values of the corresponding $\boldsymbol{y}_{n+1},\dots,\boldsymbol{y}_N$ missing, the observed values of $\boldsymbol{x}_{n+1},\dots,\boldsymbol{x}_N$ are irrelevant to the posterior analysis of $(B, \Sigma_{yy\cdot x})$. Hence, the posterior distribution of $(B, \Sigma_{yy\cdot x})$ as given by the second factor of (3.4)[3] is equivalent to that obtained by applying the result (ii) of theorem 3.1 to the posterior distribution of $(\boldsymbol{\mu}, \Sigma)$ based on the complete portion of the data, $\{(\boldsymbol{x}_i, \boldsymbol{y}_i),\, i=1,\dots,n\}$, only.

Write $\bar{\boldsymbol{w}}^* = (\bar{\boldsymbol{x}}^{*\prime}, \bar{\boldsymbol{y}}^{*\prime})'$, where

$$\bar{\boldsymbol{x}}^* = \sum_{i=1}^{n} \boldsymbol{x}_i/n, \quad \bar{\boldsymbol{y}}^* = \sum_{i=1}^{n} \boldsymbol{y}_i/n,$$

and

$$A^* = \begin{bmatrix} A^*_{xx} & A^*_{xy} \\ A^*_{yx} & A^*_{yy} \end{bmatrix} = \begin{bmatrix} \sum_{i=1}^{n} (\boldsymbol{x}_i - \bar{\boldsymbol{x}}^*)(\boldsymbol{x}_i - \bar{\boldsymbol{x}}^*)' & \sum_{i=1}^{n} (\boldsymbol{x}_i - \bar{\boldsymbol{x}}^*)(\boldsymbol{y}_i - \bar{\boldsymbol{y}}^*)' \\ \sum_{i=1}^{n} (\boldsymbol{y}_i - \bar{\boldsymbol{y}}^*)(\boldsymbol{x}_i - \bar{\boldsymbol{x}}^*)' & \sum_{i=1}^{n} (\boldsymbol{y}_i - \bar{\boldsymbol{y}}^*)(\boldsymbol{y}_i - \bar{\boldsymbol{y}}^*)' \end{bmatrix}.$$

The posterior distribution of $(\boldsymbol{\mu}, \Sigma)$, given $\{(\boldsymbol{x}_i, \boldsymbol{y}_i),\, i=1,\dots,n\}$, then is again of the normal–Wishart type, $NW_p(\boldsymbol{\theta}^*, \tau^*, \Omega^*, \nu^*)$, where now

$$\tau^* = \tau + n, \qquad \nu^* = \nu + n, \qquad \boldsymbol{\theta}^* = \tau^{*-1}(\tau\boldsymbol{\theta} + n\bar{\boldsymbol{w}}^*),$$

and

$$\Omega^* = \Omega + A^* + \frac{n\tau}{n+\tau}(\bar{\boldsymbol{w}}^* - \boldsymbol{\theta})(\bar{\boldsymbol{w}}^* - \boldsymbol{\theta})' \tag{3.5}$$

[3] Though the regression approach to the posterior distribution of $(B, \Sigma_{yy\cdot x})$ here is straightforward (cf. DeGroot, 1970, section 11.10 for the special case $s=1$), the resulting formulae for Γ^*, Ψ^* and $\Omega^*_{yy\cdot x}$ appearing in Lemma 3.1 become more complex for computation.

(cf. Dempster, 1969, p. 368). Consequently, as argued above, we have the following lemma.

Lemma 3.1. The posterior distribution of $(B, \Sigma_{yy\cdot x})$, which is independent of $(\boldsymbol{\mu}_x, \Sigma_{xx})$, a posteriori, is as follows:

(i) $\Sigma_{yy\cdot x}^{-1} \sim W_s(\Omega_{yy\cdot x}^{*-1}, \nu^*)$, and

(ii) conditional on $\Sigma_{yy\cdot x}$, $B \sim N_{s,(r+1)}(\Gamma^*, \Sigma_{yy.x}, \Psi^*)$, where $\Omega_{yy\cdot x}^*$, Γ^* and Ψ^* are counterparts of $\Omega_{yy\cdot x}$, Γ and Ψ given in (3.3) based on Ω^*, $\boldsymbol{\theta}^*$ and τ^* defined in (3.5). As a result, the marginal posterior density of B is

$$P(B) \propto \left|\Omega_{yy\cdot x}^* + (B - \Gamma^*)\Psi^{*-1}(B - \Gamma^*)'\right|^{-(\nu^* + r + 1)/2};$$

that is, B is matrix T distributed (Dickey, 1967):

$$B \sim T_{s,(r+1)}\left(\Gamma^*, \Omega_{yy\cdot x}^*, \Psi^*, \nu^* + r + 1\right). \tag{3.6}$$

Based on the foregoing analysis, the Bayesian inference for the original population parameters appearing in (2.1), except for Σ_{yy}, can now be stated explicitly as in the following theorem.

Theorem 3.2.

(1) Write

$$\bar{\boldsymbol{x}} = \sum_{i=1}^{N} \boldsymbol{x}_i/N \quad \text{and} \quad A_{xx} = \sum_{i=1}^{N} (\boldsymbol{x}_i - \bar{\boldsymbol{x}})(\boldsymbol{x}_i - \bar{\boldsymbol{x}})'.$$

Then

$$(\boldsymbol{\mu}_x, \Sigma_{xx}|\text{data}) \sim NW_n(\hat{\boldsymbol{\theta}}_x, \hat{\tau}, \hat{\Omega}_{xx}, \hat{\nu}), \tag{3.7}$$

where

$$\hat{\tau} = \tau + N,\ \hat{\nu} = \nu - s + N,$$
$$\hat{\boldsymbol{\theta}}_x = \hat{\tau}^{-1}(\tau\boldsymbol{\theta}_x + N\bar{x}),$$

and

$$\hat{\Omega}_{xx} = \Omega_{xx} + A_{xx} + \frac{N\tau}{N + \tau}(\bar{\boldsymbol{x}} - \boldsymbol{\theta}_x)(\bar{\boldsymbol{x}} - \boldsymbol{\theta}_x)'.$$

(2) Conditional on $(\boldsymbol{\mu}_x, \Sigma_{xx})$, $(\boldsymbol{\mu}_y, \Sigma_{yx})$ is matrix T distributed:

$$\begin{aligned}&\left(\boldsymbol{\mu}_y, \Sigma_{yx}|\text{data};\ \boldsymbol{\mu}_x, \Sigma_{xx}\right)\\ &\quad \sim T_{s,\,(r+1)}\left(\left(\boldsymbol{\gamma}^* + \Gamma_{yx}^*\boldsymbol{\mu}_x, \Gamma_{yx}^*\Sigma_{xx}\right), \Omega_{yy\cdot x}^*, M'\Psi^*M, \nu^* + r + 1\right),\end{aligned} \tag{3.8}$$

where

$$M = \begin{bmatrix} \mathbf{1}_s & \mathbf{0} \\ \boldsymbol{\mu}_x & \Sigma_{xx} \end{bmatrix},$$

with $\mathbf{1}_s : s \times 1$ a vector of ones.

Proof. (1) is an immediate consequence of the factorization (3.4). (2) is a result of (3.6) by noting that $(\boldsymbol{\mu}_y, \Sigma_{yx}) = BM$.

Point estimation. From (3.7), the modal estimate for $(\boldsymbol{\mu}_x, \Sigma_{xx})$ is $(\hat{\boldsymbol{\theta}}_x, \hat{\Sigma}_{xx})$, where

$$\hat{\Sigma}_{xx} = \hat{\Omega}_{xx}/\hat{\nu}.$$

The modal estimate for $(\boldsymbol{\mu}_y, \Sigma_{yx})$ evaluated at $(\boldsymbol{\mu}_x, \Sigma_{xx}) = (\hat{\boldsymbol{\theta}}_x, \hat{\Sigma}_{xx})$ is, from (3.8), $(\hat{\boldsymbol{\theta}}_y, \hat{\Sigma}_{yx})$, where

$$\hat{\boldsymbol{\theta}}_y = \boldsymbol{\gamma}^* + \Gamma^*_{yx}\hat{\boldsymbol{\theta}}_x = \boldsymbol{\theta}^*_y + \Gamma^*_{yx}(\hat{\boldsymbol{\theta}}_x - \boldsymbol{\theta}^*_x),$$

$$\hat{\Sigma}_{yx} = \Gamma^*_{yx}\hat{\Sigma}_{xx}.$$

The posterior distribution of $\Sigma_{yy} = \Sigma_{yy\cdot x} + B_{yx}\Sigma_{xx}B'_{yx}$ is difficult to derive. Intuitively, though, a sensible estimate of Σ_{yy} seems to be:

$$\hat{\Sigma}_{yy} = \Sigma^*_{yy\cdot x} + \Gamma^*_{yx}\hat{\Sigma}_{xx}\Gamma^{*\prime}_{yx},$$

where

$$\Sigma^*_{yy\cdot x} = \Omega^*_{yy\cdot x}/\nu^*.$$

These estimates for the population parameters are similar in form to the MLEs with the posterior modal estimates for $\boldsymbol{\mu}_x$, Σ_{xx}, $\boldsymbol{\alpha}$, B_{yx} and $\Sigma_{yy\cdot x}$ substituting for the MLEs, respectively.

4. Extension

The results obtained in section 3 can easily be extended to the case with multiple sets of variables whose missing data form a nested pattern. For instance, suppose $\boldsymbol{x}: r\times 1$, $\boldsymbol{y}: s\times 1$ and $\boldsymbol{z}: t\times 1$ are three sets of variables having the joint distribution $N_p(\boldsymbol{\mu}, \Sigma)$, where $p = r+s+t$:

$$\boldsymbol{\mu} = \begin{bmatrix} \boldsymbol{\mu}_x \\ \boldsymbol{\mu}_y \\ \boldsymbol{\mu}_z \end{bmatrix} \quad \text{and} \quad \Sigma = \begin{bmatrix} \Sigma_{xx} & \Sigma_{xy} & \Sigma_{xz} \\ \Sigma_{yx} & \Sigma_{yy} & \Sigma_{yz} \\ \Sigma_{zx} & \Sigma_{zy} & \Sigma_{zz} \end{bmatrix}.$$

As in (2.2), the joint density of $\boldsymbol{x}$, $\boldsymbol{y}$ and $\boldsymbol{z}$ can be written as

$$f(\boldsymbol{x}, \boldsymbol{y}, \boldsymbol{z}|\boldsymbol{\mu}, \Sigma)$$
$$= f(\boldsymbol{x}|\boldsymbol{\mu}_x, \Sigma_{xx}) f(\boldsymbol{y}|\boldsymbol{\alpha} + B_{yx}\boldsymbol{x}, \Sigma_{yy\cdot x}) f(\boldsymbol{z}|\boldsymbol{\omega} + B_{zx\cdot y}\boldsymbol{x} + B_{zy\cdot x}\boldsymbol{y}, \Sigma_{zz\cdot xy}), \tag{4.1}$$

where $\boldsymbol{\alpha}$, B_{yx} and $\Sigma_{yy\cdot x}$ are as defined in (2.3), and

$$\boldsymbol{\omega} = \boldsymbol{\mu}_z - B_{zx\cdot y}\boldsymbol{\mu}_x - B_{zy\cdot x}\boldsymbol{\mu}_y,$$

$$\left[B_{zx\cdot y} B_{zy\cdot x}\right] = \left[\Sigma_{zx}, \Sigma_{zy}\right]\begin{bmatrix}\Sigma_{xx} & \Sigma_{xy}\\ \Sigma_{yx} & \Sigma_{yy}\end{bmatrix}^{-1},$$

$$\Sigma_{zz\cdot xy} = \Sigma_{zz} - \left[\Sigma_{zx}, \Sigma_{zy}\right]\begin{bmatrix}\Sigma_{xx} & \Sigma_{xy}\\ \Sigma_{yx} & \Sigma_{yy}\end{bmatrix}^{-1}\begin{bmatrix}\Sigma_{xz}\\ \Sigma_{yz}\end{bmatrix}.$$

Under a normal–Wishart prior distribution, (3.1) and (3.2), for $(\boldsymbol{\mu}, \Sigma)$, the three distinct sets of parameters appearing on the right-hand side of (4.1) are independent of each other, as a result of theorem 3.1. Here $(\boldsymbol{\mu}_x, \Sigma_{xx})$ and $(\boldsymbol{\alpha}, B_{yx}, \Sigma_{yy\cdot x})$ have the same distributions as described by (i) and (ii) of theorem 3.1, except that the degrees of freedom for Σ_{xx}^{-1} and $\Sigma_{yy\cdot x}^{-1}$ are replaced by $\nu - s - t$ and $\nu - t$, respectively. The distribution of $(\boldsymbol{\omega}, B_{zx\cdot y}, B_{zy\cdot x}, \Sigma_{zz\cdot xy})$ is analogous to that of $(\boldsymbol{\alpha}, B_{yx}, \Sigma_{yy\cdot x})$ with, now, the pair $(\boldsymbol{x}, \boldsymbol{y})$ playing the role as $\boldsymbol{x}$, and $\boldsymbol{z}$ as $\boldsymbol{y}$. With the nested pattern of missing observations:

$$\boldsymbol{x}_1, \ldots, \boldsymbol{x}_m, \boldsymbol{x}_{m+1}, \ldots, \boldsymbol{x}_n, \boldsymbol{x}_{n+1}, \ldots, \boldsymbol{x}_N,$$
$$\boldsymbol{y}_1, \ldots, \boldsymbol{y}_m, \boldsymbol{y}_{m+1}, \ldots, \boldsymbol{y}_n,$$
$$\boldsymbol{z}_1, \ldots, \boldsymbol{z}_m,$$

the Bayesian analysis discussed in section 3 can be applied.

References

Anderson, T.W. (1957) "Maximum likelihood estimates for a multivariate normal distribution when some observations are missing", *J. of the American Statistical Association*, 52, 200–203.

Chen, C-F. (1979) "Bayesian inference for a normal dispersion matrix and its application to stochastic multiple regression analysis", *J. of the Royal Statistical Society*, B 41, 235–248.

DeGroot, M.H. (1970) *Optimal Statistical Decision* (McGraw-Hill, New York).

Dempster, A.P. (1969) *Elements of Continuous Multivariate Analysis* (Addison-Wesley, Cambridge, Mass.).

Dickey, J.M. (1967) "Matricvariate generalizations of the multivariate t distribution and the inverted multivariate t distribution", *Ann. Math. Statistics*, 38, 511–518.

Marini, M.M., A.R. Olsen and D.B. Rubin (1980) "Maximum-likelihood estimation in panel studies with missing data", in: K.F. Scheussler, ed., *Sociological Methodology* (Jossey-Bass, San Francisco).

Rubin, D.B. (1974) "Characterizing the estimation of parameters in incomplete-data problems", *J. of the American Statistical Association*, 69, 467–474.

Srivastava, M.S. and Khatri, C.G. (1979) *An Introduction to Multivariate Statistics* (North-Holland, New York).

Chapter 23

ESTIMATION OF ACTUAL REALIZATIONS IN STOCHASTIC PARAMETER MODELS

K. RAO KADIYALA

Purdue University

DENNIS OBERHELMAN

University of South Carolina

1. Introduction

The problem of simultaneous estimation of parameters has been widely studied in the literature. Bayesian solutions to the problem are well known and discussed extensively in Lindley and Smith (1972) and Rao (1975). Random coefficient approach of the type suggested by Swamy (1970, 1971, 1974) and Swami and Mehta (1975, 1977) can also be used to analyze the problem, and a recent paper by Harville (1976) provides a method of estimating the individual random coefficients, which is closely related to the Bayesian solution.

The chapter begins with a section which summarizes the RCR model and reviews the estimator for the mean coefficient vector suggested by Swamy. The next section reformulates a theorem in Harville specializing it to the present problem to yield an estimator with both classical and Bayesian interpretations. Section 1.3 reviews the Bayesian solution found in Lindley and Smith (1972) and Rao (1975) and points out the similarities between the RCR and Bayesian analyses.

Although the estimators discussed in sections 1.1–1.3 have been suggested as alternatives to OLS, there has been little study on the relative performance of these estimators. In section 2.1 Monte Carlo methods are used to compare the efficiency of an empirical Bayes estimator with that of OLS. The simulation provides some evidence supporting the use of an empirical Bayes estimator.

Bayesian Inference and Decision Techniques, edited by P. Goel and A. Zellner

1.1. A brief review of the linear RCR model

The linear RCR model is a classical econometric model used to estimate a pooled regression equation for a panel of data collected on N groups over T periods. The model allows each group to have a unique $(K \times 1)$ response vector β_i but specifies that these coefficients are independently and identically distributed about a mean vector $\bar{\beta}$. The stochastic character of the individual response vector can be accounted for by specifying that $\beta_i = \bar{\beta} + \delta_i$ and defining δ_i to be a random component distributed with a mean vector 0 and a variance–covariance matrix equal to Δ. Under this specification, the pooled regression equation for the panel is

$$\begin{Bmatrix} Y_1 \\ Y_2 \\ \vdots \\ Y_N \end{Bmatrix} = \begin{Bmatrix} X_1 \\ X_2 \\ \vdots \\ X_N \end{Bmatrix} \bar{\beta} + \begin{Bmatrix} X_1 & 0 & \dots & 0 \\ 0 & X_2 & \dots & 0 \\ \vdots & \vdots & & \vdots \\ 0 & 0 & \dots & X_N \end{Bmatrix} \begin{Bmatrix} \delta_1 \\ \delta_2 \\ \vdots \\ \delta_N \end{Bmatrix} + \begin{Bmatrix} u_1 \\ u_2 \\ \vdots \\ u_N \end{Bmatrix} \tag{1.1.1}$$

or

$$Y = X\bar{\beta} + D(X)\delta + u, \tag{1.1.2}$$

where Y_i and X_i are, respectively, the vector of endogenous variables and non-stochastic exogenous variables corresponding to the ith cross-section, and the u_i's are the general disturbance vectors for each cross-section. Each u_i is assumed to be independently distributed with a mean vector 0 and a variance–covariance matrix equal to Ω_i; in addition, it is assumed that δ_i and u_j are independent for every i and j.

Let $w_i = X_i\delta_i + u_i$. It follows from the assumptions listed above that

$$\Sigma_i = E(w_i w_i') = X_i \Delta X_i' + \Omega_i \tag{1.1.3}$$

and

$$E(w_i w_j') = 0.$$

Accordingly, the variance–covariance matrix of Y, Σ, is block diagonal with the Σ_i's as diagonals.

The Gauss–Markov theorem implies that

$$\hat{\bar{\beta}} = \left(\sum_{i=1}^{N} X_i'\Sigma_i^{-1}X_i \right)^{-1} \left(\sum_{i=1}^{N} X_i'\Sigma_i^{-1}Y_i \right) \tag{1.1.4}$$

is the best linear unbiased estimator (BLUE) of $\bar{\beta}$.

This expression can be modified to provide:

$$\hat{\bar{\beta}} = \left(\sum_{i=1}^{N} Z_i \right)^{-1} \sum_{i=1}^{N} Z_i \hat{\beta}_i, \tag{1.1.5}$$

where

$$\hat{\beta}_i = \left(X_i'\Omega_i^{-1}X_i\right)^{-1}\left(X_i'\Omega_i^{-1}Y_i\right) \quad \text{and} \quad Z_i = \left(\left(X_i'\Omega_i^{-1}X_i\right)^{-1} + \Delta\right)^{-1}, \qquad (1.1.6)$$

and where Z_i is the mean squared error of $\hat{\beta}_i$. Thus, the estimator in (1.1.5) is a matrix weighted average of the $\hat{\beta}_i$'s.

1.2. A "best" procedure for predicting the response of individual panel members

The previous section summarized a BLUE procedure for estimating $\bar{\beta}$, the mean response of N groups in a linear RCR model. The purpose of this section is to consider the estimation of the random coefficients, β_i, $i = 1, \ldots, N$.[1]

A recent paper by Harville (1976) extends the Gauss–Markov theorem to include the estimation of random effects in general mixed linear models. If theorem 1 of Harville is applied to the model specified in (1.1.2), then the best linear unbiased estimator of δ is

$$\tilde{\delta} = (I \otimes \Delta) D'(X) \Sigma^{-1}\left(Y - \hat{\bar{\beta}}\right). \qquad (1.2.1)$$

It then follows that β_i can be estimated by

$$\tilde{\beta}_i = \hat{\bar{\beta}} + \Delta Z_i \hat{\beta}_i - \Delta Z_i \hat{\bar{\beta}}, \qquad (1.2.2)$$

which can be rewritten as

$$\tilde{\beta}_i = w_{1i}\hat{\beta}_i + w_{2i}\hat{\bar{\beta}}, \qquad (1.2.3)$$

with

$$w_{1i} = \Delta Z_i,$$

$$w_{2i} = \left(X_i'\Omega_i^{-1}X_i\right)^{-1} Z_i,$$

and

$$w_{1i} + w_{2i} = I_k.$$

The prediction error of $\tilde{\beta}_i$ is

$$\begin{aligned} E(\tilde{\beta}_i - \beta_i)(\tilde{\beta}_i - \beta_i)' &= w_{2i} V w_{2i}' + \left(X_i'\Omega_i^{-1}X_i\right)^{-1} - \left(X_i'\Omega_i^{-1}X_i\right)^{-1} w_{2i}' \\ &= \left(X_i'\Omega_i^{-1}X_i\right)^{-1} w_{1i}' + w_{2i} V w_{2i}', \end{aligned} \qquad (1.2.4)$$

where

$$V = \left(\sum_{i=1}^{N} Z_i\right)^{-1}.$$

[1] Swamy and Mehta (1975) note the importance of this distinction. The Bayesian model assumes that the coefficients are fixed and the distribution of the coefficients is introduced to represent subjective prior information. While in the RCR model the distribution of the coefficients is introduced as part of the data-generating process.

Since $E(\hat{\beta}_i - \beta_i)(\hat{\beta}_i - \beta_i)' = (X_i'\Omega_i^{-1}X_i)^{-1}$, (1.2.4) implies that the difference in prediction error between $\hat{\beta}_i$ and $\tilde{\beta}_i$ is $(X_i'\Omega_i^{-1}X_i)^{-1}w_{2i}' - w_{2i}Vw_{2i}'$.

The following observations can be made about this difference:

(1) It converges to the non-negative definite matrix, $(X_i'\Omega_i^{-1}X_i)^{-1}w_{2i}'$ as N approaches ∞. This follows because V, the mean-squared error of $\hat{\bar{\beta}}$, goes to 0 as N becomes large.

(2) As T approaches ∞, the difference approaches 0. That is, $\tilde{\beta}_i$ approaches $\hat{\beta}_i$ as T becomes large.

These observations imply that it will be most beneficial to employ $\tilde{\beta}_i$ when N is large and T is small.

It is also interesting to note that a simple average of the $\tilde{\beta}_i$'s is equivalent to the estimator for $\bar{\beta}$ appearing in (1.1.6). This can be shown by summing (1.2.2) for all N:

$$\sum_{i=1}^{N} \tilde{\beta}_i = N\hat{\bar{\beta}} + \Delta \sum_{i=1}^{N} \left(\Delta + \left(X_i'\Omega_i^{-1}X_i\right)^{-1}\right)^{-1}\hat{\beta}_i - \Delta \sum_{i=1}^{N} \left(\Delta + \left(X_i'\Omega_i^{-1}X_i\right)^{-1}\right)^{-1}\hat{\bar{\beta}}.$$

When (1.1.6) is inserted for $\hat{\bar{\beta}}$ in the expression above, the last two terms cancel, and

$$\frac{1}{N}\sum_{i=1}^{N} \tilde{\beta}_i = \hat{\bar{\beta}}.$$

1.3. Summary of Bayesian results on simultaneous estimation of parameters

A Bayesian approach to the problem of simultaneously estimating coefficient vectors $\beta_1, \beta_2, \ldots, \beta_N$ from N regressions yields an estimator for β_i which is computationally equivalent to that found in (1.2.3). Detailed descriptions of the Bayesian solution can be found in Lindley and Smith (1972), Rao (1975), and the references therein. A brief summary is given below.

In the paper by Lindley and Smith, Bayesian estimators for β_i and its mean $\bar{\beta}$ are derived under the assumptions that:

(1) $Y_i \overset{d}{=} N(X_i\beta_i, \Omega_i)$,

(2) the β_i's are exchangeable and $\beta_i \overset{d}{=} N(\bar{\beta}, \Delta)$, and

(3) there is vague prior knowledge of $\bar{\beta}$. Given these assumptions, Lindley and Smith show that the mean of the posterior distribution of β_i is

$$\beta_i^* = \left((X_i'\Omega_i^{-1}X_i) + \Delta^{-1}\right)^{-1}(X'\Omega^{-1}Y + \Delta^{-1}\beta^*),$$

where

$$\beta^* = \sum_{i=1}^{N} \beta_i^*/N.$$

It can be seen that this is the same weighted average of the least-squares estimator and the estimator for $\bar{\beta}$ as suggested by $\tilde{\beta}_i$ in (1.2.3).

Moreover, the Bayesian estimator for $\bar{\beta}$ is the simple average of the Lindley–Smith estimators for the β_i's. It was shown in section 1.2 that this average is equal to the Swamy estimator for $\bar{\beta}$. Although the Bayesian solutions discussed in Lindley and Smith (1972), Novick et al. (1972) and Rao (1975) and the classical solutions given in Swamy (1970, 1971, 1974) and Swamy and Mehta (1975, 1977) lead to the same estimates, it is important to notice that the two approaches are conceptually different.

1.4. An empirical Bayes estimator for β_i

In some applied studies it is assumed that the variance–covariance matrix of each u_i is equal to a scalar times I. Accordingly, this section assumes that $E(u_iu_i') = \theta_{ii}I_T$ and discusses how the estimator in (1.2.3) can be implemented in applied studies.

Under this assumption,

$$\tilde{\beta}_i = \Delta\left(\Delta + \theta_{ii}(X_i'X_i)^{-1}\right)^{-1}\hat{\beta}_i + \theta_{ii}(X_i'X_i)^{-1}\left(\Delta + \theta_{ii}(X_i'X_i)^{-1}\right)^{-1}\hat{\bar{\beta}}, \quad (1.4.1)$$

where $\hat{\beta}_i = (X_i'X_i)^{-1}X_i'Y_i$ is the OLS estimator for the ith cross-section and

$$\hat{\bar{\beta}} = \sum_{i=1}^{N}\left[\sum_{j=1}^{N}\left(\Delta + \theta_{jj}(X_j'X_j)^{-1}\right)^{-1}\right]^{-1}\left(\Delta + \theta_{ii}(X_i'X_i)^{-1}\right)^{-1}\hat{\beta}_i \quad (1.4.2)$$

is the estimator for $\bar{\beta}$ suggested by Swamy (1970).

When the variance–covariance parameters, Δ and θ_{ii}, are unknown, they can be estimated by the estimators for Δ and θ_{ii} proposed by Swamy (1970) and Rao (1975). These are

$$\hat{\theta}_{ii} = \frac{u_i'M_iu_i}{T-K}, \quad M_i = I_T - X_i(X_i'X_i)^{-1}X_i',$$

and

$$\hat{\Delta} = \frac{1}{N-1}\left(\sum_{i=1}^{N}\hat{\beta}_i\hat{\beta}_i' - \frac{1}{N}\sum_{i=1}^{N}\hat{\beta}_i\sum_{i=1}^{N}\hat{\beta}_i'\right) - \frac{1}{N}\sum_{i=1}^{N}\hat{\theta}_{ii}(X_i'X_i)^{-1}. \quad (1.4.3)$$

The following two-stage estimator is then adopted for β_i:

$$\hat{\hat{\beta}}_i = \hat{w}_{1i}\hat{\beta}_i + \hat{w}_{2i}\hat{\bar{\beta}}, \quad (1.4.4)$$

where

$$\hat{w}_{1i} = \hat{\Delta}\left(\hat{\Delta} + \hat{\theta}_{ii}(X_i'X_i)^{-1}\right)^{-1},$$

$$\hat{w}_{2i} = \hat{\theta}_{ii}(X_i'X_i)^{-1}\left(\hat{\Delta} + \hat{\theta}_{ii}(X_i'X_i)^{-1}\right)^{-1},$$

and

$$\bar{\tilde{\beta}} = \sum_{i=1}^{N} \left[\sum_{j=1}^{N} \left(\hat{\Delta} + \hat{\theta}_{jj} \left(X_j' X_j \right)^{-1} \right)^{-1} \right]^{-1} \left(\hat{\Delta} + \hat{\theta}_{ii} \left(X_i' X_i \right)^{-1} \right)^{-1} \hat{\beta}_i .$$

The estimator in (1.4.4) is the same as the empirical Bayes estimator suggested by Rao (1975, see eq. 2.39, p. 551) with one minor difference: eq. (1.4.4) uses $\bar{\tilde{\beta}}$ as an estimator for $\bar{\beta}$ while the estimator suggested by Rao is a simple average of the OLS estimators. Accordingly, the estimator in (1.4.4) will henceforth be referred to as an empirical Bayes estimator.

Rao investigated the properties of the empirical Bayes estimator for the case where $X_1 = X_2 = \ldots = X_N$, and found that the compound loss of the empirical Bayes estimator is always less than that of OLS. With regards to estimation of parameters for an individual β_i when the X's are not replicated, the problem becomes much more complicated. We can note, however, that in the special case where the error vectors have a common dispersion matrix, $\theta^2 I_T$, $\hat{\Delta}$ in (1.4.3) and

$$\hat{\theta}^2 = \frac{1}{N} \sum_{i=1}^{N} \frac{u_i' M_i u_i}{T-K}$$

converge to Δ and θ^2 as N becomes large. Accordingly, when N becomes large and T is fixed, the difference is mean-squared error between $\hat{\beta}_i$ and $\hat{\hat{\beta}}_i$ converges to

$$\theta^2 \left(X_i' X_i \right)^{-1} \left(\Delta + \theta^2 \left(X_i' X_i \right)^{-1} \right) \theta^2 \left(X_i' X_i \right)^{-1} .$$

Since this matrix is positive definite, this implies that for sufficiently large N, $\hat{\hat{\beta}}_i$ will always dominate $\hat{\beta}_i$ (for fixed T).

In many practical situations it will not be plausible to assume that Δ and θ_{ii} are known; so it is important to ask what happens to the relative efficiency of the estimation procedures under these circumstances. Therefore, the next section uses Monte Carlo simulations to investigate the relative efficiencies of $\hat{\hat{\beta}}_i$ and $\hat{\beta}_i$.

2. Comparison of empirical Bayes and OLS estimators

In this section we report the results of a Monte Carlo investigation that compares efficiencies of OLS and empirical Bayes procedures. The results, as one would expect, are in favor of the OLS estimator for small samples. But the empirical Bayes estimator overtakes the OLS estimator for sample sizes as small as $T > 10$ and $N > 30$. The model and the results are summarized below.

2.1. Description of the model

The data used in the Monte Carlo study described in this section were simulated in the following fashion. A single variable equation

$$\underset{(20\times1)}{Y_i} = \underset{(20\times1)}{\beta_i X_i} + \underset{(20\times1)}{u_i}, \quad i=1,\dots,N,$$

was used to generate 200 samples for $N=50$ groups with:

$X_i = (20\times1)$ vector of uniform random numbers generated at the beginning of the experiment,

$\beta_i = 5.0 + \delta_i,$

$\delta_i \overset{\text{i.i.d.}}{=} N(0, \Delta),$

$u_i \overset{\text{d}}{=} N(0, \theta_{ii} I_T).$

Assigned values for w_2 in (1.2.4) were used to determine the variances of the errors, u_i, $i=1,\dots,N$. When w_2 is low, the variance of the OLS estimator is small relative to Δ and the difference in prediction error between OLS and the Bayesian estimator using true values of Δ and θ_{ii} is minimal. It is possible in this situation that the estimated values of Δ and θ_{ii} will introduce enough error into the estimation of β_i to make the empirical Bayes estimator less efficient than OLS. Therefore, any study on the performance of these estimators should consider a range for w_2. For this reason, the variances, θ_{ii}, $i=1,\dots,N$, were chosen so that the weight w_2 ranged from 0.05 to 0.50. For example, θ_{11} was selected so that w_2 was set at 0.05, θ_{22} was selected so that w_2 was set at 0.10, and θ_{1010} was selected so that w_2 was selected at 0.50. The other 40 variances were selected in a similar fashion. In addition, to study the relative performance of the estimators for different absolute levels of variance, the experiment was conducted for $\Delta = 1, 5, 25$.

Once the data were generated, two sets of OLS estimates for $\beta_1, \beta_2, \dots, \beta_{10}$ were computed: the first, using the complete vector of generated data ($T=20$), and the second, using only half the generated observations ($T=10$). In a similar fashion for each case ($T=20$) and ($T=10$) three sets of empirical Bayes estimates for $\beta_1, \beta_2, \dots, \beta_{10}$ were obtained: the first using only the first 10 groups of observations ($N=10$), the second, using the first 30 groups ($N=30$), and the third, using all 50 groups ($N=50$).

The average absolute deviation between the estimated and actual values of $\beta_1, \beta_2, \dots, \beta_{10}$ were computed, and the results for ($T=20$) and ($T=10$) are summarized in tables 23.1 and 23.2. For the sake of space, only the overall average on the results for the first five groups ($w_2 = 0.05$–0.25) and the overall average on the results for the second five groups ($w_2 = 0.30$–0.50) are reported in

Table 23.1. Average absolute deviations in 200 replications with $T = 10$.

		OLS	Empirical Bayes $N = 10$	$N = 30$	$N = 50$
$\Delta = 1$	Group 1 ($w_2 = 0.05$–0.25)	0.450744	0.438731 (2)	0.413118 (1)	0.404893 (0)
	Group 2 ($w_2 = 0.30$–0.50)	0.961022	0.720145 (0)	0.664187 (0)	0.653119 (0)
$\Delta = 5$	Group 1 ($w_2 = 0.05$–0.25)	1.033691	1.05449 (4)	0.939079 (1)	0.913153 (1)
	Group 2 ($w_2 = 0.30$–0.50)	2.266746	1.63235 (0)	1.48495 (0)	1.451614 (0)
$\Delta = 25$	Group 1 ($w_2 = 0.05$–0.25)	2.217433	2.33722 (4)	2.08086 (2)	1.97675 (1)
	Group 2 ($w_2 = 0.30$–0.50)	4.866366	3.64018 (0)	3.40976 (0)	3.32508 (0)

Table 23.2. Average absolute deviations in 200 replications with $T = 20$.

		OLS	Empirical Bayes $N = 10$	$N = 30$	$N = 50$
$\Delta = 1$	Group 1 ($w_2 = 0.05$–0.25)	0.324368	0.330858 (4)	0.313678 (0)	0.311672 (1)
	Group 2 ($w_2 = 0.30$–0.50	0.677777	0.562913 (0)	0.535114 (0)	0.529503 (0)
$\Delta = 5$	Group 1 ($w_2 = 0.05$–0.25)	0.758646	0.753429 (2)	0.716790 (0)	0.711649 (0)
	Group 2 ($w_2 = 0.30$–0.50)	1.47028	1.21822 (0)	1.16147 (0)	1.15660 (0)
$\Delta = 25$	Group 1 ($w_2 = 0.05$–0.25)	1.607949	1.61927 (4)	1.53526 (1)	1.53234 (2)
	Group 2 ($w_2 = 0.30$–0.50)	3.234086	2.65388 (0)	2.54948 (0)	2.55075 (0)

these tables. The number in parentheses represents the number of times the OLS estimator outperformed the empirical Bayes estimator in each subgroup.

2.2. *Summary of the simulation study*

Upon comparison of the average absolute deviations of the OLS and empirical Bayes estimator in tables 23.1 and 23.2, it is evident that the empirical Bayes estimator generally outperformed the least squares estimator. In fact, once the number of groups used to estimate β_i reached 30, the empirical Bayes estimator dominated OLS in all cases except for the case where the variance of the OLS estimator was on the order of 10 percent of the size of Δ (i.e. $w_2 \leq 0.10$).

When the value of w_2 reached the size of 0.30 or larger, the increase in efficiency of the Bayesian estimator was substantial. This was true even for the case where N was small. For example, looking at the column in table 23.1 corresponding to $N=10$, it is seen that when $0.30 \leq w_2 \leq 0.50$, the empirical Bayes estimator was on average 25 percent lower in mean absolute deviation (M.A.D.) than the OLS estimator. For the case where $T=20$ and $N=10$ (see table 23.2) the empirical Bayes estimator was on average 16 percent lower in M.A.D. for $0.30 \leq w_2 \leq 0.50$. These percentages increased for larger values of N.

In summary, this study demonstrates that substantial gains in efficiency over OLS can be realized through the use of the empirical Bayes estimator. This is true for moderate values of w_2 even in small samples ($N=10$, $T=10$). Furthermore, the results suggest that sample sizes of $N \geq 30$ and $T \geq 10$ are large enough to insure the dominance of the Bayesian estimator.

2.3. *An example employing* $\hat{\hat{\beta}}_i$

In order to provide an example of the efficient estimation procedure developed in section 1.2, the following methodological example was constructed involving the demand for pass-books in 10 New-England-Mid-Atlantic States. The example uses semiannual data (for 1967–1976) on per capita savings deposits at commercial banks and per capita income. It also uses data on the pass-book rate at commercial banks and the three-month treasury bill rate (as a proxy for open market rates).

The model considered here is a basic stock adjustment model in which the desired demand for pass-book accounts is a function of the rate spread between pass-book and treasury bills (SPR_{it}) and income (Y_{it}), i.e.

$$Dep^*_{it} = \alpha_{0i} + \alpha_{1i} Y_{it} + \alpha_{2i} SPR_{it} \tag{2.3.1}$$

and

$$\Delta\, Dep_{it} = \gamma\left(Dep^*_{it} - Dep_{it-1}\right) + u_{it}.$$

So, the model to be estimated is

$$\Delta\, Dep_{it} = \beta_{0i} + \beta_{1i} Y_{it} + \beta_{2i} SPR_{it} + \beta_{3i} Dep_{it-1} + u_{it}, \tag{2.3.2}$$

where it is expected that

$$\beta_{1i} > 0, \quad \beta_{2i} > 0, \quad \beta_{3i} < 0.$$

To obtain estimators for each cross-section, two choices are available – OLS or the methodology suggested in section 1.3. Both procedures were applied to the model and the results are reported in tables 23.3 and 23.4.

Referring to these tables, it is seen that the coefficients of all three variables have the proper sign and relatively low standard errors giving empirical support to the specification in (2.3.2).

Table 23.3. Estimation of eq. (2.3.2) by OLS.

Cross-section	Income	Spread	Lagged deposits
1	0.049	6.43	−0.258
	(0.016)	(2.01)	(0.130)
2	0.052	9.27	−0.099
	(0.017)	(2.00)	(0.089)
3	0.065	4.19	−0.179
	(0.019)	(1.62)	(0.090)
4	0.250	4.95	−0.868
	(0.068)	(2.44)	(0.250)
5	0.110	7.47	−0.591
	(0.020)	(1.39)	(0.118)
6	0.176	11.66	−0.442
	(0.066)	(4.34)	(0.184)
7	0.091	10.58	−0.342
	(0.026)	(1.84)	(0.102)
8	0.080	15.87	−0.381
	(0.015)	(2.47)	(0.096)
9	0.075	7.50	−0.159
	(0.033)	(1.55)	(0.112)
10	0.105	15.69	−0.168
	(0.042)	(4.291)	(0.114)

Table 23.4. Estimation of eq. (2.3.2) by eq. (1.3.4).

Cross-section	Income	Spread	Lagged deposits
1	0.049	7.00	−0.255
	(0.013)	(1.63)	(0.099)
2	0.058	9.49	−0.134
	(0.016)	(1.68)	(0.080)
3	0.075	5.55	−0.231
	(0.017)	(1.41)	(0.080)
4	0.145	6.59	−0.487
	(0.041)	(1.94)	(0.151)
5	0.093	7.08	−0.484
	(0.017)	(1.24)	(0.097)
6	0.113	9.65	−0.278
	(0.040)	(2.40)	(0.112)
7	0.085	9.83	−0.313
	(0.020)	(1.55)	(0.088)
8	0.072	12.28	−0.299
	(0.014)	(1.88)	(0.081)
9	0.083	8.09	−0.189
	(0.029)	(1.38)	(0.095)
10	0.091	12.11	−0.145
	(0.033)	(2.48)	(0.094)
$\hat{\bar{\beta}} = \frac{1}{N}\Sigma\tilde{\beta}_i$	0.086	8.77	−0.282
	(0.018)	(1.23)	(0.072)

The Monte Carlo simulations of the previous section indicated that, under the assumptions of a random coefficient regression model, the efficient estimation procedure could provide gains in efficiency over OLS. Therefore, the estimated models using this technique should be considered as a preferred alternative to those estimated by OLS.

References

Dent, W.T. and C. Hildreth (1977) "Maximum likelihood estimation in random coefficient models", *Journal of the American Statistical Association*, 72, 69–72.

Fisk, P.R. (1967) "Models of the second kind in regression analysis", *Journal of Royal Statistical Society*, Ser. B 28, 266–281.

Froelich, B.R. (1973) "Some estimators for a random coefficient regression model", *Journal of the American Statistical Association*, 68, 329–336.

Harville, D. (1976) "Extension of the Gauss–Markov theorem to include the estimation of random effects", *The Annals of Statistics*, 4, 384–395.

Harville, D. (1977) "Maximum likelihood approaches to variance component estimation and to related problems", *Journal of the American Statistical Association*, 72, 320–341.

Hildreth, C. and J.P. Houck (1968) "Some estimators for a linear model with random coefficients", *Journal of the American Statistical Association*, 63, 584–595.

Lindley, D.V. and A.F.M. Smith (1972) "Bayes estimates for the linear model", *J. Royal Statistical Society*, Ser. B 34, 1–41.

Novick, M.R., P.H. Jackson, D.T. Thayer and N.S. Cole (1972) "Estimating multiple regressions in *m* groups: A cross-validation study", *British Journal of Mathematical and Statistical Psychology*, 25, 33–50.

Raj, B. (1975) "Linear regression with random coefficients: The finite sample and convergence properties", *Journal of the American Statistical Association*, 70.

Rao, C.R. (1973) *Linear Statistical Inference and Its Applications* (John Wiley and Sons, Inc., New York).

Rao, C.R. (1975) "Simultaneous estimation of parameters in different linear models and application to biometric problems", *Biometrics*, 31, 545–554.

Rosenberg, B. (1973) "The analysis of a cross-section of time series by stochastically convergent parameter regression", *Annals of Economic and Social Measurement*, 2, 399–428.

Rubin, H. (1950) "Notes on random coefficients", in: T.C. Koopmans, ed., *Statistical Inference in Dynamic Economic Models* (Cowles Commission Monograph No. 10), pp. 419–421.

Schoenemann, P.H. (1965) "On the formal differentiation of traces and determinants", Research Memorandum No. 27, The Psychometric Laboratory, University of North Carolina.

Singh, B. and A. Ullah (1974) "Estimation of seemingly unrelated regression with random coefficients", *Journal of the American Statistical Association*, 69, 191–195.

Swamy, P.A.V.B. (1970) "Efficient inference in random coefficient regression model", *Econometrica*, 38, 311–323.

Swamy, P.A.V.B. (1971) *Statistical Inference in Random Coefficient Models* (Springer Verlag, Berlin).

Swamy, P.A.V.B. (1974) "Linear models with random coefficients", in: P. Zarembka, ed., *Frontiers in Econometrics* (Academic Press, New York).

Swamy, P.A.V.B. and J.S. Mehta (1977) "Estimation of linear models with time and cross-sectionally varying parameters", *Journal of the American Statistical Association*, 890–898.

Swamy, P.A.V.B. and J.S. Mehta (1975) "Bayesian and non-Bayesian analysis of switching regressions and of random coefficient regression models", *Journal of the American Statistical Association*, 70, 593–605.

Swamy, P.A.V.B. and P.A. Tinsley (1980) "Linear prediction and estimation methods for regression models with stationary stochastic coefficients", *Journal of Econometrics*, 103–142.

Zellner, A. (1966) "On the aggregation problem: A new approach to a troublesome problem", Report 6628, Center of Math. Studies in Business and Econ., University of Chicago published in K.A. Fox et al., eds., *Economic Models, Estimation and Risk Programming: Essays in Honor of Gerhard Tintner* (Springer-Verlag, New York, 1969), pp. 365–374.

Chapter 24

LOCAL SENSITIVITY ANALYSIS AND BAYESIAN REGRESSION DIAGNOSTICS

WOLFGANG POLASEK*

University of Vienna

1. Introduction

Local and global sensitivity analyses are general statistical concepts, discussed for example in Leamer (1978), to evaluate the stability of estimators with respect to parameters and assumptions of the underlying model. This concept can be applied to classical and Bayesian models as well. Some further application of this approach can be found in Leamer (1982).

In this chapter, first the problem of local sensitivity analysis in the regression model $\boldsymbol{y} = \boldsymbol{X\beta} + \boldsymbol{\varepsilon}$ with respect to a general covariance matrix $\boldsymbol{\Sigma} = \mathrm{Var}(\boldsymbol{\varepsilon})$ is discussed. The local sensitivity problem examines the changes in the estimates of $\boldsymbol{\beta}$ when the elements σ_{ij} in $\boldsymbol{\Sigma}$ are changed infinitesimally. A global sensitivity analysis is not meaningful for this problem, since the set of estimates for $\boldsymbol{\beta}$ is unlimited if $\boldsymbol{\Sigma}$ is allowed to vary freely. The results are derived using the matrix-by-matrix calculus derived in Balestra (1976) and Polasek (1982). The definition of proper arrangements of scalar derivatives in a derivative matrix leads to simpler interpretable results and avoids unnecessary complications with permutation (or commutation) matrices. A short overview of matrix derivatives is given in section 2. Section 3 derives the local sensitivity results for the posterior mean of the regression model $\boldsymbol{y} \sim N(\boldsymbol{X\beta}, \boldsymbol{\Sigma})$, $\boldsymbol{\beta} \sim N(\boldsymbol{b}^*, \boldsymbol{H}^{*-1})$, where $\boldsymbol{b}^*$ and $\boldsymbol{H}^{*-1}$ are the moments of the prior distribution.

It is shown how the classical results for regression diagnostics given in Belsley, Kuh and Welsch (1980) can be extended to the detection of influential observa-

*I wish to thank the referee and Prem K. Goel for valuable hints and comments.

Bayesian Inference and Decision Techniques, edited by P. Goel and A. Zellner

tions in a Bayesian linear model as well. This is demonstrated for the simple Keynesian consumption function in Austria (1954–1977).

In section 4 we show how the matrix derivatives can be successfully applied to the local sensitivity analysis of regression systems. For the seemingly unrelated regression (SUR) system the general formulae for the sensitivities of the Bayesian posterior mean are derived. As an example, we demonstrate the construction of diagnostic measures for the well-known Grünfeld data (Zellner, 1962) to check if it pays to estimate the system by a SUR estimator.

2. Regression diagnostics and matrix derivatives

In their pioneering book on regression diagnostics Belsley, Kuh and Welsch (1980) considered two basic approaches to construct diagnostic measures. One is based on the jack-knife idea of omitting successively one observation and then to compare the difference of the original estimate and the deleted version of the estimate. The second approach considers a change in weight in the covariance matrix of the residuals and derives the derivative of the estimate of this point. Fortunately, both approaches yield similar results and therefore we will use the derivative approach for the construction of Bayesian regression diagnostics. This approach relates also directly to the more general concept of local sensitivity analysis which was advocated by Leamer (1972, 1978) for a Bayesian analysis. Using the concept of matrix-by-matrix derivatives we can generalize the classical methods for the Bayesian case. Therefore we briefly summarize the essentials of these derivative concepts. They will be used for the construction of Bayesian regression diagnostics in the next sections.

The idea of matrix by matrix derivatives dates back to Dwyer and MacPhail (1948). They recognized the importance of the arrangement of matrix-by-matrix derivatives by developing an elementwise approach of the form $\partial\langle \boldsymbol{A}\rangle/\partial \boldsymbol{B}$ and $\partial \boldsymbol{A}/\partial\langle \boldsymbol{B}\rangle$. Balestra (1976) developed an overall calculus for the non-elementwise matrix-by-matrix derivatives, but only for the $\partial\langle \boldsymbol{A}\rangle/\partial \boldsymbol{B}$-type derivative, based on the work of MacRae (1974). Polasek (1982) showed that the second type of derivative, $\partial \boldsymbol{A}/\partial\langle \boldsymbol{B}\rangle$, implies a dual derivative algebra, analogous to Balestra (1976). Fortunately, both types of derivatives can be easily transformed to each other by the use of permutation matrices $\boldsymbol{P}_{m,p} = \boldsymbol{P}'_{p,m}$, which are extensively studied by Magnus and Neudecker (1979). Let $\boldsymbol{A}$ be $(m \times n)$ and $\boldsymbol{B}$ be $= (p \times q)$, then

$$\partial\langle \boldsymbol{A}\rangle/\partial \boldsymbol{B} = \boldsymbol{P}_{m,p}\,\partial \boldsymbol{A}/\partial\langle \boldsymbol{B}\rangle \boldsymbol{P}_{q,n} \tag{2.1}$$

and

$$\partial \boldsymbol{A}/\partial\langle \boldsymbol{B}\rangle = \boldsymbol{P}_{q,m}\,\partial\langle \boldsymbol{A}\rangle/\partial \boldsymbol{B} \boldsymbol{P}_{n,q}. \tag{2.2}$$

The $\partial \boldsymbol{A}/\partial\langle \boldsymbol{B}\rangle$ derivative is useful for simplifying the construction of regression diagnostics for the seemingly unrelated regression (SUR) system in section 4.

3. Bayesian regression diagnostics

Consider the normal general linear model (GLM) with general non-singular $(T \times T)$ covariance matrix:

$$\underset{(T\times 1)}{y} = \underset{(T\times K)(K\times 1)}{X\beta} + \underset{(T\times 1)}{\varepsilon}, \quad \varepsilon \sim N(0, \Sigma), \tag{3.1}$$

and prior information:

$$\beta \sim N(b^*, H^{*-1}). \tag{3.2}$$

The posterior distribution is given by $\beta|y \sim N(b^{**}, H^{-1})$ (see, for example, Smith, 1973) with the parameters:

$$b^{**} = H^{-1}(H^*b^* + X'\Sigma^{-1}y) \tag{3.3}$$

and

$$H = H^* + X'\Sigma^{-1}X. \tag{3.4}$$

In this section we consider the effects of small changes of the precision or weight matrix Σ^{-1} on the posterior mean and covariance of β, which generalises the idea of Belsley, Kuh and Welsch (1980) for the GLM. The following result has been proved in Polasek (1982):

Theorem 3.1. The local sensitivities of the posterior moments (3.3) and (3.4) with respect to Σ^{-1} in the Bayesian linear regression model (3.1)–(3.2) are given by the $(KT \times KT)$ derivative matrix:

$$\partial H^{-1}/\partial \Sigma^{-1} = -\operatorname{vec} XH^{-1} \operatorname{vec}' H^{-1}X' \tag{3.5}$$

and the $(KT \times T)$ matrix:

$$\partial b^{**}/\partial \Sigma^{-1} = (\operatorname{vec} XH^{-1})\, e', \tag{3.6}$$

where e is the residual term given by

$$e = y - Xb^{**}, \tag{3.7}$$

and vec A denotes the vectorization of the $(m \times n)$ matrix $A = (a_{.1}, \ldots, a_{.n})$ by columns to an $(mn \times 1)$ column vector. (vec A: row vector.)

The compact derivative result (3.6) shows an interesting structure in the form of a "dyadic" vec-product; the first term involves the vectorization of the XH^{-1} matrix and the second term the residual vector (3.7). The posterior variance has the role of a scaling factor, since vec $XH^{-1} = (H^{-1} \otimes I)\operatorname{vec} X$.

The compact derivative result (3.6) can be divided into elementwise derivatives which allow a simple calculation as a diagnostic measure:

$$\partial b^{**}/\partial \sigma^{ij} = H^{-1}X_i'e_j, \quad i, j = 1, \ldots, T. \tag{3.8}$$

Now σ^{ij} is the (i, j)th element of $\boldsymbol{\Sigma}^{-1}$, e_j the jth element of $\boldsymbol{e}$, and $\boldsymbol{X}_i$ is the ith $(1 \times K)$ row vector formed from the $(T \times K)$ matrix $\boldsymbol{X}$.

In practical applications not all derivatives might be interesting and these elementwise derivatives allow us to choose the appropriate ones. Note that the posterior mean $\boldsymbol{b}^{**}$ can be reduced to the GLS estimate $\hat{\boldsymbol{\beta}}_{\text{GLS}}$ by setting $\boldsymbol{H}^* = \boldsymbol{0}$.

3.1. *Local Σ-sensitivities and regression diagnostic*

In this section we show how the local sensitivities in theorem 3.1 can be used for "Bayesian regression diagnostics". For this purpose we first discuss the relationship between the general derivative results and the classical concept of regression diagnostics.

A special application of derivatives has been suggested by Belsley, Kuh and Welsch (1980) for a sensitivity analysis of the OLS estimate $\hat{\boldsymbol{\beta}}$. In order to detect observations with large influence, the weight of the ith observation is set to w_i and then the local sensitivity evaluated at $w_i = 1$ can be derived. The $\boldsymbol{\Sigma}^{-1}$ matrix is replaced by a set of diagnonal matrices of the form:

$$\boldsymbol{W}_i = \text{diag}(1,\ldots,1, w_i, 1,\ldots,1), \quad i = 1,\ldots, N, \tag{3.9}$$

and the corresponding local sensitivity for the estimator $\boldsymbol{b}(w_i) = (\boldsymbol{X}'\boldsymbol{W}_i\boldsymbol{X})^{-1}\boldsymbol{X}'\boldsymbol{W}_i\boldsymbol{y}$:

$$\left.\frac{\partial \boldsymbol{b}(w_i)}{\partial w_i}\right|_{w_i = 1} = (\boldsymbol{X}'\boldsymbol{X})^{-1}\boldsymbol{X}_i'\boldsymbol{r}_i, \quad i = 1,\ldots, N, \tag{3.10}$$

where $\boldsymbol{r} = (\boldsymbol{y} - \boldsymbol{X}\hat{\boldsymbol{\beta}})$ is the OLS residual. Since these matrices involve only one parameter, no matrix derivatives are necessary. Note that by applying a matrix inversion lemma, the GLS residuals $\boldsymbol{y} - \boldsymbol{X}\boldsymbol{b}(w_i)$ can be expressed by the OLS residuals r_i.

The local sensitivity derivative of theorem 3.1 can therefore be interpreted as a simultaneous diagnostic checking device for all elements in the covariance matrix $\boldsymbol{\Sigma}$. The basic difference lies in the calculation of the residual vector. In (3.8) we need the residuals of the posterior mean and $\boldsymbol{H}$ while (3.10) uses the OLS residual and $\boldsymbol{X}'\boldsymbol{X}$.

3.2. *Bayesian regression diagnostics for the consumption function*

Consider the simple Keynesian consumption function (Austrian data in real terms from 1954 to 1977) where C_t denotes consumption and Q_t the gross national product:

$$C_t = \beta_0 + \beta_1 Q_t + \varepsilon_t, \quad \varepsilon_t \sim N(0, \sigma_\varepsilon^2), \ \text{Cov}(\varepsilon_t, \varepsilon_s) = 0, \quad \text{for } t \neq s. \tag{3.11}$$

We compare the regression diagnostics for the following three different prior specifications with increasing confidence in the marginal propensity of consumption:

(a) $\beta_1 \sim N(0.6, 0.25)$,

(b) $\beta_1 \sim N(0.7, 0.01)$,

(c) $\beta_1 \sim N(0.8, 0.001)$.

We assume for the intercept for all three cases a clearly diffuse prior: $\beta_0 \sim N(0, 1000)$ and independence from β_1. Since we are only interested in inferences for β_1 we ignore prior specifications for β_0. The posterior means and the OLS estimate are given in Table 24.1.

As expected, the posterior means move away from the OLS (or ML) location as soon as we have tighter a priori variances for the marginal propensity of consumption, β_1.

Note that the OLS estimate coincides with the posterior mean for specification (a) and prior specification (b) hardly moves the posterior mean. Only the tight prior (c) moves the posterior mean in the direction of the prior mean 0.8. This shows that the data evidence is fairly strong against prior specifications for the marginal propensity of consumption. Only a strong prior moves the posterior mean from the OLS location.

The regression diagnostics for this consumption function are computed from (3.8). In figs. 24.1 and 24.2 we have denoted the sensitivities with respect to the intercept b_0^{**} with I and the sensitivities of the income coefficient b_1^{**} with Q.

The calculation of the regression diagnostics for the posterior means of the consumption (3.10) based on three prior specifications allows us to draw conclusions for the "sensitivity of sensitivities" as well:

(1) The sensitivities for both coefficients are almost identical for the prior specifications (a) and (b). Therefore we list in fig. 24.1 only the diagnostics for specification (a) which is also very close to the OLS regression diagnostics (3.10).

(2) Comparing the diagnostics of the regression with very diffuse prior specification (a) with the informative specification (c) we see a slight change in the patterns for I and Q. Note that for the informative prior (c) the first half of the observations is more sensitive with respect to the intercept than for the prior (a).

(3) The sensitivities with respect to the marginal propensity of consumption β_1 is scattered around the 0-line and is almost the same for all prior specifications. The most influential points for the regression estimates are the last four years (1974–1977). The reason for this behaviour lies in the fact that we observe a higher consumption for Austria in those years while the income was shrinking (1975) or expanding to a smaller degree (see table 24.2).

Table 24.1. Estimates for the consumption function.

Case	b_0^{**}	$ST \cdot D(b_0^{**})$	b_1^{**}	$ST \cdot D(b_1^{**})$
OLS	0.011	0.052	0.586	0.013
(a)	0.011	0.052	0.586	0.013
(b)	0.004	0.052	0.588	0.013
(c)	−0.109	0.048	0.617	0.012

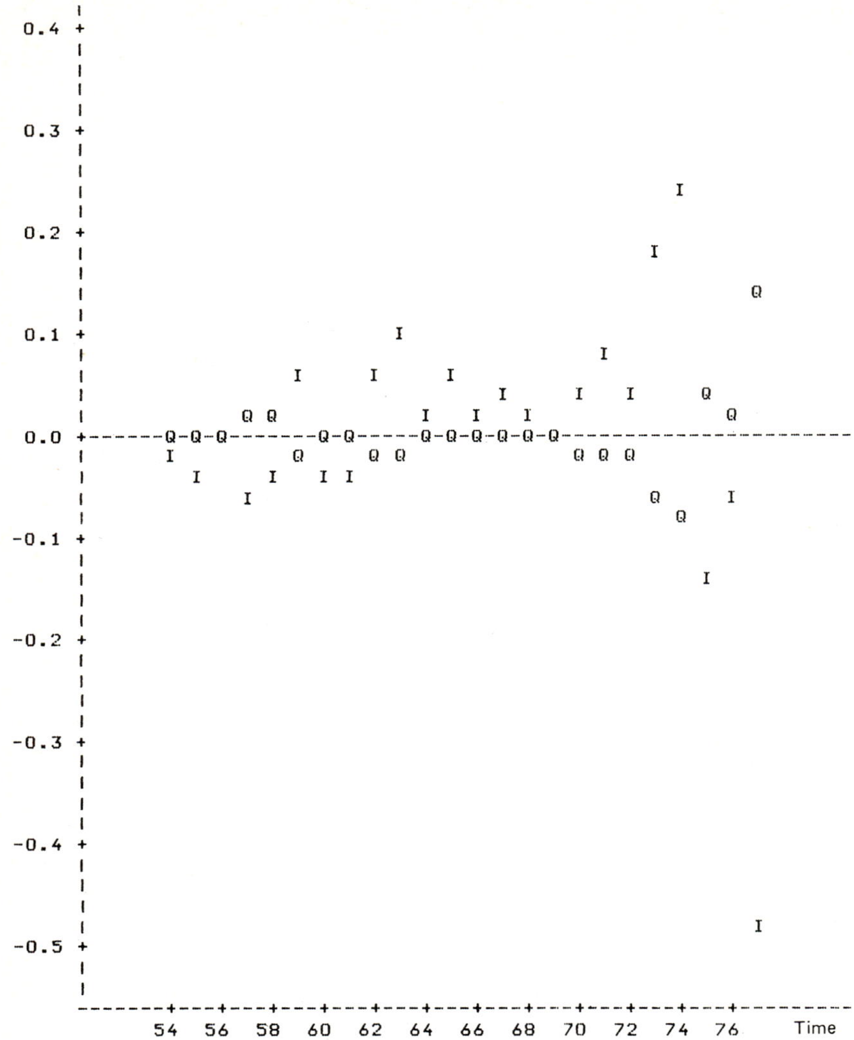

Fig. 24.1. Local sensitivities of the posterior mean for the prior specification:

$$\boldsymbol{\beta} \sim N\left[\begin{pmatrix} 0 \\ 0.6 \end{pmatrix}, \begin{pmatrix} 1000 & 0 \\ 0 & 0.25 \end{pmatrix}\right].$$

Sensitivities: I = intercept $\partial b_0^{**}/\partial\sigma^{ii}$; Q = Q-coefficient $\partial b_1^{**}/\partial\sigma^{ii}$.

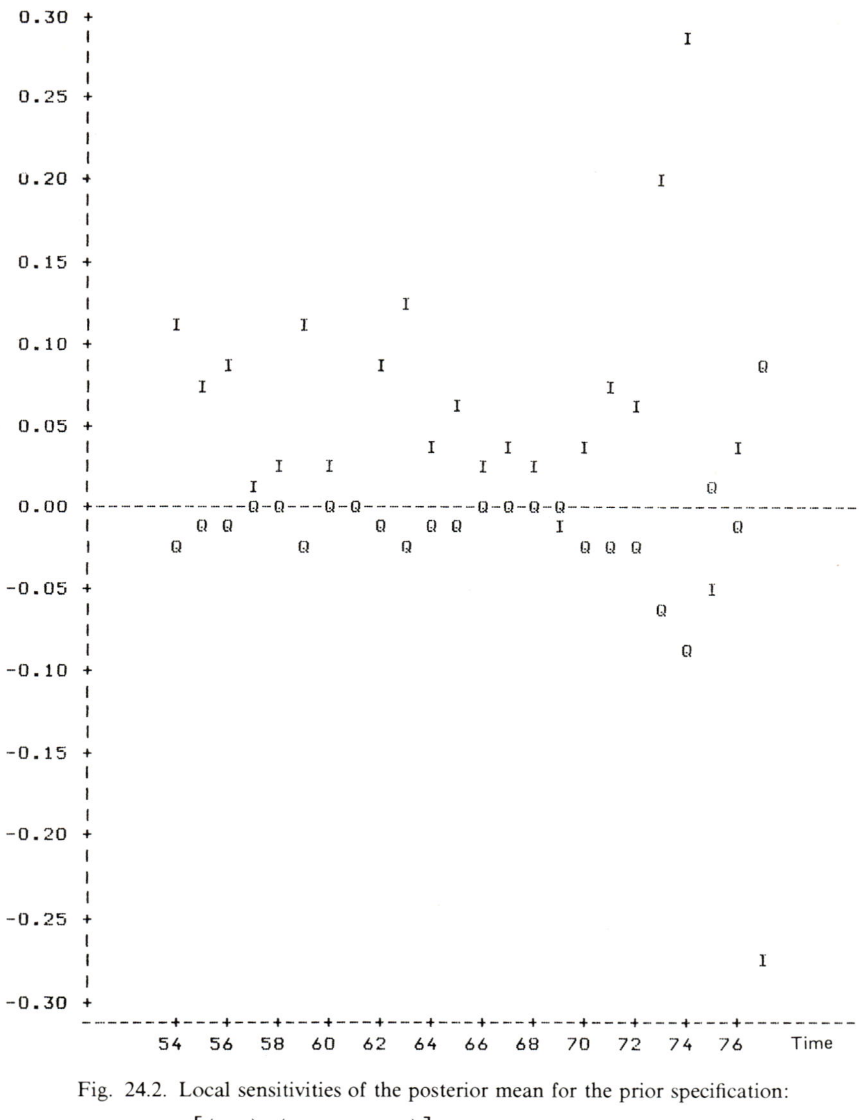

Fig. 24.2. Local sensitivities of the posterior mean for the prior specification:

$$\boldsymbol{\beta} \sim N\left[\begin{pmatrix} 0 \\ 0.8 \end{pmatrix}, \begin{pmatrix} 1000 & 0 \\ 0 & 0.001 \end{pmatrix}\right].$$

Sensitivities: I = intercept $\partial b_0^{**}/\partial \sigma^{ii}$; Q = Q-coefficient $\partial b_1^{**}/\partial \sigma^{ii}$.

Table 24.2 Austrian data for GNP (Q) and consumption (C) for the period 1954–77.

Q	C	Year
130.447	76.818	1954
144.889	84.904	1955
154.844	91.340	1956
164.317	95.579	1957
170.338	99.463	1958
175.174	104.541	1959
189.611	111.053	1960
200.142	116.746	1961
205.380	123.018	1962
213.912	129.606	1963
227.136	134.869	1964
234.943	140.818	1965
246.681	146.361	1966
252.558	151.394	1967
263.782	157.353	1968
279.218	162.467	1969
300.889	171.346	1970
316.781	182.000	1971
336.900	197.038	1972
356.406	205.169	1973
371.119	212.757	1974
363.735	218.059	1975
382.722	226.685	1976
395.941	242.317	1977

4. Local sensitivity analysis for SUR

A question which was not considered in Belsley, Kuh and Welsch (1980) is whether regression diagnostics can be extended to systems estimators in econometrics. In this section we show that these diagnostics can be extended by using matrix-by-matrix derivatives. Because the number of possible diagnostics increases with the number of dimensions, we also need new methods to summarize the sensitivities. We demonstrate our approach for the seemingly unrelated regression (SUR) system originally developed by Zellner (1962).

In a SUR system we link N regressions by an $(N \times N)$ between regression covariance matrix $\boldsymbol{\Omega}$ in the form:

$$\underset{(TN \times 1)}{\begin{pmatrix} \boldsymbol{y}_1 \\ \vdots \\ \boldsymbol{y}_N \end{pmatrix}} = \underset{(TN \times TK)}{\begin{pmatrix} \boldsymbol{X}_1 & & \boldsymbol{O} \\ & \ddots & \\ \boldsymbol{O} & & \boldsymbol{X}_N \end{pmatrix}} \underset{(KN \times 1)}{\begin{pmatrix} \boldsymbol{\beta}_1 \\ \vdots \\ \boldsymbol{\beta}_N \end{pmatrix}} + \underset{(TN \times 1)}{\begin{pmatrix} \boldsymbol{u}_1 \\ \vdots \\ \boldsymbol{u}_N \end{pmatrix}}, \tag{4.1}$$

or compactly expressed:

$$\boldsymbol{Y} = \boldsymbol{X\beta} + \boldsymbol{u}, \quad \boldsymbol{u} \sim N(\boldsymbol{O}, \boldsymbol{\Omega} \otimes \boldsymbol{I}_T). \tag{4.2}$$

Assuming a normal prior distribution $\boldsymbol{\beta} \sim N(\boldsymbol{b}^*, \boldsymbol{H}^{*-1})$ the posterior mean is given by the following matrix weighted average of $\boldsymbol{b}^*$ and the GLS estimate $\boldsymbol{b}$:

$$\boldsymbol{b}^{**} = \boldsymbol{H}^{-1}\big(\boldsymbol{H}^*\boldsymbol{b}^* + \boldsymbol{X}'(\boldsymbol{\Omega}^{-1}\otimes\boldsymbol{I}_T)\boldsymbol{X}\boldsymbol{b}\big), \quad (KN \times 1), \tag{4.3}$$

where

$$\boldsymbol{b} = \big(\boldsymbol{X}'(\boldsymbol{\Omega}^{-1}\otimes\boldsymbol{I}_T)\boldsymbol{X}\big)^{-1}\boldsymbol{X}'(\boldsymbol{\Omega}^{-1}\otimes\boldsymbol{I}_T)\boldsymbol{y}, \quad (KN \times 1), \tag{4.4}$$

$$\boldsymbol{H} = \boldsymbol{H}^* + \boldsymbol{X}'(\boldsymbol{\Omega}^{-1}\otimes\boldsymbol{I}_T)\boldsymbol{X}, \quad (KN \times KN). \tag{4.5}$$

The SUR covariance matrix $\boldsymbol{\Sigma} = \boldsymbol{\Omega}\otimes\boldsymbol{I}_T$ is now specially structured and contains many zeros. Therefore the $\partial\boldsymbol{B}/\partial\langle\boldsymbol{A}\rangle$ derivative will avoid redundant algebra in the derivative arrangement. The following result has been established in Polasek (1982):

Theorem 4.1. The local sensitivities of the posterior moments $\boldsymbol{b}^{**}$ and $\boldsymbol{H}$ of (4.3) and (4.4) of the SUR model (4.1) are given by the $(KN^2 \times KN^2)$ matrix:

$$\partial\boldsymbol{H}^{-1}/\langle\partial\boldsymbol{\Omega}\rangle = \big(\boldsymbol{I}_N\otimes\boldsymbol{H}^{-1}\boldsymbol{X}'\big)\big(\mathrm{vec}^2\boldsymbol{\Omega}^{-1}\otimes\boldsymbol{I}_T\big)\big(\boldsymbol{I}_N\otimes\boldsymbol{X}\boldsymbol{H}^{-1}\big), \tag{4.6}$$

and the $(KN^2 \times N)$ derivative matrix:

$$\partial\boldsymbol{b}^{**}/\langle\partial\boldsymbol{\Omega}\rangle = (\boldsymbol{\Omega}^{-1}\otimes\boldsymbol{H}^{-1}\boldsymbol{X}')\big(\mathrm{vec}\,\boldsymbol{I}_N\otimes\hat{\boldsymbol{U}}\boldsymbol{\Omega}^{-1}\big), \tag{4.7}$$

where the $(T \times N)$ matrix of residuals $\hat{\boldsymbol{U}}$ is built up by columns of $\hat{\boldsymbol{u}} = \mathrm{vec}\,\hat{\boldsymbol{U}} = \boldsymbol{y} - \boldsymbol{X}\boldsymbol{b}^{**}$.

The structure of derivative (4.7) is similar to the regression model in theorem 3.1. Now $\mathrm{vec}^2\boldsymbol{\Omega}^{-1}$ replaces $\mathrm{vec}^2\boldsymbol{I}_T$, because the derivative is now given with respect to $\boldsymbol{\Omega}$ (rather than $\boldsymbol{\Omega}^{-1}$).

Breaking up the sensitivity matrix (4.7) by rows, and taking into account the symmetry of $\boldsymbol{\Omega}$, we get the following elementwise sensitivities:

$$\partial\boldsymbol{b}^{**}/\partial\omega_{ij} = \begin{cases} \boldsymbol{H}^{-1}\boldsymbol{X}'(\boldsymbol{\omega}^{\cdot i}\otimes\boldsymbol{U}\boldsymbol{\omega}^{\cdot i}), & i = j, \\ \boldsymbol{H}^{-1}\boldsymbol{X}'(\boldsymbol{\omega}^{\cdot i}\otimes\boldsymbol{U}\boldsymbol{\omega}^{\cdot j} + \boldsymbol{\omega}^{\cdot j}\otimes\boldsymbol{U}\boldsymbol{\omega}^{\cdot i}), & i \neq j, \end{cases} \tag{4.8}$$

where $\boldsymbol{\Omega}^{-1} = (\boldsymbol{\omega}^{\cdot 1}, \ldots, \boldsymbol{\omega}^{\cdot N})$ is partitioned by columns. It should be noted that the residual matrix $\boldsymbol{U}$ is weighted twice by the inverse of $\boldsymbol{\Omega}^{-1}$. First an "average" residual is formed by $\boldsymbol{U}\boldsymbol{\omega}^{\cdot i}$ and then this average residual vector is weighted by $\boldsymbol{\omega}^{\cdot i}$ for the coefficient vector of every equation.

4.1. The distribution of the sensitivities

From eqs. (3.6) and (4.8) we see that the sensitivities are linear functions of the estimated residuals. Because SUR diagnostics involve a large number of observations we try to find large sensitivities by distributional approximations. The distribution of the OLS residuals has been calculated in Theil (1966) and for the

GLS residuals in Theil (1971, p. 24): If

$$y = X\beta + u, \quad u \sim N(O, \Sigma), \tag{4.9}$$

is a normal general linear model with known covariance matrix Σ, then the GLS residual

$$\hat{u} = y - Xb_{\text{GLS}} = y - X(X'\Sigma^{-1}X)X'\Sigma^{-1}y \tag{4.10}$$

is distributed as

$$\hat{u} \sim N\big(O, \Sigma - X(X'\Sigma^{-1}X)^{-1}X'\big). \tag{4.11}$$

This allows us to calculate the distribution for the SUR sensitivities for the non-informative (classical) case, i.e. for $\text{vec}\,\hat{U} = y - Xb$, where b is given as in (4.4):

$$\text{vec}\,\hat{U} \sim N\Big(O, \big(\Omega \otimes I_T - X\big(X'(\Omega^{-1} \otimes I_T)X\big)^{-1}X'\Big). \tag{4.12}$$

Note that we can write the SUR sensitivities (4.8) of the diagonal elements as

$$\partial b/\partial\omega_{ii} = H^{-1}X'\big(\omega^{\cdot i}\omega^{i\cdot} \otimes I_T\big)\text{vec}\,\hat{U}. \tag{4.13}$$

Now we can find the distribution of the sensitivities by (4.12) as

$$\partial b/\partial\omega_{ii} \sim N\big[O, H^{-1}X'(P_i\omega^{ii} - Q_i)XH^{-1}\big], \tag{4.14}$$

with $P_i = \omega^{\cdot i}\omega^{i\cdot} \otimes I_T$ and Q_i given as

$$Q_i = P_iX\big(X'(\Omega^{-1} \otimes I_T)X\big)^{-1}X'P_i, \tag{4.15}$$

$$\text{Var}\big(b/\partial\omega_{ij}\big) = H^{-1}X'\big[\omega^{jj}P_i - Q_i + \omega^{ii}P_j - Q_j\big]XH^{-1}. \tag{4.16}$$

These distributional approximations allow a better overview of the large sensitivities obtained from a local sensitivity analysis of a SUR system. To check if a certain element in the Ω-matrix has a large impact on the changes of all the regression coefficients in the SUR system, we can count the numbers of local sensitivities which are greater than the multiple of the standard error obtained from the variance in (4.14) and (4.16).

4.2. *Checking for seemingly unrelated regression estimates: The Grünfeld data*

To illustrate the sensitivity analysis we choose the well-known Grünfeld investment data of the original SUR analysis, given in Zellner (1962). The investment function for 1935–54 for the first three companies out of the 11 major U.S. corporations (General Motors, U.S. Steel, and General Electric) is given by the equation:

$$I_t = \beta_0 + \beta_1 K_{t-1} + \beta_2 F_{t-1} + \varepsilon_t,$$

where I_t is the current gross investment, K_{t-1} is the beginning-of-the-year capital

Table 24.3. OLS and SUR estimates for three regression equations.

	General Motors	U.S. Steel	General Electric
(a) OLS estimates			
INT	−149.8	−49.2	−9.96
K	0.12	0.17	0.03
F	0.37	0.39	0.15
(b) Covariance matrix			
(c) SUR estimates			
INT	−168.6	32.97	−21.9
K	0.12	0.13	0.04
F	0.38	0.43	0.13

(b) Covariance matrix

$$\hat{\Omega} = \begin{bmatrix} 8\,423.88 & (-0.30) & (0.28) \\ -2\,314.17 & 9299.6 & (0.39) \\ 714.75 & 1151.1 & 777.5 \end{bmatrix}$$

stock, and F_{t-1} is the value of outstanding shares at the beginning of the year (the firm's market value).

Various estimates are given in table 24.3, while summary statistics for the sensitivity analysis is given in table 24.4.

In table 24.4 we have calculated large sensitivities based on the variances of (4.16): $\sigma^2 = \mathrm{Var}(\partial \boldsymbol{b}/\partial \omega_{ij})$. According to the two derivative concepts of section 2 we have also two possibilities for the presentation of large ("significant") sensitivities: the third column in table 24.4 arranges the sensitivities as counts over the coefficients when the symmetric $\boldsymbol{\Omega}$ matrix is fixed: Analogous to the $\partial \boldsymbol{b}/\partial\langle\boldsymbol{\Omega}\rangle$ derivative as

$$f_{ij}(\boldsymbol{\Omega}) = \#\left(\partial \boldsymbol{b}/\partial \omega_{ij} \geq c\sigma\right), \quad i=1,\ldots,3;\; j=1,\ldots,i, \tag{4.18}$$

Table 24.4. SUR sensitivities for the Grünfeld data, two-derivative arrangements.

s	Rows	$\partial \boldsymbol{b}/\partial \omega_{ij}$, $i,j=1,\ldots,3$ $\boldsymbol{\Omega}$	Rows	$\partial \boldsymbol{b}_k/\partial \boldsymbol{\Omega}$, $k=1,\ldots,9$ G.M.	Steel	G.E.	Sum	%
		0	INT	0	0	1		
2σ	$\boldsymbol{\Omega}$	4 1	K	1	1	1	5	9.3
		0 0 0	F	0	0	1		
		0	INT	1	1	1		
1.5σ	$\boldsymbol{\Omega}$	6 2	K	2	2	2	13	24.1
		1 4 0	F	1	2	1		
		4	INT	1	1	2		
σ	$\boldsymbol{\Omega}$	6 4	K	4	4	3	21	38.9
		1 4 2	F	3	2	1		
Maximum per element		9			6		54	100

where # denotes "the number of" and the c's are chosen constants, similar to the "fence-values" in exploratory data analysis.
The fifth column contains the sensitivity counts in the form of the $\partial\langle \boldsymbol{b}\rangle/\partial\boldsymbol{\Omega}$ derivative, or as

$$f_{ki}(\boldsymbol{B}) = \#(\partial b_{ki}/\partial\boldsymbol{\Omega} \geq c\sigma), \quad i=1,\ldots,N,\ k=1,\ldots,K. \tag{4.19}$$

In the "sum" column we find the total number of significant sensitivities outside the $c\sigma$-intervals, which is simply the sum of the elements in both frequency matrices. The last column calculates the sums as percentages of the $N \cdot K \cdot N$ $(N+1)2 = 54$ possible sensitivities.

Comparing the last "percent" column with the percentages of a normal distribution we notice a flatter empirical distribution. While there are only 39 percent of the sensitivities outside the σ-interval, there are 9.3 percent outside the 2σ interval. This indicates that those sensitivities are mainly responsible for changes between the OLS and the SUR estimate.

From the $\boldsymbol{\Omega}$ arrangement of the sensitivities we see that mainly the ω_{12} element was responsible for the shifts in the estimates. Table 24.3 shows that the ω_{12} element is the biggest off-diagonal element, but the correlation coefficients (upper triangle numbers in parentheses) are almost equal in size. This could mean that the negative correlation might be the reason for the influence. The second type of arrangement shows that the five significant sensitivities only affect the capital (K) coefficients and the General Electric (G.E.) equation.

In the last row of table 24.4 we have listed the maximum number of possible frequency counts in both arrangements. There is a maximum of $K \cdot N = 9$ frequencies for every element in the $\boldsymbol{\Omega}$ arrangement and maximal $N\ (N+1)/2 = 6$ frequencies for every element in the $\boldsymbol{B}$ arrangement ($\boldsymbol{b} = \operatorname{vec} \boldsymbol{B}$).

Finally, we want to make some critical remarks about this procedure. For large systems the computation of the distribution of the sensitivities becomes technically cumbersome. Also, for the full Bayesian case these methods have to be extended and better summarizing statistics seem to be necessary.

5. Summary

In this chapter we have discussed the possibility of sensitivity analysis in Bayesian regression models. In particular we have constructed Bayesian regression diagnostics as an extension of the classical concept in Belsley, Kuh and Welsch (1980). Using the concept of matrix-by-matrix derivatives we also derived regression diagnostics for the seemingly unrelated regression (SUR) system. Regression diagnostics can be viewed as a special application of the general concept of local sensitivity analysis which was first suggested by Leamer (1972) for a prior–posterior sensitivity analysis. Local sensitivities and regression diagnostics can help to report results and indicate where more complicated models have to be considered. In this respect Bayesian methods do not differ too much from classical ones. But

Bayesian sensitivity analysis might also include the prior specification. The combination of these results, as they are discussed for example in Leamer (1982) or Polasek (1984), with regression diagnostics, is still a topic of further research. [*Note*: A SAS PROC MATRIX program for regression diagnostics is available from the author upon request.]

References

Balestra, P. (1976) *La Dérivation Matricelle* (Collection de L'IME No. 12, Sirey, Paris).

Belsley, D.A., E. Kuh and R.E. Welsch (1980) *Regression Diagnostics* (John Wiley, New York).

Box, G.E.P. and G.C. Tiao (1962) "A further look at robustness via Bayes' Theorem", *Biometrika*, 49, 419–432.

Dwyer, P.S. and M.S. MacPhail (1948) "Symbolic matrix derivatives", *Ann. Math. Stat.*, 19, 517–534.

Leamer, E.E. (1972) "A class of informative prior and distributed lag analysis", *Econometrica*, 42, 1059–1081.

Leamer, E.E. (1978) *Specification Searches* (John Wiley, New York).

Leamer, E.E. (1982) "Sets of posterior mean with bounded variance prior", *Econometrica*, 50, 725–736.

Magnus, J.R. and H. Neudecker (1979) "The commutation matrix: Some properties and applications", *Annals of Statistics*, 381–394.

Polasek, W. (1982) "A dual approach to matrix derivatives", *Metrika*, 1984, to appear.

Polasek, W. (1984) "Multivariate regression models: Estimation and sensitivity analysis for two-dimensional data", in: J. Kadane, ed., *Robustness in Bayesian Statistics* (North-Holland, Amsterdam), pp 229–309.

Theil, H. (1966) "A simplification of the BLUS procedure for analysing regression disturbances", *Journal of the American Statistical Association*, 61, 242–251.

Zellner, A. (1962) "An efficient method of estimating seemingly unrelated regression and tests of aggregation bias", *Journal of the American Statistical Association*, 57, 348–368.

PART VIII

MODELING AND FORECASTING

Chapter 25

A BAYESIAN VIEW OF STATISTICAL MODELLING

A. PHILIP DAWID

University College London

0. Introduction

Traditionally, a statistical model is a description of how certain "objective" probabilities, governing the outcomes of an experiment, are determined by the values of a set of unknown parameters. For the most part, Bayesians have been content to assume the model as an input to their analysis, merely extending it by introducing a prior distribution for the parameters. A mechanical application of Bayes' Theorem is then usually all that is required to complete the analysis.

There are a number of odd features to such a Bayesian treatment. If, as out-and-out subjectivists would aver, the only allowable probabilities are subjective, why the ready acceptance of the objectivity of the model probabilities? If, moreover, the theory of coherence is to be taken seriously, there can be no such thing as an "unknown" probability, so what is the status of a probability distribution governed by an unknown parameter?

It is my view that a Bayesian approach which is firmly based on known subjective distributions over observables can bring extra insight, for both Bayesians and non-Bayesians, into the construction, interpretation, analysis and validation of traditional statistical models. In particular, new light is shed on the role of independence in modelling, and on the relationship between probability models and real-world frequencies. One fruitful view is that a model is not so much a description of the world, as a description of the area of agreement of a bevy of Bayesians. Another is that the idea that there exists a unique "true" model is unnecessarily restrictive, since it may be perfectly self-consistent to use different models of the same set-up for different purposes.

Bayesian Inference and Decision Techniques, edited by P. Goel and A. Zellner

1. The "true" model

In a frequentist approach to statistics, one generally makes a very clear distinction between two types of quantity, *parameters* and *observables*. The observables have a joint distribution, which is regarded as having an objective physical reality, and the purpose of statistical inference is usually considered to be the framing of sensible statements about this unknown objective distribution, in the light of realized values of some of the observables. In turn, the parameters describe the class of possible distributions for the observables, with specific, but unknown, parameter values corresponding to the "true" distribution. These are not normally themselves considered to have arisen as the outcomes of any random process.

Deeper analysis of this division suggests that the distinction between parameter and observable is not as clear-cut as commonly supposed. Each such quantity has a range of possible values, and a single specific realized value. The logic which allows some of these realized values to be regarded as random outcomes, but not others, deserves more analysis than it usually gets. In particular, it is often not easy to justify and interpret the assumption of an "objective" probability distribution as underlying the unique realization of a specific observable, which is necessary to make convincing the otherwise arbitrary distinction between parameters and observables. The principal real-world distinction is often, quite simply, that a parameter is "not observable": that is not a forceful reason for denying it the status of a realization from a probability distribution. Conversely, an observable might sometimes usefully be regarded as a determinant of the joint conditional distribution of the remaining observables, given its value. This might then be better treated as a "non-random"parameter.

The Bayesian approach avoids such problematic distinctions. Observables and parameters are quantities of the same kind: all are random variables (or, better, uncertain quantities) having an overall joint subjective probability distribution which describes personal uncertainty about their actual values. Nevertheless, an important distinction remains. Observables, being just that, are quantities on which a Bayesian could lay bets, or otherwise determine numerically his subjective probability distribution. But how can one have beliefs about a parameter, something which merely serves to label the model probability distributions? We only ever obtain a single realization from the joint distribution governing all observables. So not only is a parameter (usually) intrinsically unobservable, but it may often be very unclear just what physical reality, if any, it measures. This muddiness threatens the firm operationalist foundation on which de Finetti's theory of subjective probability is built.

Now the Bayesian who uses model densities $f(\boldsymbol{x}|\boldsymbol{\theta})$ for observables $\boldsymbol{X}$ given parameters $\boldsymbol{\Theta}$, together with a prior density $f(\boldsymbol{\theta})$ for $\boldsymbol{\Theta}$, generates an implied marginal distribution for $\boldsymbol{X}$. If another Bayesian uses a different model and prior which imply the *same* marginal distribution for $\boldsymbol{X}$, the two will agree on their attitudes to all observables. If we regard $\boldsymbol{X}$ as comprising all relevant quantities which can be observed in the real world, there is no reason to distinguish between these two Bayesian descriptions. We cannot say that, if one Bayesian is using a "true" model, then the other is not – each is equally valid. The functional dependence of the model densities on the parameters, and the prior distribution, can be regarded purely as mathematical artifacts which combine to generate the desired marginal density $f(\boldsymbol{x})$. Any combination of model and prior which succeeds in this can be used for further calculations, so long as these confine themselves to questions concerning observables only. Such predictivist use of model-cum-prior is exemplified by Geisser (1982). When a choice exists, the selection of a particular combination of model and prior in a particular case will depend on considerations of mathematical tractability for the immediate purpose at hand.

Although the above analysis allows the Bayesian to pass backwards and forwards between different models as it suits him, there is no reason why this same freedom should not be extended to a frequentist statistician: a model which is suitable for Bayesian purposes is at least a prima facie candidate for frequentist use, and this can often be justified on purely frequentist terms. With this new liberty in mind, many frequentist concepts, such as sufficiency of a statistic or consistent estimability of a parameter, can be seen in a completely new light: as possible criteria one might apply to select one out of a number of otherwise equally valid candidates for the role of "model". And again, the motivation behind such a criterion should be to provide more tractable analysis.

2. Examples

In this section we present some examples to demonstrate the possibility and fruitfulness of representing one situation by more than one model.

An important consideration is that, if we have data $\boldsymbol{X}_0$, and want to predict further observables $\boldsymbol{X}_1$, it is valuable to have a model in which, given the parameters $\boldsymbol{\theta}$, $\boldsymbol{X}_1$ and $\boldsymbol{X}_0$ are independent. For in general the predictive density of $\boldsymbol{X}_1$ given $\boldsymbol{X}_0$ is $f(\boldsymbol{x}_1|\boldsymbol{x}_0) = \int f(\boldsymbol{x}_1|\boldsymbol{\theta}, \boldsymbol{x}_0)f(\boldsymbol{\theta}|\boldsymbol{x}_0)\mathrm{d}\boldsymbol{\theta}$. Under the condition above, this reduces to $\int f(\boldsymbol{x}_1|\boldsymbol{\theta})f(\boldsymbol{\theta}|\boldsymbol{x}_0)\mathrm{d}\boldsymbol{\theta}$, a straightforward expectation of the model density $f(\boldsymbol{x}_1|\boldsymbol{\theta})$ with respect to the posterior distribution of $\boldsymbol{\theta}$ given the data $\boldsymbol{x}_0$. This elimination of $\boldsymbol{x}_0$ from the first term is often a great technical advantage, although not a fundamental conceptual necessity.

Example 2.1. Consider the model for a discrete time series, (X_t), for which the (X_t) are jointly normal with $E(X_t) = \mu$, $\mathrm{var}(X_t) = \sigma^2$, $\mathrm{cov}(X_s, X_t) = \rho\sigma^2 (s \neq t)$, all the parameters being unknown. The parameters (μ, ρ, σ^2) have a joint prior distribution. A set of past observations is available, and it is desired to forecast future observables.

The analysis may be substantially eased as follows. One way of representing the (X_t) is as

$$X_t = \mu + Y + Z_t, \tag{2.1}$$

given (μ, ρ, σ^2), in which $Y \sim \mathcal{N}(0, \rho\sigma^2)$, $Z_t \sim \mathcal{N}(0, \sigma'^2)$, all mutually independent, where $\sigma'^2 = (1-\rho)\sigma^2$. Now define $\mu' = \mu + Y$. Then, given (μ', σ'^2), we can represent the distribution of the data as $X_t = \mu' + Z_t$, where now the (Z_t) are *independent* $\mathcal{N}(0, \sigma'^2)$. So, taking (μ', σ'^2) as our new parameters, we obtain as our revised model the sub-model of the original model for which $\rho = 0$. To complete the Bayesian picture, we must assign to the new parameters $(\mu', \sigma'^2) = (\mu + Y, (1-\rho)\sigma^2)$ their distribution induced from the original joint distribution of (μ, ρ, σ^2, Y). The marginal distribution of the data (X_t) using this prior with the above independence model will then be identical with that based on the original model $(\rho \neq 0)$ and its corresponding prior. Consequently, we can always, without loss of generality, assume $\rho = 0$.

This simplification is available for the Bayesian because he makes no absolute distinction between parameters and observables. By contrast, the mongrel form $\mu' = \mu + Y$ would be an outcast from any established frequentist kennel. Can a frequentist nevertheless justify replacing his original full model by its independence submodel?

One problem with the original $(\rho \neq 0)$ model is that it is *non-ergodic*. In particular, it is not possible to estimate its parameter ρ consistently from even unlimited observation on a single realization of the series (for example, the maximum likelihood estimator of ρ is 0, for any data). This should cause a frequentist to wonder what he really means by the value of ρ. One common answer is that, if we consider independent "parallel universes", each generating a realization of the series, and from each extract the values of X_s and X_t, then ρ will be the limiting value of the sample correlation between X_s and X_t over these conceptual replications. This may be termed the *external* interpretation of a model. However, it may be argued that we only have one universe, so that this interpretation is non-operational, and so meaningless. Instead, a frequentist might prefer an *internal* interpretation, in which the parameters of his model are determinable from indefinite observation on a single realization. Only the second $(\rho = 0)$ model satisfies this criterion. The general approach to be considered in sections 3 and 4 automatically leads to models with such an internal interpretation.

It is of interest that, notwithstanding the impossibility of identifying the model having $\rho \neq 0$, it is perfectly adequate for the internal purpose of predicting future values on the basis of past data, even from a sampling-theoretic viewpoint. For it implies that, given $(X_1, X_2, \ldots, X_n)$, X_{n+1} is normally distributed with mean $\overline{X}_n - [\sigma^2(1-\rho)/\{n\rho\sigma^2 + (1-\rho)\sigma^2\}](\overline{X}_n - \mu)$ and variance $\sigma^2(1-\rho)\cdot(1 + [\rho/\{1 + (n+1)\rho\}])$. For large n, $\operatorname{var}(\overline{X}_n) \doteq \rho\sigma^2$, so that $(\overline{X}_n - \mu)$ is $0(1)$, and the predictive mean may be well estimated by $\overline{X}_n$. Likewise, the predictive variance

$\doteq \sigma^2(1-\rho)$, which is consistently estimable by $\Sigma(X_i - \bar{X})^2/(n-1)$. We therefore obtain the same asymptotic form of the predictive distribution as that which we would use for the model with $\rho = 0$. This property provides a frequentist argument for regarding the two different models as essentially identical, and thus for using the simpler independence model. The asymptotic predictive equivalence of such alternative models holds in some generality.

Example 2.2 (Dawid, 1977). This time we start from a subjective joint distribution for a data-array $(X_{ij}: i \geq 1, j \geq 1)$, which we suppose to be multivariate normal, with $E(X_{ij}) = m$, $\text{var}(X_{ij}) = v + a + q(\text{all } i, j)$, $\text{cov}(X_{ij_1}, X_{ij_2}) = v + a(j_1 \neq j_2)$, $\text{cov}(X_{i_1j_1}, X_{i_2j_2}) = v(i_1 \neq i_2)$. It may be shown that, for positive-definiteness, v, a and q are non-negative. Under certain reasonable symmetry conditions this is a good model for data where i labels treatments and j replications.

One representation of this distribution is as

$$X_{ij} = \mu + \alpha_i + \varepsilon_{ij}, \tag{2.2}$$

with $\mu \sim \mathcal{N}(m, v), \alpha_i \sim \mathcal{N}(0, a), \varepsilon_{ij} \sim \mathcal{N}(0, q)$, all mutually independent.

Suppose that experimental data are available on a finite sub-array $S_1 = (X_{ij}: 1 \leq i \leq t, 1 \leq j \leq n)$. We consider two possible prediction problems.

(2.2a) Prediction for $S_2 = (X_{ij}: i = 1, \ldots, t;\ j > n)$ – that is, further replications of treatments used in the experiment.

Define $\theta_i = \mu + \alpha_i$, from (2.2). Then for all data in $S_1 \cup S_2$ we have

$$X_{ij} = \theta_i + \varepsilon_{ij}, \tag{2.3}$$

where the (ε_{ij}) are independent of each other, and of the (θ_i).

Now we nominate the θ's as "parameters" and the ε's as "errors", thus obtaining a "one-way lay-out" model in which $X_{ij} \sim \mathcal{N}(\theta_i, q)$ independently. In particular, the (X_{ij}) to be predicted are independent, in this model (i.e. given the θ's), from those in the experiment, while the same parameters are involved. This simplifies the technical prediction task: the posterior distribution of the θ's given S_1 is found, and "plugged in" to the model (2.3) for S_2.

(2.2b) Prediction for $S_3 = (X_{ij}: i > t)$ – that is, outcomes of treatments as yet untried.

Now we write

$$X_{ij} = \mu + \eta_{ij}, \tag{2.4}$$

with $\eta_{ij} = \alpha_i + \varepsilon_{ij}$, taking μ as the "parameter" and the (η_{ij}) as "errors". While the (η_{ij}) are not independent, it is still true that in this model (i.e. given μ) S_1 and S_3 are independent, so that we need only find the posterior distribution of μ based on S_1 from (2.4), and use it to average out the appropriate "sampling distribution" for S_3 (dependent on μ).

For the Bayesian, the two models (2.3) and (2.4) are trivially equivalent, each being a re-expression of (2.2); each is the more useful in its own context. A non-Bayesian statistician, who believes in a firm distinction between parameters and observables, might also like to enjoy this flexibility of being able to adjust the model to his purpose. Indeed, some frequentist approaches to problems such as recovery of interblock information in partially balanced designs suggest a willingness to bend the usual rules on such occasions.

Example 2.3 (Dawid, 1981). Suppose the $(n \times p)$ data-matrix $\boldsymbol{X}$ has the matrix-normal distribution $\mathcal{N}(\boldsymbol{I}, \boldsymbol{\Sigma})$, i.e. its rows are independently $\mathcal{N}(\boldsymbol{0}, \boldsymbol{\Sigma})$, and, in the prior distribution, $\boldsymbol{\Sigma} \sim IW_p(\delta; \boldsymbol{K})$ (i.e. $\boldsymbol{\Sigma}^{-1} \sim W(p+\delta-1; \boldsymbol{K}^{-1})$). With $\boldsymbol{X} = (\boldsymbol{X}_1 \quad \boldsymbol{X}_2)$, $\boldsymbol{X}_i$ being $(n \times p_i)$, it is desired to predict $\boldsymbol{X}_2$ from $\boldsymbol{X}_1$.

The conditional sampling distributions for $\boldsymbol{X}_2$ given $\boldsymbol{X}_1$ depend on

$$\boldsymbol{\Sigma} = \begin{pmatrix} \boldsymbol{\Sigma}_{11} & \boldsymbol{\Sigma}_{12} \\ \boldsymbol{\Sigma}_{21} & \boldsymbol{\Sigma}_{22} \end{pmatrix}$$

(where $\boldsymbol{\Sigma}_{ij}$ is $(p_i \times p_j)$) through $\boldsymbol{\Sigma}_{22\cdot 1} = \boldsymbol{\Sigma}_{22} - \boldsymbol{\Sigma}_{21}\boldsymbol{\Sigma}_{11}^{-1}\boldsymbol{\Sigma}_{12}$ and $\boldsymbol{B} = \boldsymbol{\Sigma}_{11}^{-1}\boldsymbol{\Sigma}_{12}$, while the sampling distribution of $\boldsymbol{X}_1$ involves only $\boldsymbol{\Sigma}_{11}$. In fact, $\boldsymbol{X}_2 | \boldsymbol{X}_1 \sim \boldsymbol{X}_1\boldsymbol{B} + \mathcal{N}(\boldsymbol{I}, \boldsymbol{\Sigma}_{22\cdot 1})$, while $\boldsymbol{X}_1 \sim \mathcal{N}(\boldsymbol{I}, \boldsymbol{\Sigma}_{11})$. Since it may be shown that, in the prior distribution, $\boldsymbol{\Sigma}_{11}$ is independent of $(\boldsymbol{B}, \boldsymbol{\Sigma}_{22\cdot 1})$, the posterior joint distribution for $(\boldsymbol{B}, \boldsymbol{\Sigma}_{22\cdot 1})$ given $\boldsymbol{X}_1$ is the same as its prior distribution, and this may, therefore, be found and used to average the conditional sampling distributions for $\boldsymbol{X}_2$ given $\boldsymbol{X}_1$. While this approach is straightforward, it is fairly lengthy.

Alternatively we may use an entirely different model. It may be shown that the model assumptions,

$$\underset{(n \times p)}{\boldsymbol{X}} \sim \mathcal{N}\big(\underset{(n \times n)}{\Lambda}, \underset{(p \times p)}{\boldsymbol{K}}\big) \qquad (K \text{ as before}),$$

together with the prior distribution,

$$\Lambda \sim IW(\delta; \boldsymbol{I}),$$

imply the identical marginal distribution for the observables $\boldsymbol{X}$ as the original specification. In the new "sampling distributions", $\boldsymbol{X}_2 | \boldsymbol{X}_1 \sim \boldsymbol{X}_1\boldsymbol{D} + \mathcal{N}(\Lambda, \boldsymbol{K}_{22\cdot 1})$, where $\boldsymbol{D} = \boldsymbol{K}_{11}^{-1}\boldsymbol{K}_{12}$, while $\boldsymbol{X}_1 \sim \mathcal{N}(\Lambda, \boldsymbol{K}_{11})$. The posterior distribution for Λ given $\boldsymbol{X}_1$ is easily found to be $\Lambda \sim IW(\delta + p_1; \boldsymbol{I} + \boldsymbol{S}_1)$, with $\boldsymbol{S}_1 = \boldsymbol{X}_1\boldsymbol{K}_{11}^{-1}\boldsymbol{X}_1'$. Using this to average the conditional sampling distributions immediately yields the predictive distribution for $\boldsymbol{X}_2$ given $\boldsymbol{X}_1$: $\boldsymbol{X}_2 \sim \boldsymbol{X}_1\boldsymbol{D} + T(\delta + p_1; \boldsymbol{I} + \boldsymbol{S}_1, \boldsymbol{K}_{22\cdot 1})$, involving the matrix-$t$ distribution.

Example 2.4 (Ericson, 1969). Consider a finite population of N units, each of which either possesses or does not possess a certain attribute. Let $X_i = 1$ if the ith unit possesses the attribute, $X_i = 0$ otherwise. Data are to be obtained on units $1, 2, \ldots, n$, and inference is required for the population count $T_N = \sum_{i=1}^{N} X_i$.

One way of proceeding is for the Bayesian to assess his full joint distribution for the population make-up $\boldsymbol{S}_N = (X_1, X_2, \ldots, X_N)$, and simply compute the conditional distribution for $\boldsymbol{S}_N$ given $\boldsymbol{S}_n = (X_1, X_2, \ldots, X_n)$. He can then calculate the marginal distribution for T_N given $\boldsymbol{S}_n$. This problem may be formulated as a statistical model having "parameter" $\boldsymbol{S}_N$ and data $\boldsymbol{S}_n$. The prior for $\boldsymbol{S}_N$ is just the original joint assessment, while the likelihood is $f(\boldsymbol{s}_n|\boldsymbol{s}_N)=1$, if $\boldsymbol{s}_n$ is the initial n-segment of $\boldsymbol{s}_N$, $=0$ otherwise. From a classical point of view this likelihood has been labelled "uninformative", but this is far from the case when it is combined with an informative prior for $\boldsymbol{S}_N$ which links values in $\boldsymbol{S}_n$ with the remainder, since Bayes' Theorem then provides the appropriate way of making inferences about unobserved values from those observed, and thus finding an informed posterior distribution for T_N.

Now in many contexts it would be very reasonable for the prior distribution over $\boldsymbol{S}_N$ to be *exchangeable*. It is then fairly easy to see that the posterior distribution for T_N depends on $\boldsymbol{S}_n$ only through $T_n = \Sigma_{i=1}^{n} X_i$. This may then be found from a Bayesian analysis of a new model in which T_N is the parameter, T_n the data, and the "sampling distributions" are hypergeometric $\mathscr{H}(N; n, T_N)$:

$$\Pr(T_n = t \mid T_N = u) = \binom{n}{t}\binom{N-n}{u-t}\Big/\binom{N}{u}.$$

Combined with the prior for T_N which is implied by that chosen for $\boldsymbol{S}_N$, this model again yields the correct posterior distribution for T_N. For a special prior distribution, this analysis is considered in more detail in example 5.1 below.

Lastly, if it is further supposed that $(X_1, X_2, \ldots, X_n)$ can be conceived of as the initial segment of an infinite exchangeable sequence, we know by de Finetti's theorem that we can model the X's as independent, given some parameter $\theta \in [0,1]$, with $P_\theta(X_i = 1) = \theta$. An appropriate prior distribution for θ will yield the initial prior assessment for $\boldsymbol{S}_N$. With this model and prior, we have sampling independence between T_n and $T_{N-n} = T_N - T_n$, with respective binomial distributions $\mathscr{B}(n; \theta)$ and $\mathscr{B}(N-n; \theta)$, so that the predictive distribution of T_{N-n} is found by averaging its sampling distribution with the posterior distribution of θ given T_n. Adding in the known value of T_n yields the desired distribution for T_N.

In this example we see that, under suitable assumptions, the Bayesian can choose, at will, between at least three different models. With appropriate relationships between the prior distributions for their parameters, these all imply the same joint distribution for observables, which is all that the Bayesian need concern himself with.

3. **Intersubjective models** (Dawid, 1979, section 9; Dawid, 1982a)

As shown above, a single Bayesian does not have a unique way of breaking down his subjective distribution for observables into a "model" and a "prior". Any such decomposition may be made for reasons of technical simplicity, and there is no reason to consider that the induced "sampling distributions" have any greater degree of objectivity than has the prior. However, when we consider not one but a whole bevy of Bayesians, with varying subjective distributions over the observables, we do obtain a meaningful way of separating out the model from the prior. The essential requirement is that the model distributions, if not exactly objective, are at least *intersubjective*, so that they are acceptable to every member of the bevy. The prior distribution may be allowed to vary arbitrarily with the individual Bayesian.

In some cases, there may be sufficient agreement about the nature of things for the agreed sampling distributions to take up a good deal of the structure, as when the parameter is of low dimension and the data highly multivariate. This is the kind of model treated as standard by most approaches to inference. In such a case, predictive distributions based on a substantial body of data will automatically be in very close agreement for all members of the bevy. In other applications, there may be almost total incompatibility of the various "world-views", corresponding to a trivial decomposition in which data and parameter are identified, and the "model" says no more than that the bevy will agree on the values of the data when they have observed them. Some world-views of the operation of economic systems appear to be at loggerheads in this way.

It is of importance to be able to distinguish the cases above, so that we can tell for which problems an "objective" statistical analysis is likely to be meaningful and broadly acceptable.

In example 2.4, the first model, $f(\boldsymbol{s}_n|\boldsymbol{s}_N)=1$ or 0, corresponds essentially to the degenerate case above. If we consider a bevy having an entirely unrestricted distribution over $\boldsymbol{S}_N$, this trivial model is the only one available. The amount of agreement between posterior distributions is minimal. If, however, we restrict attention to the bevy of Bayesians whose distributions over $\boldsymbol{S}_N$ are exchangeable, the hypergeometric distributions for T_n given T_N now form a universally acceptable model, the prior for T_N taking up all the subjectivity in the various joint distributions. Under further restriction to infinitely extendible exchangeability for $\boldsymbol{S}_N$, the binomial model now serves as an agreed intersubjective sampling distribution, only the prior distribution for the probability parameter θ being subjectively specified.

3.1. Sufficiency

In general, let $\mathscr{A}$ be the space of potential observables, including those for which prediction may be required as well as those to be observed in an experiment, and consider the family $\mathscr{Q}=\{Q_\lambda : \lambda\in\mathscr{L}\}$ of subjective distributions over $\mathscr{A}$, λ labelling the individual Bayesian B_λ in the bevy $\mathscr{L}$. We look for a decomposition

$$Q_\lambda(\cdot)=\int P_\theta(\cdot)\,\mathrm{d}\Pi_\lambda(\theta), \tag{3.1}$$

where the family $\mathscr{P}=\{P_\theta\}$ constitutes the intersubjective model, and Π_λ is B_λ's prior for the parameter θ.

One way to construct such a decomposition is to consider $\mathscr{Q}$ as a standard statistical model over $\mathscr{A}$, with parameter λ, and define θ to be a *sufficient statistic* for $\mathscr{Q}$ in $\mathscr{A}$, with P_θ the conditional distribution of the other observables given θ, which is independent of λ by the very definition of sufficiency. In this case the "parameter" θ is itself a function of the observables (and so has an "internal" interpretation), and Π_λ in (3.1) is just the distribution for θ induced by Q_λ. It would normally be appropriate to take θ to be *minimal* sufficient for $\mathscr{Q}$, since this maximizes the amount of agreement expressed by the model $\mathscr{P}$.

An example of this procedure is given by the hypergeometric model for example 2.4, since it is straightforward to show that T_N is the minimal sufficient statistic for the family of all exchangeable distributions over $\boldsymbol{S}_N$, and so forms the appropriate "parameter".

3.2. *Invariance*

Often the agreement expressed between the various Bayesians in a bevy may take the simple qualitative form that each member is willing to accept that his subjective distribution is invariant under some group G acting on the observable data-space: that is, all agree on certain underlying symmetries in the problem. This is exactly the case in the above example of exchangeability, with G the group of permutations of units. It then follows generally that the minimal sufficient statistic for the family of all these invariant distributions is just the maximal G-invariant function of the data. If, for example, our bevy all agree that $\boldsymbol{X}_N = (X_1, X_2, \ldots, X_N)'$ has spherical symmetry, so that, for any fixed orthogonal $(N \times N)$ matrix $\boldsymbol{U}$, $\boldsymbol{UX}$ and $\boldsymbol{X}$ have the same subjective distribution, the appropriate model has parameter $Z_N = \sum_{i=1}^N X_i^2$. If only $\boldsymbol{X}_n = (X_1, X_2, \ldots, X_n)$ is observed, the induced sampling distribution for $Z_n | Z_N$ is $Z_N \cdot \beta(\frac{1}{2}n, \frac{1}{2}(N-n))$, which may be combined with the subjective prior distribution of Z_N to yield the appropriate posterior, and further suitably combined with this posterior to provide a predictive distribution for $X_{N-n} = (X_{n+1}, X_{n+2}, \ldots, X_N)$. A special case of this analysis is considered in example 5.2 below.

4. Infinitary problems

Another fruitful way to obtain a decomposition as in (3.1) is by conceiving of the data-space $\mathscr{A}$ of interest as embedded in a larger space $\mathscr{A}^*$, carrying a family $\mathscr{Q}^* = \{Q_\lambda^* : \lambda \in \mathscr{L}\}$ of probability distributions, consistent with having margins $\mathscr{Q}$ over $\mathscr{A}$. If θ is a sufficient statistic for $\mathscr{Q}^*$ in $\mathscr{A}^*$, we obtain, in analogy to (3.1):

$$Q_\lambda^*(\cdot) = \int P_\theta^*(\cdot)\,\mathrm{d}\Pi_\lambda(\theta) \tag{4.1}$$

and, restricted to $\mathscr{A}$, this yields (3.1), with P_θ the $\mathscr{A}$-margin of the distribution

P_θ^* over $\mathscr{A}^*$. The essential difference from our previous construction is that the parameter θ no longer lives in the original space $\mathscr{A}$ of observables, and is hypothetical to the extent that $\mathscr{A}^*$ is. However, the parameter still carries an internal interpretation in so far as $\mathscr{A}^*$ describes quantities that might conceivably be observed, rather than independent replications in parallel universes.

This construction underlies the binomial model for example 2.4, where $\mathscr{A}^*$ is the space of infinite sequences $(X_1, X_2, \dots)$ and $\mathscr{Q}^*$ the family of all exchangeable distributions over $\mathscr{A}^*$. The parameter (sufficient statistic) may be taken to be $\theta = \lim_{k \to \infty} k^{-1} \sum_{i=1}^k X_i$, and we recover the binomial sampling distributions as agreed conditional distributions, given θ.

Although most real problems involve only a finite number of observables, one can often conceive of these as extended to infinity in a suitable way (which may not be unique), and models such as the above become particularly convenient. There is an intricate and extensive theory governing the decomposition (4.1) for such infinitary problems (Lauritzen, 1974, 1982; Dynkin, 1978; Dawid, 1982a). If the family $\mathscr{Q}^*$ has certain properties (suitable symmetry being sufficient, but not necessary), it will be a *simplex*, i.e. a convex set each element of which can be expressed uniquely as a mixture of its *extreme points* (those that are not non-trivial mixtures of other elements). Then in (4.1) we can take $\mathscr{P}^*$ as the set of these extreme points, and Π_λ the induced mixing distribution for Q_λ.

Application of this theory to problems with symmetry produces some familiar and unfamiliar models, including the following normal-theory models:

(a) $X_i \sim \mathscr{N}(0, v)$ independently;

(b) $X_1 \sim \mathscr{N}(u, v)$ independently; and

(c) $X_{ij} = \xi + \eta_i + \gamma_j + \varepsilon_{ij}$, where $\eta_i \sim \mathscr{N}(0, \omega^2), \gamma_j \sim \mathscr{N}(0, \tau^2), \varepsilon_{ij} \sim \mathscr{N}(0, \sigma^2)$ all independently (the parameters being $\xi, \omega^2, \tau^2, \sigma^2$).

All the above follow from suitable assumptions of spherical symmetry alone: see Dawid (1982a) for details.

One reason why such infinitary models are particularly valuable is the way in which they tend automatically to incorporate independence in the sampling distributions. The following argument, which shows independence in the extreme exchangeable distributions, may be extended to more complicated problems.

Theorem. Consider the family of exchangeable distributions for $Z = (X_1, X_2, \dots)$. If P is an extreme member of this family, then, under P, the (X_i) are independent and identically distributed.

Proof. It is easy to see that, under exchangeability, $Z^* = (X_2, X_3, \dots)$ has the same distribution as Z. But the marginal distribution of Z^* is expressible as a mixture of its conditional distributions, given X_1, each of which is invariant under the group G^* of permutations leaving unit 1 fixed. So, restricting attention to Z^*, the extremeness of P, in the family of distributions invariant under G^*, implies that the distributions of Z^* given X_1, under P, must all be identical, showing that X_1 and $(X_2, X_3, \dots)$ are independent under P. The rest of the proof is trivial.

5. Finitary problems

Some problems yield to a straightforward analysis without the introduction of an infinite extension of the observables. A class of these may be constructed as follows.

We start with X_1, X_2, X_3 independently drawn from a family $\mathscr{E}$ of distributions indexed by θ. In cases of interest, $\mathscr{E}$ will be an exponential family, or other family having a fixed-dimensional sufficient statistic. Each X_i may represent several observations. Let T_i be the sufficient statistic for θ based on X_i. Then the sufficient statistic T^2 based on $X^2 = (X_1, X_2)$ will be a function of (T_1, T_2), which we denote by $T^2 = T_1 \oplus T_2$. Similarly, for the sufficient statistic T^3 based on $X^3 = (X_1, X_2, X_3)$, $T^3 = T_1 \oplus T_2 \oplus T_3$.

We now consider X_1 as our data and X_2 as our predictands. We investigate the new model having T^2 as its parameter, and, as its sampling distributions, the conditional distributions for X^2 given T^2 implied by the original model $\mathscr{E}$. For the prior distribution of T^2, we likewise assume that this follows from $\mathscr{E}$ by conditioning on some value of T^3 (a "hyperparameter"). By sufficiency, θ is absent from these distributions.

It is enough to predict T_2, since, using the conditional independence notation of Dawid (1979), in $\mathscr{E}$: (a) $X_2 \perp\!\!\!\perp (T_1, T_3) | \theta$, whence $X_2 \perp\!\!\!\perp (T_1, T_3) | (\theta, T_2)$; and (b) $X_2 \perp\!\!\!\perp \theta | T_2$. Together these imply $X_2 \perp\!\!\!\perp (T_1, T_3, \theta) | T_2$, so that the conditional distribution of X_2 given T_2 in $\mathscr{E}$ will be unchanged by specification of either the hyperparameter or the data in our new set-up. Similarly, it may be seen that T_1 is a sufficient (adequate) summary of X_1 for prediction purposes.

Effectively the new model for our data is described by $T_1 | T^2$ (constructed from $\mathscr{E}$), and the prior by $T^2 | T^3$. We require the posterior distribution of T_2, equivalent to $T_2 | (T_1, T^3)$ in $\mathscr{E}$, or alternatively to $T_2 | (T_1, T_2 \oplus T_3)$, since $T^3 = T_1 \oplus (T_2 \oplus T_3)$. Furthermore, since $T_1 \perp\!\!\!\perp (T_2, T_3)$, this is equivalent to $T_2 | T_2 \oplus T_3$ in $\mathscr{E}$, which is easily found and will have the same general form as both model and prior.

If the posterior distribution of the parameter T^2 is required, this is simply obtained as $T^2 = T_1 \oplus T_2$, T_1 being fixed by the data and T_2 distributed as above.

In passing we note that the marginal distribution of the data T_1, for any value T^3 of the hyperparameter, is given by $T_1 | T^3$ in $\mathscr{E}$, again of the same general form.

The above abstractions are best appreciated through some examples. Where helpful, we rename T_1, T^2, T^3 as T, Λ, Ω, respectively.

Example 5.1. In $\mathscr{E}$, let X_i denote the outcomes for n_i Bernoulli trials with success probability θ. Then T_i is the number of such successes, with $T_i \sim \mathscr{B}(n_i; \theta)$, and $T_1 \oplus T_2 = T_1 + T_2$, etc.

Our *model* is then obtained as the distribution for $T = T_1$ given $\Lambda = T_1 + T_2$, which is thus hypergeometric:

$$T \sim \mathscr{H}(n_1 + n_2; n_1, \lambda), \quad \text{given } \Lambda = \lambda.$$

Likewise the *prior* is taken as

$$\Lambda \sim \mathscr{H}(n_1+n_2+n_3; n_1+n_2, \omega), \quad \text{where } \Omega=\omega.$$

From the above analysis we deduce that, given data $T=t$, the posterior distribution of $\Lambda - t$ is $\mathscr{H}(n_2+n_3; n_2, \omega - t)$. The marginal distribution for T is $\mathscr{H}(n_1+n_2+n_3; n_1, \omega)$.

Example 5.2. We start now with $T_i \sim \sigma^2\chi^2_{\nu_i}$ under $\mathscr{E}$. This yields *model*:

$$T \sim \lambda.\beta(\tfrac{1}{2}\nu_1, \tfrac{1}{2}\nu_2), \quad \text{given } \Lambda=\lambda,$$

and *prior*:

$$\Lambda \sim \omega.\beta(\tfrac{1}{2}(\nu_1+\nu_2), \tfrac{1}{2}\nu_3).$$

We deduce that the posterior distribution for $\Lambda - t$ is $(\omega - t).\beta(\frac{1}{2}\nu_2, \frac{1}{2}\nu_3)$, while the marginal distribution of T is $\omega.\beta(\frac{1}{2}\nu_1, \frac{1}{2}(\nu_2+\nu_3))$.

6. Probability models and relative frequencies

Any infinitary subjective distribution or statistical model makes assertions about the behaviour of relative frequencies. If a sequence of zero–one variables $(X_1, X_2, \ldots)$ is considered to be exchangeable, for instance, and thus to behave like a realization of a set of Bernoulli trials, for some (random) probability value, then it follows that, with subjective probability one: (a) the relative frequency $n^{-1}\sum_{i=1}^n X_i$ tends to a limit as $n \to \infty$, and (b) this limit is undisturbed by suitable selection of a subsequence: for example, that which consists of all those trials for which the previous outcome is 1.

Such an assertion is testable in the real world when the (X_i) are observed (at least conceptually, given infinite time, or to a good approximation over a long finite time). If it turns out to be false, then this discredits the original subjective assessment: self-consistent it may be, but out of touch with reality.

In Dawid (1982b) it is shown how such a falsifiable assertion about relative frequencies is implicit in any coherent subjective distribution, no matter how subtly or flexibly it attempts to take account of all eventualities. Likewise, adoption of any family of distributions, considered as a statistical model, implies similar strong conclusions: for example, the Bernoulli model with completely unspecified probability parameter again implies properties (a) and (b) above.

We are thus led to consider any subjective model, or parametric model, of the behaviour of observables as a meaningful theory in the sense of Popper: we can conceive of experimental data that would discredit it. This is true equally of those models which can be constructed by assuming nothing more than suitable symmetries in our beliefs about the world. As in physics, an assumption of symmetry has a great deal of non-trivial predictive content, and may turn out, with hindsight, to be untenable.

The moral is that, while we can use the theory of coherence to construct and interpret statistical models, we are not exempt from the responsibility of checking that they bear at least some relationship with the real world.

To an objectivistic statistician, a model is "correct" if it gives the "true values" of the objective probabilities associated with the events it models. A whole-hearted subjectivist tends to reject the idea that "objective" probabilities exist: as we have seen, he can consider them as an artifact of his subjective view of the world, and then not uniquely determined. However, even if we take the view that a subjective distribution can be falsified only on the basis of its gross predictions (such as its implications for the behaviour of limiting relative frequencies), it can be shown to follow that all non-falsified subjective distributions must eventually make essentially identical probabilistic predictions (Dawid, 1982c). Likewise, all statistical models which are not falsified must eventually imply these same predictions (in most cases, with extensive background data there will be a sharp estimate of the parameter which can be assumed correct for prediction purposes, but this is inessential). So in a certain sense, there is a unique "true probabilistic prediction", i.e. an empirically valid limiting assignment of predictive distributions; but this does not imply a unique model, since there may be different ways of setting up a parametric model (with or without a prior distribution) which all eventually yield the same predictions (as, for instance, in example 2.1). We cannot distinguish between such models empirically, so to that extent we cannot hope to identify the "true model". But, for the same reason, this lack of identification doesn't matter: any of these unfalsified contenders will serve the purpose equally well. The selection of one rather than another may then be made purely on grounds of convenience, for example, because it incorporates independence in a useful way.

7. Conclusion

We have surveyed several ways in which a Bayesian might understand a statistical model. It might be no more than a technically convenient half-way house in calculating a predictive distribution. It might be an expression of intersubjective agreement, or a reflection of perceived underlying symmetries. In any case it need not be unique, but it is empirically testable and, if valid, will yield the only appropriate probabilistic predictions. All these attitudes differ, by various degrees, from traditional frequentist understandings of models. Nevertheless, I feel that the insights gained by such Bayesian analyses have useful lessons for non-Bayesian statisticians too. It would be good to see a serious non-Bayesian attack on the fundamental philosophical questions: "What do we mean by a model?"; "How should we construct models?"; and "How can we falsify a model?"

References

Dawid, A.P. (1977) "Invariant distributions and analysis of variance models," *Biometrika*, 64, 291–297.

Dawid, A.P. (1979) "Conditional independence in statistical theory" (with discussion), *J. Roy. Statist. Soc.*, B 41, 1–31.

Dawid, A.P. (1981) "Some matrix-variate distribution theory: Notational considerations and a Bayesian application", *Biometrika*, 68, 265–274.

Dawid, A.P. (1982a) "Intersubjective statistical models, in: G. Koch and F. Spizzichino, eds., *Exchangeability in Probability and Statistics* (North-Holland Publishing Company, Amsterdam), pp. 217–232.

Dawid, A.P. (1982b) "The well-calibrated Bayesian" (with discussion) *J. Amer. Statist. Assoc.*, 77, 605–613.

Dawid, A.P. (1982c) "Objective probability forecasts", Research Report 14, Department of Statistical Science, University College London.

Dynkin, E.B. (1978) "Sufficient statistics and extreme points", *Ann. Prob.*, 6, 705–730.

Ericson, W.A. (1969) "Subjective Bayesian models in sampling finite populations" (with discussion), *J. Roy. Statist. Soc.*, B 31, 195–233.

Geisser, S. (1982) "Aspects of the predictive and estimative approaches in the determination of probabilities" (with discussion), *Biometrics*, Supplement: Current Topics in Biostatistics and Epidemiology, 75–93.

Lauritzen, S.L. (1974) "Sufficiency, prediction and extreme models", *Scand. J. Statist.*, 1, 128–134.

Lauritzen, S.L. (1982) *Statistical Models as Extremal Families* (Aalborg University Press).

Chapter 26

PRINCIPLES OF INTERACTIVE FORECASTING

JAMES Q. SMITH and R.B. GATHERCOLE*

University College London

1. Introduction

This chapter introduces a new methodology of inference based on the authors' experiences in forecasting time series. Our main concern is to build up a sensible framework in which the statistician and practitioner can cooperate and combine their separate types of knowledge to produce good forecasts. Many of these have been generated through interactive use of the Dynamic Linear Model (Harrison and Stevens, 1976).

We will begin by stating certain postulates from which our methodology has been developed. First, we believe that a model is only a convenient approximation to a way of thinking about the dynamics generating a series. The statistician's art is to utilise the information he receives from consultation with those practitioners closely concerned with a process, to build a model that will encapsulate most of the important dynamics contained in this information. In most interesting problems there is no identifiable "true" model and even when there is one the costs of identifying it will usually make it preferable to use a convenient simpler model.

Our second postulate is that a time series model is good if, and only if, it produces precise enough forecasts of the required kind. We might use a model which has been constructed under completely false assumptions about certain aspects of the underlying dynamics. Under this postulate it might still be considered adequate if its forecasts were good despite this error. More importantly, it implies that we need to know what we are required to forecast because a model may be very good for one purpose but not so good for another.

*We wish to thank the SRC for funding this research.

Bayesian Inference and Decision Techniques, edited by P. Goel and A. Zellner

Hence the idea of a "best model" is related to the practitioner's objectives. This optimality will *not* be defined outside the context of the model.

The third postulate that we wish to make is that in business applications it is optimal to use a model which promises the greatest eventual pay-off either in money or utility on money. Usually the more complicated a model is the more costly it is to use and these model costs must be taken into account in our selection procedure. For example, if we wanted to implement an ARIMA model (Box and Jenkins, 1970) then costs would normally be incurred through the employment or experts to choose a model and computer time to obtain estimates. Even when the model was set up, we would incur the running costs. Often these expenses would not be justified by the extra forecasting accuracy obtained from using the ARIMA model. On simple practical grounds it is therefore important to include such costs in any selection procedure between models.

Thus our purpose is to produce a forecasting system whose associated expected forecast losses are as small as possible. The idea of building a model around the specification of the expectation of certain important variables rather than around conditional probability statements is not new, e.g. see Whittle (1970) and Goldstein (1981). However, it is unusual to use expected model costs under optimal action as the starting point of an inferential procedure. Appropriate prior probabilities are deduced from the prior expected costs when using each model. Probabilities are not themselves regarded as existing outside the framework of a well-specified model but are just considered as useful artifacts of statements about models.

2. The proposed construction of a forecasting model

Our strategy results very naturally from the postulates in the previous section. Initially, following our second postulate, for each model in contention the statistician must assess the practitioner's total expected forecast error if he were to use that model over a certain fixed time period. Usually this information would not be obtained directly. A team of statisticians and practitioners would jointly scrutinise similar series to the one in question and take account of any other pertinent information that might be available, in order to make their assessment. For example, last year using a certain type of model, losses incurred through misforecasting might have been C_F but this year it looks likely that the market will be more stable so that our expected forecast loss when using the same model will be less than C_F. We should also be able to fix a confidence interval for the figure arrived at. Another model in contention was not used last year but it is nonetheless possible to assess how such a model would have performed had it been used. Obviously in practice one would not just consider last year's figures but also any other data available to obtain a consensus on the model's performance. This would be done in a fairly ad hoc way before any detailed and costly technical analysis had taken place.

Each model under consideration will also have implementation and running cost over a pertinent time interval. To follow our third postulate we choose the model M^* for which the sum of expected forecast and running costs is a minimum. This model is then set up in a formal Bayesian way so that the prior probability distributions of the parameters agree with the initial consensus of statements about the forecast losses when model M^* is used. An illustration of how this is done is given in section 5. Model M^* is then implemented and forecasts are obtained which are optimal conditional on M^* being correct. As data arrives so our chosen model forecast losses will be formally updated. We can also informally update expected forecast losses associated with other models (see section 7). If at any time our chosen model M^* promises less pay-off than the implementation of an alternative, then we switch models.

Thus, let $m(T)$ be the number of models considered and compared by our committee at time T, and label these models $M_1, \ldots, M_{m(T)}$. Let $E_T(M_j)$ represent the expected forecast loss at time T when using model M_j, $1 \leq j \leq m(T)$. Let $C_j^{(1)}$ represent the expected initial implementation cost of model M_j, $C_j^{(2)}(T)$ represent the expected running costs over the total remaining forecast period for M_j at time T, and write

$$C_j(T) = C_j^{(1)} + C_j^{(2)}(T).$$

In this notation, using the strategy above, we initially choose a model $M^* = M_{j^*}$ such that

$$E_0(M^*) + C_{j^*}(0) \leq E_0(M_j) + C_j(0), \quad \text{for any } 1 \leq j \leq m(0).$$

Our proposed strategy is now to continue to run M^* until the first occasion T that

$$E_T(M^*) + C_{j^*}^{(2)}(T) > E_T(M_{k^*}) + C_{k^*}(T), \quad \text{for some } 1 \leq k^* \leq m(T), \tag{2.1}$$

When this happens we implement model M_{k^*}, denote it by $M_{k^*} = M^*$, and proceed with our new model until (2.1) is satisfied when we switch models again. The sequential strategy is depicted in the flow chart given in fig. 26.1.

If the running costs of any model are negligible, then it can be seen that there is a natural inertia built into the model selection since we will certainly not change models whilst

$$E_T(M^*) \leq \min_{1 < j = j^* < m(T)} C_j^{(1)}. \tag{2.2}$$

Thus, regardless of how well we think a different model will forecast, we still retain our original model M^* is the costs of implementation of any other model does not compensate any increased effectiveness. Typically $E_T(M^*)$ will be a function of the sufficient statistics for M^* so that eq. (2.2) generates a type of hypothesis test whose size depends on model costs. In sections 4 and 6 we will investigate how our model behaves for certain combinations of models and loss functions.

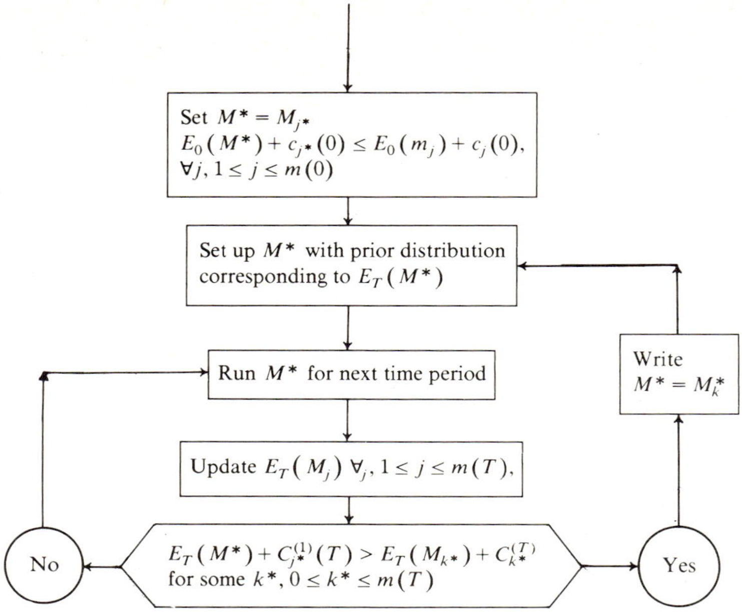

Fig. 26.1. Flow chart of sequential procedure.

3. Some of the consequences of using our strategy

3.1. The introduction of new models

One problem of using a Bayesian technique in sequential methods is that, as the data arrive, a model not considered initially, and hence allocated a prior probability of zero, suddenly becomes a possibility in your own mind. To adjust the prior to allow the consideration of this new model, we assume the subject is "incoherent" to the Bayesian if his probability space is fixed over time. He is, therefore, either forced to (a) admit that his prior distribution is just an approximation to his "true" prior. e.g. see Dickey in his discussion of Effron and Morris (1973); (b) refuse to consider this new possibility; or (c) allow his probability space to change over time. All these solutions seem to us to be rather inadequate. The first is suspicious because it makes nonsense of statements about certain techniques being "incoherent" since any deviation from "coherence" can be explained away by admitting this rather ill-defined concept of an approximation to a "true" prior that can never be specified. The second leads to bad forecasts. The third, whilst being logically sound, is not sufficiently specific to formalise into a practical procedure.

With our technique this problem with prior probabilities does not arise. If a new model suggests itself to us we just assess the expected losses and costs associated with using it. If total costs are less than those of our chosen model M^* we use it, otherwise we do not. The reason that we have no problems is that at no time have we made any statement about the *probability* of a certain model being "correct". In fact we have no interest in such probabilities. We are only interested in the prospective *performances* of the particular models under consideration. This is the essential advantage of considering prior probabilities as artifacts of a model rather than the key links between models and the real world. Events and their associated probabilities need spaces in which their respective meaning and enumeration do not depend on the model being used. However, events may have no meaning outside the model articulating them, and probability evaluations may be no longer valid if the model is possibly false, especially if we accept that no model is a complete description of reality.

3.2. The role of the statistician

There are in our opinion three levels at which modelling can take place. At the first level, an attempt is made to model a process *exactly* by isolating the precise mechanism generating our data. It is not clear that such a mechanism exists or if it does that it can be articulated. However, it is to this end that many Bayesians apply themselves. Even if this activity is justifiable such modelling needs the inspiration of a saint to discover the "correct" class of models and the tenacity of a limpet to surmount the often intractable difficulties of implementing the procedure.

At the second level of modelling, we try to find a useful mathematical analogy with the assumed underlying mechanism generating the series we observe. To do this it is essential to take account of the *context* of the model, the knowledge of people who have experienced the dynamics of the series, and be able to incorporate any theories about the system in question. A model built with these considerations in mind is in general far more robust than models picked by rules of convention.

The third level of modelling is the quantitative one where technical expertise is used to estimate and find "optimal" predictors for a given model. This is a role which most statisticians can play comfortably. However, used without the second level of modelling it is very often unrobust to changing environment since it allows us to consider only the behaviour of mathematical series isolated from its context. It follows that when this method of forecasting is used on its own it tends to have only moderate practical success.

In our proposed technique we have tried to build a system in which we can integrate these lower two levels of modelling so that the practitioner's ideas are reflected in the final choice of a model and forecast. We have relinquished the highest level of evangelical modelling to those more likely to receive divine revelation than ourselves.

4. Some examples using simple loss functions

Example 4.1. Here we consider a loss function

$$L(y, \hat{y}) = \sum_{i=1}^{n} (y_i - \hat{y}_i(M))^2,$$

where $\hat{y}_i(M)$ is the mean of y_i given $y_1, \ldots, y_{i-1}$ and model M, and n is a fixed point in the future when, for example, the forecaster will be assessed. Of course, with $n = 1$ the problem reduces to the usual estimation problem. For the purposes of this example, the running cost of all models are assumed to be negligible.

Let us now assume that $(y_i - \hat{y}_i(M_j))^2/[V(M_j)]$, $1 \le i \le n$, are independent χ_1^2 variables for all our models $M_1, \ldots, M_m$ and where $V(M_j)$ is the forecast error variance of our observations when using model M_j. If the observations $y_1, y_2, \ldots$ are jointly normally distributed and generated from a pth difference stationary model where all the parameters were known except the forecast variance $V(M_j)$ for each of the models M_j, this assumption is automatically satisfied (see Astrom, 1970, p. 231). Thus, in particular, this assumption is satisfied for most constant normal dynamic linear models (Harrison and Stevens, 1976). For the sake of illustration suppose a priori that $V(M_j)$ has the conjugate inverted gamma distribution. This choice, which is part of our model specification on M_j, makes the subsequent analysis very simple and is not a crucial factor in the ensuing arguments. In particular assume that $V(M^*)$ has initial density $f(v)$, where $f(v)$ is given by

$$f(v) = \frac{\beta_0^{\alpha_0}}{\Gamma(\alpha_0)} \exp\{-\beta_0 v^{-1}\} v^{-(\alpha_0+1)}, \quad v > 0.$$

At any time period it is easy to check (see DeGroot, 1970) that the expected loss $E_T(M^*)$ associated with the model M^* which is running initially is given by

$$E_T(M^*) = \sum_{i=1}^{T} (y_i - \hat{y}_i(M^*))^2 + (n-T)\mathbb{E}(V(M^*)|y^T),$$

where

$$\mathbb{E}(V(M^*)|y^T) = \rho_T(M^*)V_0(M^*) + (1 - \rho_T(M^*))S_T^2(M^*)$$

and where $y^T = y_1, y_2, \ldots, y_T$,

$$0 \le \rho_T(M^*) = \frac{\alpha_0 - 1}{\alpha_0 + \frac{1}{2}T - 1} \le 1,$$

$$S_T^2(M^*) = T^{-1} \sum_{i=1}^{T} (y_i - \hat{y}_i(M^*))^2,$$

and $V_0(M^*) = \beta_0(\alpha_0 - 1)^{-1}$ is the mean of V using model M^* before we take any observations. $V_0(M^*)$ will have been chosen to correspond to the initial prior expected forecast loss $E_0(M^*)$, ascertained in the way described in section 2.

If at time T we decided to switch to model M_j we would incur a total loss of

$$\sum_{i=1}^{T} (y_i - \hat{y}_i(M^*))^2 + (n-T)\mathbb{E}\big(V(M_j)|y^T\big) + C_j^{(1)},$$

where $C_j^{(1)}$ is the expected cost of implementing model M_j over the next $n-T$ time periods. Of course, we do not know what $\mathbb{E}(V(M_j)|y^T)$ is when $M_j \neq M^*$ because we have not implemented M_j. Let us suppose for the sake of argument that $\mathbb{E}(V(M_j)|y^T) = \mathbb{E}(V(M_j)) = n^{-1}E_0(M_j)$ and that our observations to date have not influenced us as to the potential performance of this model were it to be implemented. Then, we switch from model M^* to model M_j if and only if

$$(n-T)\mathbb{E}\big(V(M^*)|y^t\big) \geq \left(\frac{n-T}{n}\right)E_0(M_j) + C_j^{(1)}.$$

In particular as α_0 tends to 1, so that our prior variance on forecast loss tends to infinity, we switch from M^* to M_j only if

$$S_T^{(2)}(M^*) \geq \frac{E_0(M_j)}{n} + \frac{C_j^{(1)}}{n-T}.$$

The way that the rejection of M^* is linked to the sample variance $S_t^2(M^*)$ when the cost of implementation $C_j^{(1)}$ of the alternative models M_j is constant and equal, is illustrated in fig. 26.2. Notice that as time passes we become less inclined to drop M^* because the implementation costs for any new model will dominate the future loss arising from forecasting errors.

The next example illustrates what happens when the pertinent expected loss function does not depend only on the sufficient statistic S_T^2.

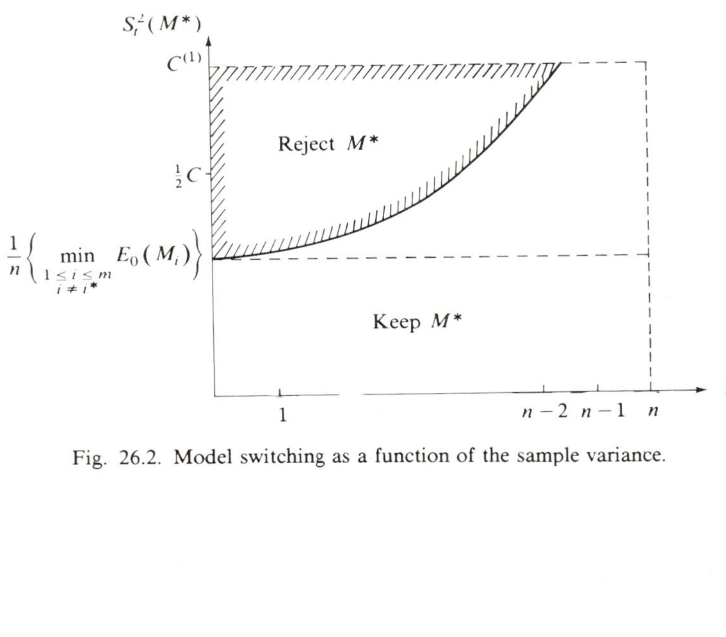

Fig. 26.2. Model switching as a function of the sample variance.

Example 4.2. Let

$$L(y, \hat{y}(M)) = \sum_{i=1}^{n} |y_i - \hat{y}_i(M)|,$$

where $\hat{y}_i(M)$ is the mean of y_i given $y_1, \ldots, y_{i-1}$ and model M. We shall again assume that

$$\frac{(Y_i - \hat{y}_i(M))^2}{V(M)} \sim \chi_1^2, \quad i = 1, 2, \ldots, n,$$

and are independent. Then the expected loss $E_T(M^*)$ at time T when using model M^* is given by

$$E_T = TR_T(M^*) + (n - T)\mathbb{E}\left(V^{1/2}(M^*)|y^T\right),$$

where $R_T = T^{-1}\sum_{i=1}^{T}|y_i - \hat{y}_i(M^*)|$, $y^T = y_1, y_2, \ldots, y_T$, and, given y^T, $V(M^*)$ has an inverted gamma distribution, $\mathrm{IG}(\alpha_1, \beta_1)$, where

$$\alpha_1 = \alpha_0 + \tfrac{1}{2}T; \qquad \beta_1 = \beta_0 + \tfrac{1}{2}TS_T^2,$$

and where α_0 and β_0 are the parameters in the prior inverted gamma distribution of $V(M^*)$, and S_T^2 is the sample "forecast" variance $T^{-1}\sum_{i=1}^{T}(y_i - \hat{y}_i(M^*))^2$ at time t. Therefore, it follows that

$$\mathbb{E}\left(V^{1/2}(M^*)|y^T\right) = \frac{\beta_1^{1/2}\Gamma\left(\alpha_1 - \frac{1}{2}\right)}{\Gamma(\alpha_1)}.$$

If we let $\alpha_0 \downarrow 1$ as before, and hold the prior expected loss associated with our original model constant, we can show that

$$\mathbb{E}\left(V^{1/2}(M^*)|y^T\right) = \tau^*(T)\left(S_T^2(M^*)\right)^{1/2},$$

where

$$\tau^*(T) = \frac{\Gamma\{\frac{1}{2}(T+1)\}\left(\frac{1}{2}T\right)^{1/2}}{\Gamma\{\frac{1}{2}(T+2)\}}.$$

In fact, Stirling's approximation implies that $\tau^*(T) \to 1$ as $T \to \infty$. Now using the same conditions on the other models similar to those in example 4.1, we change from model M^* to model M_j only if

$$\tau^*(T)\left(S_T^2\right)^{1/2} \geq \frac{E_0(M_j)}{n} + \frac{C_j^{(1)}}{n-T}.$$

Of course, usually our information about the variance of the forecast loss will not be infinite and this will adjust this condition. Notice that this example emphasises the importance of the original normality assumptions of the observations since the condition is based on S_T^2 and not R_T defined above.

In the examples 4.1 and 4.2 we have considered time series which, at least in their stochastic component, had reached their equilibrium. Often we cannot assume this. For many of the Harrison and Stevens (1976) models, however, we often find that, using a prior distribution on our level parameters,

$$Z_t = \frac{(Y_t - \hat{y}_t(M))^2}{B_t(M)} \sim \chi_1^2, \quad t = 1, 2, \ldots,$$

where the forecast variance B_t decreases to a limit B as $t \to \infty$. That is

$$Z_t = \frac{\sigma_t(M)(Y_t - \hat{y}_t(M))^2}{V(M)} \sim \chi_1^2, \quad t = 1, 2, \ldots,$$

where $0 < \sigma_t(M) < \sigma(M)$ and $\sigma_t(M)$ increases to $\sigma(M)$ as $t \to \infty$, and V is the variance of our observations Y_t given our level parameters $\boldsymbol{\theta}_t$ at that time. By putting the prior distribution on V induced by our ascertained prior distribution of loss, as in example 4.1, we can now calculate how to decide whether to keep a model or not in the light of incoming data.

Example 4.3. Let M^* imply that $Y_1, Y_2, \ldots$ came from a steady model given by

$$Y_t | \theta_t \sim n(\theta_t, V),$$
$$\theta_t | \theta_{t-1} \sim n(\theta_{t-1}, W).$$

It is then easy to show that the one step ahead estimate $\hat{y}_t$ of Y_t is given by

$$\hat{y}_t = m_{t-1} = m_{t-2} + A_t(y_{t-1} - m_{t-2}),$$

where

$$A_t = 1 - (1 + A_{t-1} + s)^{-1},$$
$$s = W/V,$$

and

$$\mathrm{Var}(\hat{Y}_t | y_1, \ldots, Y_{t-1}) = B_t = V(A_{t-1} + 1 + s).$$

Thus, if a priori V has an inverted gamma prior $\mathrm{IG}(\alpha_0, \beta_0)$ under M^*, then posterior to observing the last T readings $y_1, \ldots, y_T$, V has an inverted gamma posterior distribution $\mathrm{IG}(\alpha_T, \beta_T)$, where

$$\alpha_T = \alpha_0 + \tfrac{1}{2}T$$

and

$$\beta_T = \beta_0 + \tfrac{1}{2}TS_T'^2(M^*),$$

and where

$$S_T'^2(M^*) = T^{-1} \sum_{i=1}^{T} \sigma_i(y_i - \hat{y}_i(M^*))^2$$

and

$$\sigma_i(M^*) = (A_{i-1} + 1 + s)^{-1}, \quad 1 \le i \le n.$$

Thus $\mathbb{E}(V|y^T) = \rho_T V_0 + (1-\rho_T)S_T'^2$, where ρ_T is defined in example 4.1 and V_0 is the prior mean of V. Thus,

$$E_T(M^*) = \sum_{i=1}^{T} (y_i - \hat{y}_i(M^*))^2 + (n-T)\mathbb{E}\left\{ \sum_{i=T+1}^{n} (Y_i - \hat{y}_i(M^*))^2 | y^T \right\}$$

$$= \sum_{i=1}^{T} (y_i - \hat{y}_i)^2 + (n-T)\tau_T(\boldsymbol{\sigma})\{\rho_T V_0 + (1-\rho_T)S_T'^2(M^*)\},$$

where $\tau_T(\boldsymbol{\sigma}) = [\sigma_{T+1}^{-1} + \sigma_{T+2}^{-1} + \cdots + \sigma_n^{-1}]$, and σ_i is defined above.

In this example we have to be a little more careful in comparing M^* to M_j because we might easily expect initial losses associated with M_j to be larger than later ones for exactly the same reason as for M^*. However, since M^* and hence $\tau_T(\boldsymbol{\sigma}(M^*))$ is simple to calculate, so might be $\tau_T(\boldsymbol{\sigma}(M_j))$. If this is not the case, then we have to allow for this difficulty in some ad hoc way. Under a vague prior this analysis implies that M^* will be rejected in favour of M_j only if

$$\tau_T(\boldsymbol{\sigma}(M^*))S_T^2(M^*) \ge \tau_T(\boldsymbol{\sigma}_T(M_j))\frac{E_0(M_j)}{n} + \frac{C_j^{(1)}}{n-T}.$$

5. A simple advertising model

Here we illustrate our technique with a simplified practical case study. A firm had just spent a considerable amount of money on launching a series of new gimmicky pieces of confectionery marketed for children. Each product was to be withdrawn after six months and the firm has already set up a new product to launch in its place at that time. We show how dynamic linear models of these series are constructed and how our forecasting technique was implemented.

Let $\theta_t^{(1)}$ and $\theta_t^{(2)}$ represent the level of sales and growth in sales, respectively, at time t and write $\boldsymbol{\theta}_t = (\theta_t^{(1)}, \theta_t^{(2)})$. Then if Y_t represents the actual demand at time t we can write the process as

$$Y_t|\boldsymbol{\theta}_t \sim n(\theta_t^{(1)}, V),$$

$$\boldsymbol{\theta}_t|\boldsymbol{\theta}_{t-1} \sim n(G\boldsymbol{\theta}_{t-1}, W),$$

where v is a constant scalar, W is a covariance matrix of the form

$$W = \begin{bmatrix} w_1 & \lambda_2 w_2 \\ \lambda_2 w_2 & w_2 \end{bmatrix}, \quad w_1 > w_2,$$

and the 2×2 system matrix G is of the form

$$G = \begin{bmatrix} \lambda_1 & 1 \\ 0 & \lambda_2 \end{bmatrix}, \quad 0 \leq \lambda_1, \lambda_2 \leq 1.$$

The variance V reflects the amount of non-systematic variation in our observations y_t not modelled by our system equation, and variances w_1 and w_2 reflect the systematic variation in the level of sales and growth of sales, respectively.

λ_1 can be interpreted as the proportion of customers we retain from one week to the next and λ_2 the weekly rate of decrease of new customers enticed to try our product by the advertising campaign.

We shall assume that the objective of the analysis was to forecast the weekly sales of the product as precisely as possible. For the sake of simplicity, we will assume that the product had cost twice as much to store as the profit obtained from its sale. It is then usually appropriate to minimise the expected loss function $L(y, \hat{y})$, which is of the form

$$L(y, \hat{y}) = \sum_{i=1}^{n} \tau_a(y_i, \hat{y}_i), \tag{5.1}$$

where

$$\tau_a(y, \hat{y}) = \begin{cases} a(y - \hat{y}), & \hat{y} - y < 0, \\ 2a(\hat{y} - y), & \hat{y} - y \geq 0, \end{cases}$$

and where $a > 0$.

In previous years similar products have had values of λ_1 and λ_2 as 0.8 and 0.9, respectively. The model variances, V, w_1, and w_2, for this type of product have been related in the past by

$$\frac{V}{w_1} = \frac{w_1}{w_2} = 10, \tag{5.2}$$

and such models have worked adequately. Since all customers at the start of the process would be new customers, $\mathbb{E}(\theta_1^{(1)}) = 0$ and the initial covariance matrix was $S_1 V$, where $S_1 = \{S_1^{(i,j)}\}$ and $S_1^{(i,j)} = S_1^{(1,1)}$, $1 \leq i, \ j \leq 2$. Again from previous experience, we expected the initial number of new customers $\mathbb{E}(\theta_1^{(1)}) = 900$. We were initially reasonably certain about this and chose $S_1^{(1,1)} = 2$.

We shall call the dynamic linear model described above M_1. Again for the sake of simplicity assume there was a single alternative model M_2. This model would estimate all the parameters whose values are fixed in M_1, using information obtained from an expensive but quick detailed survey. The cost C_2 of setting up M_2 was large and at the outset the practitioner did not consider it worth the cost. From discussions with the practitioner, it was found that total forecasting losses over the 25-week period, $E_0(M_1)$ and $E_1(M_2)$, for M_1 and M_2 were given by

$$E_0(M_1) = 13{,}000a,$$
$$E_0(M_2) = 6{,}000a,$$
$$C_1 = 4{,}000a, \tag{5.3}$$

and

$$C_2 = 15{,}000a.$$

Thus, he was quite happy to work with M_1 initially. Confidence intervals were also available for the first two of these estimates.

Our first task is to use eq. (5.3) to fix the observation variance V of model $M_1 = M^*$. The one step ahead forecast errors $B_t = \sigma_t V$ at time t satisfy the following recurrence relationships (see Harrison and Stevens, 1976):

$$\sigma_t = S_t^{(1,1)} + 1,$$
$$S_t = G(S_{t-1} - S_{t-1}^*)G' + W^*, \tag{5.4}$$
$$S_t^* = \sigma_t^{-1}\begin{pmatrix} S_t^{(1,1)} \\ S_t^{(1,2)} \end{pmatrix}(S_t^{(1,1)}, S_t^{(1,2)}),$$

where $W^* = V^{-1}W$ and $S_t = \{S_t^{(i,j)}\}$, $1 \le i,\ j \le 2$.

It is easily checked that the expected forecast loss $E_0(M^*)$ is given by

$$E_0(M^*) = ak\left[\sum_{i=1}^{25} \sigma_t^{1/2}\right]\mathbb{E}(V^{1/2}) = ak\left[\sum_{i=1}^{25} \sigma_t^{1/2}\right]\frac{\beta_0^{1/2}\Gamma(\alpha_0 - \frac{1}{2})}{\Gamma(\alpha_0)}, \tag{5.5}$$

where $k = 1.086$, σ_t is defined above, and (α_0, β_0) are the prior parameters in his prior inverted gamma distribution given in example 4.1. In a similar way the variance of the forecast loss can be related to α_0 and β_0. Solving for (α_0, β_0) numerically over α_0 using (5.5) it was found that

$$(\alpha_0, \beta_0) = \left\{2.0, \left[\left(\frac{2}{\sqrt{\pi}}\right)\left(k\sum_{t=1}^{25} \sigma_t^{1/2}\right)^{-1}\left(\frac{E_0(M^*)}{a}\right)\right]^2\right\}.$$

Once the dynamic linear model M_1 was fully specified the joint distribution of observables $\{Y_t\}$ could be stated. Various implications from M_1 about the consequent forecast distribution were therefore checked with the practitioner, as outlined in section 2. He found these consequences believable. We therefore started to run $M_1 = M^*$ and did not consider implementing M_2 whilst

$$E_T^{\Delta}(M^*) = ak\left(\sum_{t=T+1}^{25} \sigma_t^{1/2}\right)\beta_T^{1/2}\frac{\Gamma(\alpha_T - \frac{1}{2})}{\Gamma(\alpha_T)} \le C_2, \tag{5.6}$$

where α_T and β_T are given in example 4.2. If the condition above were to be violated, then we would have to reassess the situation. In table 26.1 we give some suitably disguised sales figures of such a confectionery product when consumer

Table 26.1. Forecasts of some advertising data and the running expected forecast loss under an asymmetric loss function.

Week no.	Demand	Mean of $Y_T \mid y_1, y_2, \ldots, y_{T-1}$	Expected forecast loss $E_T^{\Delta}(M_1)$
0	–	–	13 000
1	812	900	10 495
2	1872	1425	10 813
3	2814	2218	11 787
4	3015	2904	10 387
5	4827	3152	18 230
6	6779	3910	29 017
7	5123	4916	26 038
8	6181	4885	24 965
9	8020	5111	28 549
10	8888	5740	30 879
11	7721	6364	28 518
12	5934	6413	25 639
13	3214	5902	25 195
14	4872	4753	22 390
15	4561	4418	19 769
16	4017	4106	17 307
17	3542	3744	14 992
18	2030	3369	13 068
19	2171	2695	10 972
20	1421	2283	9 014
21	1061	1805	7 100
22	1003	1391	5 224
23	876	1109	3 416
24	631	898	1 677
25	1074	700	0

reaction was unusual. It was clear by week 6 that the model was behaving unexpectedly poorly. By comparing $E_6^{\Delta}(M_1)$ with $E_0(M_2)$ and C_2 we decided to replace M_1 by M_2 on the basis of these 6 data points before continuing to forecast the product. It was found that $\lambda_1 = 0.8$ was far too small a value for this particular problem and we were retaining a higher proportion of old customers than usual.

Notice that in this simple example neither M_1 nor M_2 could be considered an exact description of the process generating the data. For example, both M_1 and M_2 assign positive probabilities to negative sales. It is also very probable that the reasoning behind the choice of the form of the matrix G for both M_1 and M_2 could be bettered.

However, our approach is a pragmatic one. We are interested in comparing the available alternatives M_1 and M_2 that we can afford to implement. We chose the models we expected to cost the firm least which were what they required. No scotchtape and string were needed to achieve this since the technique was built for this purpose. In addition, we were not encumbered with the sort of philosophical problems that would arise if, as in the Bayesian approach, we had postulated M_1 and M_2 as possible "correct" models. Linking the model selection problem to expected forecasts in this way gives a sound rationale for combining sequential estimation and sequential "hypothesis testing" in forecasting problems.

6. Behaviour under the action of other loss functions

6.1. Other functions of the sum of losses after a fixed time n

In the examples of the previous section we considered loss functions of the form $L=\sum_{i=1}^{n} r(y_i, \hat{y}_i(M))$. These loss functions are often unrealistic. For instance, suppose for example that last year a practitioner forecasted sales of a product at a cost A in forecast errors and implementation costs. This was considered quite adequate by his employers. This year he will be quite content to not lose more than A units. His *disutility* on loss $U(l)$ thus satisfies

$$U(l)=0, \quad \text{if } l \leq A,$$

and be increasing in l elsewhere. Let us consider the particularly simple case when

$$U_1(l)=\begin{cases} 0 & l \leq A, \\ H & l > A, \quad \text{where } H>0. \end{cases} \tag{6.1}$$

He should now minimise his expected *disutility* on loss given model M at time T with regard to the various costs of changing models. Denote his initial choice of model by M^*. Then his expected disutility on running M^* at time $T-1$, $\mathbb{E}_{T-1}(U_1|M^*)$ is given by

$$\begin{aligned} \mathbb{E}_{T-1}(U_1|M^*) &= H \operatorname{Prob}(L \geq A|M^*, y_1,\dots, y_{T-1}) \\ &= H \operatorname{Prob}\left(\sum_{i=T}^{n} r(Y_i-\hat{y}_i(M^*))^2 \geq A^*|y_1,\dots, y_{T-1}\right), \end{aligned} \tag{6.2}$$

where $A^*=A-\sum_{i=1}^{T-1}(y_i-\hat{y}_i(M^*))^2$ and r is some increasing function. Similarly, at time T:

$$\mathbb{E}_{T-1}(U_1|M_j)=H \operatorname{Prob}\left(\sum_{i=T}^{n} r(y_i-\hat{y}_i(M_j))^2 \geq A^*(y_1,\dots, y_{T-1})+C_j^{(1)}\right). \tag{6.3}$$

If the practitioner is forecasting many such products, then typically $C_j^{(1)}$, $j=1,\dots, m$, will be large compared to $E_{T-1}(U_1|M^*)$, at least initially. Note that the forecaster does not even have to consider the performance of any model other than M^* unless

$$\mathbb{E}_{T-1}(U_1|M^*) \geq C, \quad \text{where } C=\min_{\substack{1\leq j\leq m \\ j\neq j^*}} (C_j), \tag{6.4}$$

i.e. until $\operatorname{Prob}(L \geq A|M^*, y_1,\dots, y_{T-1}) \geq C/H$. This, of course, corresponds to a hypothesis test on L with M^* as the null with the size fixed provided costs and utilities remain constant over time. If inequality (6.4) is satisfied, we must assess $\mathbb{E}(U|M_j)$, $1 \le j \le m(T)$, in some informal way and decide whether or not to switch from M^*.

The interest in the inequality (6.4) is not solely that it generates a fixed size confidence test. If we consider eqs. (5.2) we see that at time T, $\mathbb{E}_T(U_1|M^*)$ is a function of:

(a) the sufficient statistics through

$$\operatorname{Prob}\left(\sum_{i=T+1}^{n} r\left(y_i - \hat{y}_i(M^*)\right)^2 \ge A^* | y_1, \dots, y_T \right);$$

(b) the observed losses up to time T through A^*.

Thus, for example, our new strategy used on the losses given in example 4.2 would not only depend on the sample variance S_T^2 up to time T but also the corresponding average *absolute* loss up to that time.

Obviously the choice of U given by eq. (6.1) will not be appropriate in many cases and usually a utility of the form

$$U_2(l) = \begin{cases} 0, & l \le A_1, \\ H\dfrac{(l - A_1)}{A_2 - A_1}, & A_1 < l \le A_2, \\ H, & A_2 < l, \end{cases}$$

would be more realistic. However, note that we would need eq. (6.4) to hold with $A = A_1$ even if we used this utility U_2 instead of U_1.

6.2. Discussion of other loss functions

Problems arise if the period over which we are assessed is not fixed. Clearly, if running costs are negligible and we allow the process to continue to run indefinitely with equal weight on all forecast errors, then forecasting errors will dominate any implementation costs. Therefore, the inertia built into our model selection procedure disappears and we will be forever adjusting our models. In fact this problem is transcended if future forecasting errors are given more weight than distant past and future errors. Our real losses will often be on the form

$$L_t(y_i, \hat{y}_i) = \sum_{i=0}^{\infty} \gamma_i(t) r(y_{t-i} - \hat{y}_{t-i}(M)) + \sum_{j=1}^{\infty} \delta_j(t) r(y_{t+j} - \hat{y}_{t+j}(M)),$$

where

$$\gamma_i(t) \to 0, \quad \text{as } i \to \infty,$$
$$\delta_j(t) \to 0, \quad \text{as } j \to \infty.$$

The loss functions we have already considered are ones where

$$\gamma_i(t) = \begin{cases} 1, & i \le t, \\ 0, & \text{otherwise,} \end{cases}$$

and

$$\delta_j(t) = \begin{cases} 1, & j \le n-t, \\ 0, & \text{otherwise.} \end{cases}$$

Another convenient loss function of this type which has been used for analogous problems is exponentially weighted in both and given by

$$L_t(y_i - \hat{y}_i) = \gamma^* \sum_{i=0}^{\infty} \gamma^i r(y_{t-i} - \hat{y}_{t-i}(M)) + \delta^* \sum_{j=0}^{\infty} \delta^j r(y_{t-j+1} - \hat{y}_{t-j+1}(M)),$$

where γ^* and δ^* are positive constants and $0 < \gamma$, $\delta < 1$. If we have linear utilities on our loss this will reduce to selecting a model under the loss

$$L_t(y_i - \hat{y}_i) = \delta^* \sum_{j=0}^{\infty} \delta^j r(y_{t-j+1} - \hat{y}_{t-j+1}(M)).$$

Returning to example 4.1 we see that with this new loss and an infinite variance on our prior expected loss, we switch from M^* to M_j at time $t = T$ only if

$$S_T^2(M^*) \frac{E(M_j | y^T) + C_j^{(1)}}{\delta^*(1-\delta)^{-1}}.$$

This is of the same basic hypothesis testing form as before. However, this result depends on the linear utility assumption which is dubious for most problems.

7. Some links with other techniques

7.1. The exogenous model change and mixture models

In 1976, Harrison and Stevens briefly outlined how a practitioner could interact with a predictive model to accommodate outside information that might be pertinent to their forecasts. They do this by introducing a system equation to describe the variation over time rather than specifying their model on the observation space. A simple such model is the steady model defined by

$$Y_t | \theta_t \sim n(\theta_t, V), \tag{7.1}$$

$$\theta_t | \theta_{t-1} \sim n(\theta_{t-1}, W), \tag{7.2}$$

as discussed in example 4.3. Let us suppose that Y_t represents the monthly sales of some toothpaste and that this month, T, there is considerable publicity about the dangers of an ingredient contained in this toothpaste. If before the information,

$$\theta_T | y_1, \ldots, y_{T-1} \sim n(m_T, C_T + W), \tag{7.3}$$

Harrison and Stevens now adjust their parameters m_T and C_T by making m_T smaller, thereby representing what they expect to sell, and C_T larger, thereby representing their increased uncertainty. They can now use eqs. (7.1) and (7.2) to produce their forecasts as before.

The methodology that we have investigated in this chapter is quite in accordance with this technique. At time $t = T$, the expected loss R^* of using this model M^* given by eqs. (7.3), (7.1) and (7.2) suddenly increases, whilst the expected losses associated with models with smaller means decrease. If implementation costs associated with changing models are very large compared with the expected losses using our original model, we hold on to M^*. If, on the other hand, we can implement a new model M_j cheaply, then we do this. In an interactive system this latter change would usually be implemented, but for marginal cases where interaction with our computer system was expensive, we can use the technique developed in previous sections to calculate whether or not to discard M^*. Thus, if we used the loss function $L_n = \sum_{i=1}^{n}(y_i - \hat{y}_i)^2$, we would look at the expected loss associated with L_n and our new distribution of $\theta_T | y_1, \ldots, y_{T-1}$.

Another model of particular interest developed by Harrison and Stevens is the multiprocess model. Such a model is a mixture of sub-models each given probability weights. The number of sub-models at each time stage is quadrupled to take into account various types of changes in underlying structure that might occur at that time stage. Obviously this would not be a practical model if the number of sub-models at each time stage were not pruned down in some way, for otherwise the sub-models would increase exponentially in number and running costs would be astonomical. They surmount this problem by collapsing 16 sub-models into 4 at each time point in a pre-assigned way. In Gathercole and Smith (1983) we show how a similar collapsing procedure can be implemented under criteria analogous to those in this chapter.

7.2. Internal information of bad performance

We have already given three examples in section 4 of how to deal with a model which is not performing as well as expected. We are essentially doing the same job as Box (1980) here. However, we do not test against undefined alternatives as Box does, but instead link our actions to the running costs of the analysis. The two techniques are different and should probably be employed in different circumstances. Ours should be preferred when we are forecasting sequentially, rather than performing a statistical analysis with a fixed data set. Nevertheless, it is interesting to note that the two techniques are similar in that they use certain functions of the observations to check whether they have reached a threshold that allows "rejection".

7.3. Akaike–Schwarz model selection

If likelihood methods are used to estimate the coefficients of a normal autoregressive process of order r, then it is found that the estimated model tends to give large forecast errors if r is too large. Techniques have not been devised to estimate coefficients that favour those autoregressive models with a small order. Thus, we choose a model M^* that minimises the quantity $k(y)$, where

$$k(y)=\sum_{i=1}^{l}(y_i-\hat{y}_i(M))^2+r(M),$$

and where $\hat{y}_i(M)$ is the mean of $Y_i|y_1,\ldots,y_{i-1}$ using model M, $r(M)$ is the number of parameters in M, and l is the number of past observations available in the series. Akaike (1973) advocates a choice of $\lambda=2$ and Schwarz (1978) advocates putting $\lambda=\log l$. Each of these techniques can be accommodated into our framework, and correspond to the stage in our procedure when we first choose a model with which to predict. The addition of the term $\lambda(M)$ just reflects our prior opinions about the performance of each of these models. Thus, for example, using a loss function of the form given in example 4.1 for the next n observations it is sufficient to choose a model which minimises

$$E(V|y^l)=(1-\rho_l(M))S_l^2+\rho_l(M)V_0(M), \tag{7.3a}$$

where

$$\rho_l(M)=\frac{2\alpha_0(M)-2}{2\alpha_0(M)-2+l},\qquad V_0(M)=\frac{\beta_0(M)}{\alpha_0(M)-1},$$

and $\alpha_0(M)$, $\beta_0(M)$ are the parameters for model M in the inverted gamma prior distribution on the forecast variance V. It is easy to check that this will give the same result as minimising $k(y)$ if we put $\beta_0(M)=\frac{1}{2}\lambda r(m)$, where $r(M)$ is defined above, and we assume that $\alpha_0(M)$ does not depend on M.

Using this prior we choose to set up the model M^* with $r(M^*)$ parameters. Once observations $y_{l+1}, y_{l+2},\ldots$ start arriving it will not usually be possible to rerun this complicated and expensive search after the arrival of each new observation. We are thus then thrown back onto our sequential methodology. Notice, however, that we will have an idea, from our preliminary search, how other models will perform on the data, especially if the driving mechanism of the data has not changed. We can use this information in our allocation of the expected losses $E_t(M_j)$.

References

Akaike, R. (1973) "Information theory and an extension of the maximum likelihood principle", in: B.N. Petrov and F. Czáki, eds., *2nd International Symposium on Information Theory* (Akademiai Kiadó, Budapest), pp. 267–281.

Astrom, K.J. (1970) *Introduction to Stochastic Control Theory* (Academic Press Inc., New York).

Box, G.E.P. (1980) "Sampling and Bayes' inference in scientific modelling and robustness", *J. Roy. Statist. Soc.*, A 143 (to appear).

Box, G.E.P. and G.M. Jenkins (1970) *Time Series Analysis Forecasting and Control* (Holden-Day, San Francisco).

DeGroot, M.H. (1970) *Optimal Statistical Decisions* (McGraw-Hill, New York).

Effron, B. and C. Morris (1973) "Combining possibly related estimation problems" (with discussion), *J. Roy. Soc.*, B 35, 379–421.

Gathercole, R. and J.Q. Smith (1983) "A dynamic forecasting model for a general class of discontinuous time series", in: O. D. Anderson, ed., *Time Series Analysis: Theory and Practice*, vol. 5 (North-Holland, Amsterdam), pp. 103–116.

Goldstein, M. (1981) "Revising previsions: A geometric interpretation" (with discussion), *J. Roy. Statist. Soc.*, B 43, 105–130.

Harrison, P.J. and C.F. Stevens (1976) "Bayesian forecasting" (with discussion), *J. Roy. Statist. Soc.*, B 38, 205–247.

Schwarz, G. (1978) "Estimating the dimension of a model", *Ann. Statist.*, 6, 461–464.

Whittle, P. (1970) *Probability* (Penguin Books, Harmondsworth).

Chapter 27

A COMPARISON OF SEVERAL MODEL SELECTION PROCEDURES

MURRAY K. CLAYTON*, SEYMOUR GEISSER** and DENNIS E. JENNINGS‡

University of Minnesota

1. Introduction

Econometric and other statistical models are often simplifications of extremely complicated phenomena, and it is a mistake to assume that any particular model is actually a true representation of the underlying process. What is hoped is that the model may be an adequate description and perhaps useful for some purpose. Hence it is often puzzling why there has been so much effort, especially in the softer social sciences, devoted to "testing" parameters of a model as if they were true entities and not, as in most instances, convenient artifices. A more substantial enterprise than testing should be model selection, i.e. selecting one of several alternative models such that the selected model (irrespective of its truth) would serve best some purpose of the investigator (descriptive or predictive). It is to be hoped that the selected model also represents some reasonable approximation to the truth, complicated as that may be.

In theory, there is no intrinsic difficulty for the Bayesian statistician who can determine a prior probability q_k for each potential model M_k that could have generated the set of data, and who can also determine, given a particular model M_k, a prior density for the parameters specified by the model. A posterior probability q_k' is obtained for each model and a model selected based on all of the q_k' and a cost or penalty for choosing an incorrect model. If prediction of a future value Y is at issue, then one calculates a predictive distribution as

$$F_p(y|D)=\sum_k q_k' F_p(y|D,M_k), \tag{1.1}$$

*Now at University of Wisconsin, Madison.

**Research supported in part by NIH grant GM 25271.

‡Now at University of Illinois, Urbana.

Bayesian Inference and Decision Techniques, edited by P. Goel and A. Zellner

where D represents the data in hand and $F_p(y|D, M_k)$ is the predictive distribution function conditional on Y having been generated by M_k. Conceptually all predictive inferences and decisions flow from (1.1). For example, a point predictor that minimizes squared error is

$$E(Y|D) = \sum_k q_k' E(Y|D, M_k). \tag{1.2}$$

Other loss functions can bring to the fore the predictive mode or median of Y.

Seldom, however, is the situation so straightforward for either the investigator or the statistician. As already indicated, the set of statistical models employed rarely includes an exact representation of the process, and even if it did, prior probabilities for either the various models or the parameters for any given M_k are rarely specifiable with any degree of confidence. Moreover, most investigators, if not statisticians, are unsure about their preference for a loss function.

With this in mind, it is of some importance to be able to devise reasonable and useful model selection procedures that do not depend on such tight specifications and evaluate them from more than one standpoint so as to make the rationale for their use as convincing as possible. Former reservations notwithstanding, it is still of some interest (of prime interest for classical statisticians, certainly) to determine the rate at which a model selection procedure yields the true model when in fact the true model is one of the alternatives considered. Another approach is to assess various selection procedures in terms of their predictive capacity, recognizing again that none of the models entertained is necessarily the true one. Here the goal is to find among competitive model selection procedures that one which possesses, in some sense, the best predictive capability over a broad spectrum of possibilities.

We studied four model selection procedures with comparisons restricted to two common distribution families. We determined, by simulation, estimates of the frequency with which each chooses a "true" model and, by using a prediction rule, compared the selection procedures on the basis of squared prediction error.

In making our comparisons we have restricted ourselves to the situation where the "true" model is among the alternatives considered in order to contrast the comparisons of correct selection rates and squared errors of prediction. Also, we decided to employ a fairly simple-minded prediction rule based only on the model selected, in contrast to what is indicated by (1.1). It is our belief that most investigators would use such a rule and so it is of interest to make our relative comparisons on this basis.

Our study indicated that although different model selection procedures can give rise to substantially disparate correct selection rates, their predictive errors are remarkably similar.

2. Selection criteria

Suppose we have available a set of data $\boldsymbol{x} = (x_1, \ldots, x_N)$, a realization of the random variable $\boldsymbol{X} = (X_1, \ldots, X_N)$. Several potential models $M_1, \ldots, M_m$ are entertained as having possibly generated the data or as offering a satisfactory explanation for the data. More directly, one of the models may be useful for predicting future values from the underlying process which generated the data.

We further assume that M_k specifies a joint density $f(\boldsymbol{x}|\boldsymbol{\theta}_k, M_k)$, where $\boldsymbol{\theta}_k$ is some set of unknown parameters.

In such situations several procedures have been put forward for choosing the most appropriate of the entertained models. Akaike (1973), using information theory, devised what has come to be known as the Akaike information criterion (AIC) which is equivalent to selecting that model which maximizes

$$\mathscr{A}_k = \mathrm{e}^{-p_k} f(\boldsymbol{x}|\hat{\boldsymbol{\theta}}_k, M_k), \tag{2.1}$$

where $\hat{\boldsymbol{\theta}}_k$ is the maximum likelihood estimator of $\boldsymbol{\theta}_k$ and p_k represents the number of unknown parameters. For regression problems this is equivalent to the C_p criterion of Mallows (1973).

Schwarz (1978) developed an asymptotic expansion to an exact Bayes procedure which, for nested situations, assigns positive probability to lower dimensional subspaces of the parameter space. He employs the two leading terms of the expansion as a basis for selecting the model which is a posteriori most probable. His work includes as special cases an earlier effort by Jeffreys (1967, ch. V) and a further development for regression situations by Zellner and Siow (1980), for large enough samples. Recently, Leonard (1982) has extended the Schwarz method to non-nested situations. The method is equivalent to selecting the model which maximizes

$$\mathscr{S}_k = \exp\left(-p_k \log\sqrt{N}\right) f(\boldsymbol{x}|\hat{\boldsymbol{\theta}}_k, M_k). \tag{2.2}$$

This large sample Bayes (LSB) criterion assumes that there is a fixed and equal penalty (with possibly minor perturbations) for choosing the wrong model. When this latter assumption is made, Schwarz points out that Akaike's AIC cannot be asymptotically optimal. Indeed, AIC is not even consistent! Although Akaike did not explicitly state what loss or penalty he considered or even if he considered one at all, it is easily shown and alluded to in Geisser and Eddy (1979) and more precisely in Geisser (1980) that AIC, for the same assumptions as Schwarz made, is also asymptotically Bayes, but with penalties that depend on the sample size and the kind of selection error made. This is immediately apparent by comparing (2.1) with (2.2) and noting

$$\mathscr{A}_k = \left(\sqrt{N}/e\right)^{p_k} \mathscr{S}_k. \tag{2.3}$$

Hence, if $M_k \subset M_{k'}$, then, in the light of (2.3), the selection of M_k over $M_{k'}$ requires that

$$\mathscr{S}_k > \left(\sqrt{N}/e\right)^{d} \mathscr{S}_{k'}$$

where $d = p_{k'} - p_k$. Hence, the relative penalty (the coefficient of $\mathscr{S}_{k'}$) increases rapidly with N for $N \geq 8$. This implies that for Akaike's procedure a severe relative penalty is incurred when a "false" lower dimensional model is selected as opposed to selecting a "false" higher dimensional model. For many prediction problems involving nested models this makes good sense.

For example, assume that n observations are taken from each of two populations, Π_1 and Π_2, where Π_i is $N(\mu_i, \sigma^2)$ and two models are entertained, namely $M_1 : \mu_1 = \mu_2$ and $M_2 : \mu_1 \neq \mu_2$. Here, when M_1 is true, it can easily be shown that the probabilities of correct selection for the criteria $\mathscr{A}_k$ and $\mathscr{S}_k$ are, respectively,

$$C(\mathscr{A}) = \Pr\left[F_{1,2(n-1)} \leq 2(n-1)(e^{1/n} - 1)\right]$$

and

$$C(\mathscr{S}) = \Pr\left[F_{1,2(n-1)} \leq 2(n-1)\left[(2n)^{1/2n} - 1\right]\right],$$

where $F_{a,b}$ is an F variable with a and b degrees of freedom. It is easily shown that $C(\mathscr{A}) \leq C(\mathscr{S})$ for all $n \geq 4$, $\lim_{n\to\infty} C(\mathscr{A}) = \Pr[\chi_1^2 \leq 2] \doteq 0.843$, and $\lim_{n\to\infty} C(\mathscr{S}) = 1$. On the other hand, it is clear that the power $1-\beta$ of these criteria is such that $1-\beta_{\mathscr{A}} \geq 1-\beta_{\mathscr{S}}$ for all $n \geq 4$ and that both tend to 1 as n grows. Moreover, when M_1 is true but M_2 is chosen by the selection process, the squared error of any sensible predictor of future values from Π_i will tend to σ^2 as n grows. For prediction purposes consistency of an error of the first kind may be largely irrelevant in this nested situation.

Geisser and Eddy (1979) put forth two model selection criteria based on predictive sample reuse (PSR) methods suitable for independently distributed variates such that X_j has density $f(x_j|\boldsymbol{\theta}_k, M_k)$. The first, termed PSR quasi-likelihood (PSRQL), selects that model which maximizes

$$\hat{L}_k = \prod_{j=1}^{N} f\left(x_j|\hat{\boldsymbol{\theta}}_{k(j)}, M_k\right), \tag{2.4}$$

where $\hat{\boldsymbol{\theta}}_{k(j)}$, is the MLE of $\boldsymbol{\theta}_k$ with x_j omitted. The second criterion, termed PSR quasi-Bayes (PSRQB), selects that model which maximizes

$$L_k = \prod_{j=1}^{N} f_p\left(x_j|\boldsymbol{x}_{(j)}, M_k\right), \tag{2.5}$$

where $\boldsymbol{x}_{(j)}$ denotes that x_j has been deleted from $\boldsymbol{x}$ and where $f_p(\cdot)$ represents a predictive density. The latter is calculated as follows:

$$f_p\left(x_j|\boldsymbol{x}_{(j)}, M_k\right) = \int f\left(x_j|\boldsymbol{\theta}_k, M_k\right) \mathrm{d}P\left(\boldsymbol{\theta}_k|\boldsymbol{x}_{(j)}, M_k\right) \tag{2.6}$$

and $P(\boldsymbol{\theta}_k|\boldsymbol{x}_{(j)}, M_k)$ is a posterior distribution of $\boldsymbol{\theta}_k$ based on $x_{(j)}$ and usually a diffuse prior on $\boldsymbol{\theta}_k$. A complete explication of these methods appears in Geisser and Eddy (1979). In comparing models in a nested situation, when a lower dimensional model M_k is assumed to be true, then for a reasonable class of priors it can be shown that $\mathscr{A}_{k'}$, $\hat{L}_{k'}$ and $L_{k'}$ are asymptotically equivalent for any $M_{k'}$ which contains M_k.

In order to have an idea of how these methods perform in practice, we shall compare them in two cases. First, we shall use the simple exponential distribution and two models. Under M_1 we assume that a dichotomously labeled random sample $\boldsymbol{X}=(\boldsymbol{X}_1, \boldsymbol{X}_2)$, $X_i=(X_{i1},\ldots, X_{iN_i})$, $i=1,2$, is a set of $N_1+N_2=N$ i.i.d. random variables each with density

$$f(x|\lambda, M_1)=\mathrm{e}^{-\lambda x},$$

i.e. the sampling distribution does not depend on the label. Under M_2 we assume that the label is relevant and X_{ij} is a random sample from

$$f(x|\lambda_i, M_2)=\lambda_i \mathrm{e}^{-\lambda_i x}, \quad i=1,2;\ j=1,\ldots, N_i.$$

The AIC criterion compares $\mathscr{A}_1$ with $\mathscr{A}_2$, where

$$-\log \mathscr{A}_1 = N+1+N\log \bar{x},$$

$$-\log \mathscr{A}_2 = N+2+N_1\log \bar{x}_1 + N_2\log \bar{x}_2,$$

and $N_i\bar{x}_i = x_{i1}+\cdots+x_{iN_i}$, $N\bar{x}=N_1\bar{x}_1+N_2\bar{x}_2$, for $i=1,2$.

The large sample Bayes procedure is given by Schwarz compares $\mathscr{S}_1$ with $\mathscr{S}_2$, where

$$-\log \mathscr{S}_1 = N+\log\sqrt{N}+N\log \bar{x},$$

$$-\log \mathscr{S}_2 = N+\log N+N_1\log \bar{x}_1+N_2\log \bar{x}_2.$$

For this situation the PSRQL method compares $\hat{L}_1$ with $\hat{L}_2$, while the PSRQB method compares L_1 with L_2, as given in Geisser and Eddy (1979, p. 155).

For the second case we assume that the data are samples from one of two normal populations with density

$$f(x|\mu_i, \sigma_i^2)=\left(2\pi\sigma_i^2\right)^{-1/2}\exp\left(-\frac{1}{2\sigma_i^2}(x-\mu_i)^2\right), \quad i=1,2. \tag{2.7}$$

We suppose that there are three possible models:

$$M_1: \mu_1=\mu_2,\ \sigma_1^2=\sigma_2^2; \qquad M_2: \mu_1\neq\mu_2,\ \sigma_1^2=\sigma_2^2; \qquad M_3: \mu_1\neq\mu_2,\ \sigma_1^2\neq\sigma_2^2. \tag{2.8}$$

Here the AIC criterion selects the model M_k according to the largest $\mathscr{A}_k$, where

$$-2\log \mathscr{A}_1 = N+4+N\log 2\pi S^2,$$

$$-2\log \mathscr{A}_2 = N+6+N\log 2\pi T^2,$$

$$-2\log \mathscr{A}_3 = N+8+N_1\log 2\pi S_1^2+N_2\log 2\pi S_2^2,$$

for $NS^2=\sum_i\sum_j(x_{ij}-\bar{x})^2$, $NT^2=\sum_i\sum_j(x_{ij}-\bar{x}_i)^2$, *and* $N_iS_i^2=\sum_j(x_{ij}-\bar{x}_i)^2$.

The LSB criterion selects M_k according to the largest of

$$-2\log \mathscr{S}_1 = 2\log N+N+N\log 2\pi S^2,$$

$$-2\log \mathscr{S}_2 = 3\log N+N+N\log 2\pi T^2,$$

$$-2\log \mathscr{S}_3 = 4\log N+N+N_1\log 2\pi S_1^2+N_2\log 2\pi S_2^2.$$

For the PSRQL and PSRQB criteria the relevant comparisons can be found in Geisser and Eddy (1979, p. 156–157).

In both the normal and exponential cases discussed, the data are differentiated by a binary label and a predicted value is required for each label. We will investigate and compare the use of the aforementioned model selection procedures with regard to correct selection rates. Also, rather than using the full Bayesian treatment alluded to prior to (1.1), we will investigate the error of prediction when a single model is chosen and used in conjunction with a predictive rule subsequently described.

3. Simple exponential models

A simulation was performed to compare model selection rates and predictive squared error of the four procedures when used in the exponential setting described in section 2. The relative size of the parameters for the two populations was varied by setting $\lambda_1 = 1$ and $\lambda_2 = 1(0.5)2.5$. For each value of λ_2 a sample of size $n(=N_1 = N_2)$ was generated for each population,[1] the model selection procedures were applied and the selected model noted. This was repeated 10 000 times and the proportion of correct selections for each procedure was recorded. The results for selected values of n ranging from 4 to 20 appear in table 27.1. Our choices of λ_2, n, and the number of replications were aided by a preliminary simulation study by Seber (1979).

As noted in table 27.1, these estimates of the probability of correct model selection have standard errors ranging from 0.004 to 0.005. When comparing different procedures for the same value of λ_2 and n, the standard error of a difference of estimated probabilities has been reduced by computing the estimates from the same data sets. In the discussion that follows, a difference will be called "significant" if it exceeds twice the standard error of that difference. We intend only to use this as a benchmark for making comparisons, and do not imply that a formal test is being made.

The behavior of the LSB procedure is distinct from the other three procedures. It is the only procedure which is consistent; that is, its correct selection rate goes to 1 as n increases under M_1. When M_1 is true, the probability of correct selection for the AIC, PSRQB and PSRQL procedures has an asymptotic value of 0.843, as noted in section 2. The first section of table 27.1 reflects this asymptotic behavior. The LSB procedure makes more correct selections under M_1 and fewer correct selections under M_2 than the other procedures. Except for some minor differences, the other three methods yield rather similar results over most parameter configurations in terms of their rates of correct selection, and there seems to be little to choose from among them.

[1] Generated by the inverse cumulative method with uniform pseudo-random numbers supplied by the University of Minnesota Computing Center's routine RAN2F.

Table 27.1. Estimated probability of correct selection between populations specified by $f_1 = e^{-x}$ and $f_2 = \lambda_2 e^{-\lambda_2 x}$ for four selection criteria.[a]

λ_2	Size of each sample	Selection criterion			
		PSRQL	PSRQB	AIC	LSB
1.0	4	0.799(0.794)[b]	0.827(0.821)[b]	0.832	0.841
	8	0.816(0.830)[b]	0.827(0.837)[b]	0.830	0.894
	12	0.834	0.827	0.835	0.920
	20	0.837	0.837	0.840	0.942
1.5	4	0.262	0.237	0.232	0.223
	8	0.291	0.284	0.282	0.201
	12	0.343	0.346	0.345	0.214
	20	0.446	0.451	0.456	0.260
2.0	4	0.342	0.334	0.339	0.327
	8	0.479	0.489	0.489	0.385
	12	0.590	0.600	0.599	0.455
	20	0.763	0.770	0.771	0.602
2.5	4	0.440	0.445	0.444	0.430
	8	0.633	0.654	0.661	0.560
	12	0.769	0.784	0.787	0.664
	20	0.918	0.926	0.928	0.834

[a]Based on 10 000 samples for each sample size with standard error of tabular entries estimated to be between 0.004 and 0.005.

[b]Values in parentheses are estimates from Geisser and Eddy (1979) using same procedure.

PSRQL: Predictive sample reuse quasi-likelihood.
PSRQB: Predictive sample reuse quasi-Bayes.
AIC: Akaike information criterion.
LSB: Large sample Bayes.

Various prediction problems could arise in this setting; the one which we consider here was discussed in section 2. The problem supposes a need to predict a future observation from each population (under M_1 they are identical populations) with loss measured as the sum of squared prediction errors. Also, we use the outcome of the model selection process to determine our predictions.

Specifically, the problem examined is to predict future observations, labeled as Y_1 and Y_2, with Y_i distributed as an observation from the ith population. If our predictors are $\hat{y}_1$ and $\hat{y}_2$, then the expected squared prediction error is $E[(Y_1 - \hat{y}_1)^2 + (Y_2 - \hat{y}_2)^2]$. Since $E(Y_i - \hat{y}_i)^2 = \operatorname{Var} Y_i + E(\hat{y}_i - EY_i)^2$, we need only estimate the last term in the simulation. Doing so increases the accuracy of the estimate and eliminates the need to generate the new observations, Y_i. Following Geisser and Eddy (1979), if M_1 is selected, then we set $\hat{y}_1 = \hat{y}_2 = \bar{x}$, the grand mean, as the predictors; if M_2 is selected, then $\hat{y}_i = \hat{x}_i$ for $i = 1,2$ are used. Table 27.2 gives the estimates of the expected squared error of prediction (SEP) for the cases examined.

Inspection of the table shows the LSB method doing better under M_1 and usually poorer under M_2. The other three procedures seem indistinguishable with regard to prediction error. Surprisingly, this is true even in those situations when

Table 27.2. Estimates of expected squared error of prediction (SEP) and their standard error based on use of four selection criteria.[a]

		Selection Criterion						
λ_2	Size of each sample	PSRQL	PSRQB	AIC	LSB	SEPC[b]	SEPI[c]	SEPK[d]
1	4	2.376(0.005)	2.383(0.005)	2.380(0.006)	2.375(0.005)	2.250	2.500	2.00
	8	2.195(0.004)	2.196(0.004)	2.196(0.004)	2.177(0.004)	2.125	2.250	
	12	2.131(0.002)	2.133(0.002)	2.133(0.002)	2.115(0.002)	2.083	2.167	
	20	2.078(0.001)	2.079(0.001)	2.079(0.001)	2.065(0.001)	2.050	2.100	
1.5	4	1.763(0.005)	1.768(0.005)	1.770(0.005)	1.768(0.005)	1.806	1.681	1.44
	8	1.620(0.002)	1.622(0.002)	1.623(0.002)	1.620(0.002)	1.625	1.590	
	12	1.573(0.001)	1.573(0.001)	1.574(0.001)	1.575(0.002)	1.565	1.560	
	20	1.529(0.001)	1.529(0.001)	1.529(0.001)	1.537(0.001)	1.517	1.536	
2.0	4	1.576(0.004)	1.583(0.005)	1.585(0.005)	1.585(0.005)	1.563	1.531	1.25
	8	1.438(0.002)	1.438(0.002)	1.438(0.002)	1.449(0.002)	1.406	1.453	
	12	1.384(0.002)	1.383(0.002)	1.382(0.002)	1.399(0.002)	1.354	1.427	
	20	1.329(0.001)	1.329(0.001)	1.328(0.001)	1.348(0.001)	1.313	1.406	
2.5	4	1.492(0.005)	1.496(0.005)	1.498(0.005)	1.500(0.005)	1.450	1.485	1.16
	8	1.346(0.002)	1.342(0.002)	1.341(0.002)	1.358(0.002)	1.305	1.413	
	12	1.283(0.002)	1.281(0.002)	1.280(0.002)	1.300(0.002)	1.257	1.388	
	20	1.226(0.001)	1.225(0.001)	1.225(0.001)	1.239(0.001)	1.218	1.369	

[a] Based on samples used in table 27.1
[b] Expected squared error of prediction (SEPC) when all selections are correct.
[c] Expected squared error of prediction (SEPI) when all selections are incorrect.
[d] Expected squared error of prediction when λ_i is known (SEPK).

one procedure clearly makes more correct model selections. Thus, only LSB yields "significantly" different prediction errors, although the actual magnitude of these differences is inconsequential. The largest relative difference in squared error is only 1.5 percent which occurs when $\lambda_2 = 2.0$ and $n = 20$. One might argue that in comparing estimates of SEP that they should be expressed in the same units as the original observations. We can do so by comparing the square roots of the estimated SEPs. In this case, the relative difference of transformed estimates is 0.8 percent. On the other hand, the largest difference in the probability of correct selection for this parameter assignment is 17 percentage points, a fairly large difference.

Three other columns are included in table 27.2 to give some guidance in judging how well we are doing or can hope to do. The expected squared error of prediction when λ_i is known (SEPK) represents the unavoidable part of the loss which occurs due to the random variability of the observations to be predicted. Consequently, it can be viewed as the asymptotic limit of the prediction error for given values of λ_1 and λ_2 as n increases. The prediction error which would occur if we knew the correct model (SEPC) but did not know the actual values of λ_1 and λ_2 represents how well a "foolproof" model selection procedure would do. The error incurred if we always use the wrong model (SEPI) is also given. In some cases ($\lambda_2 = 1.5$ with $n = 4$, 8 or 12 and $\lambda_2 = 2.0$ with $n = 4$) it is better to be always wrong than always right! Of course, it is no easier to be always wrong than always right.

Let us look more carefully at one of the cases where SEPI is less than SEPC, such as $\lambda_2 = 1.5$ and $n = 4$. In this case all of the model selection procedures have estimated squared errors of prediction which are less than SEPC and greater than SEPI. To see why this might be the case, consider using a predictor for Y_1 of the form $a\bar{x} + (1-a)\bar{x}_1$. One can show that for this parameter assignment the optimal predictor of this form has $a = 16/17$. This predictor is very close to $\bar{x}$, which is used under SEPI. This example emphasizes the fact that prediction is not coincident with model selection, and that optimal squared error predictors might better be based on the information provided by several models, rather than on the acceptance of a single model – this, of course, is well known.

4. Normal models

The correct selection rates and predictive squared error of the four selection criteria were also compared using simulated normal data described by (2.7). We consider a variety of combinations of the parameter values which conform to the possible model restrictions (2.8). For simplicity, we fix $\mu_1 = 0$, $\sigma_1 = 1$, and vary μ_2 and σ_2. Tables 27.3 and 27.4 contain the parameter configurations used.

Similar to section 3, for each given parameter assignment a sample of size $n(= N_1 = N_2)$ was generated from each population,[2] the model selection criteria were applied, and the selected model was noted. We repeated this process 2000 times[3] and noted the proportion of correct model selections by each procedure. These results for $n = 10$ and $n = 20$ appear in table 27.3. For reasons explained below, we also noted the proportion of selections of M_1; these data also appear in table 27.3.

The estimates in table 27.3 have standard errors which lie between 0.008 and 0.012. As in section 3, the estimates within a given row of table 27.3 are based on the same data sets and are thus positively correlated. The standard error of a difference in estimated probabilities is thus reduced, and typically lies between 0.002 and 0.008.

Generally speaking, the AIC is "best" in terms of frequency of correct model selection, with some interesting exceptions. Under M_1, $(\mu_2, \sigma_2) = (0, 1)$, it is the worst. In this case the LSB criterion is constructed to be consistent, and so it is not surprising that it should dominate the other selection criteria when the sample size is large. It is perhaps a little surprising that it is clearly better for n as small as 20. When n is 10, no other method is "significantly" better. The other instances where the LSB method dominates the remaining methods are when

[2] Using the University of Minnesota Computing Center's pseudo-random normal variate generator NORMAL.

[3] Except as noted in table 27.4, where 10 000 repetitions were used to reduce the standard error of the corresponding estimates.

Table 27.3. Estimated probability of correct selection (first value) and that M_1 will be selected (second value) for populations specified by

$$f_1 = \frac{1}{\sqrt{2\pi}} \exp\left[-\tfrac{1}{2}x^2\right]$$

and

$$f_2 = \frac{1}{\sigma_2\sqrt{2\pi}} \exp\left[-\frac{1}{2\sigma_2^2}(x-\mu_2)^2\right]$$

assuming that if $\mu_2 = 0$, then $\sigma_2^2 = 1$, for four selection criteria.[a]

(μ_2, σ_2)	Size of each sample	Selection criterion							
		PSRQL		PSRQB		AIC		LSB	
(0,1)	10	0.824	0.824	0.872	0.872	0.737	0.737	0.871	0.871
	20	0.812	0.812	0.873	0.873	0.767	0.767	0.917	0.917
(.25,1)	10	0.154	0.773	0.102	0.833	0.197	0.684	0.133	0.818
	20	0.215	0.701	0.152	0.775	0.245	0.649	0.136	0.841
(.50,1)	10	0.265	0.650	0.211	0.703	0.317	0.547	0.252	0.692
	20	0.436	0.448	0.355	0.538	0.463	0.402	0.359	0.602
(.75,1)	10	0.417	0.464	0.364	0.510	0.480	0.343	0.429	0.486
	20	0.684	0.193	0.609	0.262	0.697	0.146	0.638	0.315
(1.00,1)	10	0.598	0.284	0.550	0.305	0.654	0.169	0.636	0.272
	20	0.795	0.050	0.754	0.077	0.782	0.034	0.833	0.103
(2.00,1)	10	0.843	0.000	0.820	0.000	0.803	0.000	0.885	0.000
	20	0.851	0.000	0.838	0.000	0.831	0.000	0.943	0.000
(4.00,1)	10	0.854	0.000	0.818	0.000	0.802	0.000	0.877	0.000
	20	0.854	0.000	0.840	0.000	0.834	0.000	0.942	0.000
(.25,1.5)	10	0.217	0.685	0.220	0.708	0.321	0.550	0.183	0.724
	20	0.457	0.440	0.431	0.510	0.521	0.378	0.258	0.677
(.50,1.5)	10	0.231	0.603	0.238	0.630	0.344	0.461	0.190	0.641
	20	0.494	0.348	0.483	0.398	0.568	0.268	0.307	0.547
(.75,1.5)	10	0.253	0.484	0.282	0.503	0.363	0.341	0.228	0.499
	20	0.530	0.204	0.539	0.249	0.602	0.149	0.366	0.361
(1.00,1.5)	10	0.282	0.371	0.320	0.382	0.399	0.234	0.256	0.365
	20	0.549	0.091	0.580	0.129	0.629	0.057	0.393	0.202
(2.00,1.5)	10	0.354	0.044	0.412	0.038	0.452	0.009	0.312	0.027
	20	0.571	0.000	0.614	0.000	0.633	0.000	0.424	0.000
(4.00,1.5)	10	0.352	0.001	0.413	0.000	0.452	0.000	0.319	0.000
	20	0.578	0.000	0.624	0.000	0.639	0.000	0.439	0.000
(1.00,2)	10	0.535	0.294	0.602	0.282	0.699	0.165	0.552	0.289
	20	0.891	0.047	0.902	0.053	0.931	0.026	0.811	0.108
(2.00,2)	10	0.598	0.074	0.686	0.059	0.736	0.022	0.618	0.060
	20	0.907	0.002	0.937	0.002	0.942	0.001	0.870	0.003
(4.00,2)	10	0.635	0.006	0.728	0.000	0.757	0.000	0.660	0.000
	20	0.923	0.000	0.944	0.000	0.949	0.000	0.874	0.000

[a] Based on 2000 samples of each sample size and standard error of tabular entries estimated to be between 0.008 and 0.012.

Table 27.4. Estimates of expected squared error of prediction (SEP) and their standard error based on use of four selection criteria.[a]

(μ_2, σ_2)	Size of each sample	Selection criterion PSRQL	PSRQB	AIC	LSB	SEPC
(0,1)	10	2.140(0.004)	2.132(0.004)	2.156(0.004)	2.135(0.004)	2.100
	20	2.076(0.002)	2.070(0.002)	2.080(0.002)	2.066(0.002)	2.050
(.25,1)	10	2.176(0.004)	2.172(0.004)	2.185(0.004)	2.176(0.004)	2.200
	20	2.101(0.002)	2.100(0.002)	2.102(0.002)	2.098(0.002)	2.100
(.50,1)	10	2.231(0.004)	2.236(0.004)	2.223(0.004)	2.236(0.004)	2.200
	20	2.130(0.002)	2.141(0.002)	2.125(0.002)	2.148(0.002)	2.100
(.75,1)	10	2.290(0.005)	2.303(0.005)	2.260(0.005)	2.296(0.005)	2.200
	20	2.132(0.003)	2.150(0.003)	2.121(0.003)	2.163(0.003)	2.100
(1,1)	10	2.295(0.006)	2.303(0.006)	2.242(0.005)	2.287(0.006)	2.200
	20	2.116(0.003)	2.126(0.003)	2.110(0.003)	2.136(0.004)	2.100
(2,1)	10	2.210(0.006)	2.200(0.005)	2.198(0.005)	2.201(0.005)	2.200
	20	2.101(0.002)	2.101(0.002)	2.101(0.002)	2.101(0.002)	2.100
(4,1)	10	2.202(0.005)	2.202(0.005)	2.202(0.005)	2.202(0.005)	2.200
	20	2.100(0.002)	2.100(0.002)	2.100(0.002)	2.100(0.002)	2.100
(.25,1.5)	10	3.528(0.007)	3.526(0.007)	3.550(0.008)	3.528(0.007)	3.575
	20	3.408(0.004)	3.404(0.004)	3.410(0.004)	3.398(0.004)	3.413
(.50,1.5)	10	3.581(0.007)	3.584(0.007)	3.582(0.007)	3.584(0.007)	3.575
	20	3.432(0.004)	3.436(0.004)	3.426(0.004)	3.447(0.004)	3.413
(.75,1.5)	10	3.650(0.007)	3.657(0.007)	3.622(0.008)	3.655(0.007)	3.575
	20	3.445(0.004)	3.455(0.004)	3.433(0.004)	3.480(0.004)	3.413
(1,1.5)	10	3.693(0.008)	3.700(0.008)	3.636(0.008)	3.690(0.008)	3.575
	20	3.437(0.005)	3.452(0.005)	3.424(0.004)	3.480(0.005)	3.413
(2,1.5)	10*	3.645(0.005)	3.617(0.005)	3.581(0.004)	3.604(0.004)	3.575
	20	3.412(0.004)	3.412(0.004)	3.412(0.004)	3.412(0.004)	3.413
(4,1.5)	10*	3.583(0.004)	3.573(0.003)	3.573(0.003)	3.573(0.003)	3.575
	20	3.415(0.004)	3.415(0.004)	3.415(0.004)	3.415(0.004)	3.413
(1,2)	10*	5.575(0.005)	5.571(0.005)	5.530(0.006)	5.571(0.005)	5.500
	20	5.272(0.007)	5.274(0.007)	5.265(0.007)	5.291(0.007)	5.250
(2,2)	10*	5.615(0.007)	5.570(0.007)	5.516(0.006)	5.559(0.007)	5.500
	20	5.249(0.007)	5.249(0.007)	5.248(0.007)	5.251(0.007)	5.250
(4,2)	10*	5.548(0.008)	5.506(0.006)	5.506(0.006)	5.506(0.006)	5.500
	20	5.260(0.007)	5.260(0.007)	5.260(0.007)	5.260(0.007)	5.250

[a]Based on samples in table 27.3, except rows marked with an asterisk, which are based on 10 000 replications.

$\sigma_2 = 1$ ($= \sigma_1$) and μ_2 is large ($n = 10$, $\mu_2 = 2,4$ and $n = 20$, $\mu_2 = 1,2,4$). To see why this might be so, note that due to the penalty structure imposed in the LSB method it tends to choose lower dimensional models. Thus, when $\sigma_2 = 1$ and μ_2 is large, while both the AIC and LSB procedures are fairly successful in determining that M_1 is incorrect, the AIC tends to pick M_3 too frequently. This is borne out by computing the proportion of times M_3 is chosen by each method. (This can be done using the data in table 27.3.) On the other hand, when M_3 is the true model, the AIC performs better than the LSB criterion. This results because the latter method tends to prefer lower dimensional models and thus frequently avoids choosing the correct model, M_3.

The situation becomes more difficult to summarize when one attempts to rank all four selection criteria in terms of their ability to select the correct model. For the most part, the PSRQL and PSRQB methods have correct selection rates which lie between those of the AIC and LSB method. Note from table 27.3 that, while the PSRQB, PSRQL, and AIC are asymptotically equivalent, for small samples there can be large differences between their correct selection rates.

A somewhat different situation arises when we compare the four methods with respect to prediction error. As in section 3, our interest here is in the expected squared error of prediction (SEP) where the predictors are based on model selection. If, for a given data set, a model selection procedure picks M_1, then we set the predictors $\hat{y}_1 = \hat{y}_2 = \bar{x}$, the grand mean. On the other hand, if the selection procedure picks M_2 or M_3, then we set $\hat{y}_1 = \bar{x}_1$ and $\hat{y}_2 = \bar{x}_2$, the individual sample means. Also, as in section 3, we estimate the SEP by estimating the quantity $E(\hat{y}_1 - \mu_1)^2 + \sigma_1^2 + E(\hat{y}_2 - \mu_2)^2 + \sigma_2^2$. Table 27.4 contains the estimates of the SEP for each parameter combination and model selection criterion. The table also contains the standard error of each estimated SEP. Note that if, for a given parameter configuration, the selection criteria always agree to reject M_1, then $(\hat{y}_1, \hat{y}_2) = (\bar{x}_1, \bar{x}_2)$ always, and so the estimate of the SEP will be the same. This occurs, for example, when $\mu_2 = 4$, $\sigma_2 = 2$, and $n = 20$ (see tables 27.3 and 27.4).

In any case, the four procedures behave quite similarly in terms of prediction error. Generally, the AIC procedure either yields the smallest point estimate of the SEP, or that estimate is not "significantly" different from the smallest point estimate of the SEP for a given assignment of parameter values. Curiously, when this fails the AIC is in fact the worst performing procedure! In such a case there is little to distinguish between the other three criteria.

When comparing the criteria in terms of SEP it is important to note that even in those cases where the selection procedures do yield "significantly" different SEP estimates, the relative difference is small. For example, if $\mu_2 = 1$, $\sigma_2 = 1$ and $n = 10$, then the smallest point estimate of the SEP is 2.242 provided by the AIC and the largest is 2.303 provided by the PSRQB method; a relative difference of 2.7 percent. Examination of table 27.4 shows that this is an extreme case; the largest relative difference for a given parameter configuration in all other cases is less than 2.0 percent. If we compare the square roots of estimates of the SEP, thus transforming them to the same units as the original observations, we find that the largest relative difference for a given parameter configuration is less than 1.4 percent.

Since the predictive process depends only on whether M_1 is or is not chosen, it is appropriate to compare the corresponding relative differences in selection rates of M_1 for the selection criteria. For example, when $\mu_2 = 1$, $\sigma_2 = 1$, and $n = 10$, the PSRQB method picks M_1 most frequently (30.5 percent of the time) and the AIC picks M_1 least frequently (16.9 percent of the time). As in the exponential sampling situation we see here that while the selection criteria might differ considerably in their rates of selection of M_1, from a practical point of view they yield essentially the same estimates of the SEP.

The similar behavior of the selection criteria in terms of prediction seems to be a consequence of the following phenomenon. Suppose we were faced with predicting Y_1 and Y_2, and suppose we thought M_1 was correct. If σ_2 is known, then the expected prediction error is $(\sigma_2^2+1)(1+1/(2n))+\mu_2^2/2 = PE(M_1)$, say. If we thought M_1 was incorrect, then the pair $(\hat{y}_1, \hat{y}_2) = (\bar{x}_1, \bar{x}_2)$ would be used, and the expected prediction error would be $(\sigma_2^2+1)(1+1/n) = PE(\overline{M}_1)$, say. It is apparent that $PE(M_1)$ can exceed $PE(\overline{M}_1)$ by a large amount, and so if μ_2 is large, the prediction error can be large if the predicting pair $(\hat{y}_1, \hat{y}_2) = (\bar{x}, \bar{x})$ is inappropriately used. However, we see from table 27.3 that in the cases where μ_2 is large, and hence where a potentially large prediction error could be made, the selection procedures rarely, if ever, choose M_1. That is to say, in the case where a potentially large error could arise, the selection procedures are able to discriminate well enough between M_1 and the other models to avoid such an error. On the other hand, when μ_2 is close to but not equal to zero, the selection procedures generally err in picking M_1 too frequently. This does not appreciably increase the prediction error, however, for if μ_2 is small, then $PE(M_1)$ and $PE(\overline{M}_1)$ are close in value. In fact, such incorrect choices of M_1 can be helpful, since if μ_2 is close enough to zero (i.e. $\mu_2^2 < (\sigma_2^2+1)/n$), then $PE(\overline{M}_1)$ is larger than $PE(M_1)$. That is, there is a smaller predictive error incurred if $(\hat{y}_1, \hat{y}_2) = (\bar{x}, \bar{x})$ is used, even though M_1 is known not to be the correct model. Such a situation arises in the cases studied when $\mu_1 = 0.25$, with $(\sigma_2, n) = (1, 10)$, $(1.5, 10)$ or $(1.5, 20)$, and so in these cases one prefers a procedure which picks M_1 frequently. (A similar situation was noted in section 3.) This explains why the LSB, PSRQL, and PSRQB procedures dominate the AIC for those parameter assignments. Outside of these cases, however, when $\mu_2 \neq 0$, then the AIC dominates the others in terms of picking M_1 less frequently, and accordingly performs no worse in terms of estimated SEP.

5. Conclusion and remarks

One of the principal reasons for building models of statistical data is to enable one to predict new observations thought to arise from the same source. Frequently such predictions are made by first choosing a model according to some criterion, and then basing the predictions upon the chosen model. We have seen, for the models and model selection criteria examined in this study, that a particular measure of predictive capacity, namely squared error, appears to be fairly stable when predictions are made in such a manner. Under the same circumstances, the correct selection rates for those model selection procedures can vary appreciably. As a consequence, it would seem that, for predictive purposes, it might be misguided to try to find model selection procedures which have optimal error rates. Instead, it is more pertinent and sensible to devote one's time to constructing models for data and assessing the predictive adequacy of such models.

This precept makes even more sense when it is realized that the models typically considered are recognized to be artifacts, and the "true" model is not likely to be included among the model alternatives considered. In that case it seems pointless to evaluate model selection criteria on the basis of fictive error rates. Nonetheless, it is still meaningful to compare the predictive capability of various models, even if none is true.

It would therefore be of interest to see if the model selection criteria examined herein exhibit the same stability with regard to prediction error when the data are generated by a process which is not among the model alternatives considered. This would indicate how "robust" the prediction rule used in this study is to variation in both the source of the data and the model selection. It would be desirable to have a model selection procedure whose predictive ability did not depend critically on the source of the data.

One such possibility is a sample reuse procedure for low structure situations which was proposed by Geisser and Eddy (1979) and was devised to have good predictive squared error properties for the prediction rule used here.

Suppose the data are $\boldsymbol{x}=(\boldsymbol{x}_1,\boldsymbol{x}_2)$, with $\boldsymbol{x}_i=(x_{i1},\dots,x_{iN_i})$ and M_1 asserts that the label $i=1,2$ is irrelevant for prediction and M_2 asserts that it is. The low structure rule is to calculate

$$D_1=N^{-1}\sum_i\sum_j(x_{ij}-\bar{x}_{(ij)})^2;\qquad D_2=N^{-1}\sum_i\sum_j(x_{ij}-\bar{x}_{i(j)})^2,\tag{5.1}$$

and predict (Y_1,Y_2) as $(\bar{x},\bar{x})$ if $D_1<D_2$ and $(\bar{x}_1,\bar{x}_2)$ if $D_2<D_1$.

It is of interest to note for the normal models discussed in section 2 that this low structure procedure yields the probability of correct selection under M_1 as

$$\Pr\left[F_{1,2(n-1)}\le 2+\frac{1}{2(n-1)}\right],$$

where $n=N_1=N_2$. This correct selection probability is always larger than that for the AIC procedure, although it tends to the same limit. In addition to possibly being more robust, this procedure directly faces the predictive aspect of model selection. For the most part, this predictive aspect has been neglected in the development and evaluation of model selection procedures.

In summary, we feel that if the "true" or "approximately true" model is included among the alternatives considered, all reasonable model selection procedures will possess rather similar predictive capabilities. Although it is considerably more useful to devise more appropriate models for data than to devise trivially more efficient selection procedures, we also believe that the area of robust model selection approaches is still in need of further development and investigation.

References

Akaike, Hirotugu (1973) "Information theory and an extension of the maximum likelihood principle", in: B.N. Petrov and F. Czaki, eds., *Proceedings of the 2nd International Symposium on Information Theory* (Akademiai Kiado, Budapest), pp. 267–281.

Geisser, S. (1980) "Discussion on hypothesis testing", in: J.M. Bernardo et al., eds., *Bayesian Statistics* (University Press, Valencia), p. 636.

Geisser, S. and W.F. Eddy (1979) "A predictive approach to model selection", *J. Amer. Statist. Assoc.*, 74, 153–160.

Jeffreys, H. (1967) *Theory of Probability* (Clarendon Press, Oxford).

Leonard, T. (1982) "Comment on paper by M. Lejeune and G.D. Faulkenberry", *J. Amer. Statist. Assoc.*, 77, 657–658.

Mallows, C.L. (1973) "Some comments on C_p", *Technometrics*, 15, 661–675.

Schwarz, Gideon (1978) "Estimating the dimension of a model", *Annals of Statistics*, 6, 461–464.

Seber, C. (1979) "Power comparisons of the quasi-Bayes and quasi-likelihood criteria for selection of exponential models", Preliminary Report, unpublished.

Zellner, A. and A. Siow (1980) "Posterior odds ratio for selected regression hypotheses", in: J.M. Bernardo et al., eds., *Bayesian Statistics* (University Press, Valencia), 585–618.

PART IX

EXPERIMENTATION AND OPTIMAL SEARCH

Chapter 28

OPTIMAL SEARCH FOR NEW TYPES

WEN-CHEN CHEN and MORRIS H. DeGROOT*

Carnegie-Mellon University

1. Introduction

The collector's problem is well known and is discussed in many introductory texts on probability theory (see, for example, Feller, 1968, or DeGroot, 1975). Suppose that a population contains C different types of items, where C is a known positive integer. Let p_i denote the proportion of items of type i $(i=1,\dots,C)$, where $P_i > 0$ and $\sum_{i=1}^{C} p_i = 1$. Suppose that a sequential random sample is to be drawn from the population, with replacement if the population is finite. Thus, the observations in the sample are independent and identically distributed and the probability that any particular observation is of type i is p_i.

In elementary treatments of this problem, it is assumed that $p_i = 1/C$, for $i=1,\dots,C$, and the focus is on the determination of the probability distribution of the number of different items that are obtained in n observations. For example, finding the probability of obtaining a complete set of C different types of items in n observations, or finding the expected number of observations necessary to obtain a complete set of C different types of items, is a standard exercise at this level.

When $p_1,\dots,p_C$ are not all equal, the problem becomes somewhat more complicated. The standard tricks for finding the expected number of observations necessary to obtain a complete set (see, for example, Rényi, 1962; Baum and

* This research was supported in part by the National Science Foundation under grant SES-8207295. Wen-chen Chen was killed in Taiwan on 3 July 1981. We had begun working together on this topic before his death. Since the model presented here and some of the results are based on either his ideas or our joint research, I have taken the liberty of listing his name as a co-author. Perhaps I have taken a greater liberty by making myself a co-author. I am indebted to Prem K. Goel for very helpful comments and corrections.

Bayesian Inference and Decision Techniques, edited by P. Goel and A. Zellner

Billingsley, 1965; or Feller, 1968, p. 225) do not apply, and one must resort to other methods: at an arbitrary stage of the sampling process, we will be in the situation where c different types of items remain to be obtained ($c=1,\dots,C$). Let $\boldsymbol{q}=(q_1,\dots,q_c)$ denote the vector of their proportions, and let $q_+=\sum_{j=1}^{c} q_j$. Thus, the probability that the next observation will yield a new type of item is q_+ and the probability that it will yield nothing new is $1-q_+$. Furthermore, let $\boldsymbol{q}(j)$ denote the $(c-1)$-dimensional vector which is simply the vector $\boldsymbol{q}$ without its jth component. Finally, let $N(\boldsymbol{q})$ denote the expected number of observations necessary to obtain each of the remaining c different types of items. Then

$$N(\boldsymbol{q})=\frac{1}{q_+}+\sum_{j=1}^{c}\frac{q_j}{q_+}N[\boldsymbol{q}(j)]. \tag{1.1}$$

This relation holds because $1/q_+$ is the expected number of observations necessary to obtain a new type of item, q_j/q_+ is the probability that the first new type obtained will be type j, and $N[\boldsymbol{q}(j)]$ is the additional expected number of observations necessary to obtain the remaining $c-1$ types of items if the first new type obtained is type j.

If there is just one type of element remaining to be obtained, so that $\boldsymbol{q}$ has just the single component q_1, then $N(\boldsymbol{q})=1/q_1$. It follows from (1.1), that if $\boldsymbol{q}=(q_1,q_2)$, then

$$N(\boldsymbol{q})=\frac{1}{q_1+q_2}\left(1+\frac{q_1}{q_2}+\frac{q_2}{q_1}\right). \tag{1.2}$$

It follows from (1.1) and (1.2) that $N(q)$ has been determined, at least in principle, for a vector $\boldsymbol{q}$ of any dimension c.

Another way in which the collector's problem can be extended is to allow the collector to receive a bonus each time he obtains a new type of item, where the amount of the bonus may depend on the type of item obtained. The asymptotic distribution of the total bonus received in n observations, and the asymptotic distribution of the number of observations necessary to make the total bonus exceed some fixed level has been extensively studied (see, for example, Rosén, 1969, 1970, and Sen, 1979). Related problems involving sampling without replacement or in which different items in the population may have different values have been considered by Holst (1972, 1973), Lanke (1972), and Rosén (1972).

In this chapter our focus is somewhat different from this other work. We are interested in problems where there are two or more different populations, each of which contains different proportions of the different types of items. Furthermore, the collector may be uncertain about the proportions of each type in the different populations. At each stage of the sequential sampling process the collector must decide from which population he wishes to draw his next observation.

In sections 2 and 3 we consider problems in which the proportions of the different types in each population are known, and the collector must determine a sequential search procedure or sampling rule for which the expected number of observations necessary to obtain at least one item of each different type is a minimum. In section 4, optimal procedures are developed in special balanced situations.

In sections 5 and 6 the same type of problem is considered when the proportions of the different types in each population are unknown. The unknown proportions in each population are assigned a Dirichlet prior joint distribution, and the proportions in different populations are assumed to be independent. Optimal search procedures with respect to this prior distribution are studied.

In sections 7 and 8, problems of optimal stopping are considered in which the collector must pay a cost of γ units for each observation that he draws, regardless of the population from which it is drawn, and he receives a fixed bonus for each new type of item that he obtains. We discuss both problems in which the proportions of different types of items are known and problems in which these proportions are unknown and assigned Dirichlet prior distributions.

Numerical examples are presented throughout the chapter to illustrate the results that are obtained.

2. Optimal search with known proportions

Suppose that there are two different types of items ($C = 2$), and there are K different populations which contain the two types in varying proportions. Let p_{ij} denote the proportion of the items in population i that are of type j ($i = 1, \ldots, K$ and $j = 1, 2$). At each stage of the sequential process, the collector must decide from which of the K populations to draw the next observation. On any drawing from population i there is probability p_{i1} of obtaining an item of type 1 and probability $p_{i2} = 1 - p_{i1}$ of obtaining an item of type 2. This situation is represented by the following matrix:

$$\begin{array}{lcc} & & \text{types} \\ & & \begin{array}{cc} 1 & 2 \end{array} \\ \text{populations} & \begin{array}{c} 1 \\ \vdots \\ K \end{array} & \begin{bmatrix} p_{11} & p_{12} \\ \vdots & \vdots \\ p_{K1} & p_{K2} \end{bmatrix} \end{array} \tag{2.1}$$

The problem is to determine a strategy that will minimize the expected number of observations necessary to obtain both types of items.

More generally, consider a problem in which originally there may have been more than two types of items, but in the course of sampling all but two types, which we shall call types 1 and 2, have been obtained. Suppose again that there are K different populations. Then the situation, as we consider further sampling, can again be described by the matrix (2.1). Now, however, $p_{i+} = p_{i1} + p_{i2}$ may be smaller than 1, since $1 - p_{i+}$ represents the probability of observing one of the types already obtained if an observation is taken from population i.

The form of an optimal strategy in this situation is easy to describe. We will take observations from some fixed population i until an item of either type 1 or type 2 is obtained. We then take all subsequent observations from a population for which the probability of obtaining the remaining type of item is a maximum. Let

$$p_j^* = \max_{i=1,\ldots,K} p_{ij}, \quad \text{for } j=1,2, \tag{2.2}$$

so that p_j^* is the largest value in the jth column of the matrix (2.1). Then by reasoning similar to that which led to (1.1), it can be shown that the expected number N_i of observations with the above strategy is

$$N_i = \frac{1}{p_{i+}} + \frac{p_{i1}}{p_{i+}} \cdot \frac{1}{p_2^*} + \frac{p_{i2}}{p_{i+}} \cdot \frac{1}{p_1^*}. \tag{2.3}$$

We must choose the population i from which we draw the first observation so that it minimizes (2.3).

If p_{i+} has the same value for $i=1,\ldots,K$ (in particular, if $p_{i+}=1$ for $i=1,\ldots,K$), the right-hand side of (2.3) is a weighted average of $1/p_2^*$ and $1/p_1^*$, plus a constant. Hence, if $p_1^* > p_2^*$, we should choose a population i that puts the maximum weight p_{i2} on $1/p_1^*$ and the minimum weight p_{i1} on $1/p_2^*$. In other words, we should choose i such that $p_{i2} = p_2^*$. Similarly, if $p_1^* < p_2^*$, we should choose i such that $p_{i1} = p_1^*$. In summary, the optimal procedure under these conditions can be characterized as follows. Find the largest entry in the matrix (2.1), and delete the column in which it lies. Find the largest entry in the remaining column, and choose the population (the row) at which it occurs for drawing the first observation.

It is easy to see from (2.3) that if $p_1^* = p_2^*$, then we should first sample from a population i for which p_{i+} is a maximum. If p_{i+} has the same value for $i=1,\ldots,K$ and $p_1^* = p_2^*$, then it does not matter from which population the first observation is taken.

3. General problems with known proportions

Now consider a general problem and suppose that at some stage there are exactly c different types of items remaining to be obtained. This situation can be represented by the matrix:

$$\text{populations} \quad \begin{matrix} & \text{types} & \\ & \begin{matrix} 1 & \cdots & c \end{matrix} & \\ \begin{matrix} 1 \\ \vdots \\ K \end{matrix} & \begin{bmatrix} p_{11} & \cdots & p_{1c} \\ \vdots & & \vdots \\ p_{K1} & \cdots & p_{Kc} \end{bmatrix}. \end{matrix} \tag{3.1}$$

Let $p_{i+} = \sum_{j=1}^{c} p_{ij}$, for $i=1,\ldots,K$. Since there may have been more than c different types of items originally and some of them may have already been obtained, p_{i+} will typically be less than 1.

Let A denote the $K \times c$ matrix (3.1) and, for $j=1,\ldots,c$, let A_j denote the $K\times(c-1)$ matrix obtained by simply deleting the jth column of A. If an observation is drawn from any one of the K populations and an item of type j is obtained ($j=1,\ldots,c$), rather than an item of a type that had already been obtained, then the matrix A_j describes the situation for the next observation. For any matrix A, let $N(A)$ denote the expected number of observations necessary to obtain all remaining types of items when the optimal procedure is followed starting from the situation represented by A.

As before, the optimal procedure is to choose some population i, to draw observations from i until an item of one of the c required types is obtained, and then to use an optimal continuation for searching for the remaining $c-1$ types. The expected number N_i of observations with this procedure is found by reasoning similar to that which led to (1.1) or (2.3):

$$N_i = \frac{1}{p_{i+}} + \sum_{j=1}^{c} \frac{p_{ij}}{p_{i+}} N(A_j). \tag{3.2}$$

We must choose the population i that minimizes (3.2). Hence,

$$N(A) = \min_{i=1,\ldots,K} N_i. \tag{3.3}$$

It follows from (3.2) and (3.3) that if we know the optimal procedure starting from any $K\times c$ matrix A, and we know the value of $N(A)$, then we can determine the optimal procedure for any $K\times(c+1)$ matrix B and we can determine the value of $N(B)$. Since the value of $N(A)$ can be determined from (2.3) for any matrix A with two columns, we can at least in principle build up to the value of $N(A)$ for any matrix A.

Before leaving this section, we mention two results that are suggested by the nature of these problems.

Proposition 1. If the matrix A is given by (3.1), then $N(A) \ge N(A_j)$, for $j=1,\ldots,c$.

Proposition 2. Consider the situation represented by the matrix (3.1). If i and k are two different populations such that $p_{ij} \ge p_{kj}$ for $j=1,\ldots,c$, then there is an optimal sampling procedure under which no observations are drawn from population k.

4. Balanced situations

Consider the situation represented by the matrix (3.1) and suppose that $c=3$. For each column j ($j=1,2,3$), let

$$p_j^* = \max_{i=1,\ldots,K} p_{ij}, \tag{4.1}$$

and for each pair of columns a and b ($a, b = 1, 2, 3$ and $a < b$), let

$$p^*_{ab} = \max_{i=1,\ldots,K} (p_{ia} + p_{ib}). \tag{4.2}$$

We know that N_i is given by (3.2) with $c = 3$. Suppose that $p^*_1 = p^*_2 = p^*_3 = p^*$, say. Then, since each matrix A_j ($j = 1, 2, 3$) has two columns, it follows from (2.3) that

$$N(A_1) = \frac{1}{p^*_{23}} + \frac{1}{p^*}, \qquad N(A_2) = \frac{1}{p^*_{13}} + \frac{1}{p^*}, \qquad N(A_3) = \frac{1}{p^*_{12}} + \frac{1}{p^*}. \tag{4.3}$$

If we now suppose in addition that $p^*_{12} = p^*_{13} = p^*_{23} = p^{**}$, say, then it follows from (3.2) that

$$N_i = \frac{1}{p_{i+}} + \frac{1}{p^{**}} + \frac{1}{p^*}. \tag{4.4}$$

Thus, the optimal procedure is very simple to describe.

Begin sampling from the population i for which p_{i+} is a maximum, i.e. the population for which the probability of obtaining any one of the three types is a maximum. Continue sampling from this population until one of these types has been obtained. Then begin sampling from a population for which the probability is a maximum of obtaining an item that is of either one of the two remaining types. Continue sampling from this population until such an item has been obtained. Finally, then begin sampling from a population for which the probability of obtaining the remaining type of item is a maximum. In other words, the *myopic* procedure is optimal. At each stage, sample from a population where there is the largest probability of obtaining a new type.

A situation with $c = 3$ in which $p^*_1 = p^*_2 = p^*_3 = p^*$ and $p^*_{12} = p^*_{13} = p^*_{23} = p^{**}$ is called a *balanced situation*. An example is provided by the following matrix:

$$\begin{array}{cc} & \text{types} \\ & \begin{array}{ccc} 1 & 2 & 3 \end{array} \\ \text{populations} \begin{array}{c} 1 \\ 2 \\ 3 \\ 4 \end{array} & \begin{bmatrix} 0.1 & 0.2 & 0.5 \\ 0.2 & 0.5 & 0.1 \\ 0.5 & 0.1 & 0.2 \\ 0.3 & 0.3 & 0.3 \end{bmatrix} \end{array} \tag{4.5}$$

In (4.5), $p^* = 0.5$ and $p^{**} = 0.7$. Since this is a balanced situation, we begin sampling from population 4 which has the maximum probability $p_{i+} = 0.9$ of yielding one of the three types. If an item of type 1 is obtained first, then we switch to population 1, since it has the maximum probability p^{**} of yielding an item of either type 2 or type 3. If an item of type 2 or type 3 had been obtained first, then we would have switched to population 3 or population 2, respectively.

Now consider the matrix (3.1) for an arbitrary value of c. For each column j, let p^*_j be as defined in (4.1); for each pair of columns, a and b, with $a < b$, let p^*_{ab} be as defined in (4.2); and for each triple of columns, a, b, and d, with

$a < b < d$, let

$$P^*_{abd} = \max_{i=1,\ldots,K} (p_{ia} + p_{ib} + p_{id}). \tag{4.6}$$

Similar definitions should be made for every subset of four columns, every subset of five columns, etc.

Then we say that we are in a *balanced situation* if p^*_j has the same value for every column j, p^*_{ab} has the same value for every pair of columns (a, b), p^*_{abd} has the same values for every triple of columns (a, b, d), etc. In a balanced situation, the myopic procedure is optimal: at each stage, sample from a population where the probability is a maximum of obtaining a type not previously obtained.

If the situation is not balanced, the myopic procedure need not be optimal. For example, suppose that $K = 2$ and $c = 2$, and that the matrix (3.1) has the following form:

$$\begin{array}{ccc} & & \begin{array}{cc} 1 & 2 \end{array} \\ \text{populations} & \begin{array}{c} 1 \\ 2 \end{array} & \begin{bmatrix} p_{11} & p_{12} \\ 0 & p_{22} \end{bmatrix} \end{array} \tag{4.7}$$

For any values such that $p_{22} > p_{11} + p_{12}$, it is optimal to start sampling in population 1.

5. Optimal search with unknown proportions

In this section we begin our study of problems in which the proportions p_{ij} are unknown and the collector assigns a prior distribution to their values. Consider a situation represented by the matrix (3.1) with $c = 1$. In other words, there is only one type of item remaining to be found, and its proportion in population i is p_{i1}, for $i = 1,\ldots, K$. Suppose that the values of $p_{11},\ldots, p_{K1}$ are unknown and that they have independent prior beta distributions. In particular, suppose that the prior distribution of p_{i1} is a beta distribution with parameters α_i and β_i, i.e. the prior density of p_{i1} is

$$\xi(p) = \frac{\Gamma(\alpha_i + \beta_i)}{\Gamma(\alpha_i)\Gamma(\beta_i)} p^{\alpha_i - 1}(1 - p)^{\beta_i - 1}, \quad \text{for} \quad 0 < p < 1, \tag{5.1}$$

where α_i and β_i are given positive constants.

The problem is to determine a search procedure such that with respect to these prior distributions the expected number of observations necessary to obtain an item of the one type remaining will be a minimum. Let X denote the necessary number of observations. Then, by the standard formula,

$$E(X) = 1 + \sum_{n=1}^{\infty} \Pr(X > n). \tag{5.2}$$

It is well known that if we take r observations from population i, then the probability $q(i,r)$ that we will not obtain an item of the required type on any of the r observations is

$$q(i,r)=\prod_{m=0}^{r-1}\frac{\beta_i+m}{\alpha_i+\beta_i+m}. \tag{5.3}$$

Hence, if among the first n observations of a search procedure, exactly n_i are taken from population i, with $\sum_{i=1}^{K}n_i=n$, then

$$\Pr(X>n)=\prod_{i=1}^{K}q(i,n_i), \tag{5.4}$$

where $q(i,n_i)=1$, if $n_i=0$.

It can be seen from (5.3) and (5.4) that we can minimize $\Pr(X>n)$ simultaneously for all values of n, and hence minimize $E(X)$, by employing the following search procedure. For each population i, consider the increasing infinite sequence,

$$S_i=\left\{\frac{\beta_i+m}{\alpha_i+\beta_i+m};\ m=0,1,2,\ldots\right\}, \tag{5.5}$$

and merge these K sequences into a single non-decreasing infinite sequence. If the first term in this merged sequence came from the sequence S_{i_1}, then the first observation in the search procedure should be taken from the population i_1. If the second term in the merged sequence came from the sequence S_{i_2}, then the second observation should be taken from the population i_2, etc. The search continues until an item of the required type has been obtained. This optimal procedure is a myopic procedure in the sense that at each stage, an observation is taken from a population for which the marginal probability of obtaining the required type of item is a maximum.

If the value of p_{i1} is actually known for some population i, then the procedure just described must be modified as follows. Instead of being defined by (5.5), the sequence S_i for population i is defined to be the constant sequence $\{1-p_{i1},1-p_{i1},\ldots\}$. The optimal sequential procedure is then determined from the merged non-decreasing sequence just as before. It follows from this discussion that once the optimal procedure specifies sampling from a population for which the value of p_{i1} is known, then we should continue sampling from that population until an item of the required type has been obtained.

6. Two or more types with unknown proportions

In this section we consider a problem in which there are two types of items, and we are at the beginning of the search process so both types must be obtained. This situation is represented by a matrix of the form (3.1) with $c=2$, where $p_{i1}+p_{i2}=1$, for $i = 1, \ldots, K$. We will again assume that the prior distribution of p_{i1} is a beta distribution with parameters α_i and β_i, for which the density is given by (5.1), and that $p_{11}, \ldots, p_{k1}$ have independent prior distributions. It follows that $p_{i2} = 1 - p_{i1}$ will have a beta prior distribution with parameters β_i and α_i.

Under the optimal procedure, we will draw the first observation from some population i, note whether we have obtained an item of type 1 or an item of type 2, update our posterior distribution in population i, and then follow an optimal procedure of the kind developed in section 5 to obtain the one remaining type of item. Let N_i denote the expected number of observations required by this procedure and let N_{ij} denote the expected number required beyond the first observation if an item of type j is obtained on the first observation ($j = 1, 2$). Then

$$N_i = 1 + \frac{\alpha_i}{\alpha_i + \beta_i} N_{i1} + \frac{\beta_i}{\alpha_i + \beta_i} N_{i2}. \tag{6.1}$$

In accordance with section 5, the value of N_{i2} is obtained as follows. Consider the K infinite sequences $S_1, \ldots, S_K$ defined by (5.5), but with the first term $\beta_i/(\alpha_i + \beta_i)$ of the sequence S_i deleted because of the observation already drawn from population i. Order the merger of these K sequences into a single non-decreasing sequence $b_{i1} \leq b_{i2} \leq \ldots$. Then, from (5.2)–(5.4):

$$N_{i2} = 1 + b_{i1} + b_{i1}b_{i2} + b_{i1}b_{i2}b_{i3} + \ldots. \tag{6.2}$$

A similar description can be given for the value of N_{i1} with the roles of α and β interchanged. For $h = 1, \ldots, K$, let R_h denote the infinite sequence

$$R_h = \left\{ \frac{\alpha_h + m}{\alpha_h + \beta_h + m};\ m = 0, 1, 2, \ldots \right\}. \tag{6.3}$$

Again, delete the first term $\alpha_i/(\alpha_i + \beta_i)$ from the sequence R_i because we have already drawn an observation from population i, and then merge the K sequences $R_1, \ldots, R_K$ into a single non-decreasing sequence $a_{i1} \leq a_{i2} \leq \ldots$. Then

$$N_{i1} = 1 + a_{i1} + a_{i1}a_{i2} + a_{i1}a_{i2}a_{i3} + \ldots. \tag{6.4}$$

If we now let

$$a_{i0} = \frac{\alpha_i}{\alpha_i + \beta_i} \quad \text{and} \quad b_{i0} = \frac{\beta_i}{\alpha_i + \beta_i}, \tag{6.5}$$

then it follows from (6.1)–(6.3) that

$$N_i = \left(1 + \sum_{n=0}^{\infty} \prod_{t=0}^{n} a_{it}\right) + \left(1 + \sum_{n=0}^{\infty} \prod_{t=0}^{n} b_{it}\right) - 1. \tag{6.6}$$

The expression D_{1i} inside the first set of parentheses on the right-hand side of (6.6) is the minimum expected number of observations necessary to obtain an item of type 2 if we must begin the search in population i. That is to say, it is the expected value obtained from merging the sequences $R_1,\ldots, R_K$ into one decreasing sequence, and then being forced to remove the element $a_{i0}=\alpha_i/(\alpha_i+\beta_i)$ from wherever it naturally occurs in this sequence and move it to the beginning. Suppose that a_{i0} was the rth term in the merged non-decreasing sequence. Then in the summation that appears in D_{1i}, all the terms from $n=r$ onward will be the same as they would be in the summation D_1^* based strictly on the merged non-decreasing sequence with α_{i0} as the rth term. Hence we have developed the following results.

Merge the K sequences $R_1,\ldots, R_K$ as defined by (6.3) into a single non-decreasing sequence $a_1\le a_2\le\ldots$. Suppose that $\alpha_i/(\alpha_i+\beta_i)$ is the rth term in this sequence, i.e. $\alpha_i/(\alpha_i+\beta_i)=a_r$. Then

$$\begin{aligned}D_{i1}-D_1^* &= (a_r+a_ra_1+\cdots+a_ra_1\cdots a_{r-2})\\ &\quad -(a_1+a_1a_2+\cdots+a_1a_2\cdots a_{r-1})\\ &= (a_r-a_1)+a_1(a_r-a_2)+\cdots+a_1\cdots a_{r-2}(a_r-a_{r-1}).\end{aligned} \tag{6.7}$$

A similar description holds in regard to the expression D_{2i} inside the second set of parentheses on the right-hand side of (6.6). Merge the K sequences $S_1,\ldots, S_K$ as defined by (5.5) into a single non-decreasing sequence $b_1\le b_2\le\ldots$. Suppose that $\beta_i/(\alpha_i+\beta_i)$ is the sth term in this sequence, and let D_2^* denote the summation based strictly on the ordered merged sequence. Then

$$D_{i2}-D_2^*=(b_s-b_1)+b_1(b_s-b_2)+\cdots+b_1\cdots b_{s-2}(b_s-b_{s-1}). \tag{6.8}$$

Since we should take the first observation from a population i for which N_i is a minimum, it follows from (6.6)–(6.8) that we should choose a population i for which $(D_{i1}-D_1^*)+(D_{i2}-D_2^*)$ is a minimum. Both (6.7) and (6.8) are finite sums so it is possible to determine an optimal value of i by a finite amount of arithmetic.

As an example of these calculations, suppose that $K=3$ and the prior distributions are specified as follows:

		α_i	β_i
population i	1	1	1
	2	2	1
	3	4	2

(6.9)

Then

$$\begin{aligned}R_1&=\{\tfrac{1}{2},\tfrac{2}{3},\tfrac{3}{4},\ldots\},\\ R_2&=\{\tfrac{2}{3},\tfrac{3}{4},\tfrac{4}{5},\ldots\},\\ R_3&=\{\tfrac{2}{3},\tfrac{5}{7},\tfrac{3}{4},\ldots\},\end{aligned} \tag{6.10}$$

and

$$S_1 = \{\tfrac{1}{2}, \tfrac{2}{3}, \tfrac{3}{4}, \ldots\},$$
$$S_2 = \{\tfrac{1}{3}, \tfrac{1}{2}, \tfrac{3}{5}, \ldots\}, \tag{6.11}$$
$$S_3 = \{\tfrac{1}{3}, \tfrac{3}{7}, \tfrac{1}{2}, \ldots\}.$$

For $i=1$, $\alpha_1/(\alpha_1+\beta_1)=\frac{1}{2}$ is the smallest term a_1 among all three sequences R_1, R_2, R_3, so the value of (6.7) is 0. However, $\beta_1/(\alpha_1+\beta_1)=\frac{1}{2}$ is the fourth smallest term b_4 among all three sequences S_1, S_2, S_3, so the value of (6.8) is

$$(\tfrac{1}{2}-\tfrac{1}{3})+\tfrac{1}{3}(\tfrac{1}{2}-\tfrac{1}{3})+\tfrac{1}{3}\cdot\tfrac{1}{3}(\tfrac{1}{2}-\tfrac{3}{7})=0.230. \tag{6.12}$$

Hence, the sum of (6.7) and (6.8) for $i=1$ is 0.230.

For $i=2$, $\alpha_2/(\alpha_2+\beta_2)=\frac{2}{3}$ is a_2, so the value of (6.7) is $\frac{2}{3}-\frac{1}{2}=0.167$. Also, $\beta_2/(\alpha_2+\beta_2)=\frac{1}{3}$ is b_1, so the value of (6.8) is 0. Hence, the sum of (6.7) and (6.8) for $i=2$ is 0.167. The same value of 0.167 is also obtained for $i=3$. Hence, both $i=2$ and $i=3$ yield the smallest values of this sum, and we conclude that the first observation should be drawn from either population 2 or population 3. Either choice is the first stage of an optimal sequential procedure.

Now suppose that the first row of the table (6.9) is changed so that $\alpha_1=\beta_1=4$, while the other values of α_i and β_i are left unchanged. Then the sequences R_1 and S_1 become

$$R_1=S_1=\{\tfrac{1}{2}, \tfrac{5}{9}, \tfrac{3}{5}, \tfrac{7}{11}, \tfrac{2}{3}, \ldots\}, \tag{6.13}$$

while the other sequences R_i and S_i remain as in (6.10) and (6.11).

For $i=1$, the first term of R_1 is again a_1, the first term of S_1 is again b_4, and the sum of (6.7) and (6.8) is therefore again 0.230. However, for both $i=2$ and $i=3$ we now find that $\alpha_i/(\alpha_i+\beta_i)=\frac{2}{3}$ is a_5 and the value of (6.7) is

$$(\tfrac{2}{3}-\tfrac{1}{2})+\tfrac{1}{2}(\tfrac{2}{3}-\tfrac{5}{9})+\tfrac{1}{2}\cdot\tfrac{5}{9}(\tfrac{2}{3}-\tfrac{3}{5})+\tfrac{1}{2}\cdot\tfrac{5}{9}\cdot\tfrac{3}{5}(\tfrac{2}{3}-\tfrac{7}{11})=0.246. \tag{6.14}$$

The value of (6.8) is again 0 for both $i=2$ and $i=3$. Hence, it is now optimal to take the first observation from population 1.

Consider now a general problem of the type introduced in section 3. Suppose, as in that section, that at some stage there are exactly c different types of items remaining to be obtained, so the situation is again represented by the matrix A given in (3.1). Suppose now, however, that the proportions p_{ij} in A are unknown.

For $i=1,\ldots,K$, let $\boldsymbol{p}_i=(p_{i1},\ldots,p_{ic})$ denote the ith row of A. It is natural to try to extend the results that have just been obtained by assuming that the vectors $\boldsymbol{p}_1,\ldots,\boldsymbol{p}_k$ are independent and that the c components of the vector $\boldsymbol{p}_i$ have a Dirichlet distribution with parametric vector $\boldsymbol{\alpha}_i=(\alpha_{i1},\ldots,\alpha_{i(c+1)})$, as represented by the density (see, for example, DeGroot, 1970, section 5.3):

$$\xi(\boldsymbol{p}_i)=\frac{\Gamma(\alpha_{i+})}{\prod_{j=1}^{c+1}\Gamma(\alpha_{ij})}\prod_{j=1}^{c+1}p_{ij}^{(\alpha_{ij}-1)}, \tag{6.15}$$

for $0<p_{ij}<1$ $(j=1,\ldots,c)$, $p_{i+}=\sum_{j=1}^{c}p_{ij}<1$, and $p_{i(c+1)}=1-p_{i+}$. Also, in

(6.15), $\alpha_{ij} > 0$, for $j = 1, \ldots, c+1$, and $\alpha_{i+} = \sum_{j=1}^{c+1} \alpha_{ij}$.

Thus, the collector's joint prior distribution of the elements of A can be represented by the following $K \times (c+1)$ matrix:

$$H = \begin{bmatrix} \alpha_{11} & \cdots & \alpha_{1(c+1)} \\ \vdots & & \vdots \\ \alpha_{K1} & \cdots & \alpha_{K(c+1)} \end{bmatrix}. \tag{6.16}$$

Suppose next that an observation is drawn from population i. Since the unknown proportions in the other populations are distributed independently of the proportions in population i, the posterior distribution of the elements of A in any row other than row i will remain the same as their prior distribution. If the observation is of a type that has already been obtained, then the posterior distribution of the elements in the ith row of A will again be a Dirichlet distribution with the $(c+1)$st component of the parameter vector $\boldsymbol{\alpha}_i$ increased by one unit (see, for example, DeGroot, 1970, section 9.8). In other words, the joint posterior distribution of the elements of the matrix A can again be represented by a matrix of the form H, as in (6.16), but with $\alpha_{i(c+1)}$ increased to $\alpha_{i(c+1)} + 1$.

If the observation drawn from population i is of type j, for some value of j $(j = 1, \ldots, c)$, then the situation changes. The joint posterior distribution of the elements of the matrix A will still be represented by a matrix of the form H, as in (6.16), with α_{ij} increased by one unit to $\alpha_{ij} + 1$ and all other entries of H remaining unchanged. However, since items of type j will no longer need to be obtained, the $K \times (c-1)$ matrix A_j, as defined in section 3, describes the situation for the next stage of the sequential process. It follows (e.g. from DeGroot, 1970, section 5.3) that the unknown proportions of the remaining $c-1$ types of items will again have independent Dirichlet distributions in the different populations, and that the joint distribution of the elements of A_j can be represented by the following $K \times c$ matrix of parameters:

$$H_j = \begin{bmatrix} \alpha_{11} & \cdots & \alpha_{1(j-1)} & \alpha_{1(j+1)} & \cdots & \alpha_{1(c+1)} + \alpha_{1j} \\ \cdots & \cdots & \cdots & \cdots & \cdots & \cdots \\ \alpha_{i1} & \cdots & \alpha_{i(j-1)} & \alpha_{i(j+1)} & \cdots & \alpha_{i(c+1)} + \alpha_{ij} + 1 \\ \cdots & \cdots & \cdots & \cdots & \cdots & \cdots \\ \alpha_{K1} & \cdots & \alpha_{K(j-1)} & \alpha_{K(j+1)} & \cdots & \alpha_{K(c+1)} + \alpha_{Kj} \end{bmatrix}. \tag{6.17}$$

Although a few simple properties of the optimal sequential search procedure can be derived, relatively little that would be helpful in characterizing the optimal procedure is known and we shall not pursue this topic further here.

7. Optimal stopping

In this section we begin consideration of problems in which the collector must pay γ units for each observation that he draws, regardless of the population from which it is drawn, and he receives a fixed bonus for each new type of item that he obtains. Without loss of generality we assume that this bonus is one unit for each new type. The collector can stop the sampling (or search) process at any time, or not even start sampling at all. Thus, if we let Y_n denote the number of different types of items obtained among the first n observations, then the problem is to determine a sampling procedure and a stopping rule that will maximize $E(Y_n - \gamma n)$, where n is the random number of observations taken before sampling ceases.

Suppose first that there is only one population from which observations can be taken, that there are c types of items that have not yet been obtained, and that these types are present in the known proportions $p_1,\ldots, p_c$. Let $p_+ = \sum_{j=1}^{c} p_j$. If we take an observation at cost γ, then we will find a new type of item with probability p_+. Hence, the expected net gain of taking just one observation and then stopping is $p_+ - \gamma$.

If $p_+ > \gamma$, it is clearly worthwhile taking at least the next observation. Suppose, on the other hand, that $p_+ < \gamma$ and we take the next observation. If we do not obtain a new type of item, then $p_1,\ldots, p_c$ and p_+ remain unchanged. If we do obtain an item of type j ($j=1,\ldots,c$), then p_j is dropped from the list of types remaining to be obtained and p_+ is correspondingly reduced. Thus, $p_+ < \gamma$ as we consider taking subsequent observations. In summary, once we reach a stage where $p_+ < \gamma$, we will continue in stages where $p_+ < \gamma$. Thus, we are in the situation that Chow and Robbins (1961) call the monotone case (see also DeGroot, 1970, section 13.14). It follows that an optimal procedure is to continue sampling if and only if $p_+ > \gamma$.

Now suppose that in the problem just described, the values of $p_1,\ldots, p_c$ are unknown and have a prior Dirichlet distribution with parameters $\alpha_1,\ldots,\alpha_{c+1}$. Then the prior probability of obtaining a new type of item when a single observation is drawn is

$$\pi = \left(\sum_{j=1}^{c} \alpha_j\right) \Bigg/ \left(\sum_{j=1}^{c+1} \alpha_j\right).$$

Hence, the expected net gain from taking just one observation and then stopping is $\pi - \gamma$.

If $\pi > \gamma$, it is clearly optimal to continue sampling. Suppose that $\pi < \gamma$, and we take the next observation. The denominator of π will increase by one unit regardless of whether or not we obtain a new type of item. The numerator of π will remain unchanged if we do not obtain a new type, and it will be reduced by the amount α_j if we obtain an item of type j ($j=1,\ldots,c$). Hence, in any event, π will be reduced. Therefore, it is again true that once we reach a stage where $\pi < \gamma$, we will continue in stages where $\pi < \gamma$. We are again in the monotone case and an optimal procedure is to continue sampling if and only if $\pi > \gamma$. These results for optimal stopping when there is just a single population have also been described by Rasmussen and Starr (1979).

Next consider the general situation described at the end of section 6 in which there are K populations from which observations can be drawn and the proportions p_{ij} of the c different types of items in the different populations are unknown. As in section 6, assume that the vectors $\boldsymbol{p}_1, \dots, \boldsymbol{p}_K$ have independent prior Dirichlet distributions as represented by the matrix of parameters H given by (6.16). Let

$$\pi_i = \frac{\sum_{j=1}^{c} \alpha_{ij}}{\sum_{j=1}^{c+1} \alpha_{ij}}, \quad \text{for} \quad i = 1, \dots, K. \tag{7.1}$$

If $\pi_i > \gamma$ for at least one value of i, then it clearly worthwhile to continue sampling, since we can at least attain a positive expected net gain by taking just a single observation from population i and then stopping. On the other hand, if $\pi_i < \gamma$, $i = 1, \dots, K$, then it is not worthwhile taking any further observations because, as explained in this section, we can never again reach a stage where $\pi_i > \gamma$ for at least one value of i. Thus, we are in the monotone case and an optimal procedure is to continue sampling if and only if $\pi_i > \gamma$ for at least one population i.

It should be emphasized that although the result just obtained characterizes the optimal stopping rule, in the sense that it tells us exactly when to continue sampling and when to stop, it does not tell us from which population to take the next observation when we do continue. This problem is relatively difficult and we shall consider some special cases in the next section.

8. Optimal stopping with known proportions

In this, the final section, we again consider problems of the type described in section 2, in which $c = 2$ and the proportions p_{ij} are known. Suppose, therefore, that two types of items remain to be obtained and the situation is represented by the matrix (2.1), where the values of $p_{i+} = p_{i1} + p_{i2}$ may be smaller than 1.

Suppose first that both $p_1^* > \gamma$ and $p_2^* > \gamma$, where p_j^* is defined by (2.2). Under these conditions it will be worthwhile to continue sampling until both types of items have been obtained. Therefore, the expected net gain from any such sampling procedure will be

$$E(Y_n - \gamma n) = 2 - \gamma E(n). \tag{8.1}$$

It can be seen from (8.1) that a procedure that maximizes this expected net gain will minimize the expected number of observations $E(n)$ necessary to obtain both types of items. Hence, the optimal procedure described in section 2 will also be optimal here.

Next, suppose that both $p_1^* < \gamma$ and $p_2^* < \gamma$, but that

$$p_+^* = \max_{i=1,\dots,K} p_{i+} > \gamma. \tag{8.2}$$

In this case the optimal procedure is to draw observations from a population i such that $p_{i+} = p_+^*$ until an item of either type 1 or type 2 is obtained, and then to stop sampling. If $p_+^* < \gamma$, the optimal procedure is not to sample at all.

Finally, suppose that $p_j^* < \gamma$ for one value of j and $p_j^* > \gamma$ for the other value of j. To be specific, suppose that $p_1^* < \gamma$ and $p_2^* > \gamma$. The optimal procedure will have the following form. Begin sampling from population i and continue taking observations from that population until an item of either type 1 or type 2 is obtained. If the item is of type 2, then stop sampling. If the item is of type 1, then draw the remaining items from a population k such that $p_{k2} = p_2^*$ and continue sampling until an item of type 2 is obtained.

The expected gain E_i from this procedure will be

$$E_i = 1 - \frac{\gamma}{p_{i+}} + \frac{p_{i1}}{p_{i+}}\left(1 - \frac{\gamma}{p_2^*}\right). \tag{8.3}$$

Hence, it follows after some algebra that

$$\frac{1}{\gamma}(2 - E_i) = \frac{1}{p_{i+}} + \frac{p_{i1}}{p_{i+}} \cdot \frac{1}{p_2^*} + \frac{p_{i2}}{p_{i+}} \cdot \frac{1}{\gamma}. \tag{8.4}$$

The population i that maximizes E_i will minimize (8.4). It can be verified that the expression (8.4) is the same as (2.3) with $p_1^* = \gamma$. This feature leads to the following interpretation of the optimal procedure.

To the K populations with proportions represented by the matrix (2.1), we add a fictional $(K+1)$st population with proportions

$$p_{(K+1)1} = \gamma \quad \text{and} \quad p_{(K+1)2} = 0. \tag{8.5}$$

Then in this extended problem, $p_1^* = \gamma$, while p_2^* remains the same as before. Hence, in the problem of optimal stopping that we are now considering, it will be optimal to begin sampling from the population i for which N_i, as given by (2.3), is a minimum. It remains only to verify that it is not optimal to begin sampling from the fictional population $K+1$. To establish this, consider a population k such that $p_{k2} = p_2^*$. Then, from (2.3), (8.5), and the relations $p_1^* = \gamma$ and $p_{k+} = p_{k1} + p_2^*$, it follows that

$$\begin{aligned} N_{K+1} - N_k &= \frac{1}{\gamma} + \frac{1}{p_2^*} - \frac{1}{p_{k+}} - \frac{p_{k1}}{p_{k+}} \cdot \frac{1}{p_2^*} - \frac{p_2^*}{p_{k+}} \cdot \frac{1}{\gamma} \\ &= \frac{p_{k1}}{p_{k+}} \cdot \frac{1}{\gamma} \geq 0. \end{aligned} \tag{8.6}$$

Thus, sampling from population k is always at least as good as sampling from the fictional population $K+1$.

It can be seen from (8.3) that if p_{i+} has the same value for $i = 1, \ldots, K$ (in particular, if $p_{i+} = 1$ for $i = 1, \ldots, K$), then the first observation should be taken from a population i such that p_{i1} is as large as possible, i.e. such that $p_{i1} = p_1^*$.

We conclude with a simple numerical example. Suppose that $K = 2$ and the matrix (2.1) has the following numerical values:

$$\text{populations}\quad \begin{array}{c} \\ 1 \\ 2 \end{array} \overset{\displaystyle\text{types}}{\begin{array}{cc} 1 & 2 \\ \end{array}} \begin{bmatrix} 0.4 & 0.2 \\ 0 & 0.8 \end{bmatrix}. \tag{8.7}$$

Suppose also that $\gamma = 0.5$, so $p_1^* = 0.4 < \gamma < p_2^* = 0.8$. If we begin sampling from population 1 and use the optimal continuation as just described, then the expected net gain is, from (8.3):

$$E_1 = 1 - \tfrac{5}{6} + \tfrac{2}{3}\left(1 - \tfrac{5}{8}\right) = \tfrac{5}{12}. \tag{8.8}$$

If we begin sampling from population 2, the expected net gain is

$$E_2 = 1 - \tfrac{5}{8} = \tfrac{3}{8}. \tag{8.9}$$

Hence, it is optimal to begin sampling from population 1, even though $p_{1+} < p_{2+}$. In this example, the myopic procedure is not optimal.

References

Baum, L.E. and P. Billingsley (1965) "Asymptotic distribution for the coupon collector's problem", *Ann. Math. Statist.*, 36, 1835–1839.

Chow, Y.S. and H. Robbins (1961) "A martingale system theorem and applications", *Proc. Fourth Berkeley Symp. Math. Statist. Probability*, vol. 1 (University of California Press), pp. 93–104.

DeGroot, M.H. (1970) *Optimal Statistical Decisions* (McGraw-Hill Book Co., New York).

DeGroot, M.H. (1975) *Probability and Statistics* (Addison-Wesley Publishing Co., Reading, Mass.).

Feller, W. (1968) *An Introduction to Probability Theory and Its Applications*, vol. 1, 3rd edn. (John Wiley & Sons, New York).

Holst, L. (1972) "Asymptotic normality and efficiency for certain goodness-of-fit tests", *Biometrika*, 59, 137–145.

Holst, L. (1973) "Some limit theorems with applications in sampling theory", *Ann. Statist.*, 1, 644–658.

Lanke, J. (1972) "Sampling distinguishable elements with replacement", *Ann. Math. Statist.*, 43, 1329–1332.

Rasmussen, S.L. and N. Starr (1979) "Optimal and adaptive stopping in the search for new species", *J. Amer. Statist. Assoc.*, 74, 661–667.

Rényi, A. (1962) "Three new proofs and a generalization of a theorem of Irving Weiss", *Publ. Math. Inst. Hungar. Acad. Sci.*, Ser. B. 7, 203–214.

Rosén, B. (1969) "Asymptotic normality in a coupon collector's problem", *Z. Wahrscheinlichkeitstheorie und Verw. Gebiete*, 13, 256–279.

Rosén, B. (1970) "On the coupon collector's waiting time", *Ann. Math. Statist.*, 41, 1952–1969.

Rosén, B. (1972) "Asymptotic theory of successive sampling with varying probabilities without replacement, I and II", *Ann. Math. Statist.*, 43, 373–397 and 748–776.

Sen, P.K. (1979) "Invariance principles for the coupon collector's problem: A martingale approach", *Ann. Statist.*, 7, 372–380.

Chapter 29

THE RELATIONSHIP BETWEEN THE NUMBER OF FACTORS AND SIZE OF AN EXPERIMENT

DENNIS V. LINDLEY*

University of Wisconsin-Madison

1. Introduction

With the growth in computing power, statisticians have turned their attention to data sets involving a large number of factors, or variables. In this chapter we consider the relationship between n, the size of the data set, or experiment, and m, the number of factors. Intuitively it appears plausible that as the number of factors increases so should the size of the experiment in order to unravel the increasing complexity that could arise with many factors. In some sense this is not true, as the following two examples show.

In an experiment in which all the factors are at two levels and a complete, factorial arrangement is used, any contrast is a mean of 2^{m-1} values against a similar mean and has variance $2^{-(m-2)}$ times that of any unit. This value is the same for a main effect (involving only *one* factor) as it is for the highest order interaction (involving m factors). Consequently, although it is necessary to increase n with m in order to estimate all the interactions (in fact n must be at least 2^m) the increase of m does not mean the introduction of effects that are more difficult to estimate than the effect of any single factor. In the situation discussed in the present chapter this is not so, the highest-order interactions being exponentially more difficult to estimate than the main effects. (Corollary 1 gives the exact expression.)

Another example arises with the multivariate normal distribution with n observations on each of m variables. Each mean has n values to estimate it, as

*Sponsored by the United States Army under contract no. DAAG29-80-C-0041, at Mathematics Research Center, University of Wisconsin, Madison. Author's present address: 2 Periton Lane, Minehead, TA24 8AQ, U.K.

Bayesian Inference and Decision Techniques, edited by P. Goel and A. Zellner

has each variance. Equally, each covariance has n pairs of values available. Consequently, each parameter is just as easily estimated with m variables as with two. It is only when sets of parameters are considered that difficulties arise: for example, with $m > n$, the sample dispersion matrix is necessarily singular. The fact remains that each parameter is perfectly well estimated.

In this chapter a model is studied in which the increase in m does genuinely increase the complexity of the situation. The model was suggested by problems of medical diagnosis where upwards of 40 symptoms (factors) may be observed on a patient and it is required to diagnose the disease from these. Typically the presence or absence of a symptom is not too well-defined and the model allows for errors in this regard. We do not pretend to have made a contribution to actual diagnosis but merely to indicate the difficulties that might arise when many symptoms are studied. The conclusion is that in situations of a diagnostic type, in contrast to the usual normal error factorial type, it is harder to appreciate the effect of many factors.

The calculations are concerned with the likelihood function and are meaningful within any statistical philosophy. Bayes' Theorem is used in (3.1) to explain why the log-likelihood ratio is useful and for simplicity in writing down the formulae a uniform prior is used (with log-odds zero) but any prior would give the same results, which are only asymptotic. The argument leads to extensive use of the Kullback–Leibler numbers and a reader who accepts their usefulness need not depend on any Bayesian philosophy to justify the results.

2. The model

All quantities are binary, taking the values plus or minus one, or simply + or −. There are m binary factors $\xi = (\xi_1, \xi_2, \ldots, \xi_m)$ and dependent on them is a binary response η described by a response function $\eta = \delta(\xi)$. For each unit in the data set every ξ_i has chance $1/2$ of being + independently of the other factors: thus each ξ has chance 2^{-m}. Each of the factors and the response for a unit are observed with a possible error. ξ_i is observed as x_i and η as y with $p(x_i \neq \xi_i) = p(y \neq \eta) = p$, $p < \frac{1}{2}$. These are independent and independent of the factors or response. We write $x = (x_1, x_2, \ldots, x_m)$. On the basis of observations $z = (x, y)$ on each of n units it is required to make inferences about δ, the response mechanism. A few comments on the model are now offered.

The structure through the ξ's and η measured with error seems reasonable for many situations (such as medical diagnosis) where the view is that were the correct quantities to be measured the response would be determined. It therefore differs from the factorial situation mentioned above in that there are errors in the factors (economists speak of errors-in-variables models). Where the model is seriously deficient as a description of reality is in the chance structure imposed for simplicity in the subsequent analysis. It might be possible to have ξ_i (and hence x_i) equally likely to be + or − by defining the factors suitably: for example, a continuous variable could be dichotomized at the median. But it is unrealistic to assume that the factors are independent. Our reason for doing so is that we do not know of a simple way to describe adequately the dependence amongst binary factors: indeed, this is a major problem in a satisfactory statistical treatment of diagnosis. Another defect is that the errors have the same chances for each factor. In the medical situation this is not true: age is more accurately determined than blood pressure. Different chances could be studied with the penalty of a substantial increase in algebraic and computational complexity. Our defense of the model is that our interest lies primarily in the relationship between m and n as far as determining δ is concerned, and that the general form of this relationship might not be too disturbed by more realistic modelling of the chances, though strong dependence might change the details substantially. The exact numerical values calculated below should not be taken seriously: but we hope that their orders of magnitude might be.

It might be suspected that problems would arise as m increases since the number of δ's is 2^{2^m}, making the determination of δ more difficult. Notice that the model has no parameters and the argument is essentially non-parametric.

3. Discrimination between response functions

Let δ_1 and δ_2 be any two different response functions carrying ξ into η and at some stage of the experiment let $p(\delta_1)$ denote the probability attached to δ_1. If an additional unit is observed to have $z = (x, y)$, then the log-odds for δ_1 against δ_2 will become

$$\log\frac{p(\delta_1|z)}{p(\delta_2|z)} = \log\frac{p(z|\delta_1)}{p(z|\delta_2)} + \log\frac{p(\delta_1)}{p(\delta_2)}, \tag{3.1}$$

by Bayes' Theorem. Conditional on δ_1, the expected change in log-odds will be

$$\Delta(\delta_1 : \delta_2) = \sum_z p(z|\delta_1)\cdot\log\frac{p(z|\delta_1)}{p(z|\delta_2)}, \tag{3.2}$$

a Kullback–Leibler number. (Notice that in the notation the first argument of Δ is the δ for which the expectation is being calculated – the "true" value – whereas the second argument is the alternative value. In general $\Delta(\delta_1 : \delta_2) \neq \Delta(\delta_2 : \delta_1)$. We sometimes abbreviate the former to Δ_{12}.)

As more units are added the log-odds change by the addition of the appropriate log-likelihood ratio, and the expected change for n units is $n\Delta_{12}$. To achieve a prescribed level of discrimination between δ_1 and δ_2 when δ_1 obtains, we would need the log-odds to attain a prescribed level R and expect to see n units to achieve this, where $R = n\Delta_{12}$. Hence, the sample size required is proportional to Δ_{12}^{-1}. For a given δ_1 the discrimination is most difficult and requires the largest sample size for the alternative δ_2 that minimizes $\Delta(\delta_1 : \delta)$ over all δ not equal to δ_1. Since δ_1 is itself initially unknown, the size of experiment is related to the least Kullback–Leibler number $\Delta(\delta_1 : \delta_2)$ over all unequal δ_1 and δ_2.

In the discussion that follows it will be supposed that all δ's are initially equally likely so that the original log-odds are zero. If this is not so, then the appropriate log-odds should be added to $n\Delta_{12}$ before the evaluation. This will not affect the conclusions which are concerned with large n where the original log-odds have little effect.

4. Interactions

The evaluation of the Δ's for general δ presents difficulties, so we consider first an important subclass of δ's called interactions. In applications interest might center on δ's that depend only on a subclass of the m factors. If changing ξ_i has no effect on δ whatever be the values of the other factors, we say ξ_i is *irrelevant* to δ: the remaining factors are called *relevant* to δ. The set of relevant factors for δ is denoted $R(\delta)$.

δ is an *odd* (*even*), *k-factor interaction* if $R(\delta)$ contains k factors and $\eta = 1$ iff an odd (even) number of these factors are +. The odd and even interactions with the same $R(\delta)$ are called *complementary*. The 1-factor interactions may be called the *main effects* of the single, relevant factor. To complete the situation we need the two complementary zero-factor interactions in which $\eta = -(+)$ for all ξ. Notice that interactions, although similar to interactions in factorial experiments, are different in that they occur in complementary pairs, the one being obtained from the other by interchanging + and − in the response values. All the odd interactions for 3-factors are listed in the last seven columns of table 29.1.

Table 29.1. Table for three factors ($m = 3$).

				1-factor			2-factor			3
	ξ		δ	ξ_1	ξ_2	ξ_3	$\xi_1\xi_2$	$\xi_2\xi_3$	$\xi_3\xi_1$	$\xi_1\xi_2\xi_3$
−	−	−	−	−	−	−	−	−	−	−
−	−	+	−	−	−	+	−	+	+	+
−	+	−	−	−	+	−	+	+	−	+
+	−	−	−	+	−	−	+	−	+	+
+	+	−	+	+	+	−	−	+	+	−
+	−	+	+	+	−	+	+	+	−	−
−	+	+	−	−	+	+	+	−	+	−
+	+	+	+	+	+	+	−	−	−	+
			N:	7	5	5	3	5	3	3
							(5)		(5)	(5)
			w:	$\frac{3}{4}$	$\frac{1}{4}$	$\frac{1}{4}$	$\frac{1}{4}$	$\frac{1}{4}$	$\frac{1}{4}$	$\frac{1}{4}$
			E:	−0.59	−0.84			−0.77		−0.74

Note: N is the number of agreements between δ and the interaction of that column. When this is less than 2^{m-1} (here 4), N for the complementary interaction is given in parentheses. $w = 2^{-(m-1)}N - 1$ (eq. (8.2)) with $N \geq 2^{m-1}$. $E = E(\delta, \delta_k)$ for interaction δ_k (eq. (6.1)).

In an abuse of notation, a k-factor interaction will often be denoted δ_k. Two different k-factor interactions will be written δ_k and δ'_k.

The reader not interested in the proofs may at this point proceed directly to theorem 2 below: the I- and J-functions are defined immediately after the statements of lemmas 2 and 3. The mathematical techniques employed are related to those used in error-correcting codes (for example, Lin, 1970) but the development given here is self-contained.

5. Probability evaluations

For any $\zeta = (\xi, \eta)$ and δ either $\eta = \delta(\xi)$ or not. The 2^m ζ's for which $\eta = \delta(\xi)$ are said to *agree* with δ and the set of them is denoted $Z(\delta)$. Since all 2^m ξ's have the same chance,

$$p(z|\delta) = 2^{-m} \sum_{\zeta \in Z(\delta)} p(z|\zeta) \tag{5.1}$$

and

$$p(z|\zeta) = p^s q^{m+1-s},$$

where $q = 1 - p$ and s is the number of disagreements between z and ζ, a disagreement being where $\xi_i \neq x_i$ or where $\eta \neq y$. We refer to s as the (Hamming) *distance* between z and ζ.

The following simple result will be used repeatedly in the argument. If s_1 is the number of disagreements in respect of relevant factors and the response, and s_2 is the number of disagreements for irrelevant factors, so that $s = s_1 + s_2$, we may write for $\delta = \delta_k$:

$$p(z|\zeta) = p^{s_1} q^{k+1-s_1} \cdot p^{s_2} q^{m-k-s_2} \tag{5.2}$$

(s_1 is called the *relevant distance*). In summing $p(z|\zeta)$ over $\zeta \in Z(\delta)$ *all* combinations of the irrelevant factors will occur since they have no effect on η. Hence, the second factor in (5.2) will yield $(p+q)^{m-k} = 1$. Consequently, in evaluating (5.1) it is enough to consider only the relevant factors and the response. This is termed the *relevancy principle.*

We proceed to the evaluation of $p(z|\delta)$ – eq. (5.1) – when δ is a k-factor interaction. There are two cases to consider according as $z \in Z(\delta)$ or not. If $z \in Z(\delta)$ there is one ζ identical with z and distant zero. There are no ζ's distant one since a disagreement for a ξ_i will mean a disagreement either for another factor or for the response by the interaction property. There are two types of ζ distant 2: those with $\eta = y$ and two disagreements in relevant factors; and those with $\eta \neq y$ and one disagreement in relevant factors. There are $\binom{k}{2}$ of the first type and k of the second, a total of $\binom{k+1}{2}$. (The other factors are omitted by the relevancy principle.) Continuing with this line of reasoning we find that if

$z \in Z(\delta)$ there are $\binom{k+1}{s_1}$ ζ's at relevant distance s_1, provided s_1 is even, and none if s_1 is odd. If $z \notin Z(\delta)$ the same result holds with odd and even interchanged. Hence,

$$p(z|\delta_k) = 2^{-m} \sum \binom{k+1}{s_1} p^{s_1} q^{k+1-s_1},$$

where the summation is over s_1 even (odd) according as $z \in (\notin) Z(\delta_k); 0 \le s_1 \le k+1$.

Lemma 1. In n Bernoulli trials with chance p of success, the chance of an odd number of successes is

$$\alpha_n = \tfrac{1}{2} - \tfrac{1}{2}(q-p)^n.$$

This well-known result is easily established by writing down and solving a recurrence relation for α_n. Notice that since $p < \frac{1}{2}$, $p < q$, and $\alpha_n < \frac{1}{2}$; also α_n increases with n. Direct use of the lemma gives

Theorem 1. For a k-factor interaction δ_k:

$$p(z|\delta_k) = \begin{cases} 2^{-(m+1)}\{1+(q-p)^{k+1}\} = 2^{-m}(1-\alpha_{k+1}), & \text{if } z \in Z(\delta_k), \\ 2^{-(m+1)}\{1-(q-p)^{k+1}\} = 2^{-m}\alpha_{k+1}, & \text{if } z \notin Z(\delta_k). \end{cases}$$

6. The Kullback–Leibler numbers

As a first stage in their calculation we evaluate

$$E_{12} = E(\delta_1 : \delta_2) = \sum_z p(z|\delta_1) \log p(z|\delta_2). \tag{6.1}$$

Lemma 2. For a k-factor interaction δ_k:

$$E(\delta_k : \delta_k) = I(\alpha_{k+1}) - m \log 2.$$

Here $I(x) = x \log x + (1-x)\log(1-x)$ is Shannon's information function. From theorem 1, (6.1) and the consideration that both $Z(\delta_k)$ and its complement contain 2^m members:

$$\begin{aligned} E(\delta_k : \delta_k) = {} & 2^m \cdot 2^{-m}(1-\alpha_{k+1}) \log 2^{-m}(1-\alpha_{k+1}) \\ & + 2^m \cdot 2^{-m}\alpha_{k+1} \log 2^{-m}\alpha_{k+1}, \end{aligned}$$

which easily reduces to the expression given.

Lemma 3. If δ_j and δ_k are respectively j- and k-factor interactions which are neither the same nor complementary, then

$$E(\delta_j : \delta_k) = J(\alpha_{k+1}) - m \log 2.$$

Here $J(x)=\frac{1}{2}\log x(1-x)$. $\log p(z|\delta_k)$ takes only two values (theorem 1) and the multiplier of $\log 2^{-m}(1-\alpha_{k+1})$ will be $\sum_{z\in Z(\delta_k)}p(z|\delta_j)$. Each term in this summation will be either $2^{-m}(1-\alpha_{j+1})$ or $2^{-m}\alpha_{j+1}$ according as $z\in Z(\delta_j)$ or not. But since δ_j and δ_k are interactions (neither equal nor complementary) both $Z(\delta_j)\cap Z(\delta_k)$ and $\bar{Z}(\delta_j)\cap Z(\delta_k)$ contain 2^{m-1} members. Hence, the multiplier is

$$2^{m-1}\cdot 2^{-m}\alpha_{j+1}+2^{m-1}\cdot 2^{-m}(1-\alpha_{j+1})=\tfrac{1}{2}.$$

Similar considerations apply to the multiplier of $\log 2^{-m}\alpha_{k+1}$ and the result is immediate.

Theorem 2. If δ_j and δ_k are respectively j- and k-factor interactions which are neither the same nor complementary:

$$\Delta(\delta_j:\delta_k)=I(\alpha_{j+1})-J(\alpha_{k+1}). \tag{6.2}$$

If they are complementary ($\delta_k=\delta_j'$):

$$\Delta(\delta_j:\delta_j')=2\{I(\alpha_{j+1})-J(\alpha_{j+1})\} \tag{6.3}$$

(if $\delta_j=\delta_k$, the Kullback–Leibler number is clearly zero).

The theorem is obvious from lemmas 2 and 3 since $\Delta_{12}=E_{11}-E_{12}$. The special, complementary case follows easily by following an argument parallel to that used to establish lemma 3.

7. Expected size of experiment

The implications of theorem 2 are best understood by referring to fig. 29.1 which sketches the functions $I(x)$ and $J(x)$ in $0\le x\le\frac{1}{2}$. Properties of these functions that are cited are all well known and easily established by use of the differential calculus. $I(x)$ is decreasing and $J(x)$ increasing in the interval and they are equal to $-\log 2$ at $x=\frac{1}{2}$. Since α_j increases with j, $\Delta(\delta_j:\delta_k)$, which is of course positive, decreases both with j and with k: that is, the higher the orders of the interactions (or equivalently, the more factors involved in either the true or alternative response) the more difficult it is to separate them. As we saw in section 3, the expected size of the experiment is inversely proportional to the Kullback-Leibler number and with m factors the least number will be $\Delta(\delta_{m-1}:\delta_{m-1}')$ in comparing one $(m-1)$-factor interaction with another. (There are only two complementary m-factor interactions which are more easily distinguished because of the doubling of the Kullback–Leibler number that then occurs: compare (6.2) and (6.3) which differ, when $j=k$, only by a factor of 2.)

Corollary 1. In the assumed model with m factors and attention confined solely to interactions, the expected size of the experiment to distinguish adequately between all response functions is inversely proportional to

$$\Delta(\delta_{m-1}:\delta_{m-1}')=I(\alpha_m)-J(\alpha_m)=\tfrac{1}{2}(q-p)^m\cdot\log\frac{1+(q-p)^m}{1-(q-p)^m}.$$

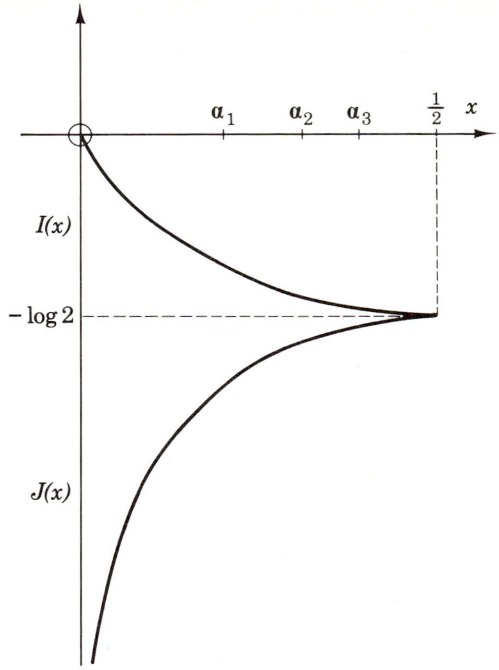

Fig. 29.1. $I(x) = x \log x + (1-x)\log(1-x)$: $J(x) = \frac{1}{2}\log x(1-x)$: $\alpha_n = \frac{1}{2} - \frac{1}{2}(q-p)^n$.

The behavior of this function with m can be appreciated by noting that for large m it is approximately $(q-p)^{2m}$, so that the increase of n with m is about exponentially fast. The following values for $p = 0.1$ underline this point:

m	2	4	8	16	32
$\Delta_{m-1,m-1}$	0.4852	0.1782	0.0284	7.925×10^{-4}	6.277×10^{-7}

so that an experiment with 32 factors (or medical symptoms) can be expected to require about 10^6 times more observations than one with only 2 factors. As explained in section 2, these numbers are not to be taken too seriously, they can only express orders of magnitude. Notice that they are exaggerated by the fact that the *worst* case is being discussed. With a distribution over δ's the sizes would fall. For example, if one was sure that only k out of the m factors were relevant (but not which k) then $\Delta_{k,k}$ not $\Delta_{m-1,m-1}$, would indicate the order. Against this, only interactions are being considered, 2^m out of the 2^{2^m} possible functions, so that even larger sizes are possible; see section 8.

If $j > k$ in theorem 2, $\Delta(\delta_j : \delta_k) > \Delta(\delta_k : \delta_j)$. Taking $k = 1$ for illustration, if a main effect is the true response it is harder to eliminate a j-factor ($j > 1$)

interaction than it would be to eliminate the main effect were the true response that j-factor interaction.

8. General response functions

A feature of interactions that greatly facilitates the computations is that $p(z|\delta)$, as z ranges over all 2^{m+1} values, assumes only two values (theorem 1) with the result that any E (eq. (6.1)) only contains two logarithmic terms. For a general response many more values will arise and the calculation of the E's, and hence the Δ's, is formidable. Some understanding of what happens with a general response can be obtained by calculating a generalized Kullback–Leibler number:

$$\Delta(\delta : \delta_1, \delta_2) = \sum_z p(z|\delta) \cdot \log \frac{p(z|\delta_1)}{p(z|\delta_2)}$$

$$= E(\delta : \delta_1) - E(\delta : \delta_2), \tag{8.1}$$

where δ_1 and δ_2 are still interactions but δ is general. This is the expected change in log-odds from one unit when comparing two interactions but when the true response is δ. Even in complete factorial experiments (section 1) it is usual to do the analysis in terms of the interactions even though the overall effect is not a pure interaction. There *any* effect is a linear combination of interactions: the same property holds here. Hence, a study of interactions with a general response is not inappropriate.

We now proceed to evaluate $E(\delta : \delta_1)$, where δ is arbitrary and δ_1 is an interaction. There are two ways to do this: either by an extension of the argument already used when δ is an interaction, or by using the theory of vector spaces in which δ is expressed as a linear combination of interactions. The second method is simpler and has the added advantage that the structure of the situation is more clearly revealed. The reader not interested in mathematical details may proceed to the statement of theorem 3, though eq. (8.2) has to be understood in order to appreciate the role of the w's in the statement of that theorem.

To express any response function as a vector in 2^m-space write out the 2^m possible ξ's in any fixed order. The first three columns of table 29.1 give such a list for $m = 3$. The values $\eta(\xi)$ form a string of 2^m pluses and minuses and provide a complete description of δ: we shall denote this vector by δ. Such a representation of a δ is given in the fourth column of the table. For any two different δ's consider the scalar product $\delta_1^T\delta_2$. The contribution to this will be $+1$ if δ_1 and δ_2 agree at a particular place, and -1 if not. Let N denote the number of agreements between δ_1 and δ_2 so that $2^m - N$ is the distance between them. Then

$$\delta_1^T\delta_2 = N - (2^m - N) = 2(N - 2^{m-1}),$$

and we write

$$2^{-m}\delta_1^{\mathrm{T}}\delta_2 = 2^{-(m-1)}(N - 2^{m-1}) = w. \tag{8.2}$$

If $\delta_1 = \delta_2$, then $2^{-m}\delta^{\mathrm{T}}\delta = 1$; if δ_1 and δ_2 are complementary, then $2^{-m}\delta^{\mathrm{T}}\delta = -1$. If δ_1 and δ_2 are any two different and non-complementary interactions, $N = 2^{m-1}$ and $w = 0$. It follows that the 2^m odd interactions form an orthogonal basis for the vector representation of δ's. (The same holds for the even interactions.) We may therefore write $\delta = \sum a_i\delta_i$, where δ is arbitrary, $\{\delta_i\}$ is the set of odd interactions, and $\{a_i\}$ is a set of numbers to be found. For interaction δ_s, $\delta^{\mathrm{T}}\delta_s = \sum a_i\delta_i\delta_s = a_s 2^m$ and consequently from (8.2) $w_s = a_s$. We therefore have the result that any response can be written as a linear combination of odd interactions with weights w equal to $2^{-(m-1)}N_s - 1$, where N_s is the number of agreements between δ and the interaction δ_s; eq. (8.2).

Consider the 2^m vectors of all odd interactions to form a square matrix. Then the number of minuses in every *row* of this matrix will be 2^{m-1} except for one row in which it will be 2^m. (Table 29.1 again illustrates for $m = 3$.) This exceptional row will be that corresponding to the ξ for which every ξ_i is $-$. If δ has also $-$ in this row (if not, then use the representation in terms of even interactions) the total number of agreements between δ and all odd interactions will be $\sum N_s = 2^m + (2^m - 1)2^{m-1}$, from which it follows that $\sum w_s = 1$.

Finally, apply the transformation $\delta' = \frac{1}{2} + (p - \frac{1}{2})\delta$ to any δ which turns each $-$ into $1 - p = q$ and each $+$ into p. Then

$$\delta' = \tfrac{1}{2} + (p - \tfrac{1}{2})\delta = \tfrac{1}{2}\sum w_s + (p - \tfrac{1}{2})\sum w_s\delta_s = \sum w_s\delta_s',$$

since $\sum w_s = 1$, and the representation is the same after transformation.

This last result enables $E(\delta : \delta_1)$ to be calculated rather easily. From (5.1):

$$p(z|\delta) = 2^{-m} \sum_{\xi \in Z(\delta)} p(z|\zeta) = 2^{-m}\sum_{\xi} p(x|\xi)p(y|\delta(\xi)),$$

but from the result just established:

$$p(y|\delta(\xi)) = \sum_s w_s p(y|\delta_s(\xi)),$$

so that

$$\begin{aligned} p(z|\delta) &= 2^{-m}\sum_{\xi} p(x|\xi)\sum_s w_s p(y|\delta_s(\xi)) \\ &= 2^{-m}\sum_s w_s \sum_{\xi} p(x|\xi)p(y|\delta_s(\xi)) \\ &= \sum w_s p(z|\delta_s). \end{aligned}$$

In words, the representation of δ in terms of interactions is also the representation of $p(z|\delta)$ in terms of $p(z|\delta_s)$. (Note that although $\sum w_s = 1$, the w's are not necessarily positive so that this is *not* a straightforward probability result.) It immediately follows from (8.1) that

$$E(\delta : \delta_1) = \sum_s w_s E(\delta_s : \delta_1)$$

and the $E(\delta_s : \delta_1)$-values are known from lemmas 2 and 3. Hence, again using $\sum w_s = 1$:

$$E(\delta : \delta_1) = w_1 I(\alpha_{i+1}) + (1 - w_1) J(\alpha_{i+1}) - m \log 2,$$

where i is the order of the interaction δ_1.

Theorem 3. For any response δ and for different and non-complementary interactions δ_s and δ_t of orders s and t, respectively, the generalized Kullback–Leibler number $\Delta(\delta : \delta_s, \delta_t)$ is

$$w_s I(\alpha_{s+1}) + (1 - w_s) J(\alpha_{s+1}) - w_t I(\alpha_{t+1}) - (1 - w_t) J(\alpha_{t+1}). \tag{8.3}$$

Here N_s is the number of agreements between δ and δ_s and $w_s = 2^{-(m-1)}(N_s - 2^{m-1})$.

The result follows directly from the evaluation of $E(\delta : \delta_s)$ and $E(\delta : \delta_t)$.

If $\delta = \delta_s$, $w_s = 1$ and $w_t = 0$ and we have theorem 2, the special case of that theorem being excluded.

Suppose $R(\delta_t)$ is not contained in $R(\delta)$, then in the representation of δ, δ_t will not appear and $w_t = 0$. If $R(\delta_t)$ is contained in $R(\delta)$, then by the relevancy principle we may confine our attention to the factors in $R(\delta)$. If these are k in number, $w_t = 2^{-(k-1)}(N_t - 2^{k-1})$, where N_t refers only to agreements in $R(\delta)$.

To understand what is happening, suppose δ is any response with $R(\delta)$ containing $k (\leq m)$ factors. If two interactions not in $R(\delta)$ are compared, (8.3) reduces to $J(\alpha_{s+1}) - J(\alpha_{t+1})$ which has the sign of $(s - t)$ so that a lower-order interaction will appear less probable than a higher-order one. If two interactions of the *same* order, $s = t$, are compared, one within $R(\delta)$ and one not, (8.3) reduces to $w_s\{I(\alpha_{s+1}) - J(\alpha_{s+1})\}$, which is positive. Hence, an interaction involving the k relevant factors will appear more probable than one of the same order involving some irrelevant factors. Of interactions within $R(\delta)$ of a given order, that with the maximum weight will predominate. It is not possible to make general statements about interactions of different orders within $R(\delta)$ since the difference between $I(\alpha_{s+1})$ and $I(\alpha_{t+1})$ (and between the J's) depends on p, whereas between w_s and w_t the difference depends solely on δ and not p.

Table 29.1 gives an example with three factors. The first three columns are the possible ξ_i's making up the eight possible ξ. The next column gives η for a δ in which the response is + if ξ_1 is + and, in addition, either ξ_2 or ξ_3, or both, are +: so $k = m = 3$. The remaining columns give η for the eight odd interactions, excluding the zero-factor interaction that always has a negative response. Under these last seven columns are listed the N's, the numbers of agreements between δ and the interaction of the column. In three cases $N < 2^{k-1} = 4$ so the number $2^{k-1} - N$ is listed below for the complementary even interaction. The resulting N's all exceed 4 and the w's all exceed 0. (This procedure of switching to complementary interactions is quite general.) The next row lists the w's and the final row provides the values of $E(\delta : \delta_s)$ for each interaction δ_s. The generalized Kullback–Leibler numbers are differences of E's and are not listed.

Consider the main effect of ξ_1 in comparison with that of ξ_2 (or ξ_3). The difference of the E's is $-0.59-(-0.84)=+0.25$ and so ξ_1 will appear more probable than the other two, and will dominate all other interactions since it has the largest E. The next most important interaction is the 3-factor one which will slightly dominate the three 2-factor ones, and all will dominate the ξ_2 and ξ_3 main effects.

Another interesting response (not tabulated) is that in which η is + only when all of $\xi_1, \xi_2, \ldots, \xi_k$ are +. In medical language the disease is only present if k (true) symptoms are present. Such a δ differs from a zero-factor interaction in only one place and hence has $2^{k-1}+1$ agreements with any other interaction (allowing for complements) giving $w_s = 2^{-(k-1)}$ for all interactions involving some or all of the $\xi_1, \xi_2, \ldots, \xi_k$. It is not difficult to show that $wI(\alpha_{s+1})+(1-w)J(\alpha_{s+1})$ increases in s if $w < \frac{1}{2}$ so that (8.3) shows that the k-factor interaction of $\xi_1, \xi_2, \ldots, \xi_k$ will dominate all other interactions in $R(\delta)$ and, by the general result, will dominate all k-factor interactions.

9. Conclusions

The main lesson to be learnt from this analysis is that there are situations in which the increasing complexity that inevitably arises from an increase in the number of factors gives rise to a need for more observations to unravel the complexity; and that this is in contrast to other familiar statistical situations in which the observational explosion does not take place. The practical consequence of this is that statisticians should be wary of data analyses involving many factors because, if a model like that studied here is reasonable, there just may not be enough observations to permit a satisfactory analysis, so that any data analysis must be a waste of time. Statisticians are often appealled to "to make sense out of this data": they should resist the temptation to do so without first checking that the extraction is possible. Rather, their talents should be directed toward sensible designs and scientists encouraged to do planned experimentation rather than idle data collection.

The analysis has also shown the usefulness of the Kullback–Leibler distance as a way of separating the possibilities and providing discrimination between them. A useful way of understanding what is happening with this model is to think of the "space" of possible δ's packed so tight that the "distances" between δ's become small. In contrast, the m-factor situation mentioned in the introduction has δ's much more loosely packed so that discrimination is easier.

Reference

Lin, Shu (1970) *An Introduction to Error-Correcting Codes* (Prentice-Hall, Englewood Cliffs)

PART X

BAYESIAN AND NON-BAYESIAN INTERACTION

Chapter 30

BAYESIAN SALESMANSHIP

JAMES BERGER*

Purdue University

1. Introduction

Bayesian statistics is growing rapidly and vigorously. By now there are extensive Bayesian methods available for use as alternatives to classical statistical techniques. However, non-Bayesian statistics also seems to be growing rapidly, and while there seems to be some trend towards the Bayesian position (cf. Zellner, 1981), this trend is not yet overpowering. Thus, although the arguments have been going on for over a century, it remains imperative for Bayesians to continue to sell their viewpoint effectively and to attempt to improve their sales techniques.

To relative newcomers in statistics the Bayesian approach seems natural and believable. (It is certainly more natural to talk about actual "probabilities" of hypotheses than about error probabilities of various types.) This chapter is addressed, not to the problem of convincing such newcomers, but instead to the far more difficult problem of changing the views of established non-Bayesians. We will argue that this can best be accomplished by: (i) arguing for the need to think conditionally on the data and follow (as much as possible) the Likelihood Principle; (ii) maintaining that, while consideration of prior distributions is necessary for meaningful conditional thought, there is no need to presume (and indeed reasons not to presume) that prior distributions can be specified completely or accurately; and (iii) admitting that, for a variety of reasons, non-Bayesian techniques are (at least for the time being) of interest to a Bayesian.

The more standard arguments for the Bayesian position are based on notions of coherency (or rationality or consistency), and on the importance of using prior information. Although these arguments can be persuasive, we will argue in section 2 that they are not sufficiently compelling to the majority of statisticians. Of

*Supported by the National Science Foundation under Grant number MCS-8101670A1.

Bayesian Inference and Decision Techniques, edited by P. Goel and A. Zellner

course, statisticians never really accept the Bayesian position until they actually try thinking in a Bayesian fashion and see the clarity and improved results that are usually obtained. It is very hard to "prove" to a non-Bayesian that Bayesian results are really better, however (or, more precisely, that in the substantial majority of problems a Bayesian analysis will give a better answer than a non-Bayesian analysis based on a similar investment of effort). Thus, the "success" of Bayesian analysis is typically the gradually perceived clinching argument for the Bayesian viewpoint, rather than the initial argument starting one on the Bayesian path. Also, the perhaps most important component of the Bayesian mode of thought is conditional thinking, which provides further justification for arguing via (i), (ii), and (iii).

There is, of course, nothing new in arguing the Bayesian viewpoint via steps (i), (ii), and (iii), and indeed there will be nothing really new anywhere in the chapter. Also, serious efforts will not be made to justify these three steps here; instead, the emphasis will be on why this is the best approach to take. Finally, very few references will be given, since the concepts discussed have been developed by a large number of people and this is not intended to be a review. Extensive bibliographies concerning some of these issues can be found in Berger (1984) and Berger and Wolpert (1984). As a final caveat, although we repeatedly refer to "non-Bayesians" as a class, there are, of course, enormous differences in view among non-Bayesians. The arguments we suggest are particularly aimed at frequentists, whose techniques form the bulk of existing statistical methodology. In particular, such theories as fiducial inference (cf. Wilkinson, 1977), structural inference (cf. Fraser, 1968), and pivotal inference (cf. Barnard, 1980) will be ignored, even though they may have major points of disagreement with the advocated form of Bayesian analysis and may have resulted in important *contributions* to the advocated form of Bayesian analysis. Finally, lack of space will also preclude discussion of various pseudo-Bayesian positions. See Berger (1984) and Berger and Wolpert (1984) for some discussion and references.

The notation that will be used will be kept simple. It will be assumed that an experiment,

$$\mathscr{E} = (X, \{f_\theta : \theta \in \Theta\}),$$

is performed, where X is the random variable observed (a particular realization of which will be denoted x), f_θ is the density of X on the sample space $\mathscr{Z}$, and θ is the unknown parameter of interest with Θ being the parameter space. (Since we seek to establish nothing rigorously, mathematical niceties will be ignored and densities will be presumed to exist.) Although θ will usually be a parameter in the usual sense, we will have occasion to allow it to be just a general index, such as when $\{f_\theta : \theta \in \Theta\}$ is to be the set of all densities on $\mathscr{Z}$ (with respect to an appropriate measure). Prior densities on Θ will be denoted $\pi(\theta)$, and $\pi(\theta|x)$ will denote the resulting posterior density.

It should be recognized that, in a fundamental Bayesian sense (such as the fundamental viewpoint of de Finetti, 1974, 1975), the more or less standard setup for statistics is deemed to be in error; parameters, for example, are mainly a figment of our imagination. Again, however, the problem addressed here is that of selling the Bayesian viewpoint to non-Bayesians, and a more effective job can be done if the non-Bayesian's framework is used, as far as possible.

2. Helpful, but non-compelling Bayesian arguments

Several arguments often used to support the Bayesian position are mentioned here, along with reasons why they seem not to be compelling to most non-Bayesians.

2.1. Prior information is important

Examples abound where prior information is a crucial element of any statistical problem. Indeed, in virtually any realistically complicated statistical setting, massive subjective choices (model assumptions, etc.) are made by any statistician. Even in such a supposedly pure non-Bayesian procedure as a significance test of a single (point) hypothesis, selection of a test statistic requires subjective consideration of the type of alternatives that are of concern. Thus, Good (1983) has said: "The people who don't know they are Bayesians are called non-Bayesians." Nevertheless, this argument fails to be persuasive to non-Bayesians. They can argue that the model often does have an "objective" reality due to theoretical considerations, and in some situations (such as that of simple Bernoulli trials) this may be the case. We will not pursue this issue, basically because of the empirically observed fact that arguments concerning the lack of objectivity of models seem to have little effect on non-Bayesians.

Non-Bayesians can also, at least, argue that their non-parametric analyses are totally objective, as in the following example (from Efron, 1982) which will be of use below.

Example 1. The experiment $\mathscr{E}$ consists of observing $X_1, \ldots, X_{15}$ which are i.i.d. observations from a completely unknown continuous density f on R^1. (Here we identify θ with the unknown f, so Θ is the set of all continuous densities on R^1.) Of interest is η, the median of the unknown density. A simple binomial calculation shows that a 96.5 percent confidence interval for η is given by $[X_{(4)}, X_{(12)}]$, where the $X_{(i)}$ are the order statistics.

2.2. Coherency

Many axiomatic systems of coherent (or consistent or rational) behavior (cf. Ramsey, 1926, 1928; de Finetti, 1974, 1975; Savage, 1954; and Rubin, 1974) have been created, which show that any coherent method of behavior corresponds to some form of Bayesian analysis. It was indeed these axiomatic developments that convinced many of today's Bayesians to become Bayesians, and they remain a cogent argument.

Those who remain unconvinced by the coherency arguments offer the following criticisms:

(i) The axioms themselves are wrong (cf. LeCam, 1977).

(ii) The axiomatic systems are only theoretically implementable, in the sense that they involve infinite numbers of comparisons or, alternatively, involve anticipation of all possible eventualities. Thus, C.A.B. Smith (in Savage et al., 1962) says: "Such an absolutely consistent person does not, of course, exist."

(iii) The coherency argument merely states that a coherent method of behavior corresponds to a Bayesian analysis. It does not say that one must explicitly use a Bayesian approach to be coherent, as the following example shows.

Example 2. Consider an invariant decision theoretic estimation problem, in which the parameter space Θ is a compact group (say, the group of rotations). For any such problem the mode of behavior "use the best invariant estimator" is completely coherent (since it happens to correspond to a Bayesian analysis with the invariant Haar measure on Θ as the prior distribution).

(iv) A Bayesian approach, even though coherent, need not necessarily be better than a non-Bayesian approach. Thus, C.A.B. Smith (in Savage et al., 1962) says: "Consistency is not *necessarily* a virtue: one can be consistently obnoxious." As an example, for parametric problems where Θ is a compact interval, the formally coherent Bayesian behavior of always using a prior distribution concentrated on the lower endpoint of the interval is clearly obnoxious. A more realistic example is the following.

Example 3. Suppose that $X \sim \mathcal{N}(\theta, 1)$ is to be observed, and that a Bayesian determines that his prior density for θ is bell-shaped with median 0 and quartiles ± 1. He then chooses the matching conjugate prior ($\mathcal{N}(0, 2.19)$) to do the analysis. If now $x = 5$ is observed, the posterior mean can be calculated to be 3.43.

As an estimate of θ this can be challenged by a non-Bayesian, since it is 1.57 (sample) standard deviations from the (MLE and MVUE) classical estimate 5. Indeed, the non-Bayesian can point out that, if the Bayesian had fitted his prior information with a Cauchy density instead, then the Bayes estimate would have been about 4.6. The non-Bayesian concludes that he could care less about coherency; he just wants to get a good estimate, and 5 looks like a pretty good estimate of θ to him.

Of course, in the above example a good Bayesian analysis would have involved a sensitivity study when the surprising (but not impossible) $x = 5$ was observed. Furthermore, in deciding between use of the normal or Cauchy prior, a Bayesian would note that the ratio of the predictive (or marginal) probabilities of $x = 5$ for the normal and Cauchy priors would be very small, indicating that the Cauchy prior is more appropriate for use. The fact that a very careful Bayesian analysis is needed is precisely the point, however; the non-Bayesian can argue that a "good" coherent Bayesian analysis is harder and more fraught with peril than a "good" non-coherent frequentist analysis. Thus, LeCam (1977) says:

> To claim that an ideal person could in principle specify such a relation is to beg the question. To claim that since an ideal person could do it, a real person should do it, is to introduce a dogma for which we have little justification.

In principle, a Bayesian might admit the validity of many of the above objections. On the other hand, it certainly seems desirable to strive for coherent behavior, and the axiomatic systems seem to indicate that the direction in which to strive is the Bayesian direction. Thus, the Bayesian can argue that most of the effort in statistics should be directed towards extending and sharpening Bayesian tools.

Another extremely valuable consequence of the coherency argument is that if any "complete" theory of statistics exists, it is the Bayesian theory. One can either be a Bayesian, or argue for some form of anarchy.

2.3. The correspondence between "good" classical procedures and Bayes procedures

When non-Bayesians attempt to define "good" classes of statistical procedures, these classes almost invariably end up corresponding to a class of Bayes procedures (or their limits). Most complete class theorems in frequentist decision theory are of this form, the first and simplest being the correspondence between most powerful tests and Bayesian tests, when testing between two simple hypotheses.

Although this correspondence could be considered to be a mathematical coincidence, it is hard to argue that a procedure should be selected from the "good" class even if it corresponds to an unreasonable prior distribution.

An example of current research interest in frequentist theory is that of the Stein effect, whereby the usual estimator of a multivariate normal mean can be improved upon (in terms of frequentist risk) in dimensions three or more. Unfortunately, the set of better estimators is enormous, and in choosing an estimator from this set it is impossible to ignore prior information, at least to the extent of deciding where to shrink towards. (See Berger, 1980, and 1982a, for more discussion.) While these arguments are certainly telling, they do not seem to be compelling to non-Bayesians.

2.4. Measuring uncertainty

A very fundamental justification for Bayesian analysis (cf. Jeffreys, 1961; de Finetti, 1974, 1975; and Jaynes, 1981, which also contain other references) is simply that the goal of statistics is to communicate evidence about uncertainty, and that the correct language to use in measuring uncertainty is probability. Only subjective probability provides a broad enough framework to encompass the types of uncertainties encountered, and Bayes' Theorem tells how to process information in the language of subjective probability. Users of statistics *want to know* the probability (after seeing the data) that a hypothesis is true, or the probability that θ is in a given interval, and yet classical statistics does not allow one to talk of such things. Instead, artificial concepts such as error probabilities and coverage probabilities are introduced as substitutes. It is ironic that non-Bayesians often claim that the Bayesians form a dogmatic unrealistic religion, when instead it is the non-Bayesian methods that are often founded on elaborate and artificial structures. Unfortunately, those who become used to these artificial structures come to view them as natural, and hence this line of argument tends to have little effect on the established non-Bayesian.

3. The conditionality-likelihood principle–robust Bayesian approach

By far the most troubling aspect of frequentist statistics is its unconditional nature, and this usually provides the easiest point of attack. We briefly outline the argument here.

3.1. Conditioning and the likelihood principle

Although most frequentists are aware of conditioning problems, the following two examples, taken from Berger and Wolpert (1984), are simple and revealing.

Example 4. Suppose X_1 and X_2 are independent and identically distributed with $P(X_i = \theta + 1) = P(X_i = \theta - 1) = \frac{1}{2}$. (Here $\theta \in \Theta = R^1$.) A 75 percent confidence set of smallest size for θ is given by

$$C(X_1, X_2) = \begin{cases} \text{the point } \frac{1}{2}(X_1 + X_2), & \text{if } X_1 \neq X_2, \\ \text{the point } X_1 - 1, & \text{if } X_1 = X_2. \end{cases}$$

Notice, however, that when we observe $x_1 \neq x_2$, we are 100 percent certain that $\theta = \frac{1}{2}(x_1 + x_2)$, while if we observe $x_1 = x_2$ the data leaves us equally uncertain as to whether θ is $x_1 + 1$ or $x_1 - 1$. Thus, the frequentist 75 percent confidence arises from the average of the *real* 100 percent confidence when $x_1 \neq x_2$ and 50 percent confidence when $x_1 = x_2$. This example forcefully shows that, at least sometimes, one must think conditionally on the actual data that occurs, rather than use frequentist long-run averages.

Example 5. Suppose that a substance to be analyzed can be sent to either a laboratory in New York or a laboratory in California. The two labs seem equally good, so a fair coin is flipped to choose between them. The coin is flipped and comes up tails, which signifies that the California lab is to be used. When the experimental results come back and a conclusion is to be made, should it be taken into account that the coin could have come up heads with the New York lab then being used?

The above example is, of course, a variant of the famous Cox (1958) example, and in symbolic form can be phrased as follows: If with probabilities 1/2 (independent of the unknown quantity θ of interest) either experiment $\mathscr{E}_1$ or experiment $\mathscr{E}_2$ (both pertaining to θ) will be performed, should the analysis depend only on the experiment actually performed or should the possibility of having done the other experiment be taken into account? (Note that we are talking about analysis with the data in hand, after one of the experiments has been done, and not about design, etc.) Uncompromising frequentist theory would suggest that the other experiment must be taken into account, since frequentist measures depend on averaging over all possible experimental outcomes, whether they happen or not. This, of course, strikes almost everyone as nonsensical, and virtually all statisticians would accept that only the experiment actually performed should matter. This has been called the *Weak Conditionality Principle* (*WCP*) (cf. Basu, 1975).

Virtually all frequentists will subscribe to the WCP and admit that conditioning must sometimes be done, but will then attempt to limit the applicability of conditioning to "relevant subsets" or "ancillary statistics" or "partitions of the sample space". (Brief descriptions and references to such efforts can be found in Berger and Wolpert, 1984.) However, Birnbaum (1962) provided a devastating, yet simple, refutation of the viability of limiting the scope of conditioning. He showed that the WCP and the Sufficiency Principle (which states that a sufficient statistic for θ contains all relevant information about θ) together imply the Likelihood Principle, which basically forces conditioning all the way down to the actual observation x. The Likelihood Principle itself traces back to ideas of G. Barnard and R.A. Fisher (see Berger and Wolpert, 1984), and in its basic form is given as follows.

Likelihood Principle. All information or evidence about θ from an experiment $\mathscr{E}$ is contained in the likelihood function for θ, namely

$$l_x(\theta) = f_\theta(x).$$

(Here x is the actual observation.) Two likelihood functions (for θ) are equivalent if they are proportional to each other.

A statistician seeking to follow the Likelihood Principle would only trust procedures or measures that depend on $\mathscr{E}$ solely through the observed likelihood function. Classical maximum likelihood estimates are of this form (as are all Bayesian procedures and measures based on the posterior distribution), but frequentist measures such as error probabilities, bias, coverage probability, and

P-values, which involve averages over unobserved x, are not of this form, and can hence not be of basic interest from the conditional viewpoint.

Frequentists can, and will, raise all sorts of objections to the Likelihood Principle and its implications. However, examples and continual application of common sense can be used to counter these objections. (See Basu, 1975, and Berger and Wolpert, 1984, for a discussion of many common objections. The latter work also shows that powerful versions of the Likelihood Principle hold in essentially complete generality, not requiring densities or even exact knowledge of the model.)

The argument for conditioning leaves one very close to the situation in section 2.4. It basically supports the argument that only the uncertainty about θ for the data, x, actually observed is of interest. For example, coverage probability of a confidence procedure $C(X)$ is something of no intrinsic interest; what is of interest is some measure of how "confident" we are that θ is in $C(x)$ for the actual observed x. (See example 4.) As Savage succinctly put it in Savage et al. (1962): "The only use I know for a confidence interval is to have confidence in it." As another example of how artificial frequentist structures can look bad in the light of conditional common sense, consider the following.

Example 6. Frequentist measures are claimed to have the very desirable property of guaranteeing "long-run" success rates. For instance, it is claimed to be scientifically desirable to test established theories (the null hypotheses) with classical tests having low probability of type I error (say $\alpha = 0.01$), since then one can be assured of incorrectly rejecting only 1 percent of correct theories. A conditionalist would argue that this is an artificial concept that does not really measure what is sought, namely the conditional chance that one is making a mistake when the null hypothesis *is* rejected. To see the difference, suppose that the observation X is either 1 or 2, with probabilities under f_0 and f_1 given by

	X = 1	X = 2
f_0	0.01	0.99
f_1	0.01001	0.98999

and it is desired to test H_0: $f = f_0$ versus H_1: $f = f_1$. The most powerful $\alpha = 0.01$ level test rejects H_0 when $X = 1$ is observed. From the conditional viewpoint, however, when $x = 1$ is actually observed (i.e. rejection of H_0 actually occurs) there is very little evidence against H_0 (because the likelihood ratio is so close to 1), so the chance of being in error is about $1/2$ (based on the data alone). Thus $\alpha = 0.01$ is providing a very misleading and false sense of security when rejection actually occurs.

One can present the case for conditioning, as outlined above, without heavy involvement of subjective prior distributions. As alternatives to coverage probability, one can present posterior probability of coverage for "objective" non-informative priors (cf. Box and Tiao, 1973). As alternatives to error probabilities in testing, one can argue simply for the use of likelihood ratios of hypotheses (at least when the hypotheses are simple). Indeed, fostering the attitude that one can often simply look at the likelihood function itself and obtain most things of interest is important. The point, of course, is to avoid introduction of subjective prior distributions until the frequentist has already accepted the conditional viewpoint. Indeed, it can be argued that acceptance of the conditional viewpoint is a far more drastic and important step than is the formal use of Bayesian methodology to implement the viewpoint. Thus L.J. Savage said (in the discussion of Birnbaum, 1962): "I, myself, came to take...Bayesian statistics...seriously only through recognition of the likelihood principle."

3.2. Robust Bayesian analysis

Although there do exist statisticians who adopt a mainly conditional viewpoint, and yet stop short of being Bayesians, most statisticians find the step from conditioning to Bayesian analysis to be a small one. The basic problem for a non-Bayesian conditionalist is that of interpreting and using the likelihood function, and cogent arguments can be presented (cf. Basu, 1975, and Berger and Wolpert, 1984) that in other than very simple applications, the likelihood function should be interpreted as a probability density with respect to a prior distribution π for the unknown θ. (The coherency arguments can be used as an aid to support the necessity of such an interpretation.) What prior π to use is a matter much debated, even among Bayesians. In Jeffreys (1961), Zellner (1971), and Box and Tiao (1973) it is argued that, for many (if not most) applications, π should be a non-informative prior, or its close relative a "reference informative prior" in testing problems (cf. Zellner, 1982). The pure subjectivist argues that π should be chosen to reflect actual subjective beliefs about θ. The robust Bayesian (cf. Berger, 1984, 1985) argues that no single prior should be chosen, but that instead one learns by passing various plausible priors, π, over the likelihood function. While not disputing the value of any of these approaches (the non-informative prior approach is very good if "objectivity" is demanded or little subjective information is available, and the single prior subjective approach rarely gives bad answers), we would argue for the robust Bayesian approach on pragmatic and scientific grounds.

The pragmatic reason for taking a robust Bayesian approach is obvious. By admitting to non-Bayesians that we cannot determine *exactly* a completely trustworthy prior distribution, and yet arguing that only by trying various plausible priors and seeing what happens can true understanding be gained, we in one stroke overcome the biggest *perceived* objection to Bayesian analysis.

A concerted effort to justify the robust Bayesian method on scientific grounds is given in Berger (1984), in which earlier references to the method can also be found. (The method is sometimes called the Doogian approach because of the long advocacy of robust Bayesian techniques by I.J. Good.) Example 3 provides a simple illustration of the need to consider robust Bayesian analysis (or sensitivity analysis). The philosophical justification for the viewpoint is clear; subjective probabilities can be specified only to a certain degree of accuracy by finite minds and, even worse, prior distributions on infinite sets require the specification of an infinite number of subjective probabilities. Indeed, the resistance of many Bayesians to the seemingly obvious conclusion that we will always be working, in reality, with prior distributions that have been arbitrarily chosen, to at least some extent, is somewhat surprising. Of course, as mentioned in example 3, good Bayesian techniques do exist for helping to select a prior that is likely to work well.)

To obtain a theory suitable for general use in practice, Bayesians must go beyond even the pure robust Bayesian position, wherein before performing the experiment a class Γ of plausible prior distributions is envisaged, which is then employed through standard Bayesian updating by the likelihood function. Among the "sins" which it seems necessary to practice are the following:

(i) Delaying at least some of the prior specification until the likelihood function (for the observed data) is available (or alternatively allowing data-based choice of the prior or class of priors). The reason for this is simply that, in realistically high-dimensional problems, it will be very difficult to specify believable prior distributions; looking first at the likelihood function can indicate where prior elicitation efforts need to be concentrated (namely, where the data provides inconclusive evidence). The viewpoint, that one is merely passing plausible priors over the likelihood function to see what happens, makes this heresy seem more palatable.

(ii) Some of the data may be ignored if its incorporation would involve considerable extra (and uncertain) prior specification, and if it seems that the posterior distribution of the quantity of interest would not be excessively changed by incorporation of this data. Thus, Hill (1975) says:

> When such a formal analysis simply cannot be made, or even when it is merely very difficult and of dubious validity, then there is little choice but to condition on that part of the data that can be effectively dealt with....

Further discussion of this point can be found in Pratt (1965), who calls it use of "insufficient statistics".

(iii) Use of non-Bayesian (even frequentist) measures may be helpful in certain situations, as will be argued in the next section.

All of the above points are discussed at considerable length in Berger (1984). When combined with an acknowledgement that, in Bayesian reporting, one must clearly report the likelihood function and the priors separately (along with the Bayesian conclusions), virtually any objection to the Bayesian approach can be dealt with.

3.3. Potential uses of frequency measures

One of the major objections to Bayesian analysis is that there are many problems that seem to be solvable in a non-Bayesian fashion (cf. example 1) and yet have no trustworthy Bayesian solution. Of course, to a large extent this is because far more research has been conducted on frequentist methods than on Bayesian methods; the existence of such problems is to be expected. Indeed, there is perhaps doubt that frequentist methods will prove to be of lasting value; thus L.J. Savage in Savage et al. (1962) states:

> I used to be bowed by critics who said, with apparent technical justification, that certain popular nonparametric techniques apply in situations where it seems meaningless even to talk of a likelihood function, but I have learned to expect that each of these techniques either has a Bayesian validation or will be found to have only illusory value as a method of inference.

Nevertheless, at the moment, certain frequentist methods are certainly of value to a Bayesian. We briefly discuss these here.

3.3.1. Design, prediction, and sequential analysis

It is indisputable that many questions in statistics involve "looking ahead" at the data that can be expected to occur. Frequentist averaging over the sample space is certainly necessary in such problems, and no Bayesian would think otherwise. Of course, these problems also have a large Bayesian component. In design, for instance, one must use subjective guesses for θ to predict what data will occur and hence what design to use. Also, the Bayesian will have the goal of obtaining good conditional performance, which may lead to a quite different design than a classical design.

3.2.2. Procedures for non-specialists

It is probably the case that most of the people who will actually be using statistics will not have been well enough trained to perform a careful conditional-robust Bayesian analysis. Hence, a major portion of theoretical work should be devoted to providing relatively simple and easy to use Bayesian procedures with built-in robustness. Since these procedures will be used repeatedly, averages over the sample space become relevant. Consider the following.

Example 3 (continued). Suppose the simple Bayesian procedure envisioned consists of eliciting the prior median and quartiles from the user, and then calculating the posterior mean for the fitted prior. If performance is measured by the squared error loss in estimation, it is easy to see that use of δ^{N}, the estimate from the conjugate normal prior π^{N}, can result in an infinite overall expected loss:

$$r(\pi, \delta^{\mathrm{N}}) = E(\theta - \delta^{\mathrm{N}}(X))^2$$

(the expectation being taken over both θ and X), when the true prior is the Cauchy prior π^{C}. On the other hand, if δ^{C}, the Bayes estimate for the fitted Cauchy prior, is used, $r(\pi^{\mathrm{N}}, \delta^{\mathrm{C}})$ is quite small. Indeed, for any reasonable prior π with median 0 and quartiles ± 1, $r(\pi, \delta^{\mathrm{C}})$ will be very satisfactory; thus, in repeated use Cauchy priors should prove more satisfactory (for this situation) than normal priors. (Even for sensible bounded losses, the overall performance of δ^{N} can be bad.)

Another much more sophisticated situation, involving the Stein effect in estimating a multivariate normal mean, can be found in Berger (1982b). Here procedures are developed whose Bayesian (conditional) performance is excellent, yet which are extremely insensitive to departures from prior assumptions.

3.3.3. Frequency measures which imply good average conditional performance

We will illustrate this notion with the concept of coverage probability. (See Pratt, 1965, for more general development and other applications.) Thus, suppose $C(X)$ is a confidence procedure at level $1-\alpha$, so that

$$P_\theta(C(X) \text{ contains } \theta) \geq 1-\alpha, \quad \text{for all } \theta \in \Theta.$$

Then, for any prior distribution π on Θ,

$$E^\pi P_\theta(C(X) \text{ contains } \theta) \geq 1-\alpha.$$

However,

$$\begin{aligned} E^\pi P_\theta(C(X) \text{ contains } \theta) &= E^m E^{\pi(\theta|X)} P(\theta \in C(X)) \\ &= E^m \lambda_\pi(X), \end{aligned}$$

where m is the marginal distribution of X and $\lambda_\pi(x)$ is clearly the posterior probability that θ is in $C(x)$ (which is the ideal conditional measure). It follows that

$$E^m \lambda_\pi(X) \geq 1-\alpha,$$

which implies that $C(x)$ has a very good chance of containing θ (if α is small) according to a good conditional measure (and no matter what π is). For the above reason, many Bayesians would accept the classical answer in example 1 as the best approximate estimate of the chance that η is in $[x_{(4)}, x_{(12)}]$ that is currently obtainable. Care should still be taken, however, to make sure that there are no obvious conditional problems (as in example 4). Thus Good (1976) states that: "... non-Bayesian methods are acceptable provided that they are not seen to contradict your honest judgments, when combined with the axioms of rationality".

3.3.4. Frequency measures as an aid to robust Bayesian analysis

When the prior elicitation process has ended, and one is left with a class Γ of possible priors, it is to be hoped that the Bayesian conditional answer is clearcut (i.e. passing the different priors in Γ over the likelihood function results in essentially the same conclusion). When this is not the case, and an answer must be produced, there is no clearcut way to proceed. A natural Bayesian possibility is to put some (arbitrary) metaprior on Γ (i.e. a prior distribution on the elements of Γ). This is merely a formal way to proceed, however, since it is being assumed that no further subjective prior elicitation is possible. As a possible alternative, one could select from among the possibilities by frequentist criteria. Discussion of this can be found in Berger (1984, 1985). It should be emphasized that there is no *foundational* reason to prefer any particular method here, since further prior elicitation is not possible.

The above view is not really a completely practical view, since it will rarely be the case that a constructed Γ is impossible to further refine, often by the data interactive process of looking at the predictive likelihoods (i.e. $m(x)$) for the various priors in Γ and observed x. It may still be the case, however, that the best *pragmatic* way to proceed is to choose among the possible priors by frequency methods. If, for instance, a prior in Γ results in a procedure with good frequency properties, then for reasons indicated in section 3.3.3 it may be reasonable to just use that prior instead of attempting further refinement of Γ. (This is especially true in those situations where Γ is very large and Bayesian calculations are very difficult.) This is not to be interpreted as an advocacy of the routine use of frequency concepts in such situations in preference to the more natural Bayesian methods, but instead is another justification for *allowing* consideration of frequency measures when a Bayesian analysis is unclear or difficult.

3.3.5. Other uses of frequency concepts in Bayesian analysis

Classical significance testing of a null hypothesis is of some value as a quick and dirty indicator of a situation needing deeper investigation (i.e. of a situation in which formulation and examination of alternatives, hopefully by Bayesian measures, is in order). As with all techniques violating the Likelihood Principle, however, care must be taken in attempting to interpret P-values literally. (See Berger and Wolpert, 1984, for a discussion and other references).

Invariance theory in statistics is closely related to the Bayesian approach with non-informative priors (cf. Berger, 1985), and indeed often suggests appropriate choices for a non-informative prior.

Asymptotics is, of course, frequently relevant, providing often needed simplifications. Indeed, there is a very substantial literature on Bayesian asymptotics. (See Berger, 1984, for references.)

The study of admissibility, with its close tie-ins with Bayesian theory, has led to a number of significant Bayesian advances, particularly in Bayesian robustness (via study of the Stein effect).

This list is by no means exhaustive, or even representative. Much of frequentist statistics has helpful things to say to Bayesians.

4. Conclusions

The essential features of the approach presented for selling the Bayesian view are (i) to break the argument up into easily digestible (and defendable) parts, and (ii) to allow a great deal of flexibility in methods. Arguing first for the adoption of a conditional viewpoint, and then for the introduction of priors to implement this viewpoint, reduces the need to deal with the initial anti-prior bias of many non-Bayesians. It also focuses the issue on the fact that the goal is to obtain good conditional probabilistic measures of uncertainty.

One can then address the concern of how best to achieve this goal. Here, the Bayesian salesman does his best to avoid being too dogmatic. Adopting a robust Bayesian position, i.e. admitting uncertainty in the prior, is only sensible and realistic. (Of course, frequentist models, etc. are every bit as uncertain, but we are not trying to sell the frequentist viewpoint). Also, admitting that there are presently many problems where believable Bayesian answers are not available and reasonable frequentist answers are, is only prudent. It must be stressed, however, that a frequentist answer is not inherently sensible; it must have some plausible relationship to a meaningful conditional measure. It would be very nice to have the arguments and debates come down to this level; namely, that of which methodology best achieves the conditional goal.

There are substantial rewards, to the statistician adopting a conditional viewpoint, that should also be pointed out. First of all, many difficult problems involving stopping rules and censoring disappear. (See Berger and Wolpert, 1984, for discussion and references.) Also, time need not be wasted on extremely difficult frequentist problems, such as solution of the Behrens–Fisher problem; conditional Bayesian solutions (including objective noninformative prior solutions) are readily available. Finally, it can lead the statistician to avoid frequentist analysis or research which is clearly ridiculous conditionally, even if the statistician is unwilling to become a Bayesian.

It is possible to object to the flexible robust Bayesian position, advocated above, on the grounds that it will frequently be incoherent, and indeed it often will be *formally* incoherent. The word "formally" is stressed, since the approach does strive for coherency to the extent obtainable. Our own view is that, as Bayesian methodology expands, there will be less and less need to leave the formal Bayesian paradigm; we will learn what types of prior distributions are inherently robust and work well for given problems. There is no need to make such a strict position a part of the Bayesian argument, however.

References

Barnard, G.A. (1980) "Pivotal inference and the Bayesian controversy" (with discussion), in: *Bayesian Statistics* (University Press, Valencia).

Basu, D. (1975) "Statistical information and likelihood" (with discussion), *Sankhyā*, Ser. A 37, 1–71.

Berger, J. (1980) "A robust generalized Bayes estimator and confidence region for a multivariate normal mean", *Ann. Statist.*, 8, 716–761.

Berger, J. (1982a) "Selecting a minimax estimator of a multivariate normal mean", *Ann. Statist.*, 10, 81–92.

Berger, J. (1982b) "Bayesian robustness and the Stein effect", *J. Amer. Statist. Assoc.*, 77, 358–368.

Berger, J. (1984) "The robust Bayesian viewpoint", in: J. Kadane, ed., *Robustness in Bayesian Statistics* (North-Holland, Amsterdam).

Berger, J. (1985) *Statistical Decision Theory and Bayesian Analysis* (Springer-Verlag, New York).

Berger, J. and R. Wolpert (1984) *The Likelihood Principle: A Review and Generalizations* (Monograph Series of the Institute of Mathematical Statistics).

Birnbaum, A. (1962) "On the foundations of statistical inference" (with discussion), *J. Amer. Statist. Assoc.*, 57, 269–326.

Box, G.E.P. and G.C. Tiao (1973) *Bayesian Inference in Statistical Analysis* (Addison-Wesley, Reading, Mass.).

Cox, D.R. (1958) "Some problems connected with statistical inference", *Ann. Math. Statist.*, 29, 357–372.

de Finetti, B. (1974, 1975) *Theory of Probability*, vols. 1 and 2 (Wiley, New York).

Efron, B. (1982) "Why isn't everyone a Bayesian?", Presented at the conference Reflections on Bayesian Approaches in Operations Research, Probability, and Statistics, Blacksburg, Virginia, 1982.

Fraser, D.A.S. (1968) *The Structure of Inference* (Wiley, New York).

Good, I.J. (1950) *Probability and the Weighing of Evidence* (Griffin, London).

Good, I.J. (1976) "The Bayesian influence, or how to sweep subjectivism under the carpet", in: W.L. Harper and C.A. Hooker, eds., *Foundations of Probability Theory, Statistical Inference, and Statistical Theories of Science*, vol. II (Reidel, Boston).

Good, I.J. (1983) "The robustness of a hierarchial model for multinomials and contingency tables", in: G.E.P. Box, T. Leonard, and C.F. Wu, eds., *Scientific Inference, Data Analysis, and Robustness* (Academic Press, New York).

Hill, B. (1975) "A simple general approach to inference about the tail of a distribution", *Ann. Statist.*, 3, 1163–1174.

Jaynes, E.T. (1981) "The intuitive inadequacy of classical statistics", Presented at the International Convention on Fundamentals of Probability and Statistics, Luino, Italy.

Jeffreys, H. (1961) *Theory of Probability*, 3rd edn. (Oxford University Press, Oxford).

LeCam, L. (1977) "A note on metastatistics or 'an essay toward stating a problem in the doctrine of chances'", *Synthese*, 36, 133–160.

Lindley, D.V. (1971) *Bayesian Statistics Review* (S.I.A.M., Philadelphia).

Pratt, J. (1965) "Bayesian interpretation of standard inference statements" (with discussion), *J. Roy Statist. Soc.*, B 27, 169–203.

Ramsey, F.P. (1926, 1928) "Truth and probability" (1926) and "Further considerations" (1928), in: *The Foundations of Mathematics and Other Logical Essays* (Harcourt, Brace, and Co., New York, 1931).

Rubin, H. (1974) "Axiomatic development of rational behavior", Technical Report, Department of Statistics, Purdue University.

Savage, L.J. (1954) *The Foundations of Statistics* (Wiley, New York).

Savage, L.J., et al. (1962) *The Foundations of Statistical Inference* (Methuen, London).

Wilkinson, G.N. (1977) "On resolving the controversy in statistical inference" (with discussion), *J. Roy. Statist. Soc.*, B 39, 119–171.

Zellner, A. (1971) *An Introduction to Bayesian Inference in Econometrics* (Wiley, New York).

Zellner, A. (1981) "The current state of Bayesian econometrics", Address at the Canadian Conference on Applied Statistics, Concordia University, Montreal, 29 April–1 May, 1981; published in A. Zellner, *Basic Issues in Econometrics* (University of Chicago Press, 1984).

Zellner, A. (1982) "Applications of Bayesian analysis in econometrics", Address at the Institute of Statisticians International Conference on Practical Bayesian Statistics, St. John's College, Cambridge, 21–24 July, 1982; published in *The Statistician*, 32 (1983) 23–34.

AUTHOR INDEX

Aigner, D.J., 157
Aitchison, J., 21
Akaike, H., 112, 113, 117, 118, 124, 422, 427
Amemiya, T., 96, 137
Anderson, T.W., 355, 356
Anderson, A., 128, 133, 134
Ando, A., 178, 191
Anscombe, F.J., 19
Arnold, B.C., 157, 158, 164, 165, 168, 169, 172
Arrow, K.J., 61, 297
Aykac, A., 1

Baldus, D., 144
Balestra, P., 375, 376
Barnard, G.A., 17, 18, 19, 25, 474
Barndorff-Nielsen, O., 20
Bartholomew, D.J., 16
Basu, D., 25, 479, 480, 481
Baum, L.E., 443, 444
Belsley, D.A., 104, 375, 376, 377, 378, 382, 386
Berger, J., 474, 477, 478, 479, 480, 481, 482, 484, 485, 486
Bernardo, J.M., 18
Bertsekas, D.P., 86
Billingsley, P., 443, 444
Birnbaum, A., 17, 479, 481
Bjerregaard, B., 248
Blackwell, D., 251, 262
Box, G.E.P., 19, 71, 97, 153, 233, 234, 326, 338, 339, 340, 352, 406, 421, 481
Bracken, J., 325
Brier, G.W., 256
Broemeling, L., 342, 344, 346, 352
Brown, B.W., 97
Brown, S.J., 148, 153, 154
Brown, R.V., 259, 267
Brumat, C., 1
Buehler, R.J., 67

Campanino, M., 21
Chamberlain, G., 221
Chow, Y.S., 22, 455
Chow, G.C., 92, 338
Cifarelli, D.M., 20, 21, 22
Cole, J., 144
Cole, N.S., 367
Coleman, A.J., 32
Cornfield, J., 178
Cornish, E.A., 178
Cox, D.R., 19, 479
Cozzolino, J.M., 168
Crosby, M.A., 328

Daboni, L., 22
Dagenais, M.G., 96
Dawid, A.P., 21, 22, 69, 179, 180, 248, 250, 267, 395, 396, 397, 400, 401, 402, 403
de Finetti, B., 14, 15, 16, 17, 20, 21, 22, 23, 24, 25, 26, 27, 32, 33, 45, 46, 60, 61, 63, 65, 197, 199, 211, 217, 218, 247, 250, 256, 265, 266, 276, 475, 477
DeGroot, M.H., 19, 120, 234, 248, 249, 250, 251, 252, 253, 255, 256, 258, 260, 262, 266, 267, 275, 276, 277, 279, 410, 443, 453, 454, 455
Dempster, A.P., 119, 127, 190, 221, 357, 359
Di Bacco, M., 26
Diaconis, P., 35
Diaz, J., 338
Dickey, J., 18, 178, 179, 180, 181, 186, 187, 192, 193, 233, 236, 288, 359
Dreze, J.H., 96
Dubins, L.E., 24, 47, 61, 70, 71, 74
Dunford, N., 66
Dunsmore, I.R., 21
Dwyer, P.S., 376
Dynkin, E.B., 400

Eddy, W.F., 427, 428, 429, 430, 431, 438
Efron, B., 408, 475
Engle, R., 128, 133, 134
Ericson, W.A., 23, 396
Eriksson, E.A., 248, 249, 250, 251, 252, 253, 258, 262, 276
Eswaren, P.K., 322

Fairley, W., 144
Farah, J., 338
Farquhar, P.H., 319
Fearn, T., 136
Feddersen, A.P., 67
Feller, W., 32, 443, 444
Ferguson, T.A., 19
Fienberg, S.E., 248, 250, 251, 253, 255, 256, 258, 260, 266, 267, 275, 276, 277
Finkelstein, M., 144
Fischhoff, B., 247, 267
Fishburn, P.C., 252, 254, 262

Fisher, R.A., 67, 16, 197
Fletcher, R., 104
Fraser, D.A.S., 68, 474
Friedman, M., 298, 328

Gani, J., 2, 6
Garthwaite, P.H., 187
Gathercole, R., 421
Geisel, M.S., 233
Geisser, S., 20, 178, 393, 427, 428, 429, 430, 431, 438
Gersch, W., 339
Geweke, J., 83
Giles, D.E.A., 235, 238
Goel, P.K., 120
Goldberger, A.S., 157
Goldfeld, S.M., 104
Goldstein, M., 198, 201, 204, 209, 406
Good, I.J., 17, 20, 117, 220, 256, 475, 482, 484
Granger, C.W.J., 128, 133, 134
Green, J.R., 289
Grunfeld, Y., 138

Habbema, J.D., 248
Hacking, I., 17
Haitovsky, Y., 120, 121, 126, 138
Hamaker, H.C., 16
Hardy, G.H., 252
Harrison, P.J., 339, 405, 410, 413, 416, 420
Hartigan, J., 64
Harvey, A.C., 350, 352
Harville, D., 363, 365
Hauser, J.R., 320, 321, 331
Heath, J.D., 60, 64, 65, 67, 68, 128, 134
Hendricks, W., 128, 134
Hewitt, E., 31
Hilden, J., 248
Hildreth, C., 17, 291, 292, 297, 298, 301, 309, 313
Hill, B.M., 23, 25, 46, 60, 70, 482
Hoerl, A.E., 79
Holst, L., 444
Howlader, H.A., 157

Jackson, P.H., 367
Jaynes, E.T., 477
Jeffreys, H., 60, 64, 71, 233, 234, 338, 427, 477, 481
Jenkins, G.M., 338, 339, 340, 352, 406

Kadane, J.B., 3, 20, 47, 56, 60, 61, 65, 66, 178, 180, 181, 186, 187, 233, 236, 267, 288
Kalman, R.E., 339
Kaufman, G.W., 178, 191
Kaufman, G.M., 193
Keefer, D.L., 322
Keelin, T., 319, 320
Keeney, R.L., 319, 320, 323, 326, 328
Kennard, R.W., 79
Keynes, J.M., 220
Khatri, C.G., 357
Kiefer, J., 178
Kirkwood, C.W., 319, 320
Klein, R.W., 148, 153, 154
Klein, L.R., 95, 97
Klein, G., 322, 328
Kloek, T., 95–96, 98, 100–105
Knowles, G.J., 291, 292, 297, 301, 316
Koch, G., 2
Koenker, R., 128, 134
Kolmogorov, A., 59, 70
Koopman, B.O., 220
Krauss, P.H., 47
Kshirsagar, A.M., 178
Kuh, E., 375, 376, 377, 378, 382, 386
Kyburg, H.E., 26, 32

Lahiff, M., 338
Land, M., 342, 346, 352
Lane, D., 46
Lanke, J., 444
Lau, L.J., 289
Lauritzen, S.L., 21, 400
LeCam, L., 476, 477
Leamer, E.E., 79, 147, 221, 233, 375, 376, 386
Leonard, T., 427
Levi, I., 60, 63, 64, 66
Lichtenstein, S., 267
Lin, S., 463
Lin, An-loh, 92
Lindley, D.V., 2, 16, 19, 22, 23, 25, 33, 60, 61, 64, 120, 124, 136, 147, 192, 193, 234, 235, 259, 267, 363, 365, 366, 367
Litterman, R.B., 79, 83, 85, 92, 338
Littlewood, J.E., 252
Lwin, T., 158, 162, 164, 172

MacPhail, M.S., 376
Magnus, J.R., 376
Makov, U.E., 339
Mallows, C.L., 427
Mardia, K.V., 169
Marini, M.M., 355
Maritz, J.S., 133
Marschak, J., 219, 223
Marshall, A.W., 258
Matheson, J.E., 269, 280
McGregor, D., 247
Meeden, G., 248, 254, 255
Meese, R., 83
Mehta, J.S., 363, 367
Meyer, R.F., 297, 320
Milgrom, P.R., 225
Monahan, J.F., 338
Morgenstern, D., 323
Morris, C.N., 117, 408

Morris, P.A., 267
Moskowitz, H., 322, 328
Muniruzzaman, A.N.M., 162
Murphy, A.H., 247, 250, 267, 268, 272
Muth, J., 233, 236

Neudecker, H., 376
Newbold, P., 352
Neyman, J., 15
Novick, M.R., 367

O'Hagan, A., 127
Olkin, I., 258
Olsen, A.R., 355

Pearson, E.S., 15
Pekelman, D., 320
Peters, S.C., 178, 180, 181, 186, 187, 233, 236, 288
Phillips, L., 22, 267
Piccinato, L., 20, 26
Poirier, D.J., 128, 134
Polasek, W., 375, 376, 377, 383, 387
Polemarchakis, H.M., 289
Pollock, S.M., 322
Polya, G., 252
Pratt, J.W., 250, 294, 297, 320, 482, 484
Press, S.J., 157, 158, 164, 165, 169, 172, 178, 192, 193
Priestley, M.B., 338, 340, 342, 343, 352

Quandt, R.E., 104

Raiffa, H., 233, 235, 319, 320, 323, 326, 328
Ramanathan, R., 128, 133, 134
Ramsey, F.P., 475, 26
Rao, C.R., 363, 365, 366, 367, 368
Rasmussen, S.L., 455
Ravindran, A., 322, 328
Rayner, A.C., 235, 238
Regazzini, E., 20, 21, 22
Renyi, A., 64, 443
Reynolds, R.A., 338, 352
Richard, J.F., 95, 96, 233
Robbins, H., 455
Roberts, H.V., 17, 20
Robinson, G.K., 66, 68, 69
Rosen, B., 444
Rothenberg, T.J., 95
Rothschild, M., 263
Rubin, H., 475
Rubin, D.B., 119, 127, 355

Salazar, D., 339
Sarin, R., 319–320
Savage, L.J., 17, 18, 25, 31, 60, 71, 219, 259, 266, 268, 271, 280, 298, 328, 475, 476, 480, 483
Scardovi, I., 16, 27
Schervish, M., 3, 47, 56, 60, 61, 65, 66, 258, 259
Schlaifer, R., 233, 235
Shoemaker, P.J.H., 320, 328
Schwartz, J.T., 66
Schwartz, G., 427
Schwarz, R., 178
Scozzafava, R., 23, 24
Seber, C., 430
Seidenfeld, T., 3, 47, 56, 60, 61, 65, 66, 67, 69
Sen, P.K., 444
Sen, S.K., 320
Shaarawy, S., 344, 346, 348, 352
Shafer, G., 147, 221
Shannon, C.E., 153
Sherman, S., 262
Shiller, R.J., 4, 79, 109, 233
Shocker, A.D., 320, 329, 330, 333
Sims, C.A., 79, 83, 85
Sinha, S.K., 157
Siow, A., 427
Smith, A.F.M., 19, 20, 23, 120, 121, 124, 126, 136, 138, 193, 339, 363, 366, 367, 377
Smith, C.A.B., 220, 221
Smith, J.Q., 421
Smith, V.L., 298
Smith, W.S., 178, 180, 181, 186, 187, 233, 236, 288
Smokler, H.E., 26, 32
Spizzichino, F., 2, 21
Srinivasan, V., 320, 329, 330, 333
Srivastava, M.S., 357
Stael von Holtstein, C.A.S., 255, 256
Starr, N., 455
Stein, C.M., 61, 79, 262
Stevens, C.F., 339, 405, 410, 413, 416, 420
Stiglitz, J.E., 263, 297
Stone, M., 69
Strassen, V., 262
Sudderth, W., 22, 32, 46, 60, 64, 65, 67, 68, 69
Swamy, P.A.V.B., 121, 363, 367

Teicher, H., 22
Testfatsion, L., 309
Thayer, D.T., 367
Theil, H., 85, 383–384
Thomson, W., 280
Tiao, G.C., 71, 153, 233, 234, 481
Tsurumi, H., 339
Tsutakawa, R.K., 119, 127
Tukey, J.W., 15
Tversky, A., 259, 267

Urban, G.L., 320, 321, 331

van Dijk, H.K., 95, 96, 98, 100, 101, 102, 103, 104, 105
Vandaele, W., 241
Vardeman, S., 248, 254, 255

Vickrey, W., 219
Vickson, R.G., 252, 254, 262
Villegas, C., 61
von Neumann, J., 323

Wachter, M.L., 137
Waid, D.C., 320, 328
Wald, A., 26
Walley, P., 220, 221
Weber, R.J., 225
Wedlin, A., 22
Welch, F., 137
Welsch, R.E., 375, 376, 377, 378, 382, 386
Whittle, P., 406
Wilkinson, G.N., 474
Williams, P.A., 220, 221
Williams, A., 233
Winkler, R.L., 20, 178, 180, 181, 186, 187, 233, 236, 247, 248, 250, 265, 267, 268, 269, 271, 273, 277, 280, 288
Wolpert, R., 474, 478, 479, 480, 481, 485, 486

Yates, J.F., 272

Zellner, A., 71, 96, 97, 147, 153, 157, 172, 178, 233, 234, 235, 236, 241, 338, 339, 341, 352, 376, 382, 384, 427, 473, 481
Zidek, J.V., 69

SUBJECT INDEX

Absolute risk aversion, 294
Admissibility, 3, 61
  Admissible rules, 16
  and finite additivity, 61
Aggregate economic behavior, 80
Almon lag estimator, 110
ARIMA models, 83, 406
ARMA Processes, 339
  Bayesian analysis, 7, 338, 341
  HPD regions, 345
Arrow's axiom system, 61
Asking prices in prob. assessment, 224
Assessment of priors
  bid–ask spread, 5, 217
  via conjugate distributions, 4, 178
  deficiency of betting games, 5, 218
  via *g*-priors, 223
  interactive computing algorithm for, 4, 178
  matrix-*t*-distribution, 4, 178, 182
  multiple linear regression model, 5, 251
  for multivariate normal sampling, 178
  multivariate-*t* distribution, 4, 178, 180
  scoring-rule procedures, 6, 280
Austrian consumption function, 7, 379
Autoregression
  univariate autoregressive process, 83, 341
  vector, 3, 79, 88
  forecasting macroeconomic vars., 3, 79
Balanced procedure (Fraser's), 68
Bayes and non-Bayes interaction, 8, 473
Bayes theorem, 13, 68, 79, 197, 341, 391, 460
Bayesian approach
  interactive forecasting, 405
  for macroeconomic forecasting, 81
  for model building, 391
  non-parametric, 19
  predictive, 20
  as model for rational beliefs, 198
  and separation of belief, 213
  threat to statistical profession, 16
Bayesian Monte Carlo integration, 4, 95, 100
  computational feasibility of, 95
Bayesian inference for
  general linear model, 7, 233, 377
  multivariate Pareto distribution, 4, 169
  nested missing-data problem, 7, 356
  Pareto distribution, 4, 158
  time series models, 7, 341
  simultaneous estimation in regression, 7, 96, 366
Bayesian regression diagnostics, 377
Bayesian reporting, 482
Bayesian robustness, 9, 481
Bayesian specification searches, 80
Bernstein polynomials, 35
Bid–ask spread, 5, 217, 224
Borel paradox, 70
Bounded rationality problem, 290
Box–Jenkins methodology
  comparison with Bayesian analysis, 7, 348
Coherence, 3, 13, 25, 62, 205, 220, 403
  Dutch books, 25, 63
  of finitely additive probability, 62
  and strong inconsistency, 62, 63
Coherent posteriors, 64
Collector's problem, 8, 443
  myopic procedures, 448
  optimal search procedures, 8, 445
Combining information, 79
Comparison of experiments
  comparing prob. forecasters, 6, 251
Complete class, 16
Concentrated likelihood function, 104
Conditional prevision, 211
Conditionality-likelihood principle, 8, 9, 478
  weak conditionality principle, 479
Conglomerability, 3, 24, 45, 46
  conditional, 66
Conglomerability and countable additivity
  equivalence of, 56
Conjugate prior distributions
  for ARMA models, 341
  for GLM, 377
  nested missing data model, 356
  for Pareto distributions, 157
Countable additivity, 45, 59
  noncoglomerability, 69
de Finetti's elicitation game, 221
de Finetti's inferential theory, 13
de Finetti's representation theorem, 3, 21
  extensions of, 31
  for finite sequence of r.v., 32, 37
  in statistical mechanics, 3, 32
  urn model approach, 32, 35
Decision problems
  Neyman–Pearson–Wald formulations, 16

Ramsey–de Finetti–Savage approach, 25
Degree of belief
probability as, 13
Discrimination cases
evidence of, 4, 141
posterior odds ratio, 4, 141
sex, 142
Disparate treatment
binomial model, 145
regression model, 147
Dispersion matrix, 120
Distributed lag estimation, 109
Distributions
Bernoulli trials, 401
beta, 338, 449
exponential, 8
gamma, 342
hypergeometric, 401
inverted gamma, 412
matrix-*t*, 4, 177, 179, 396
matrix normal, 396
multivariate normal, 355, 377, 378, 395, 459
multivariate-*t*, 4, 178, 342
normal, 8
normal-gamma, 341
normal–Wishart prior, 7, 356
Pareto, 4, 151
spherical normal, 109
symmetric under permutations, 32
uniform prior, 103
Wishart, 130
Disutility, 418
Dutch books (*see* Coherence)
Econometric modeling
(Keynesian) consumption function, 7, 378
distributed lag estimation, 109, 339
general linear model, 7, 234, 378
GLM diagnostics, 7, 378
linear hierarchical models, 119
informational variables, 3, 82
Klein–Goldberger model, 3, 95
random coefficient regression model, 364
for residential demand for electricity, 128
for return to schooling, 137
for short-run forecasting, 3, 80
SUR models
structural form model, 3, 81
time series analysis, 7, 82, 341
unemployment coefficient in wage eqn., 137
Economic variables, 92, 112
EM algorithm, 127
Empirical Bayes approach, 4, 117, 133
comparison with OLS, 7, 110
Monte Carlo comparision with OLS, 368
Employment discrimination cases, 141
Errors-in-variables, 460
Exchangeability
conference on, 2
de Finetti representation, 3, 21, 31, 33
definition of population, 22
predictivistic approach, 3, 209
Expected utility maximization, 25
Experiment
size of, 465
Federal Reserve Board, 114
Finitary problems, 401
Finite additivity, 3, 13, 23, 45, 60
statistical implications of, 59
Finite exchangeability, 33
Fair bets, 63
Flat prior, 87
Forecasting
construction of forecasting model, 406
expected forecast error, 406
interactive system, 8, 405
macroeconomic variables, 79, 82
out-of-sample, 3, 81, 352
variance, 412
Frequentist techniques
value to a Bayesian, 8, 473, 485
Full information
maximum likelihood estimates, 107
statistical analysis, 99
g-priors, 6, 233, 234
Hierarchical models
assessment of priors for, 178
linear, 4, 119
Markov property of, 121
Higher-order interactions
complexity of estimation, 8, 459
relationship with no. of factors, 8, 459
Hausdorff moment problem, 32
de Finetti representation theorem, 32
Human capital theory, 148
Hyperparameters, 4, 110, 120
empirical Bayes approach, 4, 110
Improper priors, 24, 64
Inconsistency
strong, 62, 63, 68
Weak, 68
Indecomposable sets, 51
Inductive reasoning, 2, 13
Infinitary problems, 399
Interactions, 462
Interactive forecasting system, 8, 405
postulates for, 405
Intersubjective models, 397
Invariance, 399
Invariant priors, 68, 234
Jeffrey's ignorance prior, 111
Kalman filter algorithm, 82, 86
Kinetic theory, 40
Klein–Goldberger model, 3, 95
Kullback–Leibler number, 461, 464
Laplace–de Finetti representation theorem, 34
Legendre polynomials, 35
Likelihood, 13, 65, 96
approach, 17

principle, 26, 124, 478
Linear dynamic system, 339, 405
Log-odds, 461
Logarithmic reward function, 280
Lower and upper probabilities, 5, 220
Macroeconomic variables, 90
forecasting, 3, 79, 81
Marginalization
paradox, 3, 69
Marschak's elicitation game, 223
Matrix derivatives, 376
Medical diagnosis, 460
Minimal information prior, 152
Maximum likelihood estimation
information theoretic justification, 124
nested missing-data, 355
Mixed estimation technique, 85, 96, 100
Model
external Interpretation, 394
Internal Interpretation, 394
Models
for advertising, 8, 414
ARMA, 339
for Austrian consumption function, 378
criticism, 18
definition of, 7
for distributed lag estimation, 109, 339
econometric, 107, 119, 378
infinitary, 400
intersubjective, 397
for macroeconomic forecasting, 79, 81
nonergodic, 394
nonlinear, 96
random coefficient regression, 364
robust, 409
role and limits of, 18
for socio-economic data, 4, 95, 99
simultaneous equation, 96
structural, 81, 95
time varying coefficients, 90
two-stage, 121, 367
Model selection
comparison of procedures, 425
criteria for, 426
for exponential data, 8, 430
for normal data, 8, 433
predictive error as criteria for, 8, 431
Moody's AAA bond yield, 114
MRM, 6, 234
predictive distribution for, 185
regression diagnostics, 7, 377
Neo-Bayesian approach, 13, 15, 25
Nested missing-data problem
Bayesian analysis, 7, 358
Normal–Wishart conjugate prior, 7, 356
Noncongolomerability, 3, 60
admissibility, 3
extent of, 47, 52
countable additivity, 69
curtailing, 64
disintegrability, 24, 69
of finitely additive probability, 60
strong Inconsistency, 3, 62
Nonidentifiability, 394
Non-parametric Inference
associative means, 20
Bayesian approach, 20
dirichlet process, 20
OLS estimator
comparison with empirical bayes, 7, 110, 368
Operationalist foundation, 392
Optimal search procedures
for collector's problem, 8, 445
Optimal stopping, 454, 456
Optimization problems
multiattribute utility assessment, 7, 324, 330
Orthogonal expansions, 36
Overparametrization, 87
Panel data, 119, 355, 364
Pareto distribution
censored data, 4, 161
grouped data, 4, 159
model for socio-economic data, 4, 157
mixed censored and uncensored data, 162
multivariate, 4, 169
two and three parameters, 4, 164
Philips curve, 99
Posterior mean, 4, 127, 161
computation via mixed estimation, 85, 96, 100
sampling properties of, 238
Posterior mode, 4, 127
Posterior moments
evaluation via numerical integration, 4, 96, 100
Posterior odds ratio
in discrimination cases, 4, 141
Posterior prevision, 200
Precise measurements
principle of, 18
Prediction (*see* Forecasting)
Predictive error as model selection criteria, 8, 428
Predictive density, 342, 395
Predictive distribution (*see* Predictive density)
Predictive moments, 342
Predictivistic approach, 3, 393
Prevision, 5, 198
Price of lottery tickets, 5, 221
Prior distributions
assessment of, 4, 178, 279
for ARMA models, 341
exchangeable, 397
inverted gamma, 413
of lag coefficients, 85, 109
Lwin priors, 158
minimal information, 152
for MRM via *g*-prior, 6, 234

for nested missing-data model, 7, 356
for normal regression model, 377
reference informative, 481
smoothness priors for lag estimation, 109
spherical normal, 109
for SUR model, 383
uniform prior for Klein–Goldberger model, 103
Prior beliefs (*see* Prior probability)
separation of, 5, 197, 204
Prior probability
assessment of, 4, 178, 279
via conjugate distributions, 4, 178, 341
elicitation of, 4, 178
invariant, 68
lower and upper, 5, 218
scoring-rule procedures for assessment, 6, 280
Probability
as degree of belief, 13
finitely additive, 24, 45, 60
statistical implications of, 59
frequentist definition, 14
full conditional, 46
normative, 22
subjective, 391
unconditional, 46
σ-Additive, 23, 45
Probability distributions
symmetric under permutations, 31, 32
Probability forcasting
calibration, 248, 249, 266
comparison of forecasters, 6, 247
and comparison of experiments, 6, 251
refinement of, 248, 249, 266
well-calibrated forecaster, 249, 270
Random coefficient regression model, 7, 121, 364
Rational expectation hypothesis, 6, 233
Recursive system, 96
Refinement, 248
stochastic transformation, 250
mean preserving spread, 263
Regression analysis
diagnostics in Bayesian regression, 375
for disparate treatment, 147
random coefficient model, 364
of time series, 82
RATS package, 82
Regression diagnostics, 7, 375
Relevancy principle, 463
Revising beliefs, 199
Robustness
Bayes and non-Bayes interaction, 8, 481
Schrodinger's equations, 40
Schur-convex function, 248
Scientific reporting, 230
Scoring-rule
for evaluation of probability assessors, 268
strictly proper rules, 248, 255, 284
procedures for priors assessment, 6, 280
Sensitivity analysis
global, 375
local, 377
Separating beliefs, 204, 205
σ-decomposable sets, 51
Smoothness priors, 4, 109
Statistical mechanics
N-representability, 32, 38
de Finetti representation theorem, 32
Stochastic dominance, 252
Stochastic linear system
lag coefficients, 4, 109
smoothness priors, 4, 109
Strong inconsistency, 3, 62
nonconglomerability, 3, 62
Structural form model, 3, 81
Subjective probability
assessment of, 4, 178, 197, 279
elicitation of, 4, 178, 281
interactive computing algorithm, 4, 178
price of lottery tickets, 5, 218
Subjectivity, 3, 197, 198
de Finetti's subjectivistic position, 7, 8, 15, 392
Sufficiency, 399
Sufficient statistics, 21
Supreme Court, 152
SUR models
for cross-sections of time series, 138
for Grunfeld data, 384
diagnostics checks, 7, 382
sensitivity analysis, 7, 382
Time Series
macroeconomic, 80
RATS package, 82
Regression Analysis of, 82
Uncertainty
logic of, 2, 13
Utility Assessment
certainty equivalence, 296, 325
constant risk aversion (CAR), 309, 321
farmer's utility, 6, 291
multiattribute utility, 7, 328
nonlinear programing model for, 324
parametric form, 6, 321
risk aversion, 294, 297, 325
risk neutral, 297
SCAR (SUMEX), 309, 321
scoring-rule procedure, 6, 284
Vague prior distribution, 234